**Foundations of Electromagnetic
Compatibility**

Foundations of Electromagnetic Compatibility

with Practical Applications

Bogdan Adamczyk
EMC Educational Services LLC
www.emcspectrum.com

Registered Offices
John Wiley & Sons, Inc., 111 River Street, Hoboken, NJ 07030, USA
John Wiley & Sons Ltd, The Atrium, Southern Gate, Chichester, West Sussex, PO19 8SQ, UK

Editorial Office
The Atrium, Southern Gate, Chichester, West Sussex, PO19 8SQ, UK

For details of our global editorial offices, customer services, and more information about Wiley products visit us at www.wiley.com.

Wiley also publishes its books in a variety of electronic formats and by print-on-demand. Some content that appears in standard print versions of this book may not be available in other formats.

Library of Congress Cataloging-in-Publication Data

Names: Adamczyk, Bogdan, 1960– author.
Title: Foundations of electromagnetic compatibility with practical applications / Bogdan Adamczyk.
Description: Hoboken, New Jersey : John Wiley & Sons, Inc., 2017. | Includes index.
Identifiers: LCCN 2016047894 | ISBN 9781119120780 (hardback ; cloth) |
    ISBN 1119120780 (hardback ; cloth) | ISBN 9781119120797 (Adobe PDF) |
    ISBN 1119120799 (Adobe PDF) | ISBN 9781119120803 (ePub) | ISBN 1119120802 (ePub)
Subjects: LCSH: Electromagnetic compatibility.
Classification: LCC TK7867.2 .A33 2017 | DDC 621.382/24–dc23
LC record available at https://lccn.loc.gov/2016047894

Cover design by Wiley
Front Cover: The cover shows the EMC Center at GVSU created by Prof. Adamczyk

Set in 10/12pt Warnock by SPi Global, Pondicherry, India

Contents

Preface

A few years ago when I was about to teach another EMC fundamentals course for the industry, I was contacted by some of the participants asking about a textbook for the course. Then I realized that there is no single self-contained book covering the topics of mathematic, electric circuits and electromagnetics with the focus on EMC. There is a plethora of books devoted to each of these subjects separately and each written for a general audience. It was then that the idea of writing this book was born.

This text reviews the fundamentals of mathematics, electric circuits, and electromagnetics specifically needed for the study of EMC. Each chapter reviews the material pertinent to EMC and concludes with practical EMC examples illustrating the applicability of the discussed topics. The book is intended as a reference and a refresher for both the practicing professionals and the new EMC engineers entering the field.

This book also provides a background material helpful in following the two classical texts on EMC: Clayton Paul's "Introduction to Electromagnetic Compatibility" (Wiley, 2006) and Henry Ott's "Electromagnetic Compatibility Engineering" (Wiley, 2009). Many formulas in those two books (presented without derivations) are derived from basic principles in this text.

This approach provides the reader with the understanding of the underlying assumptions and the confidence in using the final results. This insight is invaluable in the field of EMC where so many design rules and principles are based on several approximations and are only valid when the underlying assumptions are met.

The author owes a great deal of gratitude for the insight and knowledge gained from the association with colleagues from the EMC lab at Gentex Corporation (Bill Spence and Pete Vander Wel) and the EMC specialists and friends at E3 Compliance LLC (Jim Teune and Scott Mee). The author would also like to thank Mark Steffka for his guidance and help over the past ten years. Finally, the author would like to acknowledge the support of Grand Valley State University and especially its engineering dean Paul Plotkowski who was instrumental in the creation of the EMC Center, greatly contributing to the EMC education and the publication of this book.

Bogdan Adamczyk
Grand Rapids, Michigan, September 2016

Preface

Part I

Math Foundations of EMC

1

Matrix and Vector Algebra

Matrices and determinants are very powerful tools in circuit analysis and electromagnetics. Matrices are useful because they enable us to replace an array of many entries as a single symbol and perform operations in a compact symbolic form.

We begin this chapter by defining a matrix, followed by the algebraic operations and properties. We will conclude this chapter by showing practical EMC-related applications of matrix algebra.

1.1 Basic Concepts and Operations

A *matrix* is a mathematical structure consisting of rows and columns of elements (often numbers or functions) enclosed in brackets (Kreyszig, 1999, p. 305).

For example,

$$\mathbf{A} = \begin{bmatrix} 6 & 5 & 2 \\ 3 & -2 & 0 \\ 7 & 1 & 4 \end{bmatrix} \tag{1.1}$$

The entries in matrix A are real numbers. Matrices L and C in Eq. (1.2) are the matrices containing per-unit-length inductances and capacitances, respectively, representing a crosstalk model of transmission lines (Paul, 2006, p. 567). (We will discuss the details of this model later in this chapter.)

$$\mathbf{L} = \begin{bmatrix} l_G & l_m \\ l_m & l_R \end{bmatrix}, \quad \mathbf{C} = \begin{bmatrix} c_G + c_m & -c_m \\ -c_m & c_R + c_m \end{bmatrix} \tag{1.2}$$

We denote matrices by capital boldface letters. It is often convenient, especially when discussing matric operations and properties, to represent a matrix in terms of its general entry in brackets:

$$\mathbf{A} = \begin{bmatrix} a_{ij} \end{bmatrix} = \begin{bmatrix} a_{11} & a_{12} & \cdots & a_{1n} \\ a_{21} & a_{22} & \cdots & a_{2n} \\ \vdots & \vdots & \ddots & \vdots \\ a_{m1} & a_{m2} & \cdots & a_{mn} \end{bmatrix} \tag{1.3}$$

Here, $\mathbf{A}$ is an $m \times n$ matrix; that is, a matrix with m rows and n columns.

In the double-subscript notation for the entries, the first subscript always denotes the row and the second the column in which the given entry stands. Thus a_{23} is the entry in the second row and third column.

If $m = n$, we call $\mathbf{A}$ an $n \times n$ square matrix. Square matrices are particularly important, as we shall see.

A matrix that has only one column is often called a *column* vector. For example,

$$\mathbf{V}(z,t) = \begin{bmatrix} V_G(z,t) \\ V_R(z,t) \end{bmatrix}, \quad \mathbf{I}(z,t) = \begin{bmatrix} I_G(z,t) \\ I_R(z,t) \end{bmatrix} \tag{1.4}$$

Here, V and I are the column vectors representing the voltages and currents, respectively, associated with the crosstalk model of transmission lines (Paul, 2006, p. 566).

Equality of Matrices   We say that two matrices have the *same size* if they are both $m \times n$.

Two matrices $\mathbf{A} = [a_{ij}]$ and $\mathbf{B} = [b_{ij}]$ are equal, written $\mathbf{A} = \mathbf{B}$, if they are of the same size and the corresponding entries are equal; that is, $a_{11} = b_{11}$, $a_{12} = b_{12}$, and so on. For example, let

$$\mathbf{A} = \begin{bmatrix} a_{11} & a_{12} \\ a_{21} & a_{22} \end{bmatrix}, \quad \mathbf{B} = \begin{bmatrix} 7 & -4 \\ 2 & 8 \end{bmatrix} \tag{1.5}$$

Then $\mathbf{A} = \mathbf{B}$ implies that $a_{11} = 7$, $a_{12} = -4$, $a_{21} = 2$, and $a_{22} = 8$.

Matrix Addition and Scalar Multiplication   Just like the matrix equality, matrix addition and scalar multiplication are intuitive concepts, for they follow the laws of numbers. (We point this out because *matrix* multiplication, to be defined shortly, is *not* an intuitive operation.)

Addition is defined for matrices of the same size. The sum of two matrices, $\mathbf{A}$ and $\mathbf{B}$, written, $\mathbf{A} + \mathbf{B}$, is a matrix whose entries are obtained by adding the corresponding entries of $\mathbf{A}$ and $\mathbf{B}$. That is,

$$\mathbf{A} = \begin{bmatrix} a_{11} & a_{12} \\ a_{21} & a_{22} \end{bmatrix}, \quad \mathbf{B} = \begin{bmatrix} b_{11} & b_{12} \\ b_{21} & b_{22} \end{bmatrix}, \quad \mathbf{A} + \mathbf{B} = \begin{bmatrix} a_{11} + b_{11} & a_{12} + b_{12} \\ a_{21} + b_{21} & a_{22} + b_{22} \end{bmatrix} \tag{1.6}$$

The product of any matrix $\mathbf{A}$ and any scalar k, written $k\mathbf{A}$, is the matrix obtained by multiplying each element of $\mathbf{A}$ by k. That is,

$$\mathbf{A} = \begin{bmatrix} a_{11} & a_{12} \\ a_{21} & a_{22} \end{bmatrix}, \quad k\mathbf{A} = \begin{bmatrix} ka_{11} & ka_{12} \\ ka_{21} & ka_{22} \end{bmatrix} \tag{1.7}$$

From the familiar laws for numbers, we obtain similar laws for matrix addition and scalar multiplication.

$$\mathbf{A} + \mathbf{B} = \mathbf{B} + \mathbf{A} \tag{1.8a}$$

$$k(\mathbf{A} + \mathbf{B}) = k\mathbf{A} + k\mathbf{B} \tag{1.8b}$$

$$\mathbf{A} + \mathbf{0} = \mathbf{A} \tag{1.8c}$$

$$\mathbf{A} + (-\mathbf{A}) = \mathbf{0} \tag{1.8d}$$

$$1\mathbf{A} = \mathbf{A} \tag{1.8e}$$

$$0\mathbf{A} = \mathbf{0} \tag{1.8f}$$

There is one more algebraic operation: the multiplication of matrices by matrices. Since this operation does not follow the familiar rule of number multiplication we devote a separate section to it.

1.2 Matrix Multiplication

Matrix multiplication means multiplying matrices by matrices. Recall: matrices are *added* by *adding* corresponding entries, as shown in Eq. (1.6). Matrix multiplication could be defined in a similar manner:

$$\mathbf{A} = \begin{bmatrix} a_{11} & a_{12} \\ a_{21} & a_{22} \end{bmatrix}, \quad \mathbf{B} = \begin{bmatrix} b_{11} & b_{12} \\ b_{21} & b_{22} \end{bmatrix}, \quad \mathbf{AB} = \begin{bmatrix} a_{11}b_{11} & a_{12}b_{12} \\ a_{21}b_{21} & a_{22}b_{22} \end{bmatrix} \ (\textit{incorrect}) \tag{1.9}$$

But it is not. Why? Because it is not useful.

The definition of multiplication seems artificial, but it is motivated by the use of matrices in solving the systems of equations.

Matrix Multiplication   If $\mathbf{A} = [a_{ij}]$ is an $m \times n$ matrix and $\mathbf{B} = [b_{ij}]$ is an $n \times p$ matrix, then the product of A and B, $\mathbf{AB} = \mathbf{C} = [c_{ij}]$, is an $m \times p$ matrix defined by

$$c_{ij} = \sum_{k=1}^{n} a_{ik} b_{kj} = a_{i1}b_{1j} + a_{i2}b_{2j} + \cdots + a_{in}b_{nj} \tag{1.10}$$
$$i = 1, 2, \ldots, m; \quad j = 1, 2, \ldots, p$$

Note that $\mathbf{AB}$ is defined only when the number of columns of $\mathbf{A}$ is the same as the number of rows of $\mathbf{B}$. Therefore, while in some cases we can calculate the product $\mathbf{AB}$, of matrix A by matrix B, the product $\mathbf{BA}$, of matrix $\mathbf{B}$ by matrix $\mathbf{A}$, may not be defined.

We also observe that the (i,j) entry in $\mathbf{C}$ is obtained by using the ith row of $\mathbf{A}$ and the jth column of $\mathbf{B}$.

$$\begin{bmatrix} a_{11} & a_{12} & \cdots & a_{1n} \\ a_{21} & a_{22} & \cdots & a_{2n} \\ \vdots & \vdots & \ddots & \vdots \\ a_{i1} & a_{i2} & \cdots & a_{in} \\ \vdots & \vdots & \ddots & \vdots \\ a_{m1} & a_{m2} & \cdots & a_{mn} \end{bmatrix} \begin{bmatrix} b_{11} & b_{12} & \cdots & b_{1j} & \cdots & b_{1p} \\ b_{21} & b_{22} & \cdots & b_{2j} & \cdots & b_{2p} \\ \vdots & \vdots & \ddots & \vdots & \ddots & \vdots \\ b_{n1} & b_{n2} & \cdots & b_{nj} & \cdots & b_{np} \end{bmatrix} = \begin{bmatrix} c_{11} & c_{12} & \cdots & c_{1p} \\ c_{21} & c_{22} & \cdots & c_{2p} \\ \vdots & \vdots & c_{ij} & \vdots \\ c_{m1} & c_{m2} & \cdots & c_{mp} \end{bmatrix} \tag{1.11}$$

Example 1.1 Matrix multiplication

$$A = \begin{bmatrix} 7 & 2 \\ 1 & 8 \end{bmatrix}, \quad B = \begin{bmatrix} 6 & 4 \\ 5 & 2 \end{bmatrix},$$

$$AB = \begin{bmatrix} 7 & 2 \\ 1 & 8 \end{bmatrix}\begin{bmatrix} 6 & 4 \\ 5 & 2 \end{bmatrix} = \begin{bmatrix} 7\cdot6+2\cdot5 & 7\cdot4+2\cdot2 \\ 1\cdot6+8\cdot5 & 1\cdot4+8\cdot2 \end{bmatrix} = \begin{bmatrix} 52 & 32 \\ 46 & 20 \end{bmatrix}$$

∎

Example 1.2 Multiplication of a matrix and a vector

$$\begin{bmatrix} 4 & 2 \\ 1 & 8 \end{bmatrix}\begin{bmatrix} 3 \\ 7 \end{bmatrix} = \begin{bmatrix} 4\cdot3+2\cdot7 \\ 1\cdot3+8\cdot7 \end{bmatrix} = \begin{bmatrix} 26 \\ 59 \end{bmatrix}$$

whereas $\begin{bmatrix} 3 \\ 7 \end{bmatrix}\begin{bmatrix} 4 & 2 \\ 1 & 8 \end{bmatrix}$ is undefined.

∎

It is important to note that unlike number multiplication, multiplication of two square matrices is *not*, in general, commutative. That is, in general, **AB ≠ AB**

Example 1.3 Multiplication of matrices in a reverse order
Using the matrices from Example 1.1, but multiplying them in a reverse order, we get

$$A = \begin{bmatrix} 7 & 2 \\ 1 & 8 \end{bmatrix}, \quad B = \begin{bmatrix} 6 & 4 \\ 5 & 2 \end{bmatrix},$$

$$BA = \begin{bmatrix} 6 & 4 \\ 5 & 2 \end{bmatrix}\begin{bmatrix} 7 & 2 \\ 1 & 8 \end{bmatrix} = \begin{bmatrix} 6\cdot7+4\cdot1 & 6\cdot2+4\cdot8 \\ 5\cdot7+2\cdot1 & 5\cdot2+2\cdot8 \end{bmatrix} = \begin{bmatrix} 46 & 44 \\ 37 & 26 \end{bmatrix}$$

which differs from the result obtained in Example 1.1.

∎

1.3 Special Matrices

The most important special matrices are the *diagonal matrix*, the *identity matrix*, and the *inverse* of a given matrix.

Diagonal Matrix A diagonal matrix is a square matrix that can have non-zero entries only on the main diagonal. Any entry above or below the main diagonal must be zero.
For example,

$$A = \begin{bmatrix} 4 & 0 & 0 \\ 0 & -3 & 0 \\ 0 & 0 & 7 \end{bmatrix} \tag{1.12}$$

Identity Matrix   A diagonal matrix whose entries on the main diagonal are all 1 is called an *identity matrix* and is denoted by I_n or simply I.

For example,

$$I = \begin{bmatrix} 1 & 0 & 0 \\ 0 & 1 & 0 \\ 0 & 0 & 1 \end{bmatrix} \tag{1.13}$$

The identity matrix has the following important property

$$AI = IA = A \tag{1.14}$$

where A and I are square matrices of the same size.

Also, for any vector b we have

$$Ib = b \tag{1.15}$$

where the identity matrix is of the appropriate size.

1.4   Matrices and Determinants

If we were to associate a single number with a *square* matrix, what would it be? The largest element, the sum of all elements, or maybe the product? It turns out that there is one very useful single number called the *determinant*.

For a 2×2 matrix, we can obtain its determinant using the following approach:

$$\det \ A = \begin{vmatrix} a_{11} & a_{12} \\ a_{21} & a_{22} \end{vmatrix} = a_{11}a_{22} - a_{21}a_{12} \tag{1.16}$$

Note that we denote determinant by using *bars* (whereas we denote the matrices by using *brackets*).

Example 1.4   Determinant of a 2×2 matrix

$$\begin{vmatrix} 3 & 5 \\ 2 & 6 \end{vmatrix} = 3 \cdot 6 - 2 \cdot 5 = 8$$

∎

The procedure for obtaining the determinant for a 3×3 matrix is a bit more involved. Let the matrix A be specified as

$$A = \begin{bmatrix} a_{11} & a_{12} & a_{13} \\ a_{21} & a_{22} & a_{23} \\ a_{31} & a_{32} & a_{33} \end{bmatrix} \tag{1.17}$$

Its determinant

$$\det A = \begin{vmatrix} a_{11} & a_{12} & a_{13} \\ a_{21} & a_{22} & a_{23} \\ a_{31} & a_{32} & a_{33} \end{vmatrix} \tag{1.18}$$

can be obtained using the following procedure. Let's create an augmented "determinant" by rewriting the first two rows underneath the original ones:

$$\det{}_{aug} A = \begin{vmatrix} a_{11} & a_{12} & a_{13} \\ a_{21} & a_{22} & a_{23} \\ a_{31} & a_{32} & a_{33} \\ a_{11} & a_{12} & a_{13} \\ a_{21} & a_{22} & a_{23} \end{vmatrix} \tag{1.19}$$

then the value of *det* A can be obtained by adding and subtracting the triples of numbers from the augmented determinant as follows:

$$\det A = \begin{vmatrix} a_{11} & a_{12} & a_{13} \\ a_{21} & a_{22} & a_{23} \\ a_{31} & a_{32} & a_{33} \\ a_{11} & a_{12} & a_{13} \\ a_{21} & a_{22} & a_{23} \end{vmatrix}$$

$$= a_{11}a_{22}a_{33} + a_{21}a_{32}a_{13} + a_{31}a_{12}a_{23}$$
$$-a_{21}a_{12}a_{33} - a_{11}a_{32}a_{23} - a_{31}a_{22}a_{13} \tag{1.20}$$

Example 1.5   Determinant of a 3 × 3 matrix
Calculate determinant of a matrix **A** given by

$$A = \begin{bmatrix} 6 & 1 & 3 \\ 7 & -2 & 5 \\ -9 & 8 & 4 \end{bmatrix}$$

Solution: Create and evaluate the augmented determinant.

$$\det A = \begin{vmatrix} 6 & 1 & 3 \\ 7 & -2 & 5 \\ -9 & 8 & 4 \\ 6 & 1 & 3 \\ 7 & -2 & 5 \end{vmatrix}$$

$$= (6)\,(-2)\,(4) + (7)\,(8)\,(3) + (-9)\,(1)\,(5)$$
$$-(7)\,(1)\,(4) - (6)\,(8)\,(5) - (-9)\,(-2)\,(3) = -247$$

■

Why do we need to know how to obtain a second- or third-order determinant? Obviously, we could use a calculator or a software program to do that for us. There are numerous occasions when the software or a calculator would not be able to handle the calculations.

As we will later see, when discussing capacitive termination to a transmission line, we will need to obtain a symbolic solution in a proper form; even if we had access to a symbolic-calculation software, its output, in most cases, would not be in a useful form.

When discussing Maxwell's equations, we will need to evaluate a third-order determinant whose entries are vectors, vector components, and differential operators. This can only be done by hand.

1.5 Inverse of a Matrix

An inverse of a square matrix $\mathbf{A}$ (when it exists) is another matrix of the same size, denoted $\mathbf{A}^{-1}$. This new matrix, is perhaps, the most useful matrix in matrix algebra.

The inverse of a matrix has the following property of paramount importance

$$\mathbf{AA}^{-1} = \mathbf{A}^{-1}\mathbf{A} = \mathbf{I} \tag{1.21}$$

Given a square matrix of numbers we can easily obtain its inverse using a calculator or an appropriate software package. In many engineering calculations, however, we need to obtain the inverse of a 2×2 matrix in a symbolic form.

Let

$$\mathbf{A} = \begin{bmatrix} a_{11} & a_{12} \\ a_{21} & a_{22} \end{bmatrix} \tag{1.22}$$

Then the inverse of $\mathbf{A}$ can be obtained as

$$\mathbf{A}^{-1} = \frac{1}{\det \mathbf{A}} \begin{bmatrix} a_{22} & -a_{12} \\ -a_{21} & a_{11} \end{bmatrix} \tag{1.23}$$

Example 1.6   Inverse of a 2×2 matrix
Obtain the inverse of

$$\mathbf{A} = \begin{bmatrix} 4 & 3 \\ 2 & 5 \end{bmatrix}$$

Solution: According to Eq. (1.23) the inverse of $\mathbf{A}$ is

$$\mathbf{A}^{-1} = \begin{bmatrix} 4 & 3 \\ 2 & 5 \end{bmatrix}^{-1} = \frac{1}{\begin{vmatrix} 4 & 3 \\ 2 & 5 \end{vmatrix}} \begin{bmatrix} 5 & -3 \\ -2 & 4 \end{bmatrix}$$

$$= \frac{1}{20-6} \begin{bmatrix} 5 & -3 \\ -2 & 4 \end{bmatrix} = \begin{bmatrix} \dfrac{5}{14} & -\dfrac{3}{14} \\ -\dfrac{2}{14} & \dfrac{4}{14} \end{bmatrix}$$

Verification:

$$\mathbf{A}\mathbf{A}^{-1} = \begin{bmatrix} 4 & 3 \\ 2 & 5 \end{bmatrix} \begin{bmatrix} \dfrac{5}{14} & -\dfrac{3}{14} \\ -\dfrac{2}{14} & \dfrac{4}{14} \end{bmatrix} = \begin{bmatrix} \dfrac{20-6}{14} & \dfrac{-12+12}{14} \\ \dfrac{10-10}{14} & \dfrac{-6+20}{14} \end{bmatrix} = \begin{bmatrix} 1 & 0 \\ 0 & 1 \end{bmatrix}$$

∎

1.6 Matrices and Systems of Equations

We will now explain the reason behind the "unnatural" definition of matrix multiplication. Consider a system of equations:

$$\begin{aligned} a_{11}x_1 + a_{12}x_2 + a_{13}x_3 &= b_1 \\ a_{21}x_1 + a_{22}x_2 + a_{23}x_3 &= b_2 \\ a_{31}x_1 + a_{32}x_2 + a_{33}x_3 &= b_3 \end{aligned} \tag{1.24}$$

Let's define three matrices as follows:

$$\mathbf{A} = \begin{bmatrix} a_{11} & a_{12} & a_{13} \\ a_{21} & a_{22} & a_{23} \\ a_{31} & a_{32} & a_{33} \end{bmatrix}, \quad \mathbf{x} = \begin{bmatrix} x_1 \\ x_2 \\ x_3 \end{bmatrix}, \quad \mathbf{b} = \begin{bmatrix} b_1 \\ b_2 \\ b_3 \end{bmatrix} \tag{1.25}$$

Then the system of equations (1.24) can be written in compact form using matrices defined by Eq. (1.25) as

$$\mathbf{A}\mathbf{x} = \mathbf{b} \tag{1.26}$$

Since

$$\begin{aligned} \mathbf{A}\mathbf{x} &= \begin{bmatrix} a_{11} & a_{12} & a_{13} \\ a_{21} & a_{22} & a_{23} \\ a_{31} & a_{32} & a_{33} \end{bmatrix} \begin{bmatrix} x_1 \\ x_2 \\ x_3 \end{bmatrix} \\ &= \begin{bmatrix} a_{11}x_1 + a_{12}x_2 + a_{13}x_3 \\ a_{21}x_1 + a_{22}x_2 + a_{23}x_3 \\ a_{31}x_1 + a_{32}x_2 + a_{33}x_3 \end{bmatrix} = \begin{bmatrix} b_1 \\ b_2 \\ b_3 \end{bmatrix} = \mathbf{b} \end{aligned} \tag{1.27}$$

and two matrices are equal when their corresponding entries are equal. Thus, Eqs (1.24) and (1.27) are equivalent.

Equation (1.26) shows one of the benefits of using matrices: a system of linear equation can be expressed in a compact form. An even more important benefit is the fact that we can obtain the solution to the system of equations by manipulating the matrices in a symbolic form instead of the equations themselves. This will be shown in the next section.

1.7 Solution of Systems of Equations

Consider a system of equations:

$$\mathbf{Ax} = \mathbf{b} \tag{1.28}$$

If the inverse of $\mathbf{A}$ exists, then premultiplication of Eq. (1.28) by $\mathbf{A}^{-1}$ results in

$$\mathbf{A}^{-1}\mathbf{Ax} = \mathbf{A}^{-1}\mathbf{b} \tag{1.29}$$

Since $\mathbf{A}^{-1}\mathbf{A} = \mathbf{I}$, it follows

$$\mathbf{Ix} = \mathbf{A}^{-1}\mathbf{b} \tag{1.30}$$

Because $\mathbf{Ix} = \mathbf{x}$, we obtain the solution to Eq. (1.28) as

$$\mathbf{x} = \mathbf{A}^{-1}\mathbf{b} \tag{1.31}$$

Example 1.7 Solution of systems of equations using matrix inverse

Obtain the solution of

$$4x_1 + 3x_2 = 12$$
$$2x_1 + 5x_2 = -8$$

using matrix inversion.

Solution: Our system of equations in matrix form can be written as

$$\begin{bmatrix} 4 & 3 \\ 2 & 5 \end{bmatrix} \begin{bmatrix} x_1 \\ x_2 \end{bmatrix} = \begin{bmatrix} 12 \\ -8 \end{bmatrix}$$

According to Eq. (1.31), the solution, therefore, can be written as

$$\begin{bmatrix} x_1 \\ x_2 \end{bmatrix} = \begin{bmatrix} 4 & 3 \\ 2 & 5 \end{bmatrix}^{-1} \begin{bmatrix} 12 \\ -8 \end{bmatrix}$$

Utilizing the result of Example 1.6, we have

$$\begin{bmatrix} x_1 \\ x_2 \end{bmatrix} = \begin{bmatrix} 4 & 3 \\ 2 & 5 \end{bmatrix}^{-1} \begin{bmatrix} 12 \\ -8 \end{bmatrix} = \begin{bmatrix} \dfrac{5}{14} & -\dfrac{3}{14} \\ -\dfrac{2}{14} & \dfrac{4}{14} \end{bmatrix} \begin{bmatrix} 12 \\ -8 \end{bmatrix}$$

$$= \begin{bmatrix} \left(\dfrac{5}{14}\right)(12) & \left(-\dfrac{3}{14}\right)(-8) \\ \left(-\dfrac{2}{14}\right)(12) & \left(\dfrac{4}{14}\right)(-8) \end{bmatrix} = \begin{bmatrix} 6 \\ -4 \end{bmatrix}$$

■

1.8 Cramer's Rule

As we have seen, we can obtain a solution to a system of equations using matrix inversion. When dealing with 2×2 matrices, it is sometimes more expedient to use an alternative approach using *Cramer's rule.*

Let the system of equations be given by

$$a_{11}x_1 + a_{12}x_2 = b_1$$
$$a_{21}x_1 + a_{22}x_2 = b_2 \tag{1.32}$$

or in a matrix form:

$$\mathbf{Ax} = \mathbf{b} \tag{1.33}$$

where

$$\mathbf{A} = \begin{bmatrix} a_{11} & a_{12} \\ a_{21} & a_{22} \end{bmatrix}, \quad \mathbf{x} = \begin{bmatrix} x_1 \\ x_2 \end{bmatrix}, \quad \mathbf{b} = \begin{bmatrix} b_1 \\ b_2 \end{bmatrix} \tag{1.34}$$

The main determinant of the system is

$$D = \begin{vmatrix} a_{11} & a_{12} \\ a_{21} & a_{22} \end{vmatrix} \tag{1.35a}$$

Let's create two additional determinants D_1 by replacing the first column of D with the column vector $\mathbf{b}$, and the determinant D_2 by replacing the second column of D by the column vector $\mathbf{b}$. That is,

$$D_1 = \begin{vmatrix} b_1 & a_{12} \\ b_2 & a_{22} \end{vmatrix} \tag{1.35b}$$

$$D_2 = \begin{vmatrix} a_{11} & b_1 \\ a_{21} & b_2 \end{vmatrix} \tag{1.35c}$$

Then the solution of the system of equations in (1.32) is

$$x_1 = \frac{D_1}{D}, \quad x_2 = \frac{D_2}{D}, \quad D \neq 0 \tag{1.36}$$

Example 1.8 Solution of systems of equations using Cramer's rule
We will use the same system of equations as in Example 1.7.

$$4x_1 + 3x_2 = 12$$
$$2x_1 + 5x_2 = -8$$

Using Cramer's rule we obtain the solutions as

$$x_1 = \frac{\begin{vmatrix} 12 & 3 \\ -8 & 5 \end{vmatrix}}{\begin{vmatrix} 4 & 3 \\ 2 & 5 \end{vmatrix}} = \frac{(12)(5)-(-8)(3)}{(4)(5)-(2)(3)} = 6$$

$$x_2 = \frac{\begin{vmatrix} 12 & 3 \\ -8 & 5 \end{vmatrix}}{\begin{vmatrix} 4 & 3 \\ 2 & 5 \end{vmatrix}} = \frac{(4)(-8)-(2)(12)}{(4)(5)-(2)(3)} = -4$$

which, of course, agrees with the solution of the previous example.

∎

1.9 Vector Operations

In this section we define two fundamental operations on them: *scalar product* and *vector product*.

1.9.1 Scalar Product

Scalar product (or inner product, or dot product) of two vectors **A** and **B**, denoted **A** · **B**, is defined as

$$\mathbf{A} \cdot \mathbf{B} = |\mathbf{A}||\mathbf{B}|\cos\gamma \tag{1.37}$$

where $0 < \gamma < \pi$ is the angle between **A** and **B** (computed when the vectors have their initial points coinciding).

Note that the result of a scalar product, as the name indicates, is a scalar (number).

Also note that when two vectors are perpendicular to each other, their scalar product is zero.

$$if \quad \gamma = 90° \quad \Rightarrow \quad \cos\gamma = 0 \tag{1.38}$$

The order of multiplication in a scalar product does not matter, that is,

$$\mathbf{A} \cdot \mathbf{B} = \mathbf{B} \cdot \mathbf{A} \tag{1.39}$$

1.9.2 Vector Product

Vector product (or cross product) of two vectors **A** and **B**, denoted $A \times B$, is defined as a vector *V* whose length is

$$|\mathbf{V}| = |\mathbf{A}||\mathbf{B}|\sin\gamma \tag{1.40}$$

where γ is the angle between **A** and **B**, and whose direction is perpendicular to both **A** and **B** and is such that **A**, **B**, and **V**, in this order, form a right-handed triple.

Note that a vector product results in a vector. Also note that when two vectors are parallel to each other, their vector product is a zero vector.

$$if \quad \gamma = 0° \quad \Rightarrow \quad \sin\gamma = 0 \tag{1.41}$$

The order of multiplication in a vector product *does* matter, since

$$\mathbf{A} \times \mathbf{B} = -\mathbf{B} \times \mathbf{A} \tag{1.42}$$

1.10 EMC Applications

1.10.1 Crosstalk Model of Transmission Lines

In this section we will show how the matrices can be used to describe a mathematical model of the crosstalk between wires in cables or between PCB traces.

Crosstalk occurs when a signal on one pair of conductors couples to an adjacent pair of conductors, causing an unintended reception of that signal at the terminals of the second pair of conductors. Figure 1.1 shows a PCB specifically designed to produce this phenomenon.

PCB geometry is shown in Figure 1.2(a) and the corresponding circuit model is shown in Figure 1.2(b).

A pair of parallel conductors called the generator (aggressor) circuit connects a source represented by V_S and R_s to a load represented by R_L. Another pair of parallel conductors is adjacent to the generator line. These conductors, the receptor (or victim) circuit, are terminated at the near and far end. Signals in the generator circuit induce voltages across the receptor circuit terminations (Adamczyk and Teune, 2009). This is shown in Figure 1.3.

The generator and receptor circuits have per-unit-length self inductances l_G and l_R, respectively, associated with them, and a per-unit-length mutual inductance l_m between the two circuits. The per-unit-length self-capacitances between the generator conductor and the reference conductor and between the receptor conductor and the reference conductor are represented by c_G and c_R, respectively. The per-unit-length mutual

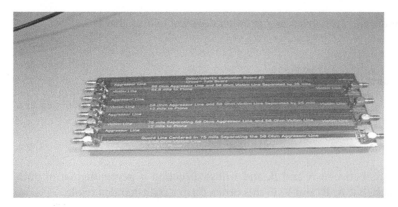

Figure 1.1 PCB used for creating crosstalk between traces.

(a)

(b)

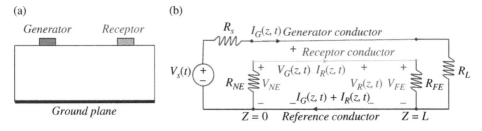

Figure 1.2 Three-conductor transmission line: (a) PCB arrangement; (b) circuit model.

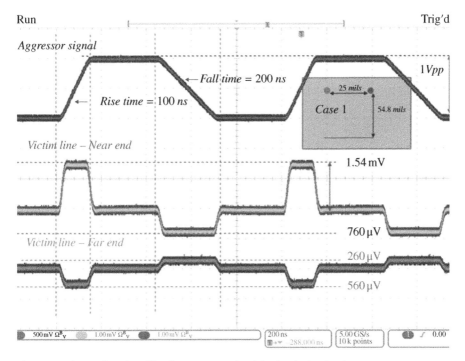

Figure 1.3 Crosstalk induced by the aggressor circuit in the victim circuit.

capacitance between the generator and receptor conductors is represented by c_m. This is shown in Figure 1.4.

Differential equations describing the model in Figure 1.4 are (Paul, 2006, pp. 565–566):

$$\frac{\partial V_G(z, t)}{\partial z} = -l_G \frac{\partial I_G(z, t)}{\partial t} - l_m \frac{\partial I_R(z, t)}{\partial t} \tag{1.43a}$$

$$\frac{\partial V_R(z, t)}{\partial z} = -l_m \frac{\partial I_G(z, t)}{\partial t} - l_R \frac{\partial I_R(z, t)}{\partial t} \tag{1.43b}$$

$$\frac{\partial I_G(z, t)}{\partial z} = -(c_G + c_m) \frac{\partial V_G(z, t)}{\partial t} + c_m \frac{\partial V_R(z, t)}{\partial t} \tag{1.43c}$$

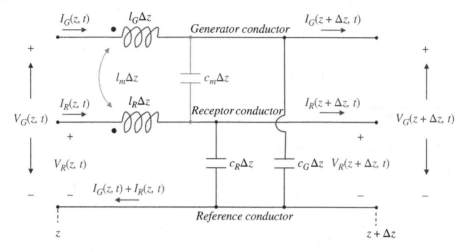

Figure 1.4 Per-unit length circuit model of three-conductor transmission line.

$$\frac{\partial I_R(z,t)}{\partial z} = c_m \frac{\partial V_G(z,t)}{\partial t} - (c_R + c_m)\frac{\partial V_R(z,t)}{\partial t} \tag{1.43d}$$

Let's introduce the following matrices:

$$\mathbf{V}(z,t) = \begin{bmatrix} V_G(z,t) \\ V_R(z,t) \end{bmatrix} \tag{1.44a}$$

$$\mathbf{I}(z,t) = \begin{bmatrix} I_G(z,t) \\ I_R(z,t) \end{bmatrix} \tag{1.44b}$$

$$\mathbf{L} = \begin{bmatrix} l_G & l_m \\ l_m & l_R \end{bmatrix} \tag{1.44c}$$

$$\mathbf{C} = \begin{bmatrix} c_G + c_m & -c_m \\ -c_m & c_R + c_m \end{bmatrix} \tag{1.44d}$$

Equations (1.43) can now be written in a matrix form as:

$$\frac{\partial}{\partial z}\mathbf{V}(z,t) = \mathbf{L}\frac{\partial}{\partial t}\mathbf{I}(z,t) \tag{1.45a}$$

$$\frac{\partial}{\partial z}\mathbf{I}(z,t) = -\mathbf{C}\frac{\partial}{\partial t}\mathbf{V}(z,t) \tag{1.45b}$$

To appreciate the usefulness of this matrix form, we will compare it to the two-conductor transmission lines equations (Paul, 2006, p. 183):

$$\frac{\partial V(z, t)}{\partial z} = -l\frac{\partial I(z, t)}{\partial t} \tag{1.46a}$$

$$\frac{\partial I(z, t)}{\partial z} = -c\frac{\partial V(z, t)}{\partial t} \tag{1.46b}$$

Notice that the equations in matrix form (1.45) have an appearance identical to that of the transmission-line equations (1.46) for a two-conductor line. *Equations of the same mathematical form have solutions of the same mathematical form.* This is a very powerful result since two-conductor transmission line theory easily provides considerable insight into the theory of multiple-line conductors.

1.10.2 Radiated Susceptibility Test

Radiated susceptibility test RS 101 of MIL STD-461-G and ISO 11452-8 standards utilizes the radiating loop fixture shown in Figure 1.5.

The fixture used can be modeled as magnetically coupled coils of a transformer shown in Figure 1.6.

Figure 1.5 Radiating loop fixture.

Figure 1.6 Coupled coils in the frequency domain.

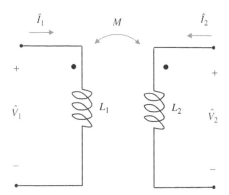

In Chapter 10 we will derive the following equations governing this circuit

$$\hat{V}_1 = j\omega L_1 \hat{I}_1 + j\omega M \hat{I}_2 \qquad (1.47a)$$

$$\hat{V}_2 = j\omega M \hat{I}_1 + j\omega L_2 \hat{I}_2 \qquad (1.47b)$$

In matrix form, Eq. (1.47) can be written as

$$\begin{bmatrix} \hat{V}_1 \\ \hat{V}_2 \end{bmatrix} = \begin{bmatrix} j\omega L_1 & j\omega M \\ j\omega M & j\omega L_2 \end{bmatrix} \begin{bmatrix} \hat{I}_1 \\ \hat{I}_2 \end{bmatrix} \qquad (1.48)$$

The solution of Eq. (1.48) can be obtained by matrix inversion or by Cramer's rule discussed in this chapter. Let's present both methods.

Matrix Inversion Approach   Using matrix inversion, the solution of Eq. (1.48) is

$$\begin{bmatrix} \hat{I}_1 \\ \hat{I}_2 \end{bmatrix} = \begin{bmatrix} j\omega L_1 & j\omega M \\ j\omega M & j\omega L_2 \end{bmatrix}^{-1} \begin{bmatrix} \hat{V}_1 \\ \hat{V}_2 \end{bmatrix} \qquad (1.49)$$

The inverse of the matrix is obtained according to Eq. (1.23) as

$$
\begin{bmatrix} j\omega L_1 & j\omega M \\ j\omega M & j\omega L_2 \end{bmatrix}^{-1} = \frac{1}{\begin{vmatrix} j\omega L_1 & j\omega M \\ j\omega M & j\omega L_2 \end{vmatrix}} \begin{bmatrix} j\omega L_2 & -j\omega M \\ -j\omega M & j\omega L_1 \end{bmatrix}
$$

$$
= \frac{1}{-\omega^2 L_1 L_2 + \omega^2 M} \begin{bmatrix} j\omega L_2 & -j\omega M \\ -j\omega M & j\omega L_1 \end{bmatrix}
$$

$$
= \begin{bmatrix} \dfrac{j\omega L_2}{-\omega^2 L_1 L_2 + \omega^2 M} & \dfrac{-j\omega M}{-\omega^2 L_1 L_2 + \omega^2 M} \\ \dfrac{-j\omega M}{-\omega^2 L_1 L_2 + \omega^2 M} & \dfrac{j\omega L_1}{-\omega^2 L_1 L_2 + \omega^2 M} \end{bmatrix} \qquad (1.50)
$$

$$
= \begin{bmatrix} \dfrac{jL_2}{\omega(M - L_1 L_2)} & \dfrac{-jM}{\omega(M - L_1 L_2)} \\ \dfrac{-jM}{\omega(M - L_1 L_2)} & \dfrac{jL_1}{\omega(M - L_1 L_2)} \end{bmatrix}
$$

And thus

$$
\begin{bmatrix} \hat{I}_1 \\ \hat{I}_2 \end{bmatrix} = \begin{bmatrix} \dfrac{jL_2}{\omega(M - L_1 L_2)} & \dfrac{-jM}{\omega(M - L_1 L_2)} \\ \dfrac{-jM}{\omega(M - L_1 L_2)} & \dfrac{jL_1}{\omega(M - L_1 L_2)} \end{bmatrix} \begin{bmatrix} \hat{V}_1 \\ \hat{V}_2 \end{bmatrix} \qquad (1.51)
$$

resulting in

$$\hat{I}_1 = \frac{jL_2}{\omega(M - L_1 L_2)}\hat{V}_1 - \frac{jM}{\omega(M - L_1 L_2)}\hat{V}_2 \tag{1.52a}$$

and

$$\hat{I}_2 = -\frac{jM}{\omega(M - L_1 L_2)}\hat{V}_1 + \frac{jL_1}{\omega(M - L_1 L_2)}\hat{V}_2 \tag{1.52b}$$

Cramer's Rule Approach   Now we will solve the system of equations

$$\begin{bmatrix} \hat{V}_1 \\ \hat{V}_2 \end{bmatrix} = \begin{bmatrix} j\omega L_1 & j\omega M \\ j\omega M & j\omega L_2 \end{bmatrix} \begin{bmatrix} \hat{I}_1 \\ \hat{I}_2 \end{bmatrix} \tag{1.53}$$

using the Cramer's rule approach.

The main determinant of the matrix in Eq. (1.53) is

$$\Delta = \begin{vmatrix} j\omega L_1 & j\omega M \\ j\omega M & j\omega L_2 \end{vmatrix} = -\omega^2 L_1 L_2 + \omega^2 M \tag{1.54a}$$

The remaining two determinants are

$$\Delta_1 = \begin{vmatrix} \hat{V}_1 & j\omega M \\ \hat{V}_2 & j\omega L_2 \end{vmatrix} = j\omega L_2 \hat{V}_1 - j\omega M \hat{V}_2 \tag{1.54b}$$

$$\Delta_2 = \begin{vmatrix} j\omega L_1 & \hat{V}_1 \\ j\omega M & \hat{V}_2 \end{vmatrix} = j\omega L_1 \hat{V}_2 - j\omega M \hat{V}_1 \tag{1.54c}$$

Therefore,

$$\hat{I}_1 = \frac{\Delta_1}{\Delta} = \frac{j\omega L_2 \hat{V}_1 - j\omega M \hat{V}_2}{-\omega^2 L_1 L_2 + \omega^2 M}$$
$$= \frac{jL_2}{\omega(M - L_1 L_2)}\hat{V}_1 - \frac{jM}{\omega(M - L_1 L_2)}\hat{V}_2 \tag{1.55a}$$

$$\hat{I}_2 = \frac{\Delta_2}{\Delta} = \frac{j\omega L_1 \hat{V}_2 - j\omega M \hat{V}_1}{-\omega^2 L_1 L_2 + \omega^2 M}$$
$$= -\frac{jM}{\omega(M - L_1 L_2)}\hat{V}_1 + \frac{jL_1}{\omega(M - L_1 L_2)}\hat{V}_2 \tag{1.55b}$$

The solutions (1.52) and (1.56) are, of course, identical.

1.10.3 *s* Parameters

To characterize high-frequency circuits, scattering parameters, or *s* parameters, are used (Ludwig and Bogdanov, 2009). Just like the other sets of parameters (*z, y, h, g,* to be discussed in Chapter 9), the *s* parameters completely describe the performance of a two-port network.

Unlike the other sets of parameters, *s* parameters do not make use of open-circuit or short-circuit measurements, but rather relate the traveling waves that are incident, reflected, and transmitted when a two-port network is inserted into a transmission line. (We will discuss travelling waves in Chapter 16 and transmission lines in Chapter 17.)

Figure 1.7 shows a two-port network (circuit or device) together with the incident, reflected, and transmitted waves.

The incident waves (a_1, a_2) and reflected waves (b_1, b_2) used to define *s* parameters for a two-port network are shown in Figure 1.8.

The linear equations describing this two-port network in terms of the *s* parameters are

$$b_1 = s_{11}a_1 + s_{12}a_2$$
$$b_2 = s_{21}a_1 + s_{22}a_2$$

$$(1.56)$$

Or in a matrix form,

$$\begin{bmatrix} b_1 \\ b_2 \end{bmatrix} = \begin{bmatrix} s_{11} & s_{12} \\ s_{21} & s_{22} \end{bmatrix} \begin{bmatrix} a_1 \\ a_2 \end{bmatrix}$$

$$(1.57)$$

where S is the *scattering matrix* given by

$$\mathbf{S} = \begin{bmatrix} s_{11} & s_{12} \\ s_{21} & s_{22} \end{bmatrix}$$

$$(1.58)$$

We will discuss *s* parameters in detail in Chapter 17.

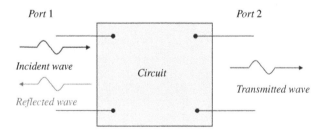

Figure 1.7 Two-port network and the travelling waves.

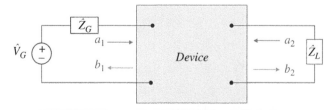

Figure 1.8 Incident and reflected waves at port 1 and port 2.

References

Adamczyk, B. and Teune, J., "EMC Crosstalk vs. Circuit Topology", ASEE North Central Section Spring Conference, Grand Rapids, MI, 2009.

Kreyszig, E., *Advanced Engineering Mathematics*, 8th ed., John Wiley and Sons, New York, 1999.

Ludwig, R. and Bogdanov, G., *RF Circuit Design*, 2nd ed., Pearson, Upper Saddle River, NJ, 2009.

Paul, C.R., *Introduction to Electromagnetic Compatibility*, 2nd ed., John Wiley and Sons, New York, 2006.

2

Coordinate Systems

In this chapter we discuss three coordinate systems frequently encountered in electromagnetics: Cartesian, cylindrical, and spherical. In each system we define the relevant operations and properties. We conclude by showing the transformations between the systems. These transformations are necessary when deriving the radiation from a Hertzian dipole, as shown in the EMC applications section at the end of this chapter.

2.1 Cartesian Coordinate System

Cartesian coordinate system is shown in Figure 2.1.

Unit vectors in this system, denoted a_x, a_y, and a_z, are usually drawn at the origin (but can be drawn at any point in space). They point in the direction of the increasing coordinate variables, and are orthogonal to each other.

A point P can be represented as a triple of numbers

$$P : (x, y, z) \tag{2.1}$$

where x, y, and z are called the *coordinates* of P.

The *ranges* of the coordinate variables are

$$-\infty < x < \infty$$

$$-\infty < y < \infty \tag{2.2}$$

$$-\infty < z < \infty$$

A *vector* **A** can be represented as a triple

$$\mathbf{A} = \left(A_x, A_y, A_z \right) \tag{2.3}$$

where A_x, A_y, and A_z are called the *components* of **A**.

A vector **A** can be decomposed into a sum of three vectors along the coordinate directions, as shown in Figure 2.2.

This decomposition can be expressed as

$$\mathbf{A} = \mathbf{A}_x + \mathbf{A}_y + \mathbf{A}_z \tag{2.4}$$

This seemingly obvious decomposition is extremely useful when evaluating line and surface integrals, as we shall see.

Foundations of Electromagnetic Compatibility with Practical Applications, First Edition. Bogdan Adamczyk.
© 2017 John Wiley & Sons Ltd. Published 2017 by John Wiley & Sons Ltd.

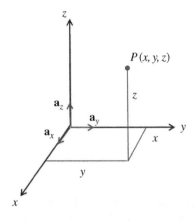

Figure 2.1 Cartesian coordinate system.

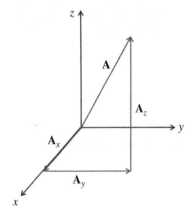

Figure 2.2 Vector decomposition.

In terms of its components, we can also write

$$\mathbf{A} = \left(A_x, A_y, A_z\right) = A_x\mathbf{a}_x + A_y\mathbf{a}_y + A_z\mathbf{a}_z \tag{2.5}$$

where the unit vectors in terms of their own components are defined as

$$\mathbf{a}_x = \left(1, 0, 0\right) \tag{2.6a}$$

$$\mathbf{a}_y = \left(0, 1, 0\right) \tag{2.6b}$$

$$\mathbf{a}_z = \left(0, 0, 1\right) \tag{2.6c}$$

Since the unit vectors are perpendicular, it follows

$$\mathbf{a}_x \cdot \mathbf{a}_y = \mathbf{a}_y \cdot \mathbf{a}_z = \mathbf{a}_z \cdot \mathbf{a}_x = 0 \tag{2.7a}$$

$$\mathbf{a}_x \cdot \mathbf{a}_x = \mathbf{a}_y \cdot \mathbf{a}_y = \mathbf{a}_z \cdot \mathbf{a}_z = 1 \tag{2.7b}$$

Figure 2.3 Cross product using cyclic permutations.

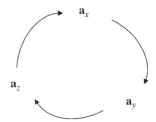

In many electromagnetics problems, we need to determine the direction of a vector resulting from a vector product of two vectors along the coordinate directions. The following equations show this result.

$$\mathbf{a}_x \times \mathbf{a}_y = \mathbf{a}_z \tag{2.8a}$$

$$\mathbf{a}_y \times \mathbf{a}_z = \mathbf{a}_x \tag{2.8b}$$

$$\mathbf{a}_z \times \mathbf{a}_x = \mathbf{a}_y \tag{2.8c}$$

These cross products can be easily obtained with the help of the cyclic permutations (Sadiku, 2010, p.15) shown in Figure 2.3.

If **A** and **B** are represented in terms of components, say, $\mathbf{A} = (A_x, A_y, A_z)$ and $\mathbf{B} = (B_x, B_y, B_z)$, their scalar product is given by

$$\mathbf{A} \cdot \mathbf{B} = A_x B_x + A_y B_y + A_z B_z \tag{2.9}$$

The vector product can be obtained by evaluating the following "determinant" (using the approach discussed in Section 1.5).

$$\mathbf{A} \times \mathbf{B} = \begin{vmatrix} \mathbf{a}_x & \mathbf{a}_y & \mathbf{a}_z \\ A_x & A_y & A_z \\ B_x & B_y & B_z \end{vmatrix} \tag{2.10}$$

Note that

$$\mathbf{A} \cdot \mathbf{A} = A_x^2 + A_y^2 + A_z^2 = |\mathbf{A}|^2 \tag{2.11}$$

and thus, the magnitude of a vector is

$$|\mathbf{A}| = \sqrt{A_x^2 + A_y^2 + A_z^2} \tag{2.12}$$

Also,

$$\mathbf{A} \times \mathbf{A} = 0 \tag{2.13}$$

2.2 Cylindrical Coordinate System

The cylindrical coordinate system is an extension of a polar system from plane to space. The cylindrical coordinate system, shown in Figure 2.4, is very convenient whenever we are dealing with problems having cylindrical symmetry (e.g. coaxial cable).

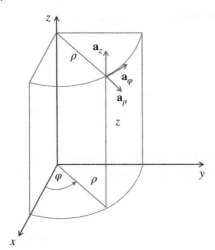

Figure 2.4 Cylindrical coordinate system.

Unit vectors in this system, denoted $\mathbf{a}_\rho$, $\mathbf{a}_\varphi$, and $\mathbf{a}_z$, are usually *not* drawn at the origin but at a convenient point in space. They point in the direction of the increasing coordinate variables, and are orthogonal to each other.

A point *P* can be represented as a triple of numbers

$$P:(\rho, \phi, z) \tag{2.14}$$

where ρ, φ, and z are called the *coordinates* of P.

The *ranges* of the coordinate variables are

$$0 \leq \rho < \infty$$
$$0 \leq \varphi < 2\pi \tag{2.15}$$
$$-\infty < z < \infty$$

A *vector* **A** can be represented as a triple

$$\mathbf{A} = (A_\rho, A_\phi, A_z) \tag{2.16}$$

where A_ρ, A_φ, and A_z are called the *components* of **A**.

A vector **A** can be represented as

$$\mathbf{A} = \mathbf{A}_\rho + \mathbf{A}_\phi + \mathbf{A}_z = A_\rho \mathbf{a}_\rho + A_\phi \mathbf{a}_\phi + A_z \mathbf{a}_z \tag{2.17}$$

The unit vectors in terms of their own components are defined as

$$\mathbf{a}_\rho = (1, 0, 0) \tag{2.18a}$$

$$\mathbf{a}_\phi = (0, 1, 0) \tag{2.18b}$$

$$\mathbf{a}_z = (0, 0, 1) \tag{2.18c}$$

Figure 2.5 Cross product using cyclic permutations in cylindrical system.

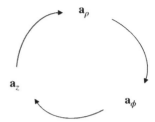

It also follows that

$$\mathbf{a}_\rho \cdot \mathbf{a}_\phi = \mathbf{a}_\phi \cdot \mathbf{a}_z = \mathbf{a}_z \cdot \mathbf{a}_\rho = 0 \qquad (2.19a)$$

$$\mathbf{a}_\rho \cdot \mathbf{a}_\rho = \mathbf{a}_\phi \cdot \mathbf{a}_\phi = \mathbf{a}_z \cdot \mathbf{a}_z = 1 \qquad (2.19b)$$

With the help of Figure 2.5 we obtain

$$\mathbf{a}_\rho \times \mathbf{a}_\phi = \mathbf{a}_z \qquad (2.20a)$$

$$\mathbf{a}_\phi \times \mathbf{a}_z = \mathbf{a}_\rho \qquad (2.20b)$$

$$\mathbf{a}_z \times \mathbf{a}_\rho = \mathbf{a}_\phi \qquad (2.20c)$$

If A and B are represented in terms of components, say, $\mathbf{A} = (A_\rho, A_\phi, A_z)$ and $\mathbf{B} = (B_\rho, B_\phi, B_z)$, their scalar product is given by

$$\mathbf{A} \cdot \mathbf{B} = A_\rho B_\rho + A_\phi B_\phi + A_z B_z \qquad (2.21)$$

The vector product can be obtained by evaluating the following "determinant":

$$\mathbf{A} \times \mathbf{B} = \begin{vmatrix} \mathbf{a}_\rho & \mathbf{a}_\phi & \mathbf{a}_z \\ A_\rho & A_\phi & A_z \\ B_\rho & B_\phi & B_z \end{vmatrix} \qquad (2.22)$$

The magnitude of the vector is

$$|\mathbf{A}| = \sqrt{A_\rho^2 + A_\phi^2 + A_z^2} \qquad (2.23)$$

2.3   Spherical Coordinate System

The spherical coordinate system, shown in Figure 2.6, is very convenient whenever we are dealing with problems having spherical symmetry (e.g. Hertzian dipole described in Section 2.5).

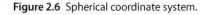

Figure 2.6 Spherical coordinate system.

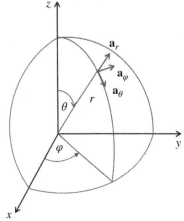

Unit vectors in this system, denoted $\mathbf{a}_r$, $\mathbf{a}_\theta$, and $\mathbf{a}_\varphi$, are usually *not* drawn at the origin but again at a convenient point in space. They point in the direction of the increasing coordinate variables, and are orthogonal to each other.

A point P can be represented as a triple of numbers

$$P:(r,\theta,\phi) \tag{2.24}$$

where r, θ, and φ, are called the *coordinates* of P.

The *ranges* of the coordinate variables are

$$0 \le r < \infty$$
$$0 \le \theta < \pi \tag{2.25}$$
$$0 \le \varphi < 2\pi$$

A *vector A* can be represented as a triple

$$\mathbf{A} = \left(A_r, A_\theta, A_\phi \right) \tag{2.26}$$

where A_r, A_θ, and A_φ are called the *components* of $\mathbf{A}$.

A vector $\mathbf{A}$ can be represented as

$$\mathbf{A} = \mathbf{A}_r + \mathbf{A}_\theta + \mathbf{A}_\phi = A_r \mathbf{a}_r + A_\theta \mathbf{a}_\theta + A_\phi \mathbf{a}_\phi \tag{2.27}$$

The unit vectors in terms of their own components are defined as

$$\mathbf{a}_r = \left(1,\, 0,\, 0 \right) \tag{2.28a}$$

$$\mathbf{a}_\theta = \left(0,\, 1,\, 0 \right) \tag{2.28b}$$

$$\mathbf{a}_\phi = \left(0,\, 0,\, 1 \right) \tag{2.28c}$$

Figure 2.7 Cross product using cyclic permutations in spherical system.

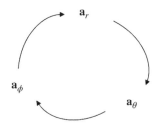

It also follows that

$$\mathbf{a}_r \cdot \mathbf{a}_\theta = \mathbf{a}_\theta \cdot \mathbf{a}_\phi = \mathbf{a}_r \cdot \mathbf{a}_r = 0 \qquad (2.29a)$$

$$\mathbf{a}_r \cdot \mathbf{a}_r = \mathbf{a}_\theta \cdot \mathbf{a}_\theta = \mathbf{a}_\phi \cdot \mathbf{a}_\phi = 1 \qquad (2.29b)$$

With the help of Figure 2.7 we obtain

$$\mathbf{a}_r \times \mathbf{a}_\theta = \mathbf{a}_\phi \qquad (2.30a)$$

$$\mathbf{a}_\theta \times \mathbf{a}_\phi = \mathbf{a}_r \qquad (2.30b)$$

$$\mathbf{a}_\phi \times \mathbf{a}_r = \mathbf{a}_\theta \qquad (2.30c)$$

If **A** and **B** are represented in terms of components, say, $\mathbf{A} = (A_r, A_\theta, A_\phi)$ and $\mathbf{B} = (B_r, B_\theta, B_\phi)$, their scalar product is given by

$$\mathbf{A} \cdot \mathbf{B} = A_r B_r + A_\theta B_\theta + A_\phi B_\phi \qquad (2.31)$$

The vector product can be obtained by evaluating the following "determinant":

$$\mathbf{A} \times \mathbf{B} = \begin{vmatrix} \mathbf{a}_r & \mathbf{a}_\theta & \mathbf{a}_\phi \\ A_r & A_\theta & A_\phi \\ B_r & B_\theta & B_\phi \end{vmatrix} \qquad (2.32)$$

The magnitude of a vector is

$$|\mathbf{A}| = \sqrt{A_r^2 + A_\theta^2 + A_\phi^2} \qquad (2.33)$$

2.4 Transformations between Coordinate Systems

In this section we discuss transformations between Cartesian and cylindrical systems, as well as the transformations between Cartesian and spherical systems.

2.4.1 Transformation between Cartesian and Cylindrical Systems

The relationships between the variables (x, y, z) of the Cartesian coordinate system and those of the cylindrical system (ρ, ϕ, z) are easily obtained from Figure 2.8.

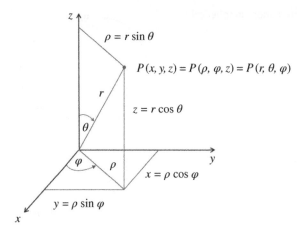

Figure 2.8 Relationship between Cartesian and cylindrical variables.

Coordinate Transformations from Cartesian to Cylindrical System

$$\rho = \sqrt{x^2 + y^2}$$

$$\phi = \tan^{-1}\frac{y}{x} \tag{2.34}$$

$$z = z$$

Coordinate Transformations from Cylindrical to Cartesian System

$$x = \rho\cos\phi$$

$$y = \rho\sin\phi \tag{2.35}$$

$$z = z$$

From Figure 2.8 we can geometrically obtain the relationships between the vector components in the two coordinate systems.

Vector Components Transformations from Cartesian to Cylindrical System (Sadiku, 2010, p. 36)

$$\begin{bmatrix} A_\rho \\ A_\phi \\ A_z \end{bmatrix} = \begin{bmatrix} \cos\phi & \sin\phi & 0 \\ -\sin\phi & \cos\phi & 0 \\ 0 & 0 & 1 \end{bmatrix} \begin{bmatrix} A_x \\ A_y \\ A_z \end{bmatrix} \tag{2.36}$$

Vector Components Transformations from Cylindrical to Cartesian System

$$\begin{bmatrix} A_x \\ A_y \\ A_z \end{bmatrix} = \begin{bmatrix} \cos\phi & -\sin\phi & 0 \\ \sin\phi & \cos\phi & 0 \\ 0 & 0 & 1 \end{bmatrix} \begin{bmatrix} A_\rho \\ A_\phi \\ A_z \end{bmatrix} \tag{2.37}$$

Example 2.1 Coordinate transformations from Cartesian to cylindrical system
Express the point $P:(1,-4,-3)$ in cylindrical coordinates.

Solution:

$$\rho = \sqrt{x^2 + y^2} = \sqrt{1+16} = 4.1231$$

$$\varphi = \tan^{-1}\frac{y}{x} = \tan^{-1}\frac{(-4)}{1} = -75.96°$$

$$z = z = -3$$

Thus,

$$P:(1,-4,-3) = P:(4.1231,-75.96°,-3)$$

∎

Note: The transformation $(A_\rho, A_\varphi, A_z) \rightarrow (A_x, A_y, A_z)$ as given by Eq. (2.37) is not complete. To complete it we need to express $\sin\phi$ and $\cos\phi$ in terms of $x, y,$ and z. From Eq. (2.35) we get

$$\cos\phi = \frac{x}{\rho} = \frac{x}{\sqrt{x^2 + y^2}} \qquad (2.38a)$$

$$\sin\phi = \frac{y}{\rho} = \frac{y}{\sqrt{x^2 + y^2}} \qquad (2.38b)$$

And thus the transformation $(A_\rho, A_\varphi, A_z) \rightarrow (A_x, A_y, A_z)$ can be expressed as

$$\begin{bmatrix} A_x \\ A_y \\ A_z \end{bmatrix} = \begin{bmatrix} \dfrac{x}{\sqrt{x^2 + y^2}} & -\dfrac{y}{\sqrt{x^2 + y^2}} & 0 \\ \dfrac{y}{\sqrt{x^2 + y^2}} & \dfrac{x}{\sqrt{x^2 + y^2}} & 0 \\ 0 & 0 & 1 \end{bmatrix} \begin{bmatrix} A_\rho \\ A_\phi \\ A_z \end{bmatrix} \qquad (2.39)$$

The cylindrical variables appearing in (A_ρ, A_φ, A_z) need to be expressed in terms of Cartesian variables before the matrix operation takes place, as illustrated by the following example.

Example 2.2 Component transformation from cylindrical to Cartesian system
Convert the following vector to Cartesian coordinates:

$$C = z\sin\phi \ \mathbf{a}_\rho - \rho\cos\phi \ \mathbf{a}_\phi + 2\rho z \ \mathbf{a}_z$$

Using Eq. (2.39) we have

$$
\begin{bmatrix} C_x \\ C_y \\ C_z \end{bmatrix} = \begin{bmatrix} \dfrac{x}{\sqrt{x^2+y^2}} & -\dfrac{y}{\sqrt{x^2+y^2}} & 0 \\ \dfrac{y}{\sqrt{x^2+y^2}} & \dfrac{x}{\sqrt{x^2+y^2}} & 0 \\ 0 & 0 & 1 \end{bmatrix} \begin{bmatrix} C_\rho \\ C_\phi \\ C_z \end{bmatrix} = \begin{bmatrix} \dfrac{x}{\sqrt{x^2+y^2}} & -\dfrac{y}{\sqrt{x^2+y^2}} & 0 \\ \dfrac{y}{\sqrt{x^2+y^2}} & \dfrac{x}{\sqrt{x^2+y^2}} & 0 \\ 0 & 0 & 1 \end{bmatrix} \begin{bmatrix} z\sin\phi \\ \rho\cos\phi \\ 2\rho z \end{bmatrix}
$$

Now that the cylindrical variables appearing in (C_ρ, C_ϕ, C_z) need to be expressed in terms of the Cartesian variables before the matrix operation takes place.

Using the relationships. (2.34) and (2.38) we obtain

$$
\begin{bmatrix} C_x \\ C_y \\ C_z \end{bmatrix} = \begin{bmatrix} \dfrac{x}{\sqrt{x^2+y^2}} & -\dfrac{y}{\sqrt{x^2+y^2}} & 0 \\ \dfrac{y}{\sqrt{x^2+y^2}} & \dfrac{x}{\sqrt{x^2+y^2}} & 0 \\ 0 & 0 & 1 \end{bmatrix} \begin{bmatrix} z\dfrac{y}{\sqrt{x^2+y^2}} \\ \left(-\sqrt{x^2+y^2}\right)\left(\dfrac{x}{\sqrt{x^2+y^2}}\right) \\ 2z\sqrt{x^2+y^2} \end{bmatrix}
$$

and thus

$$
\begin{bmatrix} C_x \\ C_y \\ C_z \end{bmatrix} = \begin{bmatrix} \dfrac{xyz}{x^2+y^2} + \dfrac{xy}{\sqrt{x^2+y^2}} \\ \dfrac{zy^2}{x^2+y^2} - \dfrac{x^2}{x^2+y^2} \\ 2z\sqrt{x^2+y^2} \end{bmatrix}
$$

2.4.2 Transformation between Cartesian and Spherical Systems

Coordinate Transformations from Cartesian to Spherical System (Sadiku, 2010, p. 36)

$$
\begin{aligned}
r &= \sqrt{x^2+y^2+z^2} \\
\theta &= \tan^{-1}\dfrac{\sqrt{x^2+y^2}}{z} \\
\phi &= \tan^{-1}\dfrac{y}{x}
\end{aligned}
\qquad (2.40)
$$

Coordinate Transformations from Spherical to Cartesian System

$$x = r\sin\theta\cos\phi$$
$$y = r\sin\theta\sin\phi \qquad (2.41)$$
$$z = r\cos\theta$$

Vector Components Transformations from Cartesian to Spherical System

$$\begin{bmatrix} A_r \\ A_\theta \\ A_\phi \end{bmatrix} = \begin{bmatrix} \sin\theta\cos\phi & \sin\theta\sin\phi & \cos\theta \\ \cos\theta\cos\phi & \cos\theta\sin\phi & -\sin\theta \\ -\sin\phi & \cos\phi & 0 \end{bmatrix} \begin{bmatrix} A_x \\ A_y \\ A_z \end{bmatrix} \qquad (2.42)$$

Vector Components Transformations from Spherical to Cartesian System

$$\begin{bmatrix} A_x \\ A_y \\ A_z \end{bmatrix} = \begin{bmatrix} \sin\theta\cos\phi & \cos\theta\cos\phi & -\sin\phi \\ \sin\theta\sin\phi & \cos\theta\sin\phi & \cos\phi \\ \cos\theta & -\sin\theta & 0 \end{bmatrix} \begin{bmatrix} A_r \\ A_\theta \\ A_\phi \end{bmatrix} \qquad (2.43)$$

2.5 EMC Applications

In this section we will show an important EMC application of the Cartesian-to-spherical systems transformations: derivations of the electric and magnetic fields radiated by a Hertzian (electric) dipole antenna.

2.5.1 Radiation Fields of an Electric Dipole Antenna

The electric or Hertzian dipole is perhaps the most fundamental antenna that facilitates the derivation of expressions for electric and magnetic field intensities of many practical antennas, like the monopole antenna used in EMC compliance testing shown in Figure 2.9.

We model the Hertzian dipole as a very short current element of length l, carrying a constant current I_0. The current element is positioned symmetrically at the origin of the coordinate system and oriented along the z axis, as shown in Figure 2.10.

In Section 6.7.3 we will show that the electric and magnetic field intensities can be calculated from the vector magnetic potential **A**, shown in Figure 2.10. At a distance r from the dipole, the vector magnetic potential is given by

$$\mathbf{A}(x, y, z) = \frac{\mu I_0 l}{4\pi r} e^{-jkr} \mathbf{a}_z \qquad (2.44a)$$

where

$$r = \sqrt{x^2 + y^2 + z^2} \qquad (2.44b)$$

Figure 2.9 Monopole antenna used in EMC compliance testing.

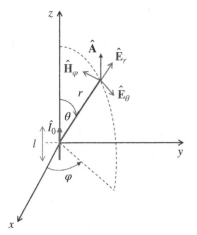

Figure 2.10 Hertzian dipole.

The vector magnetic potential in Eq. (2.44a) is expressed in Cartesian coordinate system – with the substitution of Eq. (2.44b) into it. To derive electric and magnetic field intensities at a distance r from the dipole, it is more convenient to express **A** in spherical coordinate system. This transformation is accomplished through:

$$\begin{bmatrix} A_r \\ A_\theta \\ A_\phi \end{bmatrix} = \begin{bmatrix} \sin\theta\cos\phi & \sin\theta\sin\phi & \cos\theta \\ \cos\theta\cos\phi & \cos\theta\sin\phi & -\sin\theta \\ -\sin\phi & \cos\phi & 0 \end{bmatrix} \begin{bmatrix} A_x \\ A_y \\ A_z \end{bmatrix} \tag{2.45}$$

For this problem $A_x = A_y = 0$, and thus Eq. (2.45) becomes

$$
\begin{bmatrix} A_r \\ A_\theta \\ A_\varphi \end{bmatrix} = \begin{bmatrix} \sin\theta\cos\varphi & \sin\theta\sin\varphi & \cos\theta \\ \cos\theta\cos\varphi & \cos\theta\sin\varphi & -\sin\theta \\ -\sin\varphi & \cos\varphi & 0 \end{bmatrix} \begin{bmatrix} 0 \\ 0 \\ \dfrac{\mu I_0 l}{4\pi r} e^{-jkr} \end{bmatrix} \tag{2.46}
$$

Therefore, in spherical coordinate system, the vector magnetic potential is expressed as

$$
\begin{bmatrix} A_r \\ A_\theta \\ A_\varphi \end{bmatrix} = \begin{bmatrix} \dfrac{\mu I_0 l e^{-jkr}}{4\pi r}\cos\theta \\ -\dfrac{\mu I_0 l e^{-jkr}}{4\pi r}\sin\theta \\ 0 \end{bmatrix} \tag{2.47}
$$

We will use this result in Section 6.7.4 to derive the radiation fields of a Hertzian dipole, (Paul, 2006, pp. 422–423):

$$
H_\varphi = \frac{I_0 dl}{4\pi}\beta_0^2 \sin\theta \left[j\frac{1}{\beta_0 r} + \frac{1}{\beta_0^2 r^2} \right] e^{-j\beta_0 r} \tag{2.48a}
$$

$$
E_r = 2\frac{I_0 dl}{4\pi r^2}\eta_0\beta_0^2 \cos\theta \left[\frac{1}{\beta_0^2 r^2} - j\frac{1}{\beta_0^3 r^3} \right] e^{-j\beta_0 r} \tag{2.48b}
$$

$$
E_\theta = \frac{I_0 dl}{4\pi}\eta_0\beta_0^2 \sin\theta \left[\frac{j}{\beta_0 r} + \frac{1}{\beta_0^2 r^2} - \frac{j}{\beta_0^3 r^3} \right] e^{-j\beta_0 r} \tag{2.48c}
$$

References

Sadiku, M.N.O., *Elements of Electromagnetics*, 5th ed., Oxford University Press, New York, 2010.

Paul, C.R., *Introduction to Electromagnetic Compatibility*, 2nd ed., John Wiley and Sons, New York, 2006.

3

Vector Differential Calculus

3.1 Derivatives

3.1.1 Basic Definition and Formulas

Derivatives describe the *rate* at which things change. The derivative is defined as

$$f'(x) = \lim_{\Delta x \to 0} \frac{f(x + \Delta x) - f(x)}{\Delta x} \tag{3.1}$$

(We will use this definition when deriving transmission line equations at the end of this chapter.)

Other Notations for Derivatives   Let y be the function of x, that is, $y = f(x)$. We often use the shorthand notation y' or $f'(x)$ to denote the derivative of y (Simon, 1982, p. 115). This notation does not indicate the variable with respect to which the derivative is evaluated (y could be a function not only of x but also of other variables). In many applications, it is important to identify that variable. We therefore use the alternative notation $\frac{dy}{dx}$ or $\frac{df(x)}{dx}$ to indicate that derivative of y is computed with respect to the variable x.

Derivative Formulas   Computing derivatives using the definition (3.1) can be tedious. Fortunately, such computations are usually unnecessary because there are derivative formulas that enable us to find the derivatives without computing limits.

Next we will state several useful formulas for derivatives

$$f(x) = const \;\Rightarrow\; f'(x) = 0 \tag{3.2a}$$

$$f(x) = x \;\Rightarrow\; f'(x) = 1 \tag{3.2b}$$

$$f(x) = x^a \;\Rightarrow\; f'(x) = ax^{a-1} \tag{3.2c}$$

$$f(x) = e^x \;\Rightarrow\; f'(x) = e^x \tag{3.2d}$$

$$f(x) = e^{ax} \;\Rightarrow\; f'(x) = ae^{ax} \tag{3.2e}$$

$$f(x) = \ln x \implies f'(x) = \frac{1}{x}, \quad x > 0 \tag{3.2f}$$

$$f(x) = \sin x \implies f'(x) = \cos x \tag{3.2g}$$

$$f(x) = \sin ax \implies f'(x) = a \cos ax \tag{3.2h}$$

$$f(x) = \cos x \implies f'(x) = -\sin x \tag{3.2i}$$

$$f(x) = \cos ax \implies f'(x) = -a \sin ax \tag{3.2j}$$

Derivative Properties:

$$(af)' = af' \tag{3.3a}$$

$$(f + g)' = f' + g' \tag{3.3b}$$

$$(fg)' = f'g + fg' \tag{3.3c}$$

$$\left(\frac{f}{g}\right)' = \frac{f'g - fg'}{g^2} \tag{3.3d}$$

Example 3.1   Derivative of a product

Let $f(x) = 2x^3 + 4x$ and $g(x) = x^2 - 1a$. Find the derivative of their product.

Solution A: Let's first multiply the functions out and then take the derivative.

$$(fg) = (2x^3 + 4x)(x^2 - 1) = 2x^5 - 2x^3 + 4x^3 - 4x = 2x^5 + 2x^3 - 4x$$

Thus

$$\frac{d}{dx}(fg) = \frac{d}{dx}(2x^5 + 2x^3 - 4x) = 2(5x^4) + 2(3x^2) - 4 = 10x^4 + 6x^2 - 4$$

Solution B: Let's make use of Eq. (3.3c).

$$f = 2x^3 + 4x, \quad f' = 6x^2 + 4$$
$$g = x^2 - 1, \quad g' = 2x$$

Thus,

$$(fg)' = f'g + fg' = \left(6x^2 + 4\right)\left(x^2 - 1\right) + \left(2x^3 + 4x\right)(2x)$$
$$= 6x^4 - 6x^2 + 4x^2 - 4 + 4x^4 + 8x^2 = 10x^4 + 6x^2 - 4$$

which, of course, agrees with the Solution *A*. ∎

Example 3.2   Derivative of a quotient
Let $f(x) = 2x + 4$ and $g(x) = 4x - 1$. Find the derivative of their quotient.

Solution:

$$\left(\frac{f}{g}\right)' = \frac{f'g - fg'}{g^2} = \frac{2(4x-1) - (2x+4)4}{(4x-1)^2} = \frac{8x - 2 - 8x - 16}{(4x-1)^2} = -\frac{18}{(4x-1)^2}$$ ∎

Second-Order Derivatives

It is often useful to know the derivative of f'; that is, $(f')'$. This is called the *second derivative* of f. The notation used is f'' or $\dfrac{d^2 f}{dx^2}$.

Example 3.3   Second derivative
Find the second derivative of $y = 6x^5 - 4x^3 + 2x^2$.

Solution:

$$y' = 30x^4 - 12x^2 + 4x$$
$$y'' = 120x^3 - 24x + 4$$

3.1.2   Composite Function and Chain Rule

The *composite function* of $f(x)$ and $g(x)$ is a function $f(g(x))$.
For instance:

$$f(x) = \cos x, \quad g(x) = \sqrt{x}. \text{ Then } f(g(x)) = \cos\sqrt{x}$$
$$f(x) = \sin x, \quad g(x) = x^2. \text{ Then } f(g(x)) = \sin x^2 \text{ and } g(g(f)) = \sin^2 x$$

Chain Rule – Derivative of a composite function

$$\left[f(g(x))\right]' = f'(g(x)) \cdot g'(x) \tag{3.4}$$

Example 3.4   Chain rule
Use the chain rule to differentiate $(3x^2 + 4x)^2$.

Solution:

$$\left[\left(3x^2 + 4x\right)^2\right]' = 2\left(3x^2 + 4x\right)(6x + 4)$$

3.1.3   Partial Derivative

When we have a function of several variables, say $f = f(x, y, z)$ we can obtain partial derivatives. A *partial derivative* of a function of several variables is its derivative with respect to one of those variables, with the other variables treated as constants.

A partial variable of $f(x, y, z)$ with respect to x is often denoted as $\dfrac{\partial f}{\partial x}$, while that with respect to y is denoted by $\dfrac{\partial f}{\partial y}$, and so on.

Example 3.5 Partial derivatives

Let $f(x, y, z) = 3x^2 y + 2yz$. Determine the partial derivatives of f with respect to x, y, and z.

Solution:

$$\frac{\partial f}{\partial x} = 6xy, \quad \frac{\partial f}{\partial y} = 3x^2 + 2z, \quad \frac{\partial f}{\partial z} = 2y$$

∎

3.2 Differential Elements

In the study of electromagnetics we often need to perform line, surface, and volume integrations. We will discuss these integrals in Chapter 4. The evaluation of these integrals in a particular coordinate system requires the knowledge of differential elements of length, surface, and volume. In the following subsections we describe these differential elements in each coordinate system.

3.2.1 Differential Length Element

In Section 4.1 we will introduce and learn how to evaluate line integrals of the form

$$\int_c \mathbf{F} \cdot d\mathbf{l} \tag{3.5}$$

The vector, $d\mathbf{l}$, appearing on the right-hand side of the scalar product in Eq. (3.5) is called the *differential length vector*, or the *differential displacement vector*.

Differential Length in Cartesian System Differential displacement (or length) $d\mathbf{l}$ in Cartesian coordinate system is a vector defined by

$$d\mathbf{l} = (dl_x, dl_y, dl_z) = d\mathbf{l}_x + d\mathbf{l}_y + d\mathbf{l}_z \tag{3.6}$$

and is shown in Figure 3.1. The figure shows a decomposition of a differential displacement vector $d\mathbf{l}$ into the differential displacement vectors along the coordinate axes.

Figure 3.1 Differential displacement in the Cartesian system.

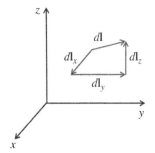

Figure 3.2 Differential displacement in cylindrical system.

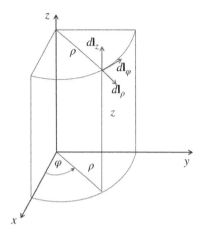

In terms of unit vectors, the differential displacement vectors along the coordinate directions can also be expressed as

$$d\mathbf{l} = dl_x \mathbf{a}_x + dl_y \mathbf{a}_y + dl_z \mathbf{a}_z \tag{3.7}$$

In Cartesian system, the differential amount of displacement, dl_x, dl_y, and dl_z, is simply dx, dy, and dz, respectively, thus we may express the differential displacement vector in (3.7) in terms of the coordinate variables as

$$d\mathbf{l} = \left(dx, dy, dz\right) = dx \mathbf{a}_x + dy \mathbf{a}_y + dz \mathbf{a}_z \tag{3.8}$$

Differential Displacement in Cylindrical System *Differential displacement (or length)* $d\mathbf{l}$ in cylindrical coordinate system is a vector defined by

$$d\mathbf{l} = \left(dl_\rho, dl_\phi, dl_z\right) = d\mathbf{l}_\rho + d\mathbf{l}_\phi + d\,\mathbf{l}_z \tag{3.9}$$

The differential displacement vectors along the coordinate directions are shown in Figure 3.2.

In terms of unit vectors, the differential displacement vectors along the coordinate directions can also be expressed as

$$d\mathbf{l} = dl_\rho \mathbf{a}_\rho + dl_\phi \mathbf{a}_\phi + dl_z \mathbf{a}_z \tag{3.10}$$

When the angle φ increases by the amount of $d\varphi$ the differential displacement dl_φ increases by $\rho d\varphi$ as shown in Figure 3.3.

Thus,

$$dl_\phi = \rho d\phi \tag{3.11}$$

Therefore, in terms of the coordinate variables, we may express the differential displacement vector in (3.10) as

$$d\mathbf{l} = \left(d\rho, \rho d\varphi, dz\right) = d\rho \mathbf{a}_\rho + \rho\, d\phi \mathbf{a}_\phi + dz \mathbf{a}_z \tag{3.12}$$

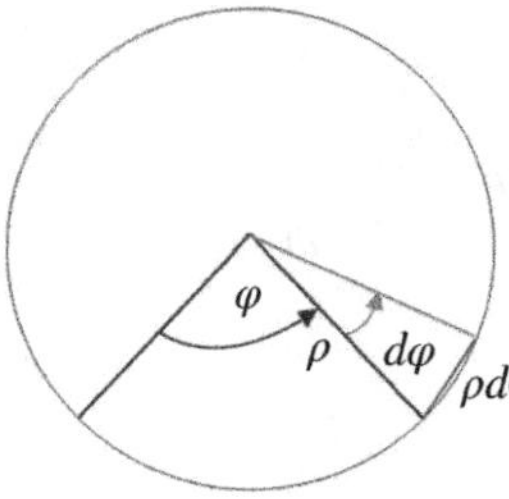

Figure 3.3 Differential displacement in φ direction.

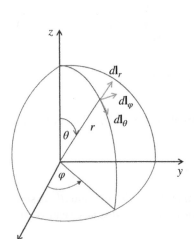

Figure 3.4 Differential displacement in spherical system.

Differential Displacement in Spherical System *Differential displacement (or length) d*l *in* spherical coordinate system is a vector defined by

$$dl = \left(dl_r, dl_\theta, dl_\phi \right) = dl_r + dl_\theta + dl_\phi \tag{3.13}$$

The differential displacement vectors along the coordinate directions are shown in Figure 3.4.

In terms of unit vectors, the differential displacement vectors along the coordinate directions can also be expressed as

$$dl = dl_r \mathbf{a}_r + dl_\theta \mathbf{a}_\theta + dl_\phi \mathbf{a}_\phi \tag{3.14}$$

In spherical coordinate system, both θ and φ are the angular coordinates, thus,

$$dl_\theta = r d\theta \tag{3.15a}$$

$$dl_\phi = \rho d\phi = r \sin\theta \, d\phi \tag{3.15b}$$

In terms of the coordinate variables, the differential displacement vector in (3.13) can be expressed as

$$dl = \left(dr, r d\theta, r \sin\theta \, d\phi \right) = dr \mathbf{a}_r + r d\theta \, \mathbf{a}_\theta + r \sin\theta \, d\phi \mathbf{a}_\phi \tag{3.16}$$

Figure 3.5 Differential surface vector.

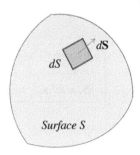

3.2.2 Differential Surface Element

Differential surface element $d\mathbf{S}$ (or differential surface area) is a vector that we will encounter when evaluating surface integrals of the form

$$\int_S \mathbf{F} \cdot d\mathbf{S} \tag{3.17}$$

Consider a surface S, and a differential amount of it, dS, as shown in Figure 3.5.

With this differential surface dS, we may associated a differential surface vector $d\mathbf{S}$ whose magnitude is equal to dS, and whose direction is perpendicular to the differential surface dS (Sadiku, 2010, p. 59)

$$d\mathbf{S} = dS\mathbf{a}_n \tag{3.18}$$

where $\mathbf{a}_n$ is a unit vector normal to the surface.

(There are, of course, two normal vectors to any such surface, so which one do we choose? In all instances when we will use such vectors, it will be clear from the problem description which perpendicular vector is of interest to us.)

In the next section we will decompose the differential surface vector into three component vectors along the coordinate directions in each of the three coordinate systems.

Differential Area in Cartesian System In Cartesian coordinate system, the differential area may, in general, be expressed as

$$d\mathbf{S} = d\mathbf{S}_x + d\mathbf{S}_y + d\mathbf{S}_z = dS_x\mathbf{a}_x + dS_y\mathbf{a}_y + dS_z\mathbf{a}_z \tag{3.19}$$

This decomposition is shown in Figure 3.6.

The differential area vector $d\mathbf{S}_x$ is perpendicular to the differential surface area dS_x which lies in the yz plane. This differential surface area is equal to the product of two differential displacements in the yz plane: dl_x and dl_z. Thus, we may write,

$$dS_x = dl_y dl_z = dydz \tag{3.20a}$$

Similarly,

$$dS_y = dl_x dl_z = dxdz \tag{3.20b}$$

$$dS_z = dl_x dl_y = dxdy \tag{3.20c}$$

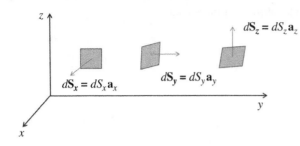

Figure 3.6 Differential surface vector decomposition.

Therefore, we may alternatively, express the differential surface area as

$$dS = (dydz, \ dxdz, \ dxdy)$$
$$= dydz\,\mathbf{a}_x + dxdz\,\mathbf{a}_y + dxdy\,\mathbf{a}_z \tag{3.21}$$

Differential Area in Cylindrical System In cylindrical coordinate system, the differential area may be expressed as

$$dS = dS_\rho + dS_\phi + dS_z = dS_\rho \mathbf{a}_\rho + dS_\varphi \mathbf{a}_\phi + dS_z \mathbf{a}_z \tag{3.22}$$

where

$$dS_\rho = dl_\phi dl_z = \rho d\phi dz \tag{3.23a}$$

$$dS_\phi = dl_\rho dl_z = d\rho dz \tag{3.23b}$$

$$dS_z = dl_\rho dl_\phi = (d\rho)(\rho d\phi) = \rho d\rho d\phi \tag{3.23c}$$

Therefore, in cylindrical system, we may alternatively, express the differential surface area as

$$dS = \left(\rho d\phi dz, d\rho dz, \rho d\phi d\rho\right)$$
$$= \rho d\phi dz\,\mathbf{a}_\rho + d\rho dz\,\mathbf{a}_\phi + \rho d\phi d\rho\,\mathbf{a}_z \tag{3.24}$$

Differential Area in Spherical System In spherical coordinate system, the differential area may be expressed as

$$dS = dS_r + dS_\theta + dS_\phi = dS_r \mathbf{a}_r + dS_\theta \mathbf{a}_\theta + dS_\phi \mathbf{a}_\phi \tag{3.25}$$

where

$$dS_r = dl_\theta dl_\phi = (rd\theta)\left(r\sin\theta d\phi\right) = r^2 \sin\theta d\theta d\phi \tag{3.26a}$$

$$dS_\theta = dl_r dl_\phi = (dr)\left(r\sin\theta d\phi\right) = r\sin\theta dr d\phi \tag{3.26b}$$

$$dS_\phi = dl_r dl_\theta = (dr)(rd\theta) = rdrd\theta \tag{3.26c}$$

Therefore, in spherical system, we may alternatively, express the differential surface area as

$$dS = \left(r^2 \sin\theta\, d\theta\, d\phi,\, r\sin\theta\, dr\, d\phi,\, r\, dr\, d\theta \right)$$
$$= r^2 \sin\theta\, d\theta\, d\phi\, \mathbf{a}_\rho + r\sin\theta\, dr\, d\phi\, \mathbf{a}_\theta + r\, dr\, d\theta\, \mathbf{a}_\phi \tag{3.27}$$

3.2.3   Differential Volume Element

We will encounter *differential volume element*, dv, in the volume integrals of the form

$$\int_v Fdv \tag{3.28}$$

The differential volume element, dv, is defined as a scalar equal to the product of three differential displacements in each coordinate system.

Differential Volume in Cartesian System   Differential volume dv, in Cartesian system is defined as the *scalar*

$$dv = dl_x dl_y dl_z \tag{3.29a}$$

or in terms of the coordinate variables:

$$dv = dxdydz \tag{3.29b}$$

Differential Volume in Cylindrical System   Differential volume dv, in cylindrical system is defined as the *scalar*

$$dv = dl_\rho dl_\phi dl_z \tag{3.30a}$$

Or in terms of the coordinate variables:

$$dv = d\rho(\rho d\phi)dz = \rho\, d\rho\, d\phi\, dz \tag{3.30b}$$

Differential Volume in Spherical System   Differential volume dv, in spherical system is defined as the *scalar*

$$dv = dl_r dl_\theta dl_\phi \tag{3.31a}$$

or in terms of the coordinate variables:

$$dv = dr\left(rd\theta\right)r\sin\theta\, d\phi$$
$$= r^2 \sin\theta\, dr\, d\theta\, d\phi \tag{3.31b}$$

3.3   Constant-Coordinate Surfaces

In this section we discuss a special class of surfaces, called *constant-coordinate surfaces*. These surfaces in Cartesian, cylindrical, or spherical coordinate systems are easily generated by keeping one of the coordinate variables constant and allowing the other two to vary.

These surfaces are extremely useful when evaluating line and surface integrals, as we will see in Chapter 4.

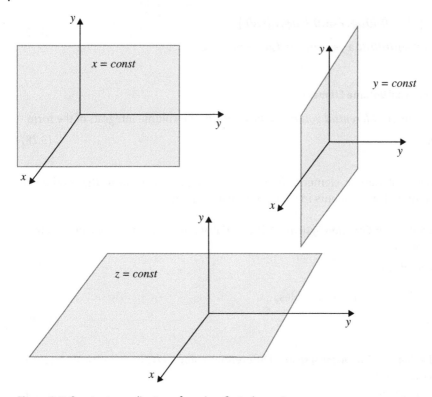

Figure 3.7 Constant-coordinate surfaces in a Cartesian system.

3.3.1 Cartesian Coordinate System

In the Cartesian system, we have three families of constant-coordinate surfaces (planes) defined by

$$x = const \quad -\infty < y < \infty, \quad -\infty < z < \infty$$
$$y = const \quad -\infty < x < \infty, \quad -\infty < z < \infty \tag{3.32}$$
$$z = const \quad -\infty < x < \infty, \quad -\infty < y < \infty$$

These surfaces are shown in Figure 3.7.

The intersection of any two such planes is a line parallel to one of the coordinate axes. For instance, $x = const$ and $y = const$ is the line parallel to the z axis. These lines are constant-coordinates lines.

3.3.2 Cylindrical Coordinate System

Orthogonal surfaces in cylindrical coordinates are described by:

$$\rho = const, \quad 0 \leq \varphi < 2\pi, \quad -\infty < z < \infty$$
$$\varphi = const, \quad 0 \leq \rho < \infty, \quad -\infty < z < \infty \tag{3.33}$$
$$z = const, \quad 0 \leq \rho < \infty, \quad 0 \leq \varphi < 2\pi$$

and are shown in Figure 3.8.

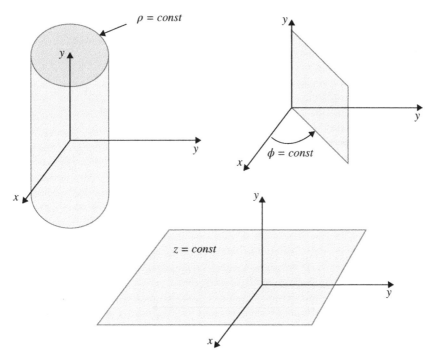

Figure 3.8 Constant-coordinate surfaces in a cylindrical system.

Note that $\rho = const$ is a circular cylinder; $\varphi = const$ is a semi-infinite plane with its edge along the z axis; $z = const$ is the same infinite plane as in a Cartesian system.

The intersection of any two surfaces is a curve – either a line or a circle: $\rho = const$ and $z = const$ is a circle of radius ρ; $z = const$ and $\varphi = const$ is a semi-infinite line originating at the z axis and passing through P; $\rho = const$ and $\varphi = const$ is an infinite line parallel to the z axis and passing through P. These curves are constant-coordinates curves.

3.3.3 Spherical Coordinate System

Orthogonal surfaces in spherical coordinates are described by:

$$
\begin{aligned}
r &= const, & 0 \le \theta &< \pi, & 0 \le \phi &< 2\pi \\
\theta &= const, & 0 \le r &< \infty, & 0 \le \phi &< 2\pi \\
\phi &= const, & 0 \le r &< \infty, & 0 \le \theta &< \pi
\end{aligned}
\tag{3.34}
$$

and are shown in Figure 3.9.

Note that $r = const$ is a sphere with its center at the origin; $\theta = const$ is a circular cone with the z axis as its axis and the origin as its vertex; $\varphi = const$ is the semi-infinite plane as in a cylindrical system.

A curve is formed by the intersection of any two surfaces; for example: $r = const$ and $\varphi = const$ is a semi-circle. These curves are constant-coordinates curves.

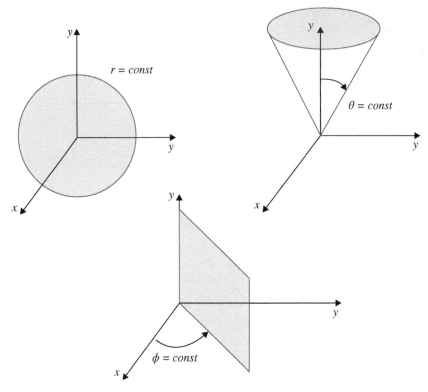

Figure 3.9 Constant-coordinate surfaces in a spherical system.

3.3.4 Differential Elements on Constant Coordinate Surfaces

Recall: Constant-coordinate surfaces in Cartesian, cylindrical, or spherical coordinate systems were obtained by keeping one of the coordinate variables constant and allowing the other two to vary.

The consequence of keeping one of the coordinate variables constant is the fact that the *differential displacement along that variable direction is zero*. Let's look at the resulting differential surface and displacement vectors in all three coordinate systems.

Cartesian Coordinate System In Cartesian, the differential surface vector is given by

$$dS = (dydz,\ dxdz,\ dxdy) \tag{3.35}$$

On the constant coordinate surfaces we have:

$$x = const \ \Rightarrow\ dx = 0 \ \Rightarrow\ dS = (dydz,\ 0,\ 0) \tag{3.36a}$$

$$y = const \ \Rightarrow\ dy = 0 \ \Rightarrow\ dS = (0,\ dxdz,\ 0) \tag{3.36b}$$

$$z = const \ \Rightarrow\ dz = 0 \ \Rightarrow\ dS = (0,\ 0,\ dxdy) \tag{3.36c}$$

Recall: the differential displacement vector, in general, is given by

$$dl = (dx, \ dy, \ dz) \tag{3.37}$$

Since the intersection of two constant-coordinate surfaces produces a constant coordinate line (parallel to the coordinate axes), it follows that on a constant-coordinate line two-out-of-three components of dl in (3.37) are zero. That is,

$$dl = (dx, 0, 0) \quad or \quad dl = (0, dy, 0) \quad or \quad dl = (0, 0, dz) \tag{3.38}$$

Cylindrical Coordinate System In cylindrical coordinate system, the differential surface vector is given by

$$dS = (\rho \, d\phi \, dz, \ d\rho \, dz, \ \rho \, d\phi \, dz) \tag{3.39}$$

On the constant coordinate surfaces we have:

$$\rho = const \quad \Rightarrow \quad d\rho = 0 \quad \Rightarrow \quad dS = (\rho \, d\phi \, dz, 0, 0) \tag{3.40a}$$

$$\phi = const \quad \Rightarrow \quad d\phi = 0 \quad \Rightarrow \quad dS = (0, \ d\rho \, dz, 0) \tag{3.40b}$$

$$z = const \quad \Rightarrow \quad dz = 0 \quad \Rightarrow \quad dS = (0, 0, \ \rho \, d\phi \, dz) \tag{3.40c}$$

Also, the differential displacement vector is, in general, given by

$$dl = (d\rho, \ \rho \, d\phi, \ dz) \tag{3.41}$$

Since the intersection of two constant-coordinate surfaces produces a constant coordinate curve, it follows that on a constant-coordinate curve two-out-of-three components of dl in (3.41) are zero. That is,

$$dl = (d\rho, 0, 0) \quad or \quad dl = (0, \ \rho \, d\phi, 0) \quad or \quad dl = (0, 0, dz) \tag{3.42}$$

Spherical Coordinate System In spherical coordinate system, the differential surface vector is given by

$$dS = (r^2 \sin\theta \, d\theta \, d\phi, \ r\sin\theta \, dr \, d\phi, \ r \, dr \, d\theta) \tag{3.43}$$

On the constant coordinate surfaces we have:

$$r = const \quad \Rightarrow \quad dr = 0 \quad \Rightarrow \quad dS = (r^2 \sin\theta \, d\theta \, d\phi, 0, 0) \tag{3.44a}$$

$$\theta = const \quad \Rightarrow \quad d\theta = 0 \quad \Rightarrow \quad dS = (0, \ r\sin\theta \, dr, 0) \tag{3.44b}$$

$$\phi = const \quad \Rightarrow \quad d\phi = 0 \quad \Rightarrow \quad dS = (0, 0, \ r \, dr \, d\theta) \tag{3.44c}$$

Also, the differential displacement vector is, in general, given by

$$dl = (dr, \ r \, d\theta, \ r\sin\theta \, d\phi) \tag{3.45}$$

Since the intersection of two constant-coordinate surfaces produces a constant coordinate curve, it follows that on a constant-coordinate curve two-out-of-three components of dl in (3.44) are zero. That is,

$$dl = (dr, 0, 0) \quad or \quad dl = (0, \ r \, d\theta, 0) \quad or \quad dl = (0, 0, \ r\sin\theta \, d\phi) \tag{3.46}$$

3.4 Differential Operators

In this section we will introduce several differential operators: gradient, divergence, curl, and Laplacian. These operators appear in Maxwell's equations and in the wave equation which we will study later in this text.

3.4.1 Gradient

Given a scalar function we can create a vector function. The operation involved is that of taking the "gradient". We will next define the gradient operation in Cartesian system, and subsequently in the cylindrical and spherical systems.

Gradient in Cartesian Coordinates The *gradient* of given scalar function $f(x, y, z)$, *grad f*, is the vector function defined by (Kreyszig, 1999, p. 446)

$$grad \quad f = \frac{\partial f}{\partial x}\mathbf{a}_x + \frac{\partial f}{\partial y}\mathbf{a}_y + \frac{\partial f}{\partial z}\mathbf{a}_z \qquad (3.47)$$

If we introduce the differential operator

$$\nabla = \frac{\partial}{\partial x}\mathbf{a}_x + \frac{\partial}{\partial y}\mathbf{a}_y + \frac{\partial}{\partial z}\mathbf{a}_z \qquad (3.48)$$

we may write

$$grad \quad f = \nabla f = \frac{\partial f}{\partial x}\mathbf{a}_x + \frac{\partial f}{\partial y}\mathbf{a}_y + \frac{\partial f}{\partial z}\mathbf{a}_z \qquad (3.49)$$

Gradient in Cylindrical Coordinates The *gradient* of a given scalar function $f(\rho, \varphi, z)$, *grad f*, is the vector function defined by (Sadiku, 2010, p. 70)

$$\nabla f = \frac{\partial f}{\partial \rho}\mathbf{a}_\rho + \frac{1}{\rho}\frac{\partial f}{\partial \phi}\mathbf{a}_\phi + \frac{\partial f}{\partial z}\mathbf{a}_z \qquad (3.50)$$

Gradient in Spherical Coordinates The *gradient* of given scalar function $f(r, \theta, \varphi)$, *grad f*, is the vector function defined by

$$\nabla f = \frac{\partial f}{\partial r}\mathbf{a}_r + \frac{1}{r}\frac{\partial f}{\partial \theta}\mathbf{a}_\theta + \frac{1}{r\sin\theta}\frac{\partial f}{\partial \phi}\mathbf{a}_\phi \qquad (3.51)$$

Example 3.6 Gradient of a scalar function
Determine the gradient of the following scalar fields:

a) $f = x^2 y + 3xyz$

b) $f = 2\rho z \sin\phi - z^2 \cos^2\phi + \rho^2$

c) $f = r^2 \cos\theta \sin\phi$

Solution:

a) $\nabla f = \dfrac{\partial f}{\partial x}\mathbf{a}_x + \dfrac{\partial f}{\partial y}\mathbf{a}_y + \dfrac{\partial f}{\partial z}\mathbf{a}_z, \quad f = x^2 y + 3xyz$

$\nabla f = (2xy + 3yz)\mathbf{a}_x + (x^2 + 3xz)\mathbf{a}_y + 3xy\mathbf{a}_z$

b) $\nabla f = \dfrac{\partial f}{\partial \rho}\mathbf{a}_\rho + \dfrac{1}{\rho}\dfrac{\partial f}{\partial \phi}\mathbf{a}_\phi + \dfrac{\partial f}{\partial z}\mathbf{a}_z, \quad f = 2\rho z \sin\phi - z^2 \cos^2\phi + \rho^2$

$\nabla f = (2z\sin\phi + 2\rho)\mathbf{a}_\rho + \dfrac{1}{\rho}\left[2\rho z\cos\phi - z^2 2\cos\phi(-\sin\phi)\right]\mathbf{a}_\phi + (2\rho\sin\phi - 2z\cos^2\phi)\mathbf{a}_z$

c) $\nabla f = \dfrac{\partial f}{\partial r}\mathbf{a}_r + \dfrac{1}{r}\dfrac{\partial f}{\partial \theta}\mathbf{a}_\theta + \dfrac{1}{r\sin\theta}\dfrac{\partial f}{\partial \phi}\mathbf{a}_\phi, \quad f = r^2\cos\theta\sin\phi$

$\nabla f = (2r\cos\theta\sin\phi)\mathbf{a}_r + \dfrac{1}{r}(r^2\sin\phi)(-\sin\theta)\mathbf{a}_\theta + \dfrac{1}{r\sin\theta}(r^2\cos\theta)\cos\phi\mathbf{a}_\phi$ ∎

3.4.2 Divergence

Given a vector function we can create a scalar function. The operation involved is that of taking the "divergence". We will next define the divergence operation in Cartesian system, and subsequently in the cylindrical and spherical systems.

Divergence in Cartesian Coordinates The *divergence* of given vector function $E(x, y, z) = (E_x, E_y, E_z)$, *div* **E**, is the scalar function defined by (Kreyszig, 1999, p. 453)

$$div\,\mathbf{E} = \frac{\partial E_x}{\partial x} + \frac{\partial E_y}{\partial y} + \frac{\partial E_z}{\partial z} \tag{3.52}$$

Another common notation for divergence is $\nabla \cdot \mathbf{E}$

$$div\,\mathbf{E} = \nabla \cdot \mathbf{E} = \left(\frac{\partial}{\partial x}\mathbf{a}_x + \frac{\partial}{\partial y}\mathbf{a}_y + \frac{\partial}{\partial z}\mathbf{a}_z\right)\cdot\left(E_x\mathbf{a}_x + E_y\mathbf{a}_y + E_z\mathbf{a}_z\right)$$

$$= \frac{\partial E_x}{\partial x} + \frac{\partial E_y}{\partial y} + \frac{\partial E_z}{\partial z} \tag{3.53}$$

Note that the divergence operation on a vector produces a scalar, while the gradient operation on a scalar produces a vector.

Divergence in Cylindrical Coordinates The *divergence* of given vector function $E(\rho, \varphi, z) = (E_\rho, E\varphi, E_z)$ is the scalar function defined by (Sadiku, 2010, p. 75)

$$\nabla \cdot \mathbf{E} = \frac{1}{\rho}\frac{\partial}{\partial \rho}\left(\rho E_\rho\right) + \frac{1}{\rho}\frac{\partial E_\phi}{\partial \phi} + \frac{\partial E_z}{\partial z} \tag{3.54}$$

Divergence in Spherical Coordinates The *divergence* of a given vector function $\mathbf{E}(r, \theta, \varphi) = (E_r, E_\theta, E_\varphi)$ is the scalar function defined by

$$\nabla \cdot \mathbf{E} = \frac{1}{r^2} \frac{\partial}{\partial r}(r^2 E_r) + \frac{1}{r \sin\theta} \frac{\partial}{\partial \theta}(E_\theta \sin\theta) + \frac{1}{r \sin\theta} \frac{\partial E_\varphi}{\partial \varphi} \qquad (3.55)$$

Example 3.7   Divergence of a vector function

Determine the divergence of the following vector fields:

a) $\mathbf{E} = e^{2xy} \mathbf{a}_x + z \sin xy \, \mathbf{a}_y + \cos^2 xz \mathbf{a}_z$

b) $\mathbf{E} = 2\rho z^2 \cos\phi \mathbf{a}_\rho - z \sin^2 \phi \mathbf{a}_z$

c) $\mathbf{E} = 2r\cos\theta \mathbf{a}_r + \dfrac{3}{r}\sin\theta \mathbf{a}_\theta + 2r^2 \sin\theta \mathbf{a}_\phi$

Solution:

a) $\nabla \cdot \mathbf{E} = \dfrac{\partial E_x}{\partial x} + \dfrac{\partial E_y}{\partial y} + \dfrac{\partial E_z}{\partial z} = 2ye^{2xy} + zx\cos xy - 2x\cos xz \sin xz$

b) $\nabla \cdot \mathbf{E} = \dfrac{1}{\rho} \dfrac{\partial}{\partial \rho}(\rho E_\rho) + \dfrac{1}{\rho} \dfrac{\partial E_\phi}{\partial \phi} + \dfrac{\partial E_z}{\partial z} = \dfrac{1}{\rho} \dfrac{\partial}{\partial \rho}(\rho 3\rho z^2 \cos\phi) + \dfrac{1}{\rho} 0 + \sin^2 \phi$

$= \dfrac{1}{\rho} \dfrac{\partial}{\partial \rho}(3\rho^2 z^2 \cos\phi) + \sin^2 \phi = \dfrac{1}{\rho}(6\rho z^2 \cos\phi) + \sin^2 \phi = 6z^2 \cos\phi + \sin^2 \phi$

c) $\nabla \cdot \mathbf{E} = \dfrac{1}{r^2} \dfrac{\partial}{\partial r}(r^2 E_r) + \dfrac{1}{r \sin\theta} \dfrac{\partial}{\partial \theta}(E_\theta \sin\theta) + \dfrac{1}{r \sin\theta} \dfrac{\partial E_\phi}{\partial \phi}$

$= \dfrac{1}{r^2} \dfrac{\partial}{\partial r}(2r^3 \cos\theta) + \dfrac{1}{r \sin\theta} \dfrac{\partial}{\partial \theta}\left(\dfrac{3}{r}\sin\theta \sin\theta\right) + \dfrac{1}{r \sin\theta} \dfrac{\partial(2r^2 \sin\theta)}{\partial \phi}$

$= \dfrac{1}{r^2}(6r^2 \cos\theta) + \dfrac{1}{r \sin\theta} \dfrac{3}{r}(2 \sin\theta \cos\theta) + 0 = 6\cos\theta + \dfrac{6}{r^2}\cos\theta$

∎

3.4.3   Curl

Given a vector function $\mathbf{H}(x, y, z)$ we can create another vector function. The operation involved is that of taking the "curl". We will next define the curl operation in Cartesian system, and subsequently in the cylindrical and spherical systems.

Curl in Cartesian Coordinates   The *curl* of given vector function $\mathbf{H}(x, y, z) = (H_x, H_y, H_z)$, *curl* **H**, is the vector function defined by (Kreyszig, 1999, p. 457)

$$\nabla \times \mathbf{H} = \begin{vmatrix} \mathbf{a}_x & \mathbf{a}_y & \mathbf{a}_z \\ \dfrac{\partial}{\partial x} & \dfrac{\partial}{\partial y} & \dfrac{\partial}{\partial z} \\ H_x & H_y & H_z \end{vmatrix} \qquad (3.56)$$

$$= \left(\dfrac{\partial H_z}{\partial y} - \dfrac{\partial H_y}{\partial z}\right)\mathbf{a}_x + \left(\dfrac{\partial H_x}{\partial z} - \dfrac{\partial H_z}{\partial x}\right)\mathbf{a}_y + \left(\dfrac{\partial H_y}{\partial x} - \dfrac{\partial H_x}{\partial y}\right)\mathbf{a}_z$$

Curl in Cylindrical Coordinates   The *curl* of given vector function $\mathbf{H}(\rho, \phi, z) = (H_\rho, H_\phi, H_z)$ is the vector function defined by (Sadiku, 2010, p. 80)

$$\nabla \times \mathbf{H} = \begin{vmatrix} \mathbf{a}_\rho & \rho \mathbf{a}_\phi & \mathbf{a}_z \\ \dfrac{\partial}{\partial \rho} & \dfrac{\partial}{\partial \phi} & \dfrac{\partial}{\partial z} \\ H_\rho & \rho H_\phi & H_z \end{vmatrix} \tag{3.57}$$

$$= \left(\frac{1}{\rho} \frac{\partial H_z}{\partial \phi} - \frac{\partial H_\phi}{\partial z} \right) \mathbf{a}_\rho + \left(\frac{\partial H_\rho}{\partial z} - \frac{\partial H_z}{\partial \rho} \right) \mathbf{a}_\phi + \frac{1}{\rho} \left[\frac{\partial}{\partial \rho} (\rho H_\phi) - \frac{\partial H_\rho}{\partial \phi} \right] \mathbf{a}_z$$

Curl in Spherical Coordinates   The *curl* of given vector function $\mathbf{H}(r, \theta, \phi) = (H_r, H_\theta, H_\phi)$ is the vector function defined by

$$\nabla \times \mathbf{H} = \frac{1}{r^2 \sin\theta} \begin{vmatrix} \mathbf{a}_r & r\mathbf{a}_\theta & r\sin\theta \, \mathbf{a}_\phi \\ \dfrac{\partial}{\partial r} & \dfrac{\partial}{\partial \theta} & \dfrac{\partial}{\partial \phi} \\ H_r & rH_\theta & r\sin\theta \, H_\phi \end{vmatrix} \tag{3.58}$$

$$= \frac{1}{r\sin\theta} \left[\frac{\partial}{\partial \theta} (H_\phi \sin\theta) - \frac{\partial H_\theta}{\partial \phi} \right] \mathbf{a}_r + \frac{1}{r} \left[\frac{1}{\sin\theta} \frac{\partial H_r}{\partial \phi} - \frac{\partial}{\partial r} (rH_\phi) \right] \mathbf{a}_\theta$$

$$+ \frac{1}{r} \left[\frac{\partial}{\partial r} (rH_\theta) - \frac{\partial H_r}{\partial \theta} \right] \mathbf{a}_\theta$$

There are two important properties of the gradient, divergence, and curl operations that we will use later in this text:

$$\nabla \times \nabla f = 0 \tag{3.59}$$

$$\nabla \cdot (\nabla \times \mathbf{H}) = 0 \tag{3.60}$$

Example 3.8   Curl of a vector function
Determine the curl of the following vector fields:

a) $\mathbf{H} = e^{xy} \mathbf{a}_x + \sin xy \, \mathbf{a}_y + \cos^2 xz \, \mathbf{a}_z$

b) $\mathbf{H} = \rho z^2 \cos\phi \, \mathbf{a}_\rho + z \sin^2 \phi \, \mathbf{a}_z$

c) $\mathbf{H} = r\cos\theta \, \mathbf{a}_r - \dfrac{1}{r}\sin\theta \, \mathbf{a}_\theta + 2r^2 \sin\theta \, \mathbf{a}_\phi$

Solution:

a) $\nabla \times \mathbf{H} = \begin{vmatrix} \mathbf{a}_x & \mathbf{a}_y & \mathbf{a}_z \\ \dfrac{\partial}{\partial x} & \dfrac{\partial}{\partial y} & \dfrac{\partial}{\partial z} \\ H_x & H_y & H_z \end{vmatrix} = \begin{vmatrix} \mathbf{a}_x & \mathbf{a}_y & \mathbf{a}_z \\ \dfrac{\partial}{\partial x} & \dfrac{\partial}{\partial y} & \dfrac{\partial}{\partial z} \\ e^{xy} & \sin xy & \cos^2 xz \end{vmatrix}$

$$= \frac{\partial}{\partial y}\left(\cos^2 xz\right)\mathbf{a}_x + \frac{\partial}{\partial x}\left(\sin xy\right)\mathbf{a}_z + \frac{\partial}{\partial z}\left(e^{xy}\right)\mathbf{a}_y$$

$$- \frac{\partial}{\partial x}\left(\cos^2 xz\right)\mathbf{a}_y - \frac{\partial}{\partial z}\left(\sin xy\right)\mathbf{a}_x - \frac{\partial}{\partial y}\left(e^{xy}\right)\mathbf{a}_z$$

$$= y\cos xy \, \mathbf{a}_z + 2z\cos xz \sin xz \, \mathbf{a}_y - xe^{xy} \, \mathbf{a}_z = 2z\cos xz \sin xz \, \mathbf{a}_y + \left(y\cos xy - xe^{xy}\right)\mathbf{a}_z$$

b) $\nabla \times \mathbf{H} = \dfrac{1}{\rho} \begin{vmatrix} \mathbf{a}_\rho & \rho\mathbf{a}_\phi & \mathbf{a}_z \\[4pt] \dfrac{\partial}{\partial\rho} & \dfrac{\partial}{\partial\phi} & \dfrac{\partial}{\partial z} \\[4pt] H_\rho & \rho H_\phi & H_z \end{vmatrix}$

$$= \frac{1}{\rho}\frac{\partial H_z}{\partial \phi}\mathbf{a}_\rho + \frac{1}{\rho}\frac{\partial\left(\rho H_\phi\right)}{\partial\rho}\mathbf{a}_z + \frac{1}{\rho}\frac{\partial H_\rho}{\partial z}\rho\mathbf{a}_\phi$$

$$-\frac{1}{\rho}\frac{\partial H_z}{\partial\rho}\rho\mathbf{a}_\phi - \frac{1}{\rho}\frac{\partial\left(\rho H_\phi\right)}{\partial z}\mathbf{a}_\rho - \frac{1}{\rho}\frac{\partial H_\rho}{\partial\phi}\mathbf{a}_z$$

$$= \frac{z\sin 2\phi}{\rho}\mathbf{a}_\rho + 2z\rho\cos\phi\,\mathbf{a}_\phi + z^2\sin\phi\,\mathbf{a}_z$$

c) $\nabla \times \mathbf{H} = \dfrac{1}{r^2\sin\theta} \begin{vmatrix} \mathbf{a}_r & r\mathbf{a}_\theta & r\sin\theta\,\mathbf{a}_\phi \\[4pt] \dfrac{\partial}{\partial r} & \dfrac{\partial}{\partial\theta} & \dfrac{\partial}{\partial\phi} \\[4pt] H_r & rH_\theta & r\sin\theta\,H_\phi \end{vmatrix}$

$$= \frac{1}{r^2\sin\theta} \begin{vmatrix} \mathbf{a}_r & r\mathbf{a}_\theta & r\sin\theta\,\mathbf{a}_\phi \\[4pt] \dfrac{\partial}{\partial r} & \dfrac{\partial}{\partial\theta} & \dfrac{\partial}{\partial\phi} \\[4pt] r\cos\theta & -\sin\theta & 2r^3\sin^2\theta \end{vmatrix}$$

$$= \frac{1}{r^2\sin\theta}\left[\frac{\partial\left(2r^3\sin^2\theta\right)}{\partial\theta}\mathbf{a}_r + \frac{\partial(-\sin\theta)}{\partial r}r\sin\theta\,\mathbf{a}_\phi + \frac{\partial(r\cos\theta)}{\partial\phi}r\mathbf{a}_\theta\right]$$

$$+ \frac{1}{r^2\sin\theta}\left[-\frac{\partial\left(2r^3\sin^2\theta\right)}{\partial r}r\mathbf{a}_\theta - \frac{\partial(-\sin\theta)}{\partial\phi}\mathbf{a}_r - \frac{\partial(r\cos\theta)}{\partial\theta}r\sin\theta\,\mathbf{a}_\phi\right]$$

$$= \frac{1}{r^2\sin\theta}\left(4r^3\sin\theta\cos\theta\,\mathbf{a}_r - 6r^3\sin^2\theta\,\mathbf{a}_\theta + r^2\sin^2\theta\,\mathbf{a}_\phi\right)$$

$$= 4r\cos\theta\,\mathbf{a}_r - 6r\sin\theta\,\mathbf{a}_\theta + \sin\theta\,\mathbf{a}_\phi$$

3.4.4 Laplacian

The remaining differential operator is the Laplacian. When the Laplacian operation is performed on a scalar function, the result is another scalar function; when it is performed on a vector function, the result is another vector function. We will next define the Laplacian of a scalar function in Cartesian system, and subsequently in the cylindrical and spherical systems.

Laplacian in Cartesian Coordinates   *Recall:* The gradient of given scalar function $f(x, y, z)$, is the vector function defined by (Sadiku, 2010, p. 88)

$$\nabla f = \frac{\partial f}{\partial x}\mathbf{a}_x + \frac{\partial f}{\partial y}\mathbf{a}_y + \frac{\partial f}{\partial z}\mathbf{a}_z \tag{3.61}$$

If we take the divergence of the resulting vector, we obtain a scalar function

$$\nabla \cdot \nabla f = \frac{\partial^2 f}{\partial x^2} + \frac{\partial^2 f}{\partial y^2} + \frac{\partial^2 f}{\partial z^2} \tag{3.62}$$

The resulting expression is called the *Laplacian* of f and is denoted by $\nabla^2 f$. Thus,

$$\nabla^2 f = \frac{\partial^2 f}{\partial x^2} + \frac{\partial^2 f}{\partial y^2} + \frac{\partial^2 f}{\partial z^2} \tag{3.63}$$

Laplacian in cylindrical coordinates   The *Laplacian* of given scalar function $f(\rho, \varphi, z)$, is the scalar function defined by

$$\nabla^2 f = \frac{1}{\rho}\frac{\partial}{\partial \rho}\left(\rho \frac{\partial f}{\partial \rho}\right) + \frac{1}{\rho^2}\frac{\partial^2 f}{\partial \phi^2} + \frac{\partial^2 f}{\partial z^2} \tag{3.64}$$

Laplacian in spherical coordinates   The *Laplacian* of given scalar function $f(r, \theta, \varphi)$, is the scalar function defined by

$$\nabla^2 f = \frac{1}{r^2}\frac{\partial}{\partial r}\left(r^2 \frac{\partial f}{\partial r}\right) + \frac{1}{r^2 \sin\theta}\frac{\partial}{\partial \theta}\left(\sin\theta \frac{\partial f}{\partial \theta}\right) + \frac{1}{r^2 \sin^2\theta}\frac{\partial^2 f}{\partial \phi^2} \tag{3.65}$$

When computing the electric and magnetic radiation fields of antennas, we will encounter the Laplacian of a vector function $\nabla^2 \mathbf{V}$. Instead of computing this Laplacian from the definition (which is quite involved), we will make use of the following identity:

$$\nabla^2 \mathbf{V} = \nabla(\nabla \cdot V) - \nabla \times \nabla \times \mathbf{V} \tag{3.66}$$

3.5   EMC Applications

3.5.1   Transmission-Line Equations

We will show the application of the concept of a derivative through the derivation of transmission line equations. (Transmission lines will be discussed in detail in Part III of this book.)

Figure 3.10 shows the per-unit-length equivalent circuit model of a transmission line; l and c represent the per-unit-length inductance and capacitance associated with the length Δz of the line (Paul, 2006, p. 182).

Writing Kirchhoff's voltage law around the outside loop (we will review the basic circuit laws in Part II of this book) gives

$$V(z + \Delta z, t) - V(z, t) = -l\Delta z \frac{\partial I(z, t)}{\partial t} \tag{3.67}$$

Dividing both sides by Δz and taking the limit as $\Delta z \to 0$ gives

$$\lim_{\Delta z \to 0} \frac{V(z + \Delta z, t) - V(z, t)}{\Delta z} = -l\frac{\partial I(z, t)}{\partial t} \tag{3.68}$$

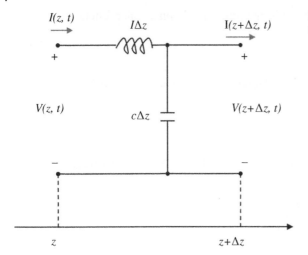

$I(z, t)$

$l\Delta z$

$I(z+\Delta z, t)$

$V(z, t)$

$c\Delta z$

$V(z+\Delta z, t)$

z

$z+\Delta z$

Figure 3.10 Equivalent circuit model of a transmission line.

We recognize that the expression on the left-hand side of Eq. (3.68) is the partial derivative of the line voltage with respect to the variable z. Thus,

$$\frac{\partial V(z,t)}{\partial z}=-l\frac{\partial I(z,t)}{\partial t} \tag{3.69}$$

This is the first transmission line equation. Similarly, writing Kirchhoff's current law at the upper node of the capacitor gives

$$I(z+\Delta z,t)-I(z,t)=-c\Delta z\frac{\partial V(z+\Delta z,t)}{\partial t} \tag{3.70}$$

Dividing both sides by Δz and taking the limit as $\Delta z \to 0$ gives

$$\lim_{\Delta z\to 0}\frac{I(z+\Delta z,t)-I(z,t)}{\Delta z}=-c\frac{\partial V(z+\Delta z,t)}{\partial t} \tag{3.71}$$

Again, we recognize that the expression on the left-hand side of Eq. (3.71) is the partial derivative of the line current with respect to the variable z. Thus,

$$\frac{\partial I(z,t)}{\partial z}=-c\frac{\partial V(z,t)}{\partial t} \tag{3.72}$$

This is the second transmission line equation. Equations (3.68) and (3.71) are called the transmission-line equations.

3.5.2 Maxwell's Equations in a Differential Form

The differential operators presented in this chapter appear throughout the study of electromagnetics and EMC. Perhaps the most important application of these operators is in Maxwell's equations.

Maxwell's equations can be expressed in several forms. Here, we present the differential time-domain version of these equations in a simple medium (Paul, 2006, pp. 899, 901).

$$\nabla\times E=-\mu\frac{\partial H}{\partial t} \tag{3.73a}$$

$$\nabla \times \mathbf{H} = \sigma \mathbf{E} + \varepsilon \frac{\partial \mathbf{E}}{\partial t} + \mathbf{J}_S \tag{3.73b}$$

$$\nabla \cdot \mathbf{E} = \frac{\rho_V}{\varepsilon} \tag{3.73c}$$

$$\nabla \cdot \mathbf{H} = 0 \tag{3.73d}$$

In Eqs (3.73) **E** denotes electric field intensity, while **H** denotes magnetic field intensity. **J** stands for volume current density, while ρ_V denotes volume charge density. We will derive and discuss these equations in Part III of this text.

3.5.3 Electromagnetic Wave Equation

The concept of a vector magnetic potential is useful in the derivation of the electric and magnetic fields radiated by an antenna. In such derivations (to be presented in Section 6.7.4), the following wave equation needs to be solved (Balanis, 2005, p. 139).

$$\nabla^2 A_z + k^2 A_z = 0 \tag{3.74}$$

where A_z is the z component of a vector magnetic potential and k is a constant.

$A_z = A_z(r)$ in spherical coordinate system (A_z is not a function of θ or φ). Thus, the Laplacian in spherical coordinate system:

$$\nabla^2 f = \frac{1}{r^2} \frac{\partial}{\partial r} \left(r^2 \frac{\partial f}{\partial r} \right) + \frac{1}{r^2 \sin\theta} \frac{\partial}{\partial \theta} \left(\sin\theta \frac{\partial f}{\partial \theta} \right) + \frac{1}{r^2 \sin^2\theta} \frac{\partial^2 f}{\partial \phi^2} \tag{3.75}$$

Applied to Eq. (3.74) this reduces to

$$\nabla^2 A_z + k^2 A_z = \frac{1}{r^2} \frac{\partial}{\partial r} \left[r^2 \frac{\partial A_z(r)}{\partial r} \right] + k^2 A_z(r) = 0 \tag{3.76}$$

which, when expanded, gives

$$\frac{1}{r^2} \left(\left[2r \frac{\partial A_z(r)}{\partial r} + r^2 \frac{\partial^2 A_z(r)}{\partial r^2} \right] \right) + k^2 A_z(r) = 0 \tag{3.77}$$

which reduces to

$$\frac{d^2 A_z(r)}{dr^2} + \frac{2}{r} \frac{d A_z(r)}{dr} + k^2 A_z(r) = 0 \tag{3.78}$$

In Chapter 6 we will discuss the solution of this differential equation.

References

Balanis, C.A., *Antenna Theory Analysis and Design*, 3rd ed., Wiley Interscience, Hoboken. New Jersey, 2005.

Paul, C.R., *Introduction to Electromagnetic Compatibility*, 2nd ed., John Wiley and Sons, New York, 2006.

Sadiku, M.N.O., *Elements of Electromagnetics*, 5th ed., Oxford University Press, New York, 2010.

Simon, A.B., *Calculus with Analytic Geometry*, Scott, Foresman and Company, Glenview, Illinois, 1982.

4

Vector Integral Calculus

4.1 Line Integrals

In this section we will define and learn how to evaluate the line integrals of the form

$$\int_c \mathbf{F} \cdot d\mathbf{l} \tag{4.1}$$

Before discussing the line integrals, however, let's review the concept of indefinite and definite integrals.

4.1.1 Indefinite and Definite Integrals

We will first introduce the indefinite integral and then use it to present the definite integral.

Indefinite Integral The *indefinite integral* can be easily defined using the concept of a derivative as follows.

Consider a function $f(x)$. If its integral exists, denoted,

$$g(x) = \int f(x)dx \tag{4.2}$$

then

$$g'(x) = f(x) \tag{4.3}$$

Several useful integral formulas are presented next:

$$f(x) = 1 \quad \int f(x)dx = \int dx = x \tag{4.4a}$$
$$(x)' = 1$$

$$f(x) = x^a \quad \int f(x)dx = \int x^a dx = \frac{1}{a+1}x^{a+1} \tag{4.4b}$$

$$\left(\frac{1}{a+1}x^{a+1}\right)' = \frac{a+1}{a+1}x^{a+1-1} = x^a$$

Foundations of Electromagnetic Compatibility with Practical Applications, First Edition. Bogdan Adamczyk.
© 2017 John Wiley & Sons Ltd. Published 2017 by John Wiley & Sons Ltd.

$$f(x) = e^x \quad \int f(x)dx = \int e^x dx = e^x \tag{4.4c}$$

$$\left(e^x\right)' = e^x$$

$$f(x) = e^{ax} \quad \int f(x)dx = \int e^{ax} dx = \frac{1}{a}e^{ex} \tag{4.4d}$$

$$\left(\frac{1}{a}e^{ex}\right)' = \frac{1}{a}(a)e^{ax} = e^{ax}$$

$$f(x) = \frac{1}{x} \quad \int f(x)dx = \int \frac{1}{x}dx = \ln x, \quad x > 0 \tag{4.4e}$$

$$\left(\ln x\right)' = \frac{1}{x}$$

$$f(x) = \sin x \quad \int f(x)dx = \int \sin x\, dx = -\cos x \tag{4.4f}$$

$$\left(-\cos x\right)' = -\left(-\sin x\right) = \sin x$$

$$f(x) = \sin ax \quad \int f(x)dx = \int \sin ax\, dx = -\frac{1}{a}\cos ax \tag{4.4g}$$

$$\left(-\frac{1}{a}\cos ax\right)' = \left(-\frac{1}{a}\right)(-\sin ax)(a) = \sin ax$$

$$f(x) = \cos x \quad \int f(x)dx = \int \cos x\, dx = \sin x \tag{4.4i}$$

$$\left(\sin x\right)' = \cos x$$

$$f(x) = \cos ax \quad \int f(x)dx = \int \cos ax\, dx = \frac{1}{a}\sin ax \tag{4.4j}$$

$$\left(\frac{1}{a}\sin ax\right)' = \left(\frac{1}{a}\right)(\cos ax)(a) = \cos ax$$

Definite Integral   If the indefinite integral is given by

$$\int f(x)dx = g(x) \tag{4.5}$$

then the *definite integral* is defined by

$$\int_a^b f(x)dx = g(x)\Big|_{x=a}^{x=b} = g(b) - g(a) \tag{4.6}$$

Integral Properties

$$\int_a^b cf(x)dx = c\int_a^b f(x)dx \tag{4.7a}$$

$$\int_a^b (f+g)(x)dx = \int_a^b f(x)dx + \int_a^b g(x)dx \tag{4.7b}$$

$$\int_a^b f(x)dx = \int_a^c f(x)dx + \int_c^b f(x)dx, \quad a < c < b \tag{4.7c}$$

$$\int_b^a f(x)dx = -\int_a^b f(x)dx \tag{4.7d}$$

4.1.2 Line Integral

The concept of a *line integral* is a simple generalization of the concept of a definite integral

$$\int_a^b f(x)dx \tag{4.8}$$

In Eq. (4.8) we integrate $f(x)$ from $x = a$ along the x axis to $x = b$. In a line integral we integrate a given function, called the integrand, along a curve C in space, or in the plane (Kreyszig, 1999, p. 464).

Consider a curve C in space extending from point a to point b, as shown in Figure 4.1. The *line integral* of a vector $\mathbf{F}$ over the curve C is defined as

$$\int_C \mathbf{F} \cdot d\mathbf{l} \tag{4.9}$$

When the curve is a closed curve (points a and b coincide) then the line integral over the curve C is defined as (Sadiku, 2010, p. 64)

$$\oint_C \mathbf{F} \cdot d\mathbf{l} \tag{4.10}$$

The evaluation of the line integral in Eqs (4.9) or (4.10) is, in general, quite difficult. However, as we will show next, if the curve is a constant-coordinate curve, this line integral reduces to the definite integral discussed earlier.

Cartesian Coordinate System In the Cartesian coordinate system, if the vector $\mathbf{F}$ and the differential displacement $d\mathbf{l}$ have the components

$$\mathbf{F} = \left(F_x, F_y, F_z \right), \qquad d\mathbf{l} = \left(dx, dy, dz \right) \tag{4.11}$$

then the line integral in Eq. (4.9) becomes

$$\int_C \mathbf{F} \cdot d\mathbf{l} = \int_C \left(F_x\, dx + F_y\, dy + F_z\, dz \right) \tag{4.12}$$

Figure 4.1 Illustration of the line integral.

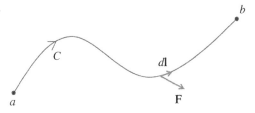

And if the line is a constant coordinate line, then this integral reduces to one of the three definite integrals:

$$\int_C \mathbf{F} \cdot d\mathbf{l} = \int_{x_1}^{x_2} F_x \, dx \tag{4.13a}$$

$$\int_C \mathbf{F} \cdot d\mathbf{l} = \int_{y_1}^{y_2} F_y \, dy \tag{4.13b}$$

$$\int_C \mathbf{F} \cdot d\mathbf{l} = \int_{z_1}^{z_2} F_z \, dz \tag{4.13c}$$

Cylindrical Coordinate System   In the cylindrical coordinate system, if the vector **F** and the differential displacement $d\mathbf{l}$ have the components

$$\mathbf{F} = \left(F_\rho, F_\phi, F_z \right), \qquad d\mathbf{l} = \left(d\rho, \rho d\phi, dz \right) \tag{4.14}$$

then the line integral in Eq. (4.9) becomes

$$\int_C \mathbf{F} \cdot d\mathbf{l} = \int_C \left(F_\rho d\rho + F_\phi \rho d\phi + F_z dz \right) \tag{4.15}$$

And if the curve is a constant coordinate curve, then this integral reduces to one of the three definite integrals:

$$\int_C \mathbf{F} \cdot d\mathbf{l} = \int_{\rho_1}^{\rho_2} F_\rho \, d\rho \tag{4.16a}$$

$$\int_C \mathbf{F} \cdot d\mathbf{l} = \int_{\phi_1}^{\phi_2} F_\phi \rho d\phi \tag{4.16b}$$

$$\int_C \mathbf{F} \cdot d\mathbf{l} = \int_{z_1}^{z_2} F_z \, dz \tag{4.16c}$$

Spherical Coordinate System   In the spherical coordinate system, if the vector **F** and the differential displacement $d\mathbf{l}$ have the components

$$\mathbf{F} = \left(F_r, F_\theta, F_\phi \right), \qquad d\mathbf{l} = \left(dr, r d\theta, r \sin\theta \, d\phi \right) \tag{4.17}$$

then the line integral in Eq. (4.9) becomes

$$\int_C \mathbf{F} \cdot d\mathbf{l} = \int_C \left(F_r \, dr + F_\theta r d\theta + F_\phi r \sin\theta \, d\varphi \right) \tag{4.18}$$

And if the curve is a constant coordinate curve, then this integral reduces to one of the three definite integrals:

$$\int_C \mathbf{F} \cdot d\mathbf{l} = \int_{r_1}^{r_2} F_r \, dr \tag{4.19a}$$

$$\int_C \mathbf{F} \cdot d\mathbf{l} = \int_{\theta_1}^{\theta_2} F_\theta r \, d\theta \tag{4.19b}$$

$$\int_C \mathbf{F} \cdot d\mathbf{l} = \int_{\phi_1}^{\phi_2} F_\phi r \sin\theta \, d\phi \tag{4.19c}$$

4.1.3 Properties of Line Integrals

From familiar properties of integrals in calculus we obtain formulas for line integrals

$$\int_C k\mathbf{F} \cdot d\mathbf{l} = k \int_C \mathbf{F} \cdot d\mathbf{l} \quad (k = const) \tag{4.20}$$

$$\int_C (\mathbf{F} + \mathbf{G}) \cdot d\mathbf{l} = \int_C \mathbf{F} \cdot d\mathbf{l} + \int_C \mathbf{G} \cdot d\mathbf{l} \tag{4.21}$$

$$\int_C \mathbf{F} \cdot d\mathbf{l} = \int_{C_1} \mathbf{F} \cdot d\mathbf{l} + \int_{C_2} \mathbf{F} \cdot d\mathbf{l} \tag{4.22}$$

where in Eq. (4.22) the path C is subdivided into two curves C_1 and C_2 that have the same orientation as C, as shown in Figure 4.2.

Line Integral Independence of Path   Consider the line integral given by

$$\int_C \mathbf{F} \cdot d\mathbf{l} = \int_C (F_x dx + F_y dy + F_z dz) \tag{4.23}$$

In Eq. (4.23) we integrate from a point a to a point b over a path C.

The value of this integral generally depends not only on a and b, but also on the path along which we integrate. This raises the question of conditions for independence of the path, so that we get the same value in integrating from a to b along any path C.

A very practical criterion for path independence is the following:

A line integral in Eq. (4.23) is independent of path if the vector **F** *is a gradient of some scalar function f.*

Figure 4.2 Illustration of Eq. (4.22).

Note that we don't need to know what that scalar function f is; we just need to know that F is the gradient of it. So, how is this useful in electromagnetics?

It is very useful, because in Part II of this book we will be evaluating the following integral

$$\int_C \mathbf{E} \cdot d\mathbf{l} \tag{4.24}$$

where E is the electric field intensity. Since

$$\mathbf{E} = -\nabla V \tag{4.25}$$

we will be free to choose any path of integration in Eq. (4.24).

Integration along Closed Curves   When discussing electrostatic fields we will make use of the following property of line integrals:

The line integral

$$\int_C \mathbf{E} \cdot d\mathbf{l} = \int_C \left(E_x dx + E_y dy + E_z dz \right) \tag{4.26}$$

is independent of path if its value around every closed path is zero.

On the other hand, if we know that the line integral is independent of path, then

$$\nabla \times \mathbf{E} = 0 \tag{4.27}$$

Example 4.1   Evaluation of a line integral
Let F be given by

$$\mathbf{A} = 4\rho \sin\phi \, \mathbf{a}_\rho + 3\cos\phi \, \mathbf{a}_\phi$$

and the curve C, in the xy plane, be defined and oriented as shown in Figure 4.3. Evaluate the line integral $\int_C \mathbf{F} \cdot d\mathbf{l}$

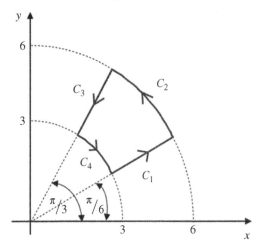

Figure 4.3 Line integral example.

Solution: To evaluate this integral we make use of Eq. (4.22)

$$\oint_C \mathbf{A} \cdot d\mathbf{l} = \int_{C_1} \mathbf{A} \cdot d\mathbf{l} + \int_{C_2} \mathbf{A} \cdot d\mathbf{l} + \int_{C_2} \mathbf{A} \cdot d\mathbf{l} + \int_{C_2} \mathbf{A} \cdot d\mathbf{l}$$

Along C_1 have

$$C_1 : \rho = 3 \to 6, \quad \varphi = \pi/6 = const, \quad z = 0 = const$$
$$d\mathbf{l} = (d\rho, \rho d\varphi, dz) = (d\rho, 0, 0)$$

$$\int_{C_1} \mathbf{A} \cdot d\mathbf{l} = \int_{C_1} (4\rho \sin\varphi, 3\cos\varphi, 0) \cdot (d\rho, 0, 0) = \int_{\substack{\rho=3 \\ \varphi=\pi/6}}^{\rho=6} 4\rho \sin\varphi \, d\rho$$

$$= 4\sin\frac{\pi}{6} \int_{\rho=3}^{6} \rho \, d\rho = (4)\left(\frac{1}{2}\right)\left(\frac{\rho^2}{2}\bigg|_{\rho=3}^{6}\right) = 2\left(\frac{36}{2} - \frac{9}{2}\right) = 25$$

Along C_2:

$$C_2 : \rho = 6, \quad \varphi = \pi/6 \to \pi/3, \quad z = const$$
$$d\mathbf{l} = (d\rho, \rho d\varphi, dz) = (0, \rho d\varphi, 0)$$

Along C_3:

$$C_3 : \rho = 6 \to 3, \quad \varphi = \pi/3 = const, \quad z = 0 = const$$
$$d\mathbf{l} = (d\rho, \rho d\varphi, dz) = (d\rho, 0, 0)$$

$$\int_{C_3} \mathbf{A} \cdot d\mathbf{l} = \int_{C_3} (4\rho \sin\phi, 3\cos\phi, 0) \cdot (d\rho, 0, 0) = \int_{\substack{\rho=6 \\ \varphi=\pi/3}}^{\rho=3} 4\rho \sin\varphi \, d\rho$$

$$= 4\sin\frac{\pi}{3} \int_{\rho=6}^{3} \rho \, d\rho = (4)\left(\frac{\sqrt{3}}{2}\right)\left(\frac{\rho^2}{2}\bigg|_{\rho=6}^{3}\right) = 2\sqrt{3}\left(\frac{9}{2} - \frac{36}{2}\right) = -25\sqrt{3}$$

Along C_4:

$$C_4 : \rho = 3, \quad \varphi = \pi/3 \to \pi/6, \quad z = const$$
$$d\mathbf{l} = (d\rho, \rho d\varphi, dz) = (0, \rho d\varphi, 0)$$

$$\int_{C_4} \mathbf{A} \cdot d\mathbf{l} = \int_{C_4} (4\rho \sin\varphi, 3\cos\varphi, 0) \cdot (0, \rho d\varphi, 0) = \int_{\substack{\rho=3 \\ \varphi=\pi/3}}^{\varphi=\pi/6} 3\rho \cos\phi \, d\phi$$

$$= (3)(3) \int_{\varphi=\pi/3}^{\varphi=\pi/6} \cos\varphi \, d\varphi = 9\left(\sin\varphi\big|_{\varphi=\pi/3}^{\varphi=\pi/6}\right) = 9\left(\frac{1}{2} - \frac{\sqrt{3}}{2}\right) = 4.5\left(1 - \sqrt{3}\right)$$

Therefore

$$\oint_C \mathbf{A} \cdot d\mathbf{l} = \int_{C_1} \mathbf{A} \cdot d\mathbf{l} + \int_{C_2} \mathbf{A} \cdot d\mathbf{l} + \int_{C_2} \mathbf{A} \cdot d\mathbf{l} + \int_{C_2} \mathbf{A} \cdot d\mathbf{l}$$

$$= 25 + 9\left(\sqrt{3} - 1\right) - 25\sqrt{3} + 4.5\left(1 - \sqrt{3}\right) = 20.5\left(1 - \sqrt{3}\right)$$

∎

4.2 Surface Integrals

The concept of a *surface integral* is a simple generalization of the concept of a double integral, which we will define next.

4.2.1 Double Integrals

In a definite integral $\int_a^b f(x)dx$, we integrate a function $f(x)$ over an interval of the x axis. In a *double integral*, we integrate a function $f(x, y)$ over a region R in the xy plane (Kreyszig, 1999, p. 480).

$$\iint_R f(x, y)dxdy \tag{4.27}$$

Double integrals have properties similar to those of definite integrals.

$$\iint_R kf(x, y)dxdy = k\iint_R f(x, y)dxdy \tag{4.28a}$$

$$\iint_R (f + g)dxdy = \iint_R fdxdy + \iint_R gdxdy \tag{4.28b}$$

$$\iint_R f(x, y)dxdy = \iint_{R_1} f(x, y)dxdy + \iint_{R_2} f(x, y)dxdy \tag{4.28c}$$

where the region R in Eq. (2.28c) is subdivided into two regions R_1 and R_2, as shown in Figure 4.4.

In many electromagnetics problems the region R is a rectangular region described by $a \le x \le b$ and $c \le y \le d$, and the double integral over a region R in Eq. (4.27) may be evaluated by two successive integrations:

$$\iint_R f(x, y)dxdy = \int_a^b \left[\int_c^d f(x, y)dy \right]dx \tag{4.29}$$

$$\iint_R f(x, y)dxdy = \int_c^d \left[\int_a^b f(x, y)dx \right]dy \tag{4.30}$$

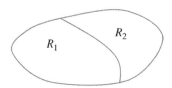

Figure 4.4 Subdivision of the region R in Eq. (4.28c).

In Eq. (4.29) we first integrate the inner integral with respect to y, treating x as a constant. Then, we integrate this result with respect to x.

In Eq. (4.30) we first integrate the inner integral with respect to x, treating y as a constant. Then, we integrate this result with respect to y.

Both (4.29) and (4.30) produce the same result, as illustrated by the following example.

Example 4.2   Evaluation of a double integral

Evaluate

$$\int\limits_{x=0}^{2}\int\limits_{y=1}^{3}\left(2xy+3y^2\right)dydx$$

and then reverse the order of integration and reevaluate.

Solution:

$$\int\limits_{x=0}^{2}\int\limits_{y=1}^{3}\left(2xy+3y^2\right)dydx=\int\limits_{x=0}^{2}\left[\int\limits_{y=1}^{3}\left(2xy+3y^2\right)dy\right]dx=\int\limits_{x=0}^{2}\left[\left(2x\frac{y^2}{2}+3\frac{y^3}{3}\right)\Bigg|_{y=1}^{y=3}\right]dx$$

$$=\int\limits_{x=0}^{2}\left[\left(xy^2+y^3\right)\Big|_{y=1}^{y=3}\right]dx=\int\limits_{x=0}^{2}(9x+27)-(x+1)dx=\int\limits_{x=0}^{2}(8x+26)dx$$

$$=\left(8\frac{x^2}{2}+26x\right)\Bigg|_{x=0}^{x=2}=16+52=68$$

Now, let's reverse the limits of integration.

$$\int\limits_{y=1}^{3}\int\limits_{x=0}^{2}\left(2xy+3y^2\right)dxdy=\int\limits_{y=1}^{3}\left[\int\limits_{x=0}^{2}\left(2xy+3y^2\right)dx\right]dy=\int\limits_{y=1}^{3}\left[\left(2\frac{x^2}{2}y+3y^2x\right)\Bigg|_{x=0}^{x=2}\right]dy$$

$$=\int\limits_{y=1}^{3}\left[\left(x^2y+3y^2x\right)\Big|_{x=0}^{x=2}\right]dy=\int\limits_{y=1}^{3}\left(4y+6y^2\right)dy=\left(4\frac{y^2}{2}+6\frac{y^3}{3}\right)\Bigg|_{y=1}^{y=3}$$

$$=\left(2y^2+2y^3\right)\Big|_{y=1}^{y=3}=(18+54)-(2+2)=68$$

■

Useful Application of Double Integrals   The *area A* of a region R in the xy plane is given by a double integral

$$A=\iint\limits_{R}dxdy \tag{4.31}$$

4.2.2   Surface Integrals

For a given vector function **F**, the surface integral is defined by

$$\iint\limits_{S}\mathbf{F}\cdot d\mathbf{S} \tag{4.32}$$

This integral, in many applications, is called the flux of **F** through **S**. When the surface is a closed surface, we denote the integral in (4.32) by

$$\oiint\limits_{S}\mathbf{F}\cdot d\mathbf{S} \tag{4.33}$$

The evaluation of the surface integral in Eqs (4.33) or (4.34) is, in general, quite difficult. However, as we will show next, if the surface is a constant-coordinate surface, this surface integral reduces to the double integral discussed in the previous section.

Cartesian Coordinate System In Cartesian coordinate system, if the vector **F** and the differential surface vector $d\mathbf{S}$ have the components

$$\mathbf{F} = \left(F_x, F_y, F_z \right), \quad d\mathbf{S} = \left(dydz, dxdz, dxdy \right) \tag{4.34}$$

then the surface integral in Eq. (4.33) becomes

$$\iint_S \mathbf{F} \cdot d\mathbf{S} = \iint_S \left(F_x dydz + F_y dxdz + F_z dxdy \right) \tag{4.35}$$

And if the surface is a constant coordinate surface, then this integral reduces to one of the three double integrals:

$$\iint_S \mathbf{F} \cdot d\mathbf{S} = \int_{z_1}^{z_2} \int_{y_1}^{y_2} F_x dydz \tag{4.36a}$$

$$\iint_S \mathbf{F} \cdot d\mathbf{S} = \int_{z_1}^{z_2} \int_{x_1}^{x_2} F_x dxdz \tag{4.36b}$$

$$\iint_S \mathbf{F} \cdot d\mathbf{S} = \int_{y_1}^{y_2} \int_{x_1}^{x_2} F_z dxdy \tag{4.36c}$$

Cylindrical Coordinate System In cylindrical coordinate system, if the vector **F** and the differential surface vector $d\mathbf{S}$ have the components

$$\mathbf{F} = \left(F_\rho, F_\phi, F_z \right), \quad d\mathbf{S} = \left(\rho d\phi dz, d\rho dz, \rho d\phi d\rho \right) \tag{4.37}$$

then the surface integral in Eq. (4.33) becomes

$$\iint_S \mathbf{F} \cdot d\mathbf{S} = \iint_S \left(F_\rho \rho d\phi dz + F_\phi d\rho dz + F_z \rho d\phi d\rho \right) \tag{4.38}$$

And if the surface is a constant coordinate surface, then this integral reduces to one of the three double integrals:

$$\iint_S \mathbf{F} \cdot d\mathbf{S} = \int_{z_1}^{z_2} \int_{\phi_1}^{\phi_2} F_\rho \rho d\phi dz \tag{4.39a}$$

$$\iint_S \mathbf{F} \cdot d\mathbf{S} = \int_{z_1}^{z_2} \int_{\rho_1}^{\rho_2} F_\phi d\rho dz \tag{4.39b}$$

$$\iint_S \mathbf{F} \cdot d\mathbf{S} = \int_{\rho_1}^{\rho_2} \int_{\phi_1}^{\phi_2} F_z \rho d\phi d\rho \tag{4.39c}$$

Spherical Coordinate System In spherical coordinate system, if the vector **F** and the differential surface vector $d\mathbf{S}$ have the components

$$\mathbf{F} = \left(F_r, F_\theta, F_\phi \right) \tag{4.40}$$

$$d\mathbf{S} = \left(r^2 \sin\theta \, d\theta \, d\phi, r \sin\theta \, dr \, d\phi, r \, dr \, d\theta \right)$$

then the surface integral in Eq. (4.33) becomes

$$\iint_S \mathbf{F} \cdot d\mathbf{S} = \iint_S \left(F_r r^2 \sin\theta \, d\theta \, d\phi + F_\theta r \sin\theta \, dr \, d\phi + F_\phi r \, dr \, d\theta \right) \tag{4.41}$$

And if the surface is a constant coordinate surface, then this integral reduces to one of the three double integrals:

$$\iint_S \mathbf{F} \cdot d\mathbf{S} = \int_{\phi_1}^{\phi_2} \int_{\theta_1}^{\theta_2} F_r r^2 \sin\theta \, d\theta \, d\phi \tag{4.42a}$$

$$\iint_S \mathbf{F} \cdot d\mathbf{S} = \int_{\phi_1}^{\phi_2} \int_{r_1}^{r_2} F_\theta r \sin\theta \, dr \, d\phi \tag{4.42b}$$

$$\iint_S \mathbf{F} \cdot d\mathbf{S} = \int_{\theta_1}^{\theta_2} \int_{r_1}^{r_2} F_\phi r \, dr \, d\theta \tag{4.42c}$$

Example 4.3 Evaluation of a Surface Integral
Evaluate the surface integral when the function A is given by

$$\mathbf{F} = 3\rho z^2 \mathbf{a}_\rho - 4z \sin\phi \, \mathbf{a}_\phi + \rho z \mathbf{a}_z$$

and the closed surface shown in Figure 4.5 is given by

$$S: \rho = 2, \quad 0 \le \varphi \le 2\pi, \quad 1 \le z \le 4$$

Figure 4.5 Closed surface in Example 4.3.

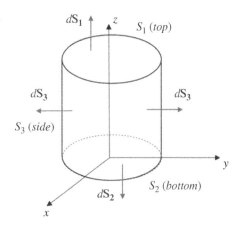

Solution:

$$\oint \mathbf{F} \cdot d\mathbf{S} = \iint_{S_1} \mathbf{F} \cdot d\mathbf{S}_1 + \iint_{S_2} \mathbf{F} \cdot d\mathbf{S}_2 + \iint_{S_3} \mathbf{F} \cdot d\mathbf{S}_3$$

On S_1 we have,

$$S_1 : 0 < \rho < 2, \quad 0 < \varphi < 2\pi, \quad z = 4$$

$$\mathbf{F} = \left(3\rho z^2, -4z\sin\varphi, \rho z\right)$$

$$d\mathbf{S}_1 = (0, 0, dS_1) = (0, 0, \rho\, d\rho\, d\varphi)$$

$$\iint_{S_1} \mathbf{F} \cdot d\mathbf{S}_1 = \iint_{\substack{0 < \rho < 2 \\ 0 < \varphi < 2\pi \\ z=4}} \left(3\rho z^2, -4z\sin\varphi, \rho z\right) \cdot (0, 0, \rho\, d\rho\, d\varphi)$$

$$= \iint_{\substack{0 < \rho < 2 \\ 0 < \varphi < 2\pi \\ z=4}} z\rho^2\, d\rho\, d\varphi = (4) \int_{\varphi=0}^{\varphi=2\pi} d\varphi \int_{\rho=0}^{\rho=2} \rho^2\, d\rho = (4)(2\pi)\left(\frac{\rho^3}{3}\right)\Big|_0^2 = \frac{64\pi}{3}$$

On S_2:

$$S_2 : 0 < \rho < 2, \quad 0 < \varphi < 2\pi, \quad z = 1$$

$$\mathbf{F} = \left(3\rho z^2, -4z\sin\varphi, \rho z\right)$$

$$d\mathbf{S}_2 = (0, 0, -dS_1) = (0, 0, -\rho\, d\rho\, d\varphi)$$

$$\iint_{S_2} \mathbf{F} \cdot d\mathbf{S}_2 = \iint_{\substack{0 < \rho < 2 \\ 0 < \varphi < 2\pi \\ z=1}} \left(3\rho z^2, -4z\sin\varphi, \rho z\right) \cdot (0, 0, -\rho\, d\rho\, d\varphi)$$

$$= -\iint_{\substack{0 < \rho < 2 \\ 0 < \varphi < 2\pi \\ z=1}} z\rho^2\, d\rho\, d\varphi = -(1) \int_{\varphi=0}^{\varphi=2\pi} d\varphi \int_{\rho=0}^{\rho=2} \rho^2\, d\rho = -(2\pi)\left(\frac{\rho^3}{3}\right)\Big|_0^2 = -\frac{16\pi}{3}$$

On S_3:

$$S_3 : \rho = 2, \quad 0 < \varphi < 2\pi, \quad 1 < z < 4$$

$$\mathbf{F} = \mathbf{F} = \left(3\rho z^2, -4z\sin\varphi, \rho z\right)$$

$$d\mathbf{S}_3 = (dS_\rho, 0, 0) = (\rho\, d\varphi\, dz,\ 0,\ 0)$$

$$\iint_{S_3} \mathbf{F} \cdot d\mathbf{S}_3 = \iint_{\substack{\rho=2 \\ 0 < \varphi < 2\pi \\ 1 < z < 4}} \left(3\rho z, -4z\sin\varphi, \rho z\right) \cdot (\rho\, d\varphi\, dz,\ 0,\ 0)$$

$$= \iint_{\substack{\rho=2 \\ 0 < \varphi < 2\pi \\ 1 < z < 4}} 3\rho z^2\, d\varphi\, dz = (3)(2)\int_0^{2\pi} d\varphi \int_1^4 z^2\, dz$$

$$= (6)(2\pi)\left(\frac{z^3}{3}\right)\Big|_1^4 = \frac{12\pi}{3}(64-1) = \frac{756\pi}{3}$$

Therefore,

$$\oint_S \mathbf{F} \cdot d\mathbf{S} = \iint_{S_1} \mathbf{F} \cdot d\mathbf{S}_1 + \iint_{S_2} \mathbf{F} \cdot d\mathbf{S}_2 + \iint_{S_3} \mathbf{F} \cdot d\mathbf{S}_3 = \frac{64\pi}{3} - \frac{16\pi}{3} + \frac{756\pi}{3} = 268$$

∎

4.3 Volume Integrals

The *volume integral* is a generalization of the triple integral and is denoted by

$$\iiint_V f(x, y, z)\,dxdydz \tag{4.43}$$

Triple integrals can be evaluated by three successive integrations.

Example 4.4 Evaluation of a Triple Integral
Let $f(\rho,\phi,z) = 2z\rho$. Then

$$\int_{z=0}^{2}\int_{\rho=0}^{2}\int_{\varphi=0}^{2\pi} 2z\rho\,d\varphi d\rho dz = 2\int_{z=0}^{2}\int_{\rho=0}^{2}\left(\int_{\varphi=0}^{2\pi} d\varphi\right)z\rho d\rho dz = 4\pi\int_{z=0}^{2}\left(\int_{\rho=0}^{2}\rho d\rho\right)zdz$$

$$= \left(4\pi\right)\left(\frac{4}{2}\right)\int_{z=0}^{2} zdz = \left(8\pi\right)\left(\frac{4}{2}\right) = 16\pi$$

∎

4.4 Divergence Theorem of Gauss

The *divergence theorem of Gauss* states that the total outward flux of a vector field $\mathbf{F}$ through the closed surface S equals the volume integral of the divergence of $\mathbf{F}$ (Sadiku, 2010, p. 76).

$$\oint_S \mathbf{F} \cdot d\mathbf{S} = \iiint \nabla \cdot \mathbf{F} dv \tag{4.44}$$

where the closed surface S defines the volume captured inside it.

4.5 Stokes's Theorem

Stokes's theorem states that the line integral of vector $\mathbf{F}$ along closed surface curve C equals the surface integral of the curl of $\mathbf{F}$.

$$\oint_C \mathbf{F} \cdot d\mathbf{r} = \iint_S (\nabla \times \mathbf{F}) \cdot d\mathbf{S} \tag{4.45}$$

where the closed curve C defines the surface S as shown in Figure 4.6

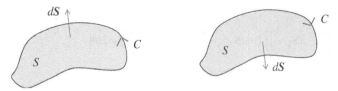

Figure 4.6 Illustration of Stokes's Theorem.

4.6 EMC Applications

4.6.1 Maxwell's Equations in an Integral Form

In Section 3.5.1 we presented Maxwell's equations in a differential form. The equivalent, integral for of these equations in simple medium is (Paul, 2006, Pgs. 899 and 901).

$$\oint_C \mathbf{E} \cdot d\mathbf{l} = -\mu \int_S \frac{\partial \mathbf{H}}{\partial t} \cdot d\mathbf{S} \tag{4.46a}$$

$$\oint_C \mathbf{H} \cdot d\mathbf{l} = \int_S \left(\sigma \mathbf{E} + \varepsilon \frac{\partial \mathbf{E}}{\partial t} \right) \cdot d\mathbf{S} + \int_S \mathbf{J}_S \cdot d\mathbf{S} \tag{4.46b}$$

$$\oint_S \mathbf{E} \cdot d\mathbf{S} = \frac{1}{\varepsilon} \int_v \rho_v \, dv \tag{4.46c}$$

$$\oint_S \mathbf{H} \cdot d\mathbf{S} = 0 \tag{4.46d}$$

In (4.46) $\mathbf{E}$ denotes electric field intensity, while $\mathbf{H}$ denotes magnetic field intensity. $\mathbf{J}$ stands for volume current density, while ρ_v denotes volume charge density. We will derive and discuss these equations in Part III of this book.

4.6.2 Loop and Partial Inductance

The concept of partial inductance is very powerful in EMC, for among other phenomena, allows one to explain the ground bounce and power rail collapse (Paul, 2006, p.779; Ott, 2009, p. 770).

Consider a current flowing out of the source, through a forward path arriving at the load, and then going back to the source through a return path, as shown in Figure 4.7.

According to the Biot-Savart's law, current I flowing in the loop produces a *magnetic flux density* $\mathbf{B}$.

The surface integral of the magnetic flux density over the surface of the loop gives the *magnetic flux* crossing the surface S

$$\Phi = \int_S \mathbf{B} \cdot d\mathbf{S} \tag{4.47}$$

The *loop self inductance* is defined as the ratio of the magnetic flux (due to the current I) crossing the surface S of this loop to the current I flowing in the same loop:

$$L = \frac{\Phi}{I} \tag{4.48}$$

Figure 4.7 Magnetic flux as a surface integral.

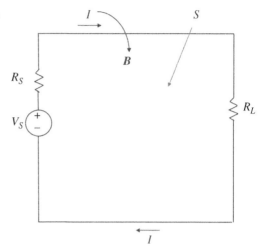

or

$$L = \frac{\displaystyle\int_S \mathbf{B} \cdot d\mathbf{S}}{I} \qquad (4.49)$$

In Part II, we will show that the magnetic flux density, **B**, is related to the *magnetic vector potential*, **A**, by

$$\mathbf{B} = \nabla \times \mathbf{A} \qquad (4.50)$$

By using Stokes's theorem, the surface integral in (4.47) over the surface area S can be transformed into a line integral of the vector magnetic potential **A** over the circumference C of the surface area

$$\int_S \mathbf{B} \cdot d\mathbf{S} = \oint_C \mathbf{A} \cdot d\mathbf{l} \qquad (4.51)$$

and therefore the magnetic flux crossing the loop can be obtained from

$$\Phi = \oint_C \mathbf{A} \cdot d\mathbf{l} \qquad (4.52)$$

Thus, the loop self inductance can alternatively be obtained from

$$L = \frac{\displaystyle\oint_C \mathbf{A} \cdot d\mathbf{l}}{I} \qquad (4.53)$$

The closed curve C can be broken into four line segments C_1, C_2, C_3, and C_4, as shown in Figure 4.8.

Thus the line integral in Eq. (4.52) can be evaluated as the sum of the line integrals along each segment of the loop:

$$\oint_C \mathbf{A} \cdot d\mathbf{l} = \int_{C_1} \mathbf{A} \cdot d\mathbf{l} + \int_{C_2} \mathbf{A} \cdot d\mathbf{l} + \int_{C_3} \mathbf{A} \cdot d\mathbf{l} + \int_{C_4} \mathbf{A} \cdot d\mathbf{l} \qquad (4.54)$$

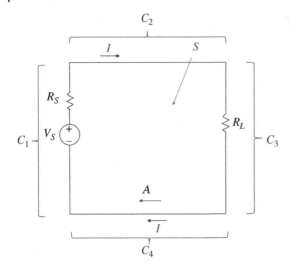

Figure 4.8 Magnetic flux as a line integral.

Using (4.54) in (4.53) we obtain

$$L = \frac{\int_{C_1} \mathbf{A} \cdot d\mathbf{l}}{I} + \frac{\int_{C_2} \mathbf{A} \cdot d\mathbf{l}}{I} + \frac{\int_{C_3} \mathbf{A} \cdot d\mathbf{l}}{I} + \frac{\int_{C_4} \mathbf{A} \cdot d\mathbf{l}}{I} \qquad (4.55)$$
$$= L_1 + L_2 + L_3 + L_4$$

Thus, the loop self inductance equals the sum of inductances attributed to each segment of the loop. The inductances L_1, L_2, L_3, and L_4 are called the *partial self inductances* (Paul, p. 780).

4.6.3   Ground Bounce and Power Rail Collapse

Let's start our discussion with a CMOS inverter logic gate in a totem-pole configuration, shown in Figure 4.9.

In a high-speed digital circuits we often encounter the cascaded CMOS configuration shown in Figure 4.10.

A simplified model of this configuration is shown in Figure 4.11.

Let's investigate the operation of this configuration on the low-to-high and high-to-low transition of the input to the first inverter.

First, assume that the load capacitors C_{GP} and C_{GN} are initially uncharged. When the input signal IN = Low, the upper transistor is ON and the lower is OFF. The current flows though the upper transistor, signal trace, and the capacitor C_{GN} to ground. This is shown in Figure 4.12.

Eventually the capacitor C_{GN} is charged to (approximately) V_{CC} and the current flow stops, as shown in Figure 4.13.

Now, the driver inverter transitions from low-to-high. Subsequently, the upper transistor turns OFF and the lower transistor turns ON, as shown in Figure 4.14.

At this point we have two sources of current:

1) current supplied by C_{GN} as it discharges (dashed arrow)
2) current supplied by V_{CC} as it charges the upper load capacitor

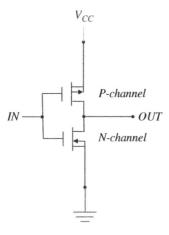

Figure 4.9 CMOS inverter logic gate.

C_{GP} & C_{GN} - *Gate capacitances*

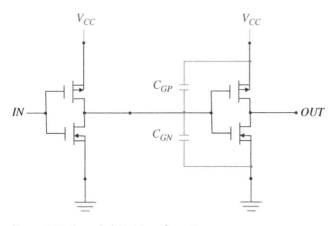

Figure 4.10 Cascaded CMOS configuration.

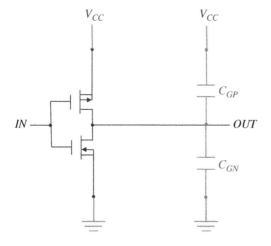

Figure 4.11 Cascaded CMOS inverters.

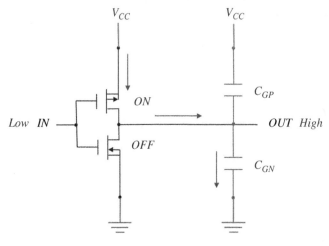

Figure 4.12 Input signal is Low.

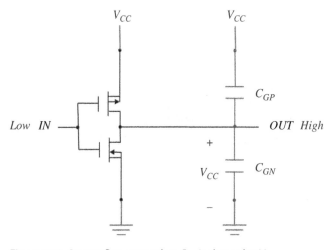

Figure 4.13 Current flow stops when C_{GN} is charged to V_{CC}.

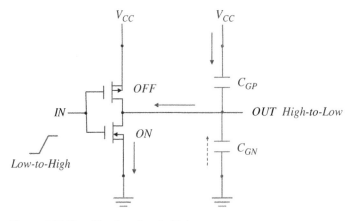

Figure 4.14 Transition from low-to-high.

The current then flows along the trace towards the driver and through the lower transistor to ground. Eventually the current flow stops, and the voltage across capacitor C_{GP} is V_{CC}. This is shown in Figure 4.15.

Now, the driver inverter transitions from high to low. Subsequently, the upper transistor turns ON and the lower transistor turns OFF, as shown in Figure 4.16.

At this point we have two sources of current:

1) current supplied by C_{GP} as it discharges (dashed arrow)
2) current supplied by V_{CC} flowing through the upper transistor, along the trace, and through the lower load capacitor, eventually charging it to V_{CC}

Let's focus on the currents supplied by V_{CC} on both transitions. These are the currents that affect the ground bounce and power rail collapse.

Now let's turn our attention to a more complete circuitry that shows the power distribution system that includes the source, V_S, and power and ground traces. This is shown in Figure 4.17.

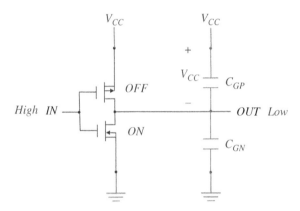

Figure 4.15 Current flow stops when C_{GP} is charged to V_{CC}.

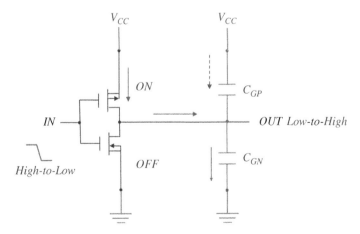

Figure 4.16 Transition from high to low.

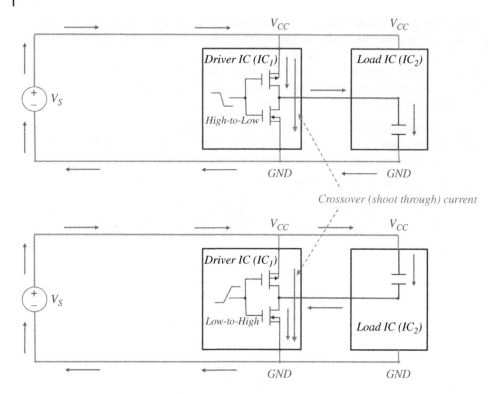

Figure 4.17 Power distribution system and the current flow.

In addition to the currents previously discussed, Figure 4.17 also shows the crossover currents which flow when both transistor are briefly ON at the same time.

When a CMOS gate switches, a current transient is drawn from the power distribution system. This current transient flows through both the power and ground traces. Both of these traces possess (partial) inductance, as shown in Figure 4.18, for a low-to-high transition, and in Figure 4.19 for a high-to-low transition.

Note: The models shown in Figures 4.17 and 4.18 can be applied at frequencies where the impedance of the short PCB traces connecting the ICs is low enough compared to the impedance of the long supply traces and thus can be neglected (Hubing et al., 1995).

The impedance of the long supply traces cannot be neglected and is modeled as the power trace inductance L_P and the ground trace inductance L_G.

When IC_1 switches (and also during the crossover event) the current is drawn from the source, resulting in the voltages V_P and V_G across the power and ground trace inductances. These voltages are often referred to as *power rail collapse*, and *ground bounce*, respectively.

The important consequence of these voltages is the fact that the voltage V_{IC} at the IC_1 power and ground pins is no longer equal to V_S, potentially causing signal integrity issues.

$$V_{IC} = V_S - \left(L_p + L_G\right)\frac{di(t)}{dt} \tag{4.56}$$

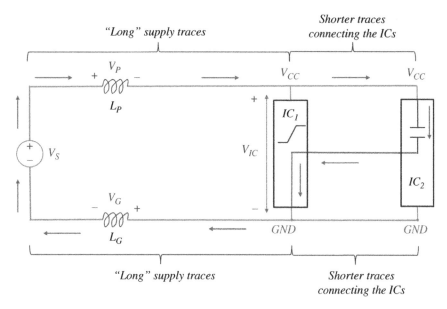

Figure 4.18 Partial inductance: transition from low to high.

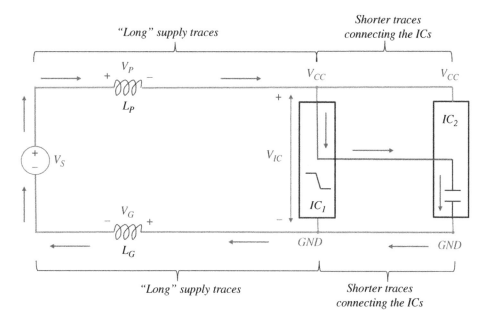

Figure 4.19 Partial inductance: transition from high to low.

References

Kreyszig, E., *Advanced Engineering Mathematics*, 8th ed., John Wiley and Sons, New York, 1999.

Paul, C.R., *Introduction to Electromagnetic Compatibility*, 2nd ed., John Wiley and Sons, New York, 2006.

Sadiku, M.N.O., *Elements of Electromagnetics*, 5th ed., Oxford University Press, New York, 2010.

Hubing, T.H., Drewniak, J.L., Van Doren, T.P. & Hockanson, D.M. – *Power Bus Decoupling on Multilayer Printed Circuit Boards*, IEEE Trans. on EMC, Vol. 37, No. 2, May 1995.

Ott, H.W., *Electromagnetic Compatibility Engineering*, John Wiley and Sons, Hoboken, New Jersey, 2009.

5

Differential Equations

Differential equations are of fundamental importance in electromagnetics because many electromagnetic laws and EMC concepts are mathematically described in the form of differential equations.

We will focus on a selected sample of the differential equations relevant to the subject of EMC. We begin by discussing the first-order RC and RL circuits and their solutions, and then focus on the second-order RLC circuits.

We conclude this chapter by presenting several EMC applications described by the differential equations.

5.1 First Order Differential Equations – RC and RL Circuits

5.1.1 RC Circuit

A typical time-domain RC circuit configuration is shown in Figure 5.1. (In Section 9.3 we will learn how to transform any linear circuit into such a configuration using the Thévenin theorem approach).

At $t = 0$, the switch closes and a dc voltage source, V_T, is connected to a capacitor with an initial voltage of V_0. R_T represents the Thévenin resistance of the circuitry connected to the capacitor.

The differential equation governing the capacitor voltage in this circuit is (Alexander & Sadiku, 2009, p. 274).

$$R_T C \frac{dv_C(t)}{dt} + v_C(t) = V_T, \quad v_C(0) = V_0 \quad t > 0 \tag{5.1}$$

Mathematics provides a number of approaches to solving this equation. We will solve it using the separation of variables. Rearranging terms gives

$$\frac{dv_C}{dt} = \frac{V_T - v_C}{R_T C} = -\frac{v_C - V_T}{R_T C} \tag{5.2}$$

or

$$\frac{dv_C}{v_C - V_T} = -\frac{dt}{R_T C} \tag{5.3}$$

Foundations of Electromagnetic Compatibility with Practical Applications, First Edition. Bogdan Adamczyk.
© 2017 John Wiley & Sons Ltd. Published 2017 by John Wiley & Sons Ltd.

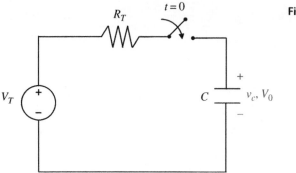

Figure 5.1 RC circuit.

Integrating both sides and using the definite integral approach we get

$$\int_{V_0}^{v_C(t)} \frac{dv_C}{v_C - V_T} = -\int_0^t \frac{dt}{R_T C} \tag{5.4}$$

Thus

$$\ln\left(v_C - V_T\right)\Big|_{V_0}^{v_C(t)} = -\frac{t}{R_T C}\Big|_0^t \tag{5.5}$$

and subsequently

$$\ln\left[v_C(t) - V_T\right] - \ln\left(V_0 - V_T\right) = -\frac{t}{R_T C} \tag{5.6}$$

or

$$\ln\frac{v_C - V_T}{V_0 - V_T} = -\frac{t}{R_T C} \tag{5.7}$$

Therefore

$$e^{\ln\frac{v_C - V_T}{V_0 - V_T}} = e^{-\frac{t}{R_T C}} \tag{5.8}$$

or

$$\frac{v_C - V_T}{V_0 - V_T} = e^{-t/R_T C} \tag{5.9}$$

leading to

$$v_C - V_T = \left(V_0 - V_T\right)e^{-t/R_T C} \tag{5.10}$$

Figure 5.2 RC circuit step response.

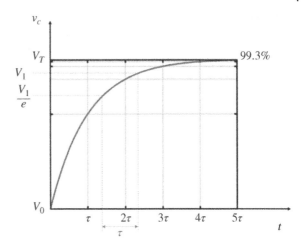

Figure 5.3 RL circuit.

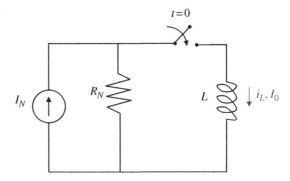

And finally the step response of the capacitor voltage is given by

$$v_C(t) = V_T + (V_0 - V_T)e^{-t/\tau}, \quad \tau = R_T C, \quad t \geq 0 \tag{5.11}$$

This response is shown in Figure 5.2, where $\tau = R_T C$ is the time constant of the circuit.

5.1.2 RL Circuit

A typical time-domain RL circuit configuration is shown in Figure 5.3.

At $t = 0$, the switch closes and a dc current source, I_N, is connected to an inductor with an initial current of I_0. R_N represents the Norton resistance of the circuitry connected to the inductor (discussed in Section 9.4).

The differential equation governing the inductor in this circuit is (Nilsson & Riedel, 2015, p. 225).

$$\frac{L}{R_N}\frac{di_L}{dt} + i_L = I_N, \quad i_L(0) = I_0, \quad t \geq 0 \tag{5.12}$$

We will solve it using the separation of variables, but this time we will use the *indefinite* integral approach. Rearranging terms gives

$$\frac{di_L}{dt} = \frac{I_N - i_L}{L/R_N} = -\frac{i_L - I_N}{L/R_N} \tag{5.13}$$

or

$$\frac{di_L}{i_L - I_N} = -\frac{dt}{L/R_N} \tag{5.14}$$

Integrating both sides and using the *indefinite* integral approach we get

$$\int \frac{di_L}{i_L - I_N} = -\int \frac{dt}{L/R_N} \tag{5.15}$$

Thus

$$\ln\left[i_L(t) - I_N\right] = -\frac{t}{L/R_N} + A \tag{5.16}$$

Therefore

$$i_L(t) - I_N = e^{-\frac{t}{L/R_N} + A} = e^A = Be^{-\frac{t}{L/R_N}} \tag{5.17}$$

or

$$i_L(t) = I_N + Be^{-\frac{t}{L/R_N}} \tag{5.18}$$

Evaluating Eq. (5.18) at $t = 0$ gives

$$B = I_0 - I_N \tag{5.19}$$

Substituting Eq. (5.19) in Eq. (5.18) and rearranging produces the final result, i.e. the step response of the inductor current

$$i_L(t) = I_N + (I_0 - I_N)e^{-t/\tau}, \quad \tau = \frac{L}{R_N}, \quad t \geq 0 \tag{5.20}$$

This response is shown in Figure 5.4, where $\tau = L/R_N$ is the time constant of the circuit.

Compare Eq. (5.20) with Eq. (5.11). They have the same mathematical form! This is not a coincidence. Look at Eq. (5.12) and Eq. (5.1). They also have the same mathematical form!

This leads to a very important observation that we will use on several occasions. Mathematical equations of the same form have the solutions of the same form.

Figure 5.4 RL circuit step response.

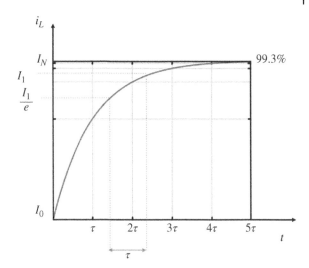

5.2 Second-Order Differential Equations – Series and Parallel RLC Circuits

5.2.1 Series RLC Circuit

A typical series RLC circuit configuration is shown in Figure 5.5.

At $t = 0$, the switch closes and a dc voltage source, V_T, is connected to a *series LC* configuration. The capacitor has an initial voltage of V_0 and the inductor has an initial current of I_0.

The differential equation governing the capacitor voltage in this circuit is (Nilsson & Riedel, 2015, p. 287)

$$\frac{d^2 v_C}{dt^2} + \frac{R_T}{L} \frac{dv_C}{dt} + \frac{1}{LC} v_C = \frac{V_T}{LC}, \quad t \geq 0$$
$$v_C(0) = V_0,$$
$$\frac{dv_C(0)}{dt} = \frac{I_0}{C}$$

(5.21)

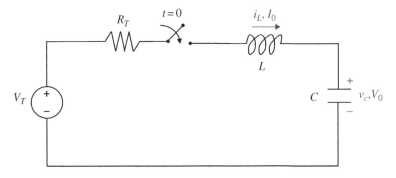

Figure 5.5 Series RLC circuit.

Note that this equation could be written as

$$\frac{d^2 v_C}{dt^2} + a\frac{dv_C}{dt} + bv_C = \frac{V_T}{LC}, \quad t \geq 0$$
$$v_C(0) = V_0,$$
$$\frac{dv_C(0)}{dt} = \frac{I_0}{C}$$

(5.22)

Instead of solving Eq. (5.22) directly, we will modify it by introducing two new parameters ζ and ω_0, instead of a and b. Why? Because these two new constants have a physical meaning (unlike a and b), are very descriptive (we will see that, when looking at the solution of Eq. (5.21)) and they are indispensable when analyzing and designing second-order systems.

The new parameters are indirectly defined by:

$$2\zeta\omega_0 = a$$

(5.23a)

$$\omega_0^2 = b$$

(5.23b)

Using these two parameters, Eq. (5.22) can be written as

$$\frac{d^2 v_C}{dt^2} + 2\zeta\omega_0 \frac{dv_C}{dt} + \omega_0^2 v_C = \frac{V_T}{LC}, \quad t \geq 0$$
$$v_C(0) = V_0,$$
$$\frac{dv_C(0)}{dt} = \frac{I_0}{C}$$

(5.24)

There are many ways of solving Eq. (5.24). The choice of approach strongly depends of the functional form of the forcing function V_T.

When the forcing function is identically equal to zero, $V_T = 0$, the circuit is driven by the initial conditions only: the initial capacitor voltage and the initial inductor current (at least one of these two values must be non-zero for the circuit response to be non-zero).

In this case, the Eq. (5.24) describing the series RLC circuit becomes

$$\frac{d^2 v_C}{dt^2} + 2\zeta\omega_0 \frac{dv_C}{dt} + \omega_0^2 v_C = 0, \quad t \geq 0$$
$$v_C(0) = V_0,$$
$$\frac{dv_C(0)}{dt} = \frac{I_0}{C}$$

(5.25)

The response of the circuit is termed the *natural* response.

We will subsequently obtain the solution of Eq. (5.24) in the time domain for the case when the forcing function is a dc (step) voltage. Also, we will set the initial conditions to zero.

$$\frac{d^2 v_C}{dt^2} + 2\zeta\omega_0 \frac{dv_C}{dt} + \omega_0^2 v_C = \frac{V_T}{LC}, \quad t \geq 0$$

$$v_C(0) = 0, \tag{5.26}$$

$$i_L(0) = 0$$

Such a response is termed the *forced* response.

Because the circuit is linear, we can solve for these responses separately and superimpose them to get the total response. (This important property is called *superposition* and will be explained in Part II.)

Natural Response of the Series RLC Circuit   The natural response is governed by Eq. (5.26), which requires the capacitor voltage plus a constant times its first derivative, plus another constant times its second derivative to add to zero for all ≥ 0. The only way this can happen is for $v_C(t)$, its first derivative, and its second derivative to have the same functional form.

There is only one such mathematical function: an exponential function. No matter how many times we differentiate an exponential function, the result is another exponential function.

This observation plus experience with first-order circuits, suggests that we try a solution of the form

$$v_C(t) = Ae^{st} \tag{5.27}$$

where A and s are constants yet to be determined.

If $v_C(t)$, as defined by Eq. (5.27), is to be the solution of Eq. (5.26) then it must satisfy Eq. (5.26). Let's see where this reasoning leads us. First, let's obtain the first and second derivative of $v_C(t)$, so that we can substitute them in Eq. (5.26).

$$\frac{dv_C(t)}{dt} = sAe^{st} \tag{5.28a}$$

$$\frac{d^2 v_C(t)}{dt^2} = s^2 Ae^{st} \tag{5.28b}$$

Substituting Eq. (5.27) and Eq. (5.28) into Eq. (5.26) leads to

$$s^2 Ae^{st} + 2\zeta\omega_0 sAe^{st} + \omega_0^2 Ae^{st} = 0 \tag{5.29}$$

Or

$$Ae^{st}\left(s^2 + 2\zeta\omega_0 s + \omega_0^2\right) = 0 \tag{5.30}$$

The function e^{st} cannot be zero for all $t \geq 0$. The condition $A = 0$ is not considered because it is a trivial solution of no interest to us. That leaves us with the requirement that

$$s^2 + 2\zeta\omega_0 s + \omega_0^2 = 0 \tag{5.31}$$

This quadratic equation is known as the characteristic equation of Eq. (5.24), since the roots of the equation dictate the character of the solution.

The two roots are

$$s_1 = -\zeta\omega_0 + \omega_0\sqrt{\zeta^2 - 1} \tag{5.32a}$$

$$s_2 = -\zeta\omega_0 - \omega_0\sqrt{\zeta^2 - 1} \tag{5.32b}$$

The two values of s in Eq. (5.32) indicate that there are two possible solutions for $v_C(t)$, each of which is in the form of the assumed solution in Eq. (5.27):

$$v_{C1} = A_1 e^{s_1 t} \tag{5.33a}$$

$$v_{C2} = A_2 e^{s_2 t} \tag{5.33b}$$

Since Eq. (5.25) is a linear equation, any linear combination of the two distinct solutions $v_{C1}(t)$ and $v_{C2}(t)$ is also a solution of Eq. (5.25). The general solution of Eq. (5.25), therefore, is

$$v_C(t) = A_1 e^{s_1 t} + A_2 e^{s_2 t} \tag{5.34}$$

where the constants s_1 and s_2 are given by Eq. (5.31) and the constants A_1 and A_2 are determined from the initial capacitor voltage and inductor current, as follows.

At $t = 0$, Eq. (5.34) becomes

$$v_C(0) = V_0 = A_1 + A_2 \tag{5.35}$$

To use the initial condition on the inductor current, we differentiate Eq. (5.34) to obtain

$$\frac{dv_C(t)}{dt} = A_1 s_1 e^{s_1 t} + A_2 s_2 e^{s_2 t} \tag{5.36}$$

At $t = 0$, Eq. (5.36) becomes

$$\frac{dv_C(0)}{dt} = A_1 s_1 + A_2 s_2 \tag{5.37}$$

Since

$$\frac{dv_C(0)}{dt} = \frac{I_0}{C} \tag{5.38}$$

we have

$$\frac{I_0}{C} = A_1 s_1 + A_2 s_2 \tag{5.39}$$

The solution of Eq. (5.35) and Eq. (5.39) yields the constants A_1 and A_2.

Examining the roots of the characteristic equation (5.31) we notice that the roots can be of three different types:

1) If $\zeta > 1$ we have *two distinct real roots* (the voltage response is said to be *overdamped*).
2) If $\zeta = 1$, we have *two equal real roots* (the voltage response is said to be *critically damped*).
3) If $\zeta < 1$, we have *two complex conjugate roots* (the voltage response is said to be *underdamped*).

Each type of the roots leads to a different mathematical form of the solution and hence to a different circuit behavior identified as *overdamped, critically damped,* or *underdamped.*

These three cases will be addressed next. (It is the underdamped case that is of most concern to an EMC engineer, as we shall see.)

Case 1 – Overdamped Response   When $\zeta > 1$, we have two distinct real roots, both of which are negative and real

$$s_1 = -\zeta\omega_0 + \omega_0\sqrt{\zeta^2 - 1} \tag{5.40a}$$

$$s_2 = -\zeta\omega_0 - \omega_0\sqrt{\zeta^2 - 1} \tag{5.40b}$$

The *overdamped* response is given by

$$v_C(t) = A_1 e^{s_1 t} + A_2 e^{s_2 t} \tag{5.41}$$

Case 2 – Critically Damped Response   When $\zeta = 1$, we have two equal real negative roots.

$$s_1 = s_2 = -\zeta\omega_0 \tag{5.42}$$

The *critically damped* response is given by

$$v_C(t) = B_1 e^{st} + B_2 t e^{st} \tag{5.43}$$

Constants B_1 and B_2 are evaluated using the initial conditions, as follows. At $t = 0$, Eq. (5.43) becomes

$$v_C(0) = B_1 \tag{5.44}$$

Upon differentiation Eq. (5.43) produces

$$\frac{dv_C(t)}{dt} = B_1 s e^{st} + B_2 e^{st} + B_2 t s e^{st} \tag{5.45}$$

Evaluating Eq. (5.45) at $t = 0$ results in

$$\frac{dv_C(0)}{dt} = B_1 s + B_2 \tag{5.46}$$

Now using Eq. (5.22) in Eq. (5.46) we arrive at

$$\frac{I_0}{C} = B_1 s + B_2 \tag{5.47}$$

Solving Eq. (5.44) and Eq. (5.47) simultaneously produces the unknown constants B_1 and B_2.

Case 3 – Underdamped Response When $\zeta < 1$, we have two complex conjugate roots.

$$s_1 = -\zeta\omega_0 + \omega_0\sqrt{\zeta^2 - 1} = -\zeta\omega_0 + \omega_0\sqrt{(-1)(1 - \zeta^2)} \tag{5.48a}$$
$$= -\zeta\omega_0 + \omega_0\sqrt{-1}\sqrt{1 - \zeta^2} = -\zeta\omega_0 + j\omega_0\sqrt{1 - \zeta^2}$$

$$s_2 = -\zeta\omega_0 - j\omega_0\sqrt{1 - \zeta^2} \tag{5.48b}$$

It is convenient (and practical from the design standpoint) to define a new frequency ω_d, called the *damped natural frequency*

$$\omega_d = \omega_0\sqrt{1 - \zeta^2} \tag{5.49}$$

Then the complex roots in Eq. (5.48) can be written as

$$s_1 = -\zeta\omega_0 + \omega_0\sqrt{\zeta^2 - 1} \tag{5.50a}$$

$$s_2 = -\zeta\omega_0 - \omega_0\sqrt{\zeta^2 - 1} \tag{5.50b}$$

The *underdamped* circuit response is given by

$$v_C(t) = D_1 e^{-\zeta\omega_0 t} \cos\omega_d t + D_2 e^{-\zeta\omega_0 t} \sin\omega_d t, \quad t \geq 0 \tag{5.51}$$

The constants D_1 and D_2 are evaluated from the initial conditions as follows. Evaluating Eq. (5.50) at $t = 0$ results in

$$v_C(0) = D_1 \tag{5.52}$$

That was easy. Obtaining D_2 will take some work. Differentiate Eq. (5.51) to obtain

$$\frac{dv_C(t)}{dt} = D_1 \frac{d}{dt}\left(e^{-\zeta\omega_0 t}\cos\omega_d t\right) + D_2 \frac{d}{dt}\left(e^{-\zeta\omega_0 t}\sin\omega_d t\right)$$
$$= D_1\left[(-\zeta\omega_0)e^{-\zeta\omega_0 t}\cos\omega_d t + e^{-\zeta\omega_0 t}(-\omega_d\sin\omega_d t)\right] \tag{5.53}$$
$$+ D_2\left[(-\zeta\omega_0)e^{-\zeta\omega_0 t}\sin\omega_d t + e^{-\zeta\omega_0 t}(\omega_d\cos\omega_d t)\right]$$

Now, evaluate Eq. (5.53) at $t = 0$ to get

$$\frac{dv_C(0)}{dt} = D_1(-\zeta\omega_0) + D_2\omega_d \tag{5.54}$$

Finally, make use of Eq. (5.22) in Eq. (5.54) to arrive at

$$D_1\left(-\zeta\omega_0\right)+D_2\omega_d = \frac{I_0}{C} \tag{5.55}$$

Use Eq. (5.52) in Eq. (5.55) to solve for D_2.

Note: Using the trigonometric identities (or phasors), the underdamped response, as given by Eq. (5.51), can be expressed in a much more useful from:

$$v_C\left(t\right) = De^{-\zeta\omega_0 t}\left(\sin\omega_d t + \theta\right), \qquad t \geq 0 \tag{5.56}$$

The new constants D and θ are determined from the knowledge of the constants D_1 and D_2.

Forced Response of the Series RLC Circuit Consider the series RLC Circuit, shown in Figure 5.6 with zero initial conditions and driven by a step input.

This circuit is governed by the differential equation

$$\frac{d^2 v_C}{dt^2} + \frac{R_T}{L}\frac{dv_C}{dt} + \frac{1}{LC}v_C = \frac{V_T}{LC}, \qquad t \geq 0 \tag{5.57}$$

Or using the damping ratio and undamped natural frequency

$$\frac{d^2 v_C}{dt^2} + 2\zeta\omega_0\frac{dv_C}{dt} + \omega_0^2 v_C = \omega_0^2 V_T \qquad t \geq 0 \tag{5.58}$$

where

$$2\zeta\omega_0 = \frac{R}{L} \tag{5.59a}$$

$$\omega_0^2 = \frac{1}{LC} \tag{5.59b}$$

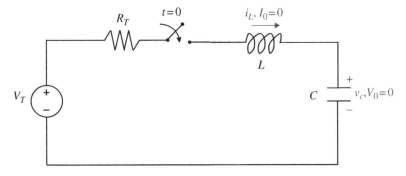

Figure 5.6 Series RLC circuit with zero initial conditions.

and the initial capacitor voltage and the initial inductor current are zero.

$$v_C(0) = V_0 = 0 \tag{5.60a}$$

$$i_L(0) = I_0 = 0 \tag{5.60b}$$

Under these assumptions the solution of Eq. (5.58) is called the *forced* response.

In order to obtain the solution of Eq. (5.58) we need the value of the initial capacitor voltage (which is zero in this case) as well as the value of the derivative of the capacitor voltage at $t = 0$.

The current through the capacitor is related to the voltage across the capacitor by:

$$i_C(t) = C\frac{dv_C(t)}{dt} \tag{5.61}$$

Since the current through the inductor is the same as the capacitor current (they are in series), we have

$$i_L(t) = C\frac{dv_C(t)}{dt} \tag{5.62}$$

Evaluating Eq. (5.62) at $t = 0$ results in

$$i_L(0) = C\frac{dv_C(0)}{dt} \tag{5.63}$$

Since the initial inductor current is zero, it follows that

$$\frac{dv_C(0)}{dt} = 0 \tag{5.64}$$

This the differential equation we need to solve; subject to its initial condition, it can be alternatively stated as

$$\begin{aligned}
&\frac{d^2v_C}{dt^2} + 2\zeta\omega_0\frac{dv_C}{dt} + \omega_0^2 v_C = \omega_0^2 V_T \quad t \geq 0 \\
&v_C(0) = 0 \\
&\frac{dv_C(0)}{dt} = 0
\end{aligned} \tag{5.65}$$

It can be shown (recall the discussion regarding the natural response) that the forced solution of Eq. (5.65) assumes one of the following forms.

Overdamped response $(\zeta > 1)$

$$v_C(t) = V_T + E_1 e^{s_1 t} + E_2 e^{s_2 t} \tag{5.66a}$$

Critically damped response $(\zeta = 1)$

$$v_C(t) = V_T + (E_1 + E_2 t)e^{st} \tag{5.66b}$$

Underdamped response $(\zeta < 1)$

$$v_C(t) = V_T + E_1 e^{-\zeta\omega_0 t}\cos\omega_d t + E_2 e^{-\zeta\omega_0 t}\sin\omega_d t \tag{5.66c}$$

The constants E_1 and E_2 are evaluated using the initial conditions following the procedure discussed for the natural response.

Just as for the natural response, it is more desirable from the engineering standpoint to express the underdamped case in the form

$$v_C(t) = V_T + Ee^{-\zeta\omega_0 t}\sin(\omega_d t + \theta) \tag{5.67}$$

instead of the form in Eq. (5.66c). The new constants E and θ are determined from the knowledge of the constants E_1 and E_2.

The forced responses are shown in Figure 5.7.

Figure 5.8 shows the underdamped responses for different value of the damping ratio ζ. It is important to note that for small values of ζ the response is highly oscillatory. This fact manifests itself in the EMC phenomenon known as *ringing*, which will be described at the end of this chapter.

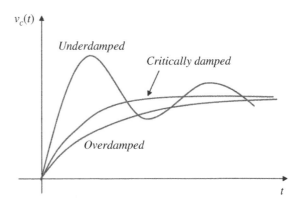

Figure 5.7 Forced responses of a series RLC circuit.

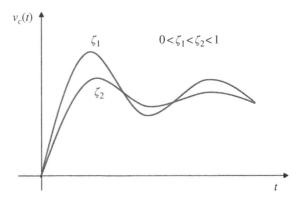

Figure 5.8 Underdamped responses for different values of ζ.

Total Response of the Series RLC Circuit Total response of the series RLC circuit is due to both the forcing function and the non-zero initial conditions. Based on the property of superposition (which will be discussed in Part II) the total response is the sum of the natural and forced responses.

The total response assumes one of the following forms.

Overdamped response $(\zeta > 1)$

$$v_C(t) = V_T + F_1 e^{s_1 t} + F_2 e^{s_2 t} \tag{5.68a}$$

Critically damped response $(\zeta = 1)$

$$v_C(t) = V_T + (F_1 + F_2 t) e^{st} \tag{5.68b}$$

Underdamped response $(\zeta < 1)$

$$v_C(t) = V_T + F_1 e^{-\zeta \omega_0 t} \cos \omega_d t + F_2 e^{-\zeta \omega_0 t} \sin \omega_d t \tag{5.68c}$$

$$v_C(t) = V_T + F e^{-\zeta \omega_0 t} \sin(\omega_d t + \theta) \tag{5.68d}$$

All the constants in Eq. (5.68) are evaluated in a similar manner to that discussed for the natural response.

5.2.2 Parallel RLC Circuit

A typical parallel RLC circuit configuration is shown in Figure 5.9.

At $t = 0$, the switch closes and a dc current source, I_N, is connected to a *parallel LC* configuration. The capacitor has an initial voltage of V_0 and the inductor has an initial current of I_0.

The differential equation governing the inductor current in this circuit is (Nilsson & Riedel, 2015, p. 280)

$$\frac{d^2 i_L}{dt^2} + \frac{1}{R_N C} \frac{di_L}{dt} + \frac{1}{LC} i_L = \frac{I_N}{LC}, \quad t \geq 0$$

$$i_L(0) = I_0, \tag{5.69}$$

$$\frac{di_L(0)}{dt} = \frac{V_0}{L}$$

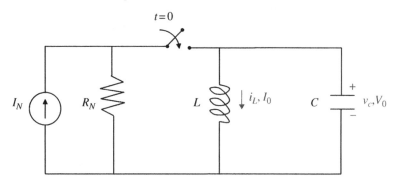

Figure 5.9 Parallel RLC circuit.

or

$$\frac{d^2 i_L}{dt^2} + 2\zeta\omega_0 \frac{di_L}{dt} + \omega_0^2 i_L = \omega_0^2 I_N, \quad t \geq 0$$
$$i_L(0) = I_0,$$
$$\frac{di_L(0)}{dt} = \frac{V_0}{L}$$

(5.70)

where

$$\omega_0^2 = \frac{1}{LC}$$

(5.71a)

$$2\zeta\omega_0 = \frac{1}{RC}$$

(5.71b)

Comparing Eqs (5.70) and (5.24) we note that they have the same mathematical form. It follows that the solutions of these equations will have the same mathematical form. Therefore the natural, forced, and the total responses will have the following forms.

Overdamped response $(\zeta > 1)$

$$i_L(t) = I_N + F_1 e^{s_1 t} + F_2 e^{s_2 t}$$

(5.72a)

Critically damped response $(\zeta = 1)$

$$i_L(t) = I_N + (F_1 + F_2 t) e^{st}$$

(5.72b)

Underdamped response $(\zeta < 1)$

$$i_L(t) = I_N + F_1 e^{-\zeta\omega_0 t} \cos\omega_d t + F_2 e^{-\zeta\omega_0 t} \sin\omega_d t$$

(5.72c)

$$i_L(t) = I_N + F e^{-\zeta\omega_0 t} \cos(\omega_d t + \theta)$$

(5.72d)

All the constants in Eq. (5.72) are evaluated in a similar manner to the one discussed for the series RLC circuit.

5.3   Helmholtz Wave Equations

In Section 6.7.4 we will derive the formulas for the radiated fields of the electric dipole of the antenna. In our derivations we will utilize the results derived in this section.

In this section we will present the solution of the inhomogeneous Helmholtz equation (Balanis, 2005, p. 139)

$$\nabla^2 \hat{\mathbf{A}} + k^2 \hat{\mathbf{A}} = -\mu \hat{\mathbf{J}}$$

(5.73)

In order to solve this equation for the vector magnetic potential A, we will proceed as follows.

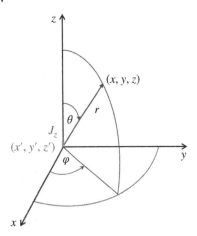

Let's assume that an infinitesimal source with current density $\mathbf{J} = \left(0,0,J_z\right)$ is placed at the origin of the coordinate system as shown in Figure 5.10.

To distinguish between the location where the source exists and the location of the observation point, we will use the prime coordinates for the source and the unprimed coordinates for the observation point, as shown in Figure 5.10.

Since the current density vector $\mathbf{J}$ has only a z component, then the vector $\mathbf{A}$ will only have a z component, and the Eq. (5.73) can be rewritten as

$$\nabla^2 A_z + k^2 A_z = -\mu J_z \tag{5.74}$$

At the observation point, the current density $J_z = 0$, and Eq. (5.74) becomes

$$\nabla^2 A_z + k^2 A_z = 0 \tag{5.75}$$

In the limit, the source is a point, and therefore A_z is not a function of θ or φ; it will only be a function of the distance from the origin, $A_z = A_z(r)$. Due to the apparent symmetry, we will choose spherical coordinate system for evaluation of the Laplacian in Eq. (5.75).

The Laplacian in spherical coordinate system is given by

$$\nabla^2 A_z = \frac{1}{r^2}\frac{\partial}{\partial r}\left(r^2\frac{\partial A_z}{\partial r}\right) + \frac{1}{r^2\sin\theta}\frac{\partial}{\partial\theta}\left(\sin\theta\frac{\partial A_z}{\partial\theta}\right) + \frac{1}{r^2\sin^2\theta}\frac{\partial^2 A_z}{\partial\phi^2} \tag{5.76}$$

Thus, Eq. (5.75) can be written as

$$\nabla^2 A_z + k^2 A_z = \frac{1}{r^2}\frac{\partial}{\partial r}\left[r^2\frac{\partial A_z(r)}{\partial r}\right] + k^2 A_z(r) = 0 \tag{5.77}$$

Taking the derivative of the term in brackets leads to

$$\nabla^2 A_z + k^2 A_z = \frac{1}{r^2}\frac{\partial}{\partial r}\left[r^2\frac{\partial A_z(r)}{\partial r}\right] + k^2 A_z(r) = 0 \tag{5.78}$$

or

$$\frac{1}{r^2}\left[2r\frac{\partial A_z(r)}{\partial r}+r^2\frac{\partial^2 A_z(r)}{\partial r^2}\right]+k^2 A_z(r)=0 \tag{5.79}$$

Equation (5.79) has two independent solutions (Balanis, 1989, p. 277)

$$A_{z1}=C_1\frac{e^{-jkr}}{r} \tag{5.80a}$$

and

$$A_{z2}=C_2\frac{e^{+jkr}}{r} \tag{5.80b}$$

Equation (5.80a) represents an outwardly (in the radial direction) traveling wave and Eq. (5.80b) describes an inwardly traveling wave. Since the source is placed at the origin, giving rise to an outwardly traveling wave, we chose the solution (5.80a) and discard the solution (5.80b). Thus,

$$A_z=A_{z1}=C_1\frac{e^{-jkr}}{r} \tag{5.81}$$

In the static case $\omega=0$, and thus

$$k^2=\omega^2\mu\varepsilon=0 \tag{5.82}$$

and Eq. (5.81) simplifies to

$$A_z=\frac{C_1}{r} \tag{5.83}$$

which is the solution to Eq. (5.73) or Eq. (5.75) when $k=0$.

Thus, at the locations away from the source the time-varying solution (5.81) and the static solution (5.83) differ only by the e^{-jkr} factor.

Thus, the time-varying solution can be obtained by multiplying the static solution by e^{-jkr}.

Now, at the locations when the source is present, the Helmholtz equation is that of Eq. (5.74), repeated here

$$\nabla^2 A_z+k^2 A_z=-\mu J_z \tag{5.84}$$

which in the static case becomes

$$\nabla^2 A_z=-\mu J_z \tag{5.85}$$

This is a well-known Poisson's equation. The most familiar version of Poisson's equations is that from electrostatics

$$\nabla^2 V=-\frac{\rho}{\varepsilon} \tag{5.86}$$

where V is the scalar electric potential and ρ is the electric charge density. The solution of Eq. (5.86) is known to be

$$V = \frac{1}{4\pi\varepsilon} \int\limits_{v} \frac{\rho}{r} dv' \tag{5.87}$$

The single prime for the variable of integration indicates that we integrate over the volume where the source is present.

Since Eqs (5.85) and (5.86) have the same mathematical forms, their solutions have the same mathematical forms. Thus the solution of Eq. (5.84) for the static case is

$$A_z = \frac{\mu}{4\pi} \int\limits_{v} \frac{J_z}{r} dv' \tag{5.88}$$

By analogy to Eqs (5.81) and (5.83), the time-varying solution of Eq. (5.84) can be obtained by multiplying the static solution (5.88) by e^{-jkr}.

Thus,

$$A_z = \frac{\mu}{4\pi} \int\limits_{v} J_z \frac{e^{-jkr}}{r} dv' \tag{5.89}$$

is the solution of Eq. (5.84).

If the current densities were in the x and y directions (J_x and J_y respectively), the wave equation

$$\nabla^2 \hat{\mathbf{A}} + k^2 \hat{\mathbf{A}} = -\mu \hat{\mathbf{J}} \tag{5.90}$$

for each direction would reduce to

$$\nabla^2 A_x + k^2 A_x = -\mu J_x \tag{5.91a}$$

$$\nabla^2 A_y + k^2 A_y = -\mu J_y \tag{5.91b}$$

with corresponding solutions of the form as in Eq. (5.89) given by

$$A_x = \frac{\mu}{4\pi} \int\limits_{v} J_x \frac{e^{-jkr}}{r} dv' \tag{5.92a}$$

$$A_y = \frac{\mu}{4\pi} \int\limits_{v} J_y \frac{e^{-jkr}}{r} dv' \tag{5.92b}$$

Thus, the solution to Eq. (5.90) is

$$\mathbf{A} = \frac{\mu}{4\pi} \int\limits_{v} \mathbf{J} \frac{e^{-jkr}}{r} dv' \tag{5.93}$$

where the solution vector $\mathbf{A}$ has the components A_x, A_y, A_z, given by Eqs (5.92a), (5.92b), and (5.89), respectively.

In the discussion so far, we have assumed that an infinitesimal source with current density $\mathbf{J} = (0, 0, J_z)$ was placed at the origin of the coordinate system.

Figure 5.11 Current density source located away from the origin.

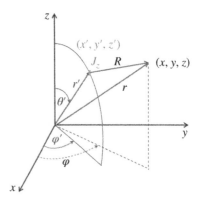

Let's now place the source away from the origin, as shown in Figure 5.11.

Note that the source is represented by the primed coordinates, while the observation point by the unprimed coordinates.

Now, the solution (5.90) can be written as

$$A(x,y,z) = \frac{\mu}{4\pi} \int_v J(x',y',z') \frac{e^{-jkR}}{R} dv' \tag{5.94}$$

where R is the distance from any point on the source to the observation point.

The solution (5.94) was derived for the volume current density (in A/m²). If **J** represents surface current density (in A/m), then the solution for **A** is given in terms of the surface integral

$$A(x,y,z) = \frac{\mu}{4\pi} \int_S J_S(x',y',z') \frac{e^{-jkR}}{R} dS' \tag{5.95}$$

and if **J** represents electric current (in A), then the solution for **A** is given in terms of the line integral

$$A(x,y,z) = \frac{\mu}{4\pi} \int_c I_e(x',y',z') \frac{e^{-jkR}}{R} dl' \tag{5.96}$$

5.4 EMC Applications

5.4.1 Inductive Termination of a Transmission Line

In this section we will show the application of the RL circuit differential equation discussed in Section 5.1.2 to the transmission line terminated by an inductive load. We will discuss transmission lines in detail in Part III.

Consider the circuit shown in Figure 5.12.

A line of length d is terminated by an inductor L with zero initial current. A constant voltage source with internal resistance equal to the characteristic impedance Z_C of the line is connected to the line at $t = 0$.

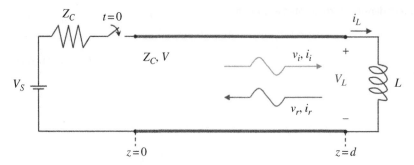

Figure 5.12 Transmission line terminated by an inductive load.

As we will learn in Chapter 17, the moment the switch closes at $t = 0$, the voltage and current waves (v_i and i_i) originate at $z = 0$ and travel down the line to reach the load end at time T.

Upon arriving at the load the reflected voltage and current waves (v_r and i_r) are created. The differential equation (we will derive it in Section 17.1.3) governing the reflected voltage wave is

$$\frac{L}{Z_C}\frac{dv_r}{dt} + v_r = -\frac{V_S}{2}, \quad v_r(0) = \frac{V_S}{2}, \quad t > T \tag{5.97}$$

Rearranging Eq. (5.97) results in

$$\frac{dv_r}{dt} + \frac{v_r}{L/Z_C} = -\frac{V_S}{2}\frac{Z_C}{L} \tag{5.98}$$

or

$$\frac{dv_r}{dt} + \frac{v_r}{\tau} = K \tag{5.99}$$

where

$$\tau = \frac{L}{Z_C} \tag{5.100a}$$

$$K = -\frac{V_S}{2}\frac{Z_C}{L} \tag{5.100b}$$

Rearranging Eq. (5.99) we get

$$\frac{dv_r}{dt} = \frac{v_r}{\tau} + K \tag{5.101}$$

or

$$\frac{dv_r}{dt} = -\frac{\left(v_r - K\tau\right)}{\tau} \tag{5.102}$$

Separating the variables we get

$$\frac{dv_r}{v_r - K\tau} = -\frac{1}{\tau} dt \tag{5.103}$$

Now, integrating Eq. (5.103) we obtain

$$\int_{v_r(t_0=T)}^{v_r(t)} \frac{dv_r}{v_r - K\tau} = -\int_{t_0=T}^{t} \frac{1}{\tau} dt \tag{5.104}$$

resulting in

$$\ln\left(v_r - K\tau\right)\Big|_{v_r(T)}^{v_r(t)} = -\frac{1}{\tau}\Big|_{T}^{t} \tag{5.105}$$

or

$$\ln\frac{v_r(t) - K\tau}{v_r(T) - K\tau} = -\frac{1}{\tau}(t - T) \tag{5.106}$$

and thus

$$\frac{v_r(t) - K\tau}{v_r(T) - K\tau} = e^{-\frac{1}{\tau}(t-T)} \tag{5.107}$$

leading to

$$v_r(t) = K\tau + \left[v_r(T) - K\tau\right] e^{-\frac{1}{\tau}(t-T)} \tag{5.108}$$

Utilizing (5.100) in (5.108) we obtain

$$v_r(t) = \left(-\frac{V_S}{2}\frac{Z_C}{L}\right)\left(\frac{L}{Z_C}\right) + \left(\frac{V_S}{2}\right) - \left(-\frac{V_S}{2}\frac{Z_C}{L}\right)\left(\frac{L}{Z_C}\right) e^{-\frac{Z_C}{L}(t-T)} \tag{5.109}$$

and finally

$$v_r(t) = -\frac{V_S}{2} + V_S e^{-\frac{Z_C}{L}(t-T)} \tag{5.110}$$

More specifically, the general solution for the differential equation (5.97) is

$$v_r(d, t) = -\frac{V_S}{2} + V_S e^{-(Z_0/L)(t-T)}, \quad \text{for} \quad t > T \tag{5.111}$$

The total voltage across the inductor, and the total current through the inductor are obtained by adding the incident and reflected waves.

$$v(l, t) = v_i + v_r = \frac{V_S}{2} + v_r(d, t)$$

$$= \begin{cases} 0, & for \quad t < T \\ V_S e^{-(Z_c/L)(t-T)}, & for \quad t > T \end{cases} \tag{5.112}$$

Figure 5.13 shows a circuit schematic of a transmission line driven by a 5 V CMOS, and terminated in an inductive load.

The driver voltage and the voltage across the inductor are displayed in Figure 5.14.

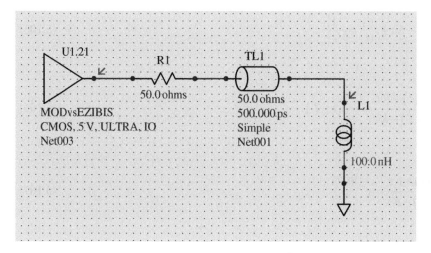

Figure 5.13 HyperLynx circuit model of a transmission line terminated by an inductive load.

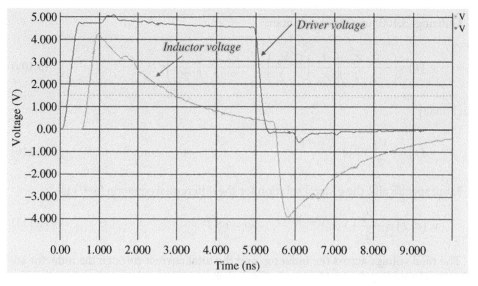

Figure 5.14 Driver voltage and the voltage across the inductor.

5.4.2 Ringing on a Transmission Line

In this section we will the application of the series RLC circuit model to the phenomenon of ringing on the PCB traces.

Many pulse circuits can be represented by the lumped equivalent circuit shown in Figure 5.15.

This is our familiar series RLC circuit that we discussed in Section 5.2.1. When ringing occurs, the voltage across the capacitors exhibits sinusoidal oscillations, i.e. the circuit is underdamped.

The underdamped response was obtained as

$$v_C(t) = V_T + Ee^{-\zeta\omega_0 t}\sin(\omega_d t + \theta) \tag{5.113}$$

This response occurs when the roots s_1 and s_2 of the characteristic equation (5.31) are complex.

$$s_1 = -\zeta\omega_0 + \omega_0\sqrt{\zeta^2 - 1} \tag{5.114a}$$

$$s_1 = -\zeta\omega_0 - \omega_0\sqrt{\zeta^2 - 1} \tag{5.114b}$$

This occurs when $\zeta < 1$.
Recall: In series RLC circuit we have

$$2\zeta\omega_0 = \frac{R}{L} \tag{5.115a}$$

$$\omega_0^2 = \frac{1}{LC} \tag{5.115b}$$

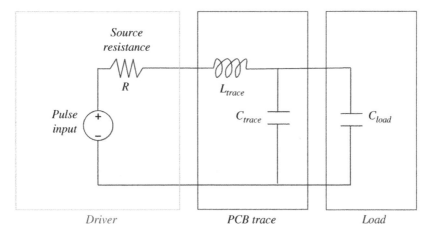

Figure 5.15 Lumped parameter model of a pulse circuit.

Solving (5.115a) for ζ we obtain

$$\zeta = \frac{R}{2L\omega_0} \tag{5.116}$$

Substituting for ω_0 from Eq. (5.115b) into Eq. (5.116), results in

$$\zeta = \frac{R}{2L\omega_0} = \frac{R}{2L\dfrac{1}{\sqrt{LC}}} = \frac{R\sqrt{LC}}{2L} = \sqrt{\frac{R^2 LC}{4L^2}} \tag{5.117}$$

or

$$\zeta = \sqrt{\frac{R^2 C}{4L}} \tag{5.118}$$

To avoid ringing we need $\zeta \geq 1$, or equivalently

$$\frac{R^2 C}{4L} \geq 1 \tag{5.119}$$

Thus the loop inductance would have to satisfy the inequality,

$$L \leq \frac{R^2 C}{4} \tag{5.120}$$

Let's use some typical values for R and C and let $R = 20\,\Omega$, $C = 10\,\text{pF}$. Then the loop inductance would have to satisfy

$$L \leq \frac{(20)^2 \times 10 \times 10^{-12}}{4} = 1\,\text{nH} \tag{5.121}$$

It is easy to see that it is very easy to create a condition of ringing in an electronic circuit.

Ringing Measurements   The laboratory test setup to measure ringing is shown in Figure 5.16.

The circuit diagram showing all *intentional* circuit components is shown in Figure 5.17.

The function generator produces a 1 MHz trapezoidal 1 V_{pp} pulse train with adjustable rise and fall time.

With the rise and fall times set to 10 ns, the voltage measured at the input to the PCB trace is shown in Figure 5.18.

There is no noticeable ringing present at the input to the trace.

Figure 5.19(a) shows the waveform when the rise time has been changed to 2.5 ns, while the fall time stayed at 10 ns. In Figure 5.19(b), both the rise and fall time are at 2.5 ns.

With the rise and/or fall time changed to 2.5 ns we observe a significant ringing present in the system.

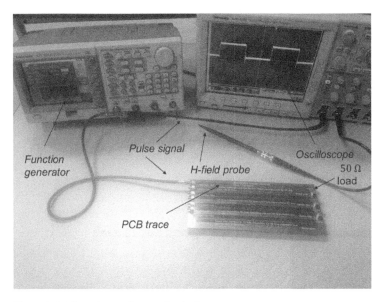

Figure 5.16 Experimental setup for ringing measurements.

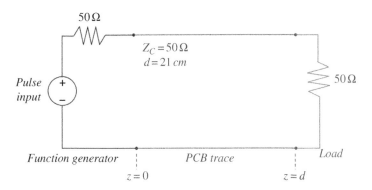

Figure 5.17 Circuit model of the experimental setup.

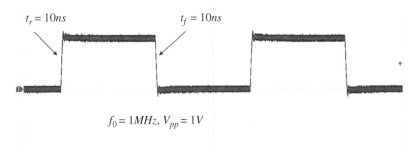

Figure 5.18 Trapezoidal pulse train produced by a function generator.

(a)

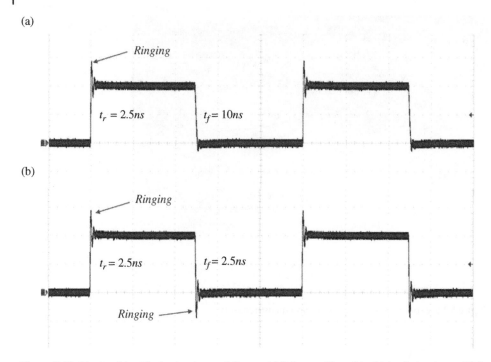

(b)

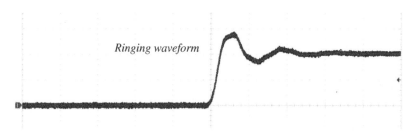

Figure 5.19 Ringing (a): with the rise time at 2.5 ns and fall time at 10 ns, (b) with both the rise and fall time at 2.5 ns.

Figure 5.20 Ringing waveform on the rising edge.

Figure 5.20 shows the expanded view of the ringing waveform on the rising edge.

It is evident from Figure 5.20 that the ringing waveform resembles an underdamped sinusoid described by Eq. (5.113), repeated here.

$$v_C(t) = V_T + Ee^{-\zeta\omega_0 t}\sin(\omega_d t + \theta) \tag{5.122}$$

At this point we may pose a question: why was the ringing not present (negligible) with the 10 ns rise/fall time but very pronounced with the 2.5 ns rise/fall time? This question can be answered when we compare the physical size of the PCB trace with the electromagnetic wave wavelength at the highest frequency present in the signal. We will discuss this topic in Chapter 15.

It is very instructive and revealing to look at the current waveforms that can be captured using an H-field probe, as shown in Figure 5.21.

These waveforms are shown in Figure 5.22 for the 10 ns rise time case, and in Figure 5.23 for the 2.5 ns rise time case.

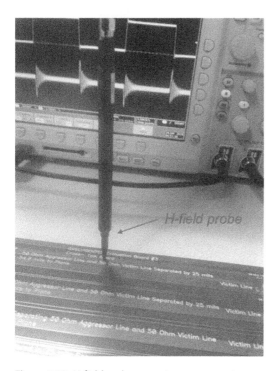

Figure 5.21 H-field probe current measurements.

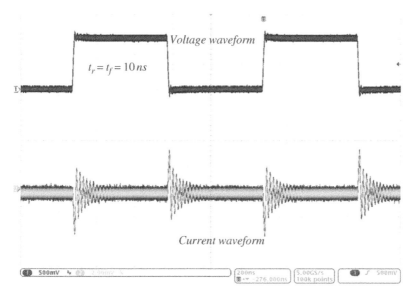

Figure 5.22 Voltage and current waveforms for the 10 ns rise time case.

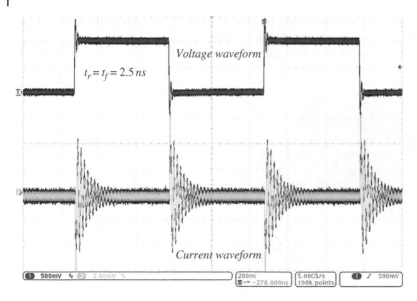

Figure 5.23 Voltage and current waveforms for the 2.5 ns rise time case.

Note that the current waveform exhibits ringing in both cases, while the voltage waveform in the 10 ns case exhibits minimal or no noticeable ringing.

References

Alexander, C.K. and Sadiku, N.O., *Fundamentals of Electric Circuits*, 4th ed., McGraw Hill, New York, 2009.

Balanis, C.A., *Advanced Engineering Electromagnetics*, 3rd ed., John Wiley and Sons, 1989.

Balanis, C.A., *Antenna Theory Analysis and Design*, 3rd ed., Wiley Interscience, Hoboken. New Jersey, 2005.

Nilsson, J.W. and Riedel, S.A., *Electric Circuits*, 10th ed., Pearson, Upper Saddle River, NJ, 2015.

Nise, N.S., *Control Systems Engineering*, 5th ed., Wiley Interscience, Hoboken. New Jersey, 2008.

6

Complex Numbers and Phasors

6.1 Definitions and Forms

A *complex number z* is a number that can be expressed as

$$z = x + jy \tag{6.1}$$

where x and y are real numbers and

$$j^2 = -1 \tag{6.2}$$

x is called a *real part* of z, and we write $x = \text{Re}(z)$.
y is called an *imaginary part* of z, and we write $y = \text{Im}(z)$.
j is called an *imaginary unit*, sometimes expressed as $j = \sqrt{-1}$.

As we will soon see, there are several equivalent representations of complex numbers. The representation in Eq. (6.1) is called the *rectangular form*.

Example 6.1 Rectangular form of a complex number

$$z_1 = 3 + j8, \quad \text{Re}(z_1) = 3, \quad \text{Im}(z_1) = 8$$
$$z_2 = 2 - j5, \quad \text{Re}(z_2) = 2, \quad \text{Im}(z_2) = -5$$
$$z_3 = -4 + j0 = -4, \quad \text{Re}(z_3) = -4, \quad \text{Im}(z_3) = 0$$
$$z_4 = 0 + j7 = j7, \quad \text{Re}(z_4) = 0, \quad \text{Im}(z_4) = 7$$

∎

When dealing with complex numbers, it is often expedient to represent them graphically. We can easily do that in a complex plane using a Cartesian coordinate system, where the x axis is the real axis and the y axis is the imaginary axis, as shown in Figure 6.1.

Example 6.2 Complex plane
Let $z = 4 - 3j$. Determine the location of this number in the complex plane.

Solution: The location is shown in Figure 6.2. ∎

Foundations of Electromagnetic Compatibility with Practical Applications, First Edition. Bogdan Adamczyk.
© 2017 John Wiley & Sons Ltd. Published 2017 by John Wiley & Sons Ltd.

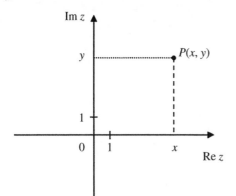

Figure 6.1 Complex plane.

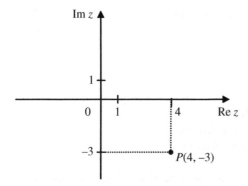

Figure 6.2 Solution of Example 6.2.

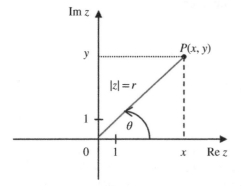

Figure 6.3 Complex number representation as a directed line segment.

In electromagnetics problems, it is often practical to represent a complex number z as a directed line segment from the origin to the point P in the complex plane, as shown in Figure 6.3.

Note that r and θ are the familiar polar coordinates of P defined by

$$x = r\cos\theta = |z|\cos\theta$$
$$y = r\sin\theta = |z|\sin\theta$$

(6.3)

r is called the magnitude of z, denoted $|z|$ and

$$|z| = r = \sqrt{x^2 + y^2} \tag{6.4}$$

Geometrically, $|z|$ is the distance of the point P from the origin, θ is called the angle of z, and

$$\theta = \tan^{-1} \frac{y}{x} \quad (\pm \pi) \tag{6.5}$$

Geometrically, θ is the directed angle from the positive x axis and is positive in the counterclockwise sense.

By substituting Eq. (6.3) into the rectangular form of z expressed by Eq. (6.1) we obtain another form of a complex number z

$$z = x + jy = r\cos\theta + jr\sin\theta$$
$$= r(\cos\theta + j\sin\theta) = |z|(\cos\theta + j\sin\theta) \tag{6.6}$$

or

$$z = |z|(\cos\theta + j\sin\theta) \tag{6.7}$$

The form in Eq. (6.7) is often denoted as

$$z = |z| \angle \theta \tag{6.8}$$

The representations in Eqs (6.7) and (6.8) are called the *polar form* of a complex number z.

All three forms are equivalent.

$$z = x + jy = |z|\cos\theta + j|z|\sin\theta = |z|(\cos\theta + j\sin\theta) = |z| \angle \theta \tag{6.9}$$

Example 6.3   Polar form of z

Let $z = 1 + j$. Obtain the polar form of z.

Solution: The magnitude of z is

$$|z| = \sqrt{x^2 + y^2} = \sqrt{1 + 1} = \sqrt{2}$$

while the angle of z is

$$\theta = \tan^{-1} \frac{1}{1} = \frac{\pi}{4} = 45°$$

The polar form of complex number z is, therefore:

$$z = \sqrt{2} \angle \frac{\pi}{4}$$

6.2   Complex Conjugate

Given a complex number

$$z = x + jy \tag{6.10}$$

we can create another related (and very useful) complex number

$$z^* = x - jy \tag{6.11}$$

This new complex number is called the *complex conjugate* of z.

Example 6.4 Complex conjugate

$$z_1 = 3 + j8, \quad z_1^* = 3 - j8$$
$$z_2 = 2 - j5, \quad z_2^* = 2 + j5$$
$$z_3 = -4 + j0 = -4, \quad z_3^* = -4 - j0 = -4$$
$$z_4 = 0 + j7 = j7, \quad z_4^* = 0 - j7 = -j7$$

∎

If z is expressed as:

$$z = |z|\cos\theta + j|z|\sin\theta = |z|\angle\theta \tag{6.12}$$

then

$$z^* = |z|\cos\theta - j|z|\sin\theta = |z|\cos\theta + j|z|\sin(-\theta)$$
$$= |z|\cos(-\theta) + j|z|\sin(-\theta) = |z|\angle-\theta \tag{6.13}$$

or

$$\text{If} \quad z = |z|\angle\theta \quad \text{then} \quad z^* = |z|\angle-\theta \tag{6.14}$$

The graphical representation of the above two complex numbers is shown in Figure 6.4.

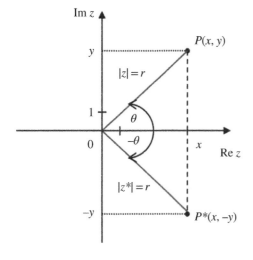

Figure 6.4 Complex number and its complex conjugate.

6.3   Operations on Complex Numbers

Equality of Complex Numbers in Rectangular Form   Two complex numbers

$$z_1 = x_1 + jy_1 \qquad z_2 = x_2 + jy_2 \tag{6.15}$$

are equal if their real parts are equal and their imaginary parts are equal. Thus,

$$z_1 = z_2 \quad \Leftrightarrow \quad x_1 = x_2 \quad \text{and} \quad y_1 = y_2 \tag{6.16}$$

Equality of Complex Numbers in Polar Form   Two complex numbers

$$z_1 = |z_1|\angle\theta_1, \quad z_2 = |z_2|\angle\theta_2 \tag{6.17}$$

are equal if their magnitudes are equal and their angles are equal. Thus,

$$z_1 = z_2 \quad \Leftrightarrow \quad |z_1| = |z_2| \quad \text{and} \quad \theta_1 = \theta_2 \tag{6.18}$$

Addition of Complex Numbers   Let

$$z_1 = x_1 + jy_1 \qquad z_2 = x_2 + jy_2 \tag{6.19}$$

Then their sum is defined as,

$$z_1 + z_2 = (x_1 + jy_1) + (x_2 + jy_2) = (x_1 + x_2) + j(y_1 + y_2) \tag{6.20}$$

Therefore, two complex numbers in rectangular form are added by adding the real parts and the imaginary parts separately.

Addition in polar form cannot be performed (except for the trivial cases) and therefore is not defined.

Example 6.5   Addition of complex numbers

$$z_1 = 5 + j2, \quad z_2 = 4 + j3$$
$$z_1 + z_2 = 5 + j2 + 4 + j3 = (5+4) + j(2+3) = 9 + j5$$

∎

Multiplication by a Real Number in Rectangular Form   Let $z = x + jy$ and a be a real number. Then the product of a and z is a complex number defined as

$$az = a(x + jy) = ax + jay \tag{6.21}$$

Multiplication of Complex Numbers in Rectangular Form   Let

$$z_1 = x_1 + jy_1 \qquad z_2 = x_2 + jy_2 \tag{6.22}$$

Then the product of z_1 and z_2 is a complex number obtained as follows

$$z_1 z_2 = (x_1 + jy_1)(x_2 + jy_2) = x_1 x_2 + jx_1 y_2 + jy_1 x_2 + j^2 y_1 y_2 \tag{6.23}$$

Since $j^2 = -1$, we have

$$z_1 z_2 = (x_1 x_2 - y_1 y_2) + j(x_1 y_2 + y_1 x_2)$$ (6.24)

Example 6.6   Multiplication of Complex Numbers in Rectangular Form

$$z_1 = 3 + j8, \quad z_2 = 4 - j5,$$
$$z_1 z_2 = (3 + j8)(4 - j5) = 12 - j15 + j32 - j^2 40 = 52 + j17$$

∎

Multiplication of Complex Numbers in Polar Form   Let

$$z_1 = r_1(\cos\theta_1 + j\sin\theta_1) = r_1\cos\theta_1 + jr_1\sin\theta_1$$ (6.25a)

$$z_2 = r_2(\cos\theta_2 + j\sin\theta_2) = r_2\cos\theta_2 + jr_2\sin\theta_2$$ (6.25b)

Then the product of z_1 and z_2 is a complex number that can be obtained as follows

$$
\begin{aligned}
z_1 z_2 &= (r_1\cos\theta_1 + jr_1\sin\theta_1)(r_2\cos\theta_2 + jr_2\sin\theta_2) \\
&= (r_1\cos\theta_1)(r_2\cos\theta_2) + j(r_1\cos\theta_1)(r_2\sin\theta_2) \\
&\quad + j(r_1\sin\theta_1)(r_2\cos\theta_2) + j^2 r_1 r_2(\sin\theta_1)(\sin\theta_2) \\
&= r_1 r_2(\cos\theta_1\cos\theta_2 - \sin\theta_1\sin\theta_2) + jr_1 r_2(\sin\theta_1\cos\theta_2 + \cos\theta_1\sin\theta_2) \\
&= r_1 r_2[\cos(\theta_1 + \theta_2) + j\sin(\theta_1 + \theta_2)] = r_1 r_2\angle(\theta_1 + \theta_2)
\end{aligned}
$$

or

$$|z_1 z_2| = |z_1||z_2|, \quad \angle(z_1 z_2) = \angle z_1 + \angle z_2$$ (6.26)

Product of a Complex Number and Its Complex Conjugate   This is one of the most important properties, which we will utilize often.
Let

$$z = x + jy, \quad z^* = x - jy$$ (6.27)

then

$$zz^* = (x + jy)(x - jy) = x^2 - jxy + jyx - j^2 y^2 = x^2 + y^2$$ (6.28)

or

$$zz^* = x^2 + y^2$$ (6.29)

Alternatively, if the complex number and its conjugate are in polar form

$$z = |z|\angle\theta, \quad z^* = |z|\angle-\theta$$ (6.30)

then

$$zz^* = (|z|\angle\theta)(|z|\angle-\theta) = |z|^2 \angle(\theta-\theta) = |z|^2 \angle 0 = |z|^2 \tag{6.31}$$

or

$$zz^* = |z|^2 \tag{6.32}$$

Note that the product of a complex number and its complex conjugate is a *real* number.

Example 6.7 Multiplication by j in rectangular form
Let $z = x + jy$. Then

$$jz = j(x + jy) = jx + j^2 y = -y + jx \tag{6.33}$$

∎

Example 6.8 Multiplication by j in polar form
Let $z = |z|\angle\theta$. Then since $j = 0 + 1j = 1\angle90°$, we get

$$jz = (1\angle90°)(|z|\angle\theta) = |z|\angle(\theta+90°) \tag{6.34}$$

∎

Therefore, multiplication by *j* is equivalent to counterclockwise rotation 90°, as shown in Figure 6.5.

Example 6.9 Negative number in polar form
Let $z = -5$. Then

$$-5 = (-1)(5) = (1\angle180°)(5\angle0°) = 5\angle180°$$

∎

Figure 6.5 Multiplication by j.

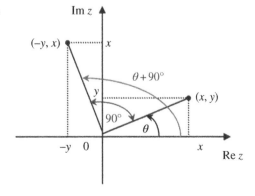

Division of Complex Numbers in Rectangular Form   Let

$$z_1 = x_1 + jy_1 \qquad z_2 = x_2 + jy_2 \tag{6.35}$$

Then the *division* of z_1 and z_2 results in a complex number z that can be obtained as follows

$$
\begin{aligned}
z &= \frac{z_1}{z_2} = \frac{x_1 + jy_1}{x_2 + jy_2} = \frac{(x_1 + jy_1)(x_2 - jy_2)}{(x_2 + jy_2)(x_2 - jy_2)} \\
&= \frac{x_1 x_2 - jx_1 y_2 + jy_1 x_2 - j^2 y_1 y_2}{x_2^2 + y_2^2} \\
&= \frac{x_1 x_2 + y_1 y_2}{x_2^2 + y_2^2} + j\frac{x_2 y_1 - x_1 y_2}{x_2^2 + y_2^2}
\end{aligned}
\tag{6.36}
$$

or

$$\frac{x_1 + jy_1}{x_2 + jy_2} = \frac{x_1 x_2 + y_1 y_2}{x_2^2 + y_2^2} + j\frac{x_2 y_1 - x_1 y_2}{x_2^2 + y_2^2} \tag{6.37}$$

Example 6.10   Division of complex numbers in rectangular form

$$z_1 = 3 + j8, \quad z_2 = 4 - j5,$$

$$\frac{z_1}{z_2} = \frac{3 + j8}{4 - j5} = \frac{(3 + j8)(4 + j5)}{(4 - j5)(4 + j5)} = \frac{12 + j15 + j32 + j^2 40}{4^2 + 5^2} = \frac{-28 + j37}{41} = -\frac{28}{41} + j\frac{37}{41}$$

∎

Division of Complex Numbers in Polar Form   Let

$$z_1 = |z_1| \angle \theta_1, \quad z_2 = |z_2| \angle \theta_2 \tag{6.38}$$

The quotient $z = \dfrac{z_1}{z_2}$ is the complex number satisfying $zz_2 = z_1$. Therefore, we have

$$|zz_2| = |z||z_2| = |z_1| \quad \Rightarrow \quad |z| = \frac{|z_1|}{|z_2|} \tag{6.39}$$

and

$$
\begin{aligned}
&\angle(zz_2) = \angle(z) + \angle(z_2) = \angle(z_1) \\
&\Rightarrow \ \angle(z) = \angle(z_1) - \angle(z_2)
\end{aligned}
\tag{6.40}
$$

Thus

$$\frac{z_1}{z_2} = \frac{|z_1| \angle \theta_1}{|z_2| \angle \theta_2} = \frac{|z_1|}{|z_2|} \angle(\theta_1 - \theta_2) \tag{6.41}$$

Example 6.11 Division of complex numbers in polar form

$$z_1 = 6\angle 30°, \quad z_2 = 2\angle 45°$$

$$\frac{z_1}{z_2} = \frac{6\angle 30°}{2\angle 45°} = 3\angle -15°$$

■

Example 6.12 Division by j in rectangular form

Let $z = x + jy$. Then

$$\frac{z}{j} = \frac{x+jy}{j} = \frac{(x+jy)(-j)}{j(-j)} = \frac{y-jx}{1} = y - jx \tag{6.42}$$

■

Example 6.13 Division by j in polar form

Let $z = |z|\angle \theta$. Then

$$\frac{z}{j} = \frac{(|z|\angle \theta)}{(1\angle 90°)} = |z|\angle(\theta - 90°) \tag{6.43}$$

■

Therefore, division by j is equivalent to clockwise rotation by 90°, as shown in Figure 6.6.

Powers of Complex Numbers Let $z = |z|\angle \theta$. Then

$$z^2 = zz = (|z|\angle \theta)(|z|\angle \theta) = |z||z|\angle(\theta + \theta) = |z|^2 \angle 2\theta \tag{6.44}$$

More generally,

$$z^n = (z\angle \theta)^n = |z|^n \angle n\theta \tag{6.45}$$

Figure 6.6 Division by *j*.

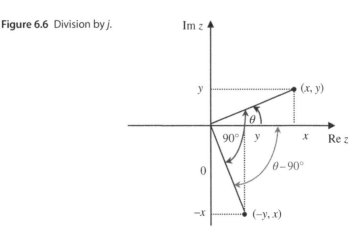

6.4   Properties of Complex Numbers

In this section we list some of the basic properties of complex numbers.

$$z_1 + z_2 = z_2 + z_1 \tag{6.46a}$$

$$z_1 z_2 = z_2 z_1 \tag{6.46b}$$

$$(z_1 + z_2) + z_3 = z_1 + (z_2 + z_3) \tag{6.46c}$$

$$(z_1 z_2) z_3 = z_1 (z_2 z_3) \tag{6.46d}$$

$$z_1 (z_2 + z_3) = z_1 z_2 + z_1 z_3 \tag{6.46e}$$

$$0 + z = z + 0 = z \tag{6.46f}$$

$$z + (-z) = (-z) + z = 0 \tag{6.46g}$$

$$z \cdot 1 = z \tag{6.46h}$$

Complex Conjugate Properties

$$(z_1 + z_2)^* = z_1^* + z_2^* \tag{6.47a}$$

$$(z_1 - z_2)^* = z_1^* - z_2^* \tag{6.47b}$$

$$(z_1 \cdot z_2)^* = z_1^* \cdot z_2^* \tag{6.47c}$$

$$\left(\frac{z_1}{z_2}\right)^* = \frac{z_1^*}{z_2^*} \tag{6.47d}$$

$$(z^*)^* = z \tag{6.47e}$$

Useful Identity

$$\frac{1}{j} = -j \tag{6.48}$$

Let's prove it.

$$\frac{1}{j} = \left(\frac{1}{j}\right)\left(\frac{-j}{-j}\right) = \frac{-j}{-j^2} = \frac{-j}{-(-1)} = -j \tag{6.49}$$

6.5   Complex Exponential Function

Let $z = x + jy$ be a complex number. The complex exponential function, e^z, is defined as (Kreyszig, 1999, p. 679)

$$e^z = e^x (\cos y + j \sin y) \tag{6.50}$$

where e^x, $\cos y$, and $\sin y$ are real functions.

Since x and y can be any real numbers, let us set $x = 0$. Then,

$$e^z = e^{0+jy} = e^0 e^{jy} = e^{jy} \tag{6.51}$$

On the other hand, using Eq. (6.50)

$$e^z = e^x \left(\cos y + j \sin y \right) = e^0 \left(\cos y + j \sin y \right) = \cos y + j \sin y \tag{6.52}$$

Comparing Eqs (6.50) and (6.52) we obtain the *Euler formula*

$$e^{jy} = \cos y + j \sin y \tag{6.53}$$

or in terms of θ

$$e^{j\theta} = \cos\theta + j \sin\theta \tag{6.54}$$

Now, since a complex number z can be expressed as

$$z = |z|\left(\cos\theta + j \sin\theta \right) \tag{6.55}$$

using Eq. (6.54), it can also be expresses as

$$z = |z| e^{j\theta} \tag{6.56}$$

This form of a complex number is called an *exponential form*. This form, perhaps, is the most useful form of a complex number in electromagnetic compatibility literature. Euler formula expressed by Eq. (6.54) leads to two very useful results, as shown next.

$$e^{-j\theta} = e^{j(-\theta)} = \cos(-\theta) + j \sin(-\theta) = \cos\theta - j \sin\theta \tag{6.57}$$

Thus, we have,

$$e^{j\theta} = \cos\theta + j \sin\theta \tag{6.58a}$$

$$e^{-j\theta} = \cos\theta - j \sin\theta \tag{6.58b}$$

Adding both sides we get

$$\cos\theta = \frac{e^{j\theta} + e^{-j\theta}}{2} \tag{6.59}$$

Subtracting both sides we get

$$\sin\theta = \frac{e^{j\theta} - e^{-j\theta}}{2j} \tag{6.60}$$

6.6 Sinusoids and Phasors

6.6.1 Sinusoids

Consider a single frequency sinusoidal signal

$$v(t) = V \cos\omega t \tag{6.61}$$

where V is the amplitude of the sinusoid and ω is the angular frequency in radians per second, rad/s.

The period T and the angular frequency ω are related by

$$T = \frac{\omega}{2\pi} \tag{6.62}$$

The reciprocal of the period is the (cyclic) frequency (in Hz)

$$f = \frac{1}{T} \tag{6.63}$$

The angular frequency ω and the cyclic frequency f are obviously related by

$$\omega = 2\pi f \tag{6.64}$$

Let us now consider a more general expression for a sinusoid,

$$v(t) = V\cos\left(\omega t + \phi\right) \tag{6.65}$$

where $(\omega t + \varphi)$ is called the argument of the cosine function, and φ is its phase.

The sinusoidal functions may, in general, be expressed in any of the four different forms: either as a sine or a cosine function, with either positive or negative amplitude.

For example,

$$v_1(t) = 2\cos\left(\omega t + 30°\right) \tag{6.66a}$$

$$v_2(t) = -3\cos\left(\omega t - 60°\right) \tag{6.66b}$$

$$v_3(t) = 4\sin\left(\omega t + 45°\right) \tag{6.66c}$$

$$v_4(t) = -5\sin\left(\omega t - 15°\right) \tag{6.66d}$$

As we will see in the next section, we often need the sinusoid to be expressed as a cosine function with positive amplitude, as shown in Eq. (6.66a).

Therefore, we need to be able to transform the other three forms into the positive cosine form. To accomplish that, we could use the following trigonometric identities:

$$-\cos\left(\omega t + \theta\right) = \cos\left(\omega t + \theta \pm 180°\right) \tag{6.67a}$$

$$\sin\left(\omega t + \theta\right) = \cos\left(\omega t + \theta - 90°\right) \tag{6.67b}$$

$$-\sin\left(\omega t + \theta\right) = \cos\left(\omega t + \theta + 90°\right) \tag{6.67c}$$

Therefore, Eqns. (5.64b–d) can be expressed as

$$v_2(t) = -3\cos\left(\omega t - 60°\right) = 3\cos\left(\omega t + 120°\right) \tag{6.68a}$$

Figure 6.7 Trigonometric relations.

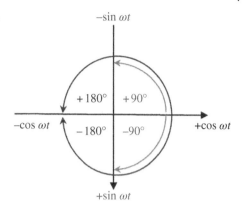

$$v_3(t) = 4\sin\left(\omega t + 45°\right) = 4\cos\left(\omega t - 45°\right) \tag{6.68b}$$

$$v_4(t) = -5\sin\left(\omega t - 15°\right) = 5\cos\left(\omega t + 75°\right) \tag{6.68c}$$

Alternatively, we may use the graphical approach (Alexander and Sadiku, 2009, p. 374) as follows.

Consider the set of axis shown in Figure 6.7. The horizontal axis represents the cosine, while the vertical axis (pointing down) denotes the sine. Angles are measured positively counterclockwise from the horizontal, as usual in polar coordinates.

This figure can be used to obtain positive cosine out of the other three forms, as follows.

Negative cosine is equivalent to positive cosine plus or minus 180°. Positive sine is equivalent to positive cosine minus 90°. Negative sine is equivalent to positive cosine plus 90°.

6.6.2 Phasors

Consider a positive cosine function of the form

$$v(t) = V\cos\left(\omega t + \theta\right) \tag{6.69}$$

We could use its amplitude and phase to create a related complex number

$$V\angle\theta = Ve^{j\theta} \tag{6.70}$$

Obviously, the complex number in expression (6.70) is *related* to the sinusoid in expression (6.69). We often say that this complex number *represents* the respective sinusoid.

Is this representation useful? Extremely! Instead of performing mathematical operations on sinusoids in the time domain (which is often difficult to do), we can perform the operations on complex numbers related to these sinusoids, in the complex domain (which is relatively easy to do).

Note that the sinusoid exists in the time domain, while the complex number representing it exists in the complex domain. Therefore, they are *not* equal; they *correspond* to each other.

$$V \cos\left(\omega t + \theta\right) \leftrightarrow V e^{j\theta} \tag{6.71}$$

When a complex number represents a sinusoid, we call it a *phasor*. By representing the sinusoid as a phasor we transform the sinusoid from the time domain to the phasor or frequency domain.

So what is the difference between a phasor and a complex number? Every phasor is a complex number, but not every complex number is a phasor. Only when the complex number represents a sinusoid is it referred to as a phasor.

In electromagnetic compatibility we often deal with complex voltages and currents. These complex expressions represent sinusoids in the time domain, and therefore they are phasors. We also encounter complex impedance, but the impedance does not represent a time-domain sinusoid, so it is not a phasor, but just a complex expression.

To distinguish between the time domain variables and the complex variables, we will adopt the notation from Paul (2006, p. 261). A complex variable will always have a "hat" above it.

$$\hat{V} = V\angle\theta = V e^{j\theta} \tag{6.72}$$

In the above expression, the magnitude V and the angle θ are real, thus they do not have "hats", but the phasor is complex.

Given a phasor, in polar or exponential form, we can easily determine the time-domain sinusoid corresponding to it. For instance, if the phasor is given by

$$\hat{I} = I e^{j\theta} \tag{6.73}$$

then the sinusoid corresponding to it is simply

$$\hat{I} = I e^{j\theta} \quad \Leftrightarrow \quad i(t) = I \cos\left(\omega t + \theta\right) \tag{6.74}$$

Alternatively, the time-domain form of phasor quantities may be obtained by multiplying the phasor form by $e^{j\omega t}$ and taking the real part of the result.

$$\mathrm{Re}\{\hat{I} e^{j\omega t}\} = I \cos(\omega t + \theta) = i(t) \tag{6.75}$$

We will show this operation in the next section, when presenting the phasor form of Maxwell's equations.

Derivative in the Phasor Domain   Let the time-domain sinusoid be expressed as

$$v(t) = V \cos\left(\omega t + \theta\right) \tag{6.76}$$

Its corresponding phasor is

$$\hat{V} = V e^{j\theta} \tag{6.77}$$

If we take the derivative of $v(t)$ in expression (6.76), we will obtain another sinusoid; a negative sine function, to be exact. That negative sine function can be expressed as a positive cosine using the transformations discussed earlier.

Therefore, we could create a phasor representing it. The question we pose is as follows: what is the relationship between the original phasor representing $v(t)$ and the phasor representing its derivative?

To answer this questions let's take the derivative of $v(t)$:

$$\frac{dv(t)}{dt} = -\omega V \sin(\omega t + \theta) = \omega V \cos(\omega t + \theta + 90°) \tag{6.78}$$

Thus the phasor representing the derivative of $v(t)$ is

$$\omega V \cos(\omega t + \theta + 90°) \leftrightarrow \omega V e^{j(\theta + 90°)} \tag{6.79}$$

Let's have a closer look at this phasor.

$$\omega V e^{j(\theta + 90°)} = \omega V e^{j\theta} e^{j90°} \tag{6.80}$$

However,

$$e^{j90°} = \cos 90° + j \sin 90° = j \tag{6.81}$$

and therefore

$$\omega V e^{j\theta} e^{j90°} = j\omega V e^{j\theta} = j\omega \hat{V} \tag{6.82}$$

We have arrived at a very important observation.

$$\begin{aligned} v(t) &\leftrightarrow \hat{V} \\ \frac{dv(t)}{dt} &\leftrightarrow j\omega \hat{V} \end{aligned} \tag{6.83}$$

That is, to obtain the phasor representing the derivative of a (sinusoidal) function, we simply take the phasor representing that function and multiply it by $j\omega$.

6.7 EMC Applications

6.7.1 Maxwell's Equations in a Phasor Form

Of major interest in EMC are the sinusoidal electromagnetic fields and current and charge densities. That is, the time-domain vectors and scalar expressions are sinusoidal functions of time and space.

For instance, the electric field intensity vector $\mathbf{E}$ in the time domain is given by

$$\mathbf{E}(x, y, z, t) = \left[E_x(x, y, z, t), E_y(x, y, z, t), E_z(x, y, z, t) \right] \tag{6.84}$$

where each of its components is a sinusoidal function

$$E_x(x, y, z, t) = E_{xm} \cos(\omega t + \theta_x) \tag{6.85a}$$

$$E_y(x, y, z, t) = E_{ym} \cos(\omega t + \theta_y) \tag{6.85b}$$

$$E_z(x, y, z, t) = E_{zm} \cos(\omega t + \theta_z) \tag{6.85c}$$

The corresponding phasors are

$$\hat{E}_x(x, y, z) = E_{xm} \angle \theta_x = E_{xm} e^{j\theta_x} \tag{6.86a}$$

$$\hat{E}_y(x, y, z) = E_{ym} \angle \theta_y = E_{ym} e^{j\theta_y} \tag{6.86b}$$

$$\hat{E}_z(x, y, z) = E_{zm} \angle \theta_z = E_{zm} e^{j\theta_z} \tag{6.86c}$$

Thus, the phasor form of the **E** vector in Eq. (6.84) is

$$\hat{\mathbf{E}}(x, y, z) = \left[\hat{E}_x(x, y, z), \hat{E}_y(x, y, z), \hat{E}_z(x, y, z) \right] \tag{6.87}$$

The time-domain form of phasor quantities may be obtained by multiplying the phasor form by $e^{j\omega}$ and taking the real part of the result. For example

$$E_x(x, y, z, t) = \text{Re}\left\{ \hat{E}_x(x, y, z) e^{j\omega t} \right\} = \text{Re}\left\{ E_{xm} e^{j\theta_x} e^{j\omega t} \right\}$$

$$= \text{Re}\left\{ E_{xm} e^{j(\omega t + \theta_x)} \right\} = \text{Re}\left\{ E_{xm} \cos(\omega t + \theta_x) + j E_{xm} \sin(\omega t + \theta_x) \right\} \tag{6.88}$$

$$= E_{xm} \cos(\omega t + \theta_x)$$

The phasor form of the derivative of the *E* field in (5.84) is

$$\frac{\partial \mathbf{E}}{\partial t} \quad \leftrightarrow \quad j\omega \hat{\mathbf{E}} \tag{6.89}$$

Thus, in order to obtain Maxwell's equations for sinusoidal excitation, we replace the field vectors and functions with their phasor forms, and their *time* derivatives with the phasor forms multiplied by $j\omega$. In a simple medium these equations in a phasor form become (Paul, 2006, p. 908) the following

Differential form of Maxwell's equations

$$\nabla \times \hat{\mathbf{E}} = -j\omega\mu\hat{\mathbf{H}} \tag{6.90a}$$

$$\nabla \times \hat{\mathbf{H}} = (\sigma + j\omega\varepsilon)\hat{\mathbf{E}} + \hat{\mathbf{J}}_S \tag{6.90b}$$

$$\nabla \cdot \hat{\mathbf{E}} = \frac{\hat{\rho}_V}{\varepsilon} \tag{6.90c}$$

$$\nabla \cdot \hat{\mathbf{H}} = 0 \tag{6.90d}$$

Integral form of Maxwell's equations

$$\oint_C \hat{\mathbf{E}} \cdot d\mathbf{l} = -j\omega\mu \int_S \hat{\mathbf{H}} \cdot d\mathbf{S} \tag{6.91a}$$

$$\oint_C \hat{\mathbf{H}} \cdot d\mathbf{l} = (\sigma + j\omega\varepsilon) \int_S \hat{\mathbf{E}} \cdot d\mathbf{S} + \int_S \hat{\mathbf{J}}_S \cdot d\mathbf{S} \tag{6.91b}$$

$$\oint_S \hat{\mathbf{E}} \cdot d\mathbf{S} = \frac{1}{\varepsilon} \int_v \hat{\rho}_v dv \tag{6.91c}$$

$$\oint_s \hat{\mathbf{H}} \cdot d\mathbf{S} = 0 \tag{6.91d}$$

6.7.2 Transmission Line Equations in a Phasor Form

In Section 3.5.1 we obtained the transmission line equations (for a lossless line) as

$$\frac{\partial V(z, t)}{\partial z} = -l \frac{\partial I(z, t)}{\partial t} \tag{6.92a}$$

$$\frac{\partial I(z, t)}{\partial z} = -c \frac{\partial V(z, t)}{\partial t} \tag{6.92b}$$

The phasor transmission line equations are obtained by replacing the circuit variables with the corresponding phasors, and replacing the time derivatives with $j\omega$.

$$\frac{d\hat{V}(z)}{dz} = -j\omega l \hat{I}(z) \tag{6.93a}$$

$$\frac{d\hat{I}(z)}{dz} = -j\omega l \hat{V}(z) \tag{6.93b}$$

6.7.3 Magnetic Vector Potential

We are now ready to utilize the knowledge gained in these first six chapters to study the vector magnetic potential vector. This is one of the most useful concepts in the study of radiation from antennas and the concept of the partial inductance.

We will begin with Maxwell's divergence equation for magnetic fields:

$$\nabla \cdot \hat{\mathbf{B}} = 0 \tag{6.94}$$

Now, let's recall the following vector identity (true for any vector):

$$\nabla \cdot \nabla \times \hat{\mathbf{A}} = 0 \tag{6.95}$$

Thus, we could define the new vector **A**, called the *magnetic vector potential*, as a vector related to the magnetic flux density vector **B** by

$$\hat{\mathbf{B}} = \nabla \times \hat{\mathbf{A}} \tag{6.96}$$

Even though the concept of a magnetic vector potential is a purely mathematical invention, it proves to be very useful, as we shall see.

Since

$$\hat{\mathbf{B}} = \mu \hat{\mathbf{H}} \tag{6.97}$$

then, in terms of the magnetic field intensity **H**, the vector magnetic potential **A** is defined as

$$\mu \hat{\mathbf{H}} = \nabla \times \hat{\mathbf{A}} \tag{6.98}$$

or, alternatively

$$\hat{H} = \frac{1}{\mu}\nabla \times \hat{A} \qquad (6.99)$$

Now, recall Maxwell's curl equation

$$\nabla \times \hat{E} = -j\omega\mu\hat{H} \qquad (6.100)$$

Substituting Eq. (6.98) into Eq. (6.100) we obtain

$$\nabla \times \hat{E} = -j\omega\nabla \times \hat{A} \qquad (6.101)$$

which can be written as

$$\nabla \times \left(\hat{E} + j\omega\hat{A}\right) = 0 \qquad (6.102)$$

Now, we will use another vector identity

$$\nabla \times \left(-\nabla V\right) = 0 \qquad (6.103)$$

This identity holds for any arbitrary scalar function V. Comparing Eqs (6.102) and (6.103) we get

$$\hat{E} + j\omega\hat{A} = -\nabla V \qquad (6.104)$$

or

$$\hat{E} = -\nabla V - j\omega\hat{A} \qquad (6.105)$$

The scalar function V in Eq. (6.102) represents *electric scalar potential*.
Now, let's take the curl of both sides of Eq. (6.99) to get

$$\nabla \times \hat{H} = \nabla \times \left(\frac{1}{\mu}\nabla \times \hat{A}\right) \qquad (6.106)$$

The right-hand side of (6.106) can be written as

$$\nabla \times \left(\frac{1}{\mu}\nabla \times \hat{A}\right) = \frac{1}{\mu}\left(\nabla \times \nabla \times \hat{A}\right) \qquad (6.107)$$

Comparing (6.106) and (6.107) we can write

$$\mu\nabla \times \hat{H} = \nabla \times \nabla \times \hat{A} \qquad (6.108)$$

Next, we will use another vector identity

$$\nabla \times \nabla \times A = \nabla\left(\nabla \cdot A\right) - \nabla^2 A \qquad (6.109)$$

Combining Eq. (6.108) and Eq. (6.109) we arrive at

$$\mu\nabla \times \hat{H} = \nabla\left(\nabla \cdot \hat{A}\right) - \nabla^2\hat{A} \qquad (6.110)$$

Using Maxwell's curl equation for magnetic field in the region away from a conduction current ($\sigma = 0$)

$$\nabla \times \hat{\mathbf{H}} = j\omega\varepsilon\hat{\mathbf{E}} + \hat{\mathbf{J}}_S \tag{6.111}$$

We rewrite Eq. (6.111) as

$$j\omega\mu\varepsilon\,\hat{\mathbf{E}} + \mu\hat{\mathbf{J}} = \nabla(\nabla \cdot \hat{\mathbf{A}}) - \nabla^2\hat{\mathbf{A}} \tag{6.112}$$

Now, we will make use of Eq. (6.105) to obtain

$$j\omega\mu\varepsilon\left(-\nabla V - j\omega\hat{\mathbf{A}}\right) + \mu\hat{\mathbf{J}} = \nabla(\nabla \cdot \hat{\mathbf{A}}) - \nabla^2\hat{\mathbf{A}} \tag{6.113a}$$

Thus

$$\mu\hat{\mathbf{J}} - \nabla j\omega\mu\varepsilon\,V + \omega^2\mu\varepsilon\hat{\mathbf{A}} = \nabla(\nabla \cdot \hat{\mathbf{A}}) - \nabla^2\hat{\mathbf{A}} \tag{6.113b}$$

or

$$\nabla^2\hat{\mathbf{A}} + \omega^2\mu\varepsilon\hat{\mathbf{A}} = -\mu\hat{\mathbf{J}} + \nabla(\nabla \cdot \hat{\mathbf{A}}) + \nabla j\omega\mu\varepsilon\,V \tag{6.113c}$$

or

$$\nabla^2\hat{\mathbf{A}} + \omega^2\mu\varepsilon\hat{\mathbf{A}} = -\mu\hat{\mathbf{J}} + \nabla\left(\nabla \cdot \hat{\mathbf{A}} + j\omega\mu\varepsilon V\right) \tag{6.113d}$$

Introducing a new constant,

$$k^2 = \omega^2\mu\varepsilon \tag{6.114}$$

Eq. (6.113d) can be rewritten as

$$\nabla^2\hat{\mathbf{A}} + k^2\hat{\mathbf{A}} = -\mu\hat{\mathbf{J}} + \nabla\left(\nabla \cdot \hat{\mathbf{A}} + j\omega\mu\varepsilon V\right) \tag{5.115}$$

In Eq. (6.98), repeated here, we implicitly defined the vector magnetic potential **A** by its curl:

$$\mu\hat{\mathbf{H}} = \nabla \times \hat{\mathbf{A}} \tag{6.116}$$

In order to uniquely define a vector, we need to define it by both the curl and the divergence. The definition of the divergence of **A** is independent of its curl. Thus, we are free to choose a convenient definition.

In order to simplify Eq. (6.115) we choose

$$\nabla \cdot \hat{\mathbf{A}} = -j\omega\mu\varepsilon V \quad \Rightarrow \quad V = -\frac{1}{j\omega\mu\varepsilon}\nabla \cdot \hat{\mathbf{A}} \tag{6.117}$$

which is known as the *Lorentz condition* (Balanis, 2005, p. 136)

Substituting Eq. (6.117) into Eq. (6.115) leads to

$$\nabla^2\hat{\mathbf{A}} + k^2\hat{\mathbf{A}} = -\mu\hat{\mathbf{J}} \tag{6.118}$$

Additionally, Eq. (6.105), repeated here,

$$\hat{\mathbf{E}} = -\nabla V - j\omega\hat{\mathbf{A}} \tag{6.119}$$

reduces to

$$\hat{\mathbf{E}} = -\nabla\phi_e - j\omega\hat{\mathbf{A}} = -j\omega\hat{\mathbf{A}} - \nabla\left(-\frac{1}{j\omega\mu\varepsilon}\nabla\cdot\hat{\mathbf{A}}\right)$$

$$= -j\omega\hat{\mathbf{A}} + \nabla\left(\frac{1}{j\omega\mu\varepsilon}\nabla\cdot\hat{\mathbf{A}}\right) \tag{6.120}$$

or

$$\hat{\mathbf{E}} = -j\omega\hat{\mathbf{A}} - j\frac{1}{\omega\mu\varepsilon}\nabla\left(\nabla\cdot\hat{\mathbf{A}}\right) \tag{6.121}$$

Thus, once **A** is known, **E** can be obtained from Eq. (6.121) and **H** from

$$\hat{\mathbf{H}} = \frac{1}{\mu}\nabla\times\hat{\mathbf{A}} \tag{6.122}$$

Alternatively, **E** can be found form Maxwell's equation

$$\nabla\times\hat{\mathbf{H}} = j\omega\varepsilon\hat{\mathbf{E}} + \hat{\mathbf{J}}_S \tag{6.123}$$

with **J** = **0**:

$$\nabla\times\hat{\mathbf{H}} = j\omega\varepsilon\hat{\mathbf{E}} \tag{6.124}$$

That is, **E** can be obtained from

$$\hat{\mathbf{E}} = \frac{1}{j\omega\varepsilon}\nabla\times\hat{\mathbf{H}} \tag{6.125}$$

6.7.4 Radiated Fields of an Electric Dipole

Electric dipole, often referred to as Hertzian dipole, shown in Figure 6.8, consists of a short thin wire of length l, carrying a phasor current $\hat{I}$, positioned symmetrically at the origin of the coordinate system and oriented along the z axis.

Ideally the wire is infinitely short, and practically a wire of the length $l \ll \lambda/50$ (λ = wavelength) can be considered a Hertzian dipole. Although Hertzian dipoles are not very practical, they are utilized as building blocks of more complex geometries.

Since the current element is very short, we may assume the current to be constant

$$\hat{\mathbf{I}}(z') = I_0\mathbf{a}_z, \quad I_0 = const \tag{6.126}$$

To find the fields radiated by the current element, we will use the two-step procedure. First, we will determine **A** from the solution of the Helmholtz equation (see Eq. (5.96))

$$\mathbf{A}(x, y, z) = \frac{\mu}{4\pi}\int_c \mathbf{I}_e\left(x', y', z'\right)\frac{e^{-jkR}}{R}dl' \tag{6.127}$$

Note that R is the distance from the source location to the observation point, and r is the distance from the origin to the observation point. Since the source is at the origin, $r = R$.

Figure 6.8 Hertzian dipole.

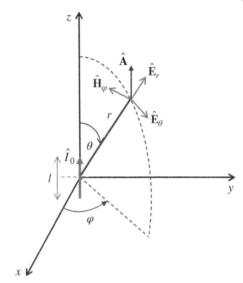

The next step is to determine **H** and **E** from

$$\hat{\mathbf{H}} = \frac{1}{\mu} \nabla \times \hat{\mathbf{A}}$$ (6.128)

$$\hat{\mathbf{E}} = -j\omega\hat{\mathbf{A}} - j\frac{1}{\omega\mu\varepsilon}\nabla(\nabla \cdot \hat{\mathbf{A}})$$ (6.129)

or alternatively, once **H** is computed from **A** by Eq. (6.128), **E** can be obtained from **H** as

$$\hat{\mathbf{E}} = \frac{1}{j\omega\varepsilon} \nabla \times \hat{\mathbf{H}}$$ (6.130)

For the Hertzian dipole shown, we have

$$\mathbf{I}_e(x', y', z') = I_0 \mathbf{a}_z$$ (6.131)

Since

$$R = r = \sqrt{x^2 + y^2 + z^2}$$ (6.132)

$$dl' = dz'$$ (6.133)

we rewrite Eq. (6.127) as

$$\mathbf{A}(x, y, z) = \frac{\mu}{4\pi} \int_c \mathbf{I}_e(x', y', z') \frac{e^{-jkR}}{R} dl' = \frac{\mu}{4\pi} \int_{-l/2}^{l/2} I_0 \mathbf{a}_z \frac{e^{-jkr}}{r} dz'$$

$$= \frac{\mu I_0}{4\pi r} e^{-jkr} \left(\int_{-l/2}^{l/2} dz' \right) \mathbf{a}_z = \frac{\mu I_0 l}{4\pi r} e^{-jkr} \mathbf{a}_z$$ (6.134)

or

$$\mathbf{A}(x, y, z) = \frac{\mu I_0 l}{4\pi r} e^{-jkr} \mathbf{a}_z \qquad (6.135)$$

The next step is to determine H, using Eq. (6.128). We will perform this operation in spherical coordinates.

The transformation between rectangular and spherical components is given by:

$$\begin{bmatrix} A_r \\ A_\theta \\ A_\phi \end{bmatrix} = \begin{bmatrix} \sin\theta\cos\phi & \sin\theta\sin\phi & \cos\theta \\ \cos\theta\cos\phi & \cos\theta\sin\phi & -\sin\theta \\ -\sin\phi & \cos\phi & 0 \end{bmatrix} \begin{bmatrix} A_x \\ A_y \\ A_z \end{bmatrix} \qquad (6.136)$$

For this problem $A_x = A_y = 0$, thus,

$$\begin{bmatrix} A_r \\ A_\theta \\ A_\phi \end{bmatrix} = \begin{bmatrix} \sin\theta\cos\varphi & \sin\theta\sin\varphi & \cos\theta \\ \cos\theta\cos\varphi & \cos\theta\sin\varphi & -\sin\theta \\ -\sin\varphi & \cos\varphi & 0 \end{bmatrix} \begin{bmatrix} 0 \\ 0 \\ \dfrac{\mu I_0 l}{4\pi r} e^{-jkr} \end{bmatrix} \qquad (6.137)$$

or

$$\begin{bmatrix} A_r \\ A_\theta \\ A_\phi \end{bmatrix} = \begin{bmatrix} \dfrac{\mu I_0 l e^{-jkr}}{4\pi r} \cos\theta \\ -\dfrac{\mu I_0 l e^{-jkr}}{4\pi r} \sin\theta \\ 0 \end{bmatrix} \qquad (6.138)$$

The curl of $\mathbf{A}$ in spherical coordinates is

$$\nabla \times \mathbf{A} = \frac{1}{r\sin\theta} \left[\frac{\partial}{\partial\theta}(A_\phi \sin\theta) - \frac{\partial A_\theta}{\partial\phi} \right] \mathbf{a}_r$$

$$+ \frac{1}{r} \left[\frac{1}{\sin\theta} \frac{\partial A_r}{\partial\phi} - \frac{\partial}{\partial r}(rA_\phi) \right] \mathbf{a}_\theta + \frac{1}{r} \left[\frac{\partial}{\partial r}(rA_\theta) - \frac{\partial A_r}{\partial\theta} \right] \mathbf{a}_\phi \qquad (6.139)$$

Since $A_\varphi = 0$ and there are no φ variations in $\mathbf{A}$, we have

$$\nabla \times \mathbf{A} = \frac{1}{r} \left[\frac{\partial}{\partial r}(rA_\theta) - \frac{\partial A_r}{\partial\theta} \right] \mathbf{a}_\phi \qquad (6.140)$$

and thus

$$\hat{\mathbf{H}} = \frac{1}{\mu} \nabla \times \hat{\mathbf{A}} = \frac{1}{\mu r} \left[\frac{\partial}{\partial r}(rA_\theta) - \frac{\partial A_r}{\partial\theta} \right] \mathbf{a}_\phi \qquad (6.141)$$

Now, we use Eq. (6.138) in Eq. (6.141) to produce

$$
\hat{\mathbf{H}} = \frac{1}{\mu r}\left[\frac{\partial}{\partial r}(rA_\theta) - \frac{\partial A_r}{\partial \theta}\right]\mathbf{a}_\varphi
$$

$$
= \frac{1}{\mu r}\left[\frac{\partial}{\partial r}\left(-r\frac{\mu I_0 l e^{-jkr}}{4\pi r}\sin\theta\right) - \frac{\partial}{\partial \theta}\left(\frac{\mu I_0 l e^{-jkr}}{4\pi r}\cos\theta\right)\right]\mathbf{a}_\varphi
$$

$$
= \frac{1}{\mu r}\left[-\frac{\mu I_0 l \sin\theta}{4\pi}\frac{\partial}{\partial r}\left(e^{-jkr}\right) - \frac{\mu I_0 l e^{-jkr}}{4\pi r}\frac{\partial}{\partial \theta}(\cos\theta)\right]\mathbf{a}_\varphi
$$

$$
= \frac{1}{r}\left[-\frac{I_0 l \sin\theta}{4\pi}\left(-jke^{-jkr}\right) + \frac{I_0 l e^{-jkr}}{4\pi r}(\sin\theta)\right]\mathbf{a}_\varphi \tag{6.142}
$$

$$
= \left[\frac{I_0 l \sin\theta}{4\pi r}\left(jke^{-jkr}\right) + \frac{I_0 l e^{-jkr}}{4\pi r^2}(\sin\theta)\right]\mathbf{a}_\varphi
$$

$$
= j\frac{kI_0 l \sin\theta}{4\pi r}\left[1 + \frac{1}{jkr}\right]e^{-jkr}\mathbf{a}_\varphi
$$

Thus, the components of the magnetic field intensity **H** at a distance r from a Hertzian dipole are

$$
H_r = 0 \tag{6.143a}
$$

$$
H_\theta = 0 \tag{6.143b}
$$

$$
H_\varphi = j\frac{kI_0 l \sin\theta}{4\pi r}\left[1 + \frac{1}{jkr}\right]e^{-jkr} \tag{6.143c}
$$

The electric field **E** can now be found using Eq. (6.129) or Eq. (6.130). Let's use the latter approach first. That is, let's calculate **E** from

$$
\hat{\mathbf{E}} = \frac{1}{j\omega\varepsilon}\nabla\times\hat{\mathbf{H}} \tag{6.144}
$$

The curl of **H** in spherical coordinates is

$$
\nabla\times\mathbf{H} = \frac{1}{r\sin\theta}\left[\frac{\partial}{\partial\theta}(H_\phi\sin\theta) - \frac{\partial H_\theta}{\partial\phi}\right]\mathbf{a}_r + \frac{1}{r}\left[\frac{1}{\sin\theta}\frac{\partial H_r}{\partial\phi} - \frac{\partial}{\partial r}(rH_\phi)\right]\mathbf{a}_\theta
$$
$$
+ \frac{1}{r}\left[\frac{\partial}{\partial r}(rH_\theta) - \frac{\partial H_r}{\partial\theta}\right]\mathbf{a}_\phi \tag{6.145}
$$

Since $H_r = H_\theta = 0$ and there are no φ variations in **H**, we have

$$
\mathbf{E} = \frac{1}{j\omega\varepsilon}\nabla\times\mathbf{H} = \frac{1}{j\omega\varepsilon}\left\{\frac{1}{r\sin\theta}\left[\frac{\partial}{\partial\theta}\left(H_\varphi\sin\theta\right)\right]\mathbf{a}_r + \frac{1}{r}\left[-\frac{\partial}{\partial r}\left(rH_\varphi\right)\right]\mathbf{a}_\theta\right\}
$$

$$
= \frac{1}{j\omega\varepsilon}\frac{1}{r\sin\theta}\left[\frac{\partial}{\partial\theta}\left(j\frac{kI_0l\sin\theta}{4\pi r}\left[1+\frac{1}{jkr}\right]e^{-jkr}\sin\theta\right)\right]\mathbf{a}_r \tag{6.146}
$$

$$
+ \frac{1}{j\omega\varepsilon}\frac{1}{r}\left[-\frac{\partial}{\partial r}\left(rj\frac{kI_0l\sin\theta}{4\pi r}\left[1+\frac{1}{jkr}\right]e^{-jkr}\right)\right]\mathbf{a}_\theta
$$

Let's evaluate the r component first:

$$
E_r = \frac{1}{j\omega\varepsilon}\frac{1}{r\sin\theta}\left[\frac{\partial}{\partial\theta}\left(j\frac{kI_0l\sin\theta}{4\pi r}\left[1+\frac{1}{jkr}\right]e^{-jkr}\sin\theta\right)\right]
$$

$$
= \frac{1}{\omega\varepsilon}\frac{1}{\sin\theta}\left[1+\frac{1}{jkr}\right]\frac{kI_0l}{4\pi r^2}e^{-jkr}\left[\frac{\partial}{\partial\theta}\left(\sin^2\theta\right)\right] \tag{6.147}
$$

$$
= \frac{1}{\omega\varepsilon}\frac{1}{\sin\theta}\left[1+\frac{1}{jkr}\right]\frac{kI_0l}{4\pi r^2}e^{-jkr}2\sin\theta\cos\theta
$$

$$
= \frac{1}{\omega\varepsilon}\frac{kI_0l\cos\theta}{2\pi r^2}\left[1+\frac{1}{jkr}\right]e^{-jkr}
$$

and now the θ-component:

$$
E_\theta = \frac{1}{j\omega\varepsilon}\frac{1}{r}\left[-\frac{\partial}{\partial r}\left(rj\frac{kI_0l\sin\theta}{4\pi r}\left[1+\frac{1}{jkr}\right]e^{-jkr}\right)\right]
$$

$$
= \frac{1}{\omega\varepsilon}\frac{1}{r}\left[-\frac{\partial}{\partial r}\left(\frac{kI_0l\sin\theta}{4\pi}\left[1+\frac{1}{jkr}\right]e^{-jkr}\right)\right] \tag{6.148}
$$

$$
= \frac{1}{\omega\varepsilon}\frac{kI_0l\sin\theta}{4\pi r}\left[-\frac{\partial}{\partial r}\left(\left[1+\frac{1}{jkr}\right]e^{-jkr}\right)\right]
$$

Let's evaluate the derivative term:

$$
\frac{\partial}{\partial r}\left(\left[1+\frac{1}{jkr}\right]e^{-jkr}\right) = \frac{\partial}{\partial r}\left(e^{-jkr}+\frac{e^{-jkr}}{jkr}\right)
$$

$$
= -jke^{-jkr}+\frac{\left(-jke^{-jkr}\right)\left(jkr\right)-e^{-jkr}jk}{\left(jkr\right)^2}
$$

$$
= -jke^{-jkr}-\frac{\left(ke^{-jkr}\right)\left(kr\right)-jke^{-jkr}}{\left(kr\right)^2} \tag{6.149}
$$

$$
= \left[-j-\frac{\left(kr\right)-j}{\left(kr\right)^2}\right]ke^{-jkr} = \left[-j-\frac{1}{kr}+\frac{j}{\left(kr\right)^2}\right]ke^{-jkr}
$$

$$
= \left[1+\frac{1}{jkr}-\frac{1}{\left(kr\right)^2}\right]\left(-jke^{-jkr}\right)
$$

Substitute Eq. (6.149) into Eq. (6.148):

$$E_\theta = \frac{1}{\omega\varepsilon}\frac{kI_0l\sin\theta}{4\pi r}\left[-\frac{\partial}{\partial r}\left(\left[1+\frac{1}{jkr}\right]e^{-jkr}\right)\right]$$

$$= -\frac{1}{\omega\varepsilon}\frac{kI_0l\sin\theta}{4\pi r}\left[\frac{\partial}{\partial r}\left(\left[1+\frac{1}{jkr}\right]e^{-jkr}\right)\right]$$

$$= -\frac{1}{\omega\varepsilon}\frac{kI_0l\sin\theta}{4\pi r}\left[1+\frac{1}{jkr}-\frac{1}{(kr)^2}\right]\left(-jke^{-jkr}\right) \qquad (6.150)$$

$$= \frac{1}{\omega\varepsilon}\frac{kI_0l\sin\theta}{4\pi r}\left[1+\frac{1}{jkr}-\frac{1}{(kr)^2}\right]jke^{-jkr}$$

Thus, the electric field intensity at a distance r from a Hertzian dipole is given by

$$E_r = \frac{1}{\omega\varepsilon}\frac{kI_0l\cos\theta}{2\pi r^2}\left[1+\frac{1}{jkr}\right]e^{-jkr} \qquad (6.151a)$$

$$E_\theta = \frac{1}{\omega\varepsilon}\frac{kI_0l\sin\theta}{4\pi r}\left[1+\frac{1}{jkr}-\frac{1}{(kr)^2}\right]jke^{-jkr} \qquad (6.151b)$$

$$E_\phi = 0 \qquad (6.151c)$$

Whereas the magnetic field intensity at a distance r form a Hertzian dipole was derived earlier, and is repeated here:

$$H_r = 0 \qquad (6.152a)$$

$$H_\theta = 0 \qquad (6.152b)$$

$$H_\phi = j\frac{kI_0l\sin\theta}{4\pi r}\left[1+\frac{1}{jkr}\right]e^{-jkr} \qquad (6.152c)$$

Note that these two sets of equations (6.151) and (6.152) are equivalent to the set of equations (7.1) on page 423 of Paul 2006.

We will show this equivalence using the alternative approach to computing the electric field intensity from Eq. (6.129), repeated here:

$$\hat{\mathbf{E}} = -j\omega\hat{\mathbf{A}} - j\frac{1}{\omega\mu\varepsilon}\nabla\left(\nabla\cdot\hat{\mathbf{A}}\right) \qquad (6.153)$$

Recall that the vector magnetic potential in spherical coordinates was given by Eq. (6.138), repeated here:

$$\begin{bmatrix} A_r \\ A_\theta \\ A_\phi \end{bmatrix} = \begin{bmatrix} \dfrac{\mu I_0 l e^{-jkr}}{4\pi r}\cos\theta \\ -\dfrac{\mu I_0 l e^{-jkr}}{4\pi r}\sin\theta \\ 0 \end{bmatrix} \qquad (6.154)$$

First, we need to compute the divergence of A. In spherical coordinates we have

$$\nabla \cdot \mathbf{A} = \frac{1}{r^2} \frac{\partial}{\partial r} \left(r^2 A_r \right) + \frac{1}{r \sin \theta} \frac{\partial}{\partial \theta} \left(A_\theta \sin \theta \right) + \frac{1}{r \sin \theta} \frac{\partial A_\phi}{\partial \phi} \tag{6.155}$$

Since $A_\varphi = 0$ and there are no φ variations in $\mathbf{A}$, we have

$$
\begin{aligned}
\nabla \cdot \mathbf{A} &= \frac{1}{r^2} \frac{\partial}{\partial r} \left(r^2 A_r \right) + \frac{1}{r \sin \theta} \frac{\partial}{\partial \theta} \left(A_\theta \sin \theta \right) \\
&= \frac{1}{r^2} \frac{\partial}{\partial r} \left(r^2 \frac{\mu I_0 l e^{-jkr}}{4\pi r} \cos \theta \right) + \frac{1}{r \sin \theta} \frac{\partial}{\partial \theta} \left(-\frac{\mu I_0 l e^{-jkr}}{4\pi r} \sin \theta \sin \theta \right) \\
&= \frac{1}{r^2} \frac{\partial}{\partial r} \left(r \frac{\mu I_0 l e^{-jkr}}{4\pi} \cos \theta \right) + \frac{1}{r \sin \theta} \frac{\partial}{\partial \theta} \left(-\frac{\mu I_0 l e^{-jkr}}{4\pi r} \sin^2 \theta \right) \\
&= \frac{1}{r^2} \frac{\mu I_0 l \cos \theta}{4\pi} \frac{\partial}{\partial r} \left(r e^{-jkr} \right) - \frac{1}{r \sin \theta} \frac{\mu I_0 l e^{-jkr}}{4\pi r} \frac{\partial}{\partial \theta} \left(\sin^2 \theta \right) \\
&= \frac{1}{r^2} \frac{\mu I_0 l \cos \theta}{4\pi} \left(e^{-jkr} - rjk e^{-jkr} \right) - \frac{1}{r \sin \theta} \frac{\mu I_0 l e^{-jkr}}{4\pi r} \left(2 \sin \theta \cos \theta \right) \\
&= \frac{\mu I_0 l \cos \theta}{4\pi r^2} \left(e^{-jkr} - rjk e^{-jkr} \right) - \frac{\mu I_0 l e^{-jkr}}{4\pi r^2} \left(2 \cos \theta \right) \\
&= \frac{\mu I_0 l \cos \theta}{4\pi r^2} e^{-jkr} - j \frac{k \mu I_0 l \cos \theta}{4\pi r} e^{-jkr} - \frac{2 \mu I_0 l \cos \theta}{4\pi r^2} e^{-jkr} \\
&= -\frac{\mu I_0 l}{4\pi} \left(\frac{1}{r^2} + \frac{jk}{r} \right) e^{-jkr} \cos \theta
\end{aligned}
\tag{6.156}
$$

Now, the gradient of the scalar function f in spherical coordinates is

$$\nabla f = \frac{\partial f}{\partial r} \mathbf{a}_r + \frac{1}{r} \frac{\partial f}{\partial \theta} \mathbf{a}_\theta + \frac{1}{r \sin \theta} \frac{\partial f}{\partial \phi} \mathbf{a}_\phi \tag{6.157}$$

Thus,

$$
\begin{aligned}
\nabla(\nabla \cdot \mathbf{A}) &= \frac{\partial f}{\partial r} \mathbf{a}_r + \frac{1}{r} \frac{\partial f}{\partial \theta} \mathbf{a}_\theta \\
&= \frac{\partial}{\partial r} \left[-\frac{\mu I_0 l}{4\pi} \left(\frac{1}{r^2} + \frac{jk}{r} \right) e^{-jkr} \cos \theta \right] \mathbf{a}_r \\
&\quad + \frac{1}{r} \frac{\partial}{\partial \theta} \left[-\frac{\mu I_0 l}{4\pi} \left(\frac{1}{r^2} + \frac{jk}{r} \right) e^{-jkr} \cos \theta \right] \mathbf{a}_\theta
\end{aligned}
\tag{6.158}
$$

Let's start with the r component.

$$
\begin{aligned}
\frac{\partial}{\partial r} \left[-\frac{\mu I_0 l}{4\pi} \left(\frac{1}{r^2} + \frac{jk}{r} \right) e^{-jkr} \cos \theta \right] &= -\frac{\mu I_0 l}{4\pi} \cos \theta \frac{\partial}{\partial r} \left[\left(\frac{1}{r^2} + \frac{jk}{r} \right) e^{-jkr} \right] \\
&= -\frac{\mu I_0 l}{4\pi} \cos \theta \left[\frac{\partial}{\partial r} \left(\frac{1}{r^2} + \frac{jk}{r} \right) e^{-jkr} + \left(\frac{1}{r^2} + \frac{jk}{r} \right) \frac{\partial}{\partial r} \left(e^{-jkr} \right) \right] \\
&= -\frac{\mu I_0 l}{4\pi} \cos \theta \left[\left(-\frac{2}{r^3} - \frac{jk}{r^2} \right) e^{-jkr} + \left(\frac{1}{r^2} + \frac{jk}{r} \right) (-jk) \left(e^{-jkr} \right) \right]
\end{aligned}
\tag{6.157}
$$

$$= \frac{\mu I_0 l}{4\pi} \cos\theta \left[\left(\frac{2}{r^3} + \frac{jk}{r^2} \right) e^{-jkr} + \left(\frac{1}{r^2} + \frac{jk}{r} \right) (jk)\left(e^{-jkr} \right) \right]$$

$$= \frac{\mu I_0 l}{4\pi} \cos\theta \left[\frac{2}{r^3} + \frac{jk}{r^2} + \frac{jk}{r^2} - \frac{k^2}{r} \right] e^{-jkr} = \frac{\mu I_0 l}{4\pi} \cos\theta \left[\frac{2}{r^3} + \frac{2jk}{r^2} - \frac{k^2}{r} \right] e^{-jkr}$$

Next, the θ-component

$$\frac{1}{r} \frac{\partial}{\partial\theta} \left[-\frac{\mu I_0 l}{4\pi} \left(\frac{1}{r^2} + \frac{jk}{r} \right) e^{-jkr} \cos\theta \right]$$

$$= -\frac{1}{r} \frac{\mu I_0 l}{4\pi} \left(\frac{1}{r^2} + \frac{jk}{r} \right) e^{-jkr} \frac{\partial}{\partial\theta} [\cos\theta] \tag{6.158}$$

$$= \frac{\mu I_0 l}{4\pi} \left(\frac{1}{r^3} + \frac{jk}{r^2} \right) e^{-jkr} \sin\theta$$

Next, substitute Eqs (6.154), (6.157), and (6.158) in (6.153), repeated here:

$$\hat{\mathbf{E}} = -j\omega\hat{\mathbf{A}} - j\frac{1}{\omega\mu\varepsilon} \nabla(\nabla\cdot\hat{\mathbf{A}}) \tag{6.159}$$

Again, let's start with the r component.

$$E_r = -j\omega \frac{\mu I_0 l e^{-jkr}}{4\pi r} \cos\theta - j\omega \frac{\mu I_0 l e^{-jkr}}{4\pi r} \cos\theta$$

$$- j\frac{1}{\omega\mu\varepsilon} \frac{\mu I_0 l}{4\pi} \cos\theta \left[\frac{2}{r^3} + \frac{2jk}{r^2} - \frac{k^2}{r} \right] e^{-jkr}$$

$$= -j\omega \frac{\mu I_0 l e^{-jkr}}{4\pi r} \cos\theta - j\frac{1}{\omega\mu\varepsilon} \frac{\mu I_0 l}{4\pi} \cos\theta \left[\frac{2}{r^3} + \frac{2jk}{r^2} \right] e^{-jkr}$$

$$+ j\frac{1}{\omega\mu\varepsilon} \frac{\mu I_0 l}{4\pi} \cos\theta \left[\frac{k^2}{r} \right] e^{-jkr} \tag{6.160}$$

Let's look at the sum of the first and the last term:

$$-j\omega \frac{\mu I_0 l e^{-jkr}}{4\pi r} \cos\theta + j\frac{1}{\omega\mu\varepsilon} \frac{\mu I_0 l}{4\pi} \cos\theta \left[\frac{k^2}{r} \right] e^{-jkr}$$

$$= \left(-j\omega + j\frac{k^2}{\omega\mu\varepsilon} \right) \frac{\mu I_0 l e^{-jkr}}{4\pi r} \cos\theta \tag{6.161}$$

$$= \left(-j\omega + j\frac{\omega^2\mu\varepsilon}{\omega\mu\varepsilon} \right) \frac{\mu I_0 l e^{-jkr}}{4\pi r} \cos\theta = 0$$

thus,

$$E_r = -j\frac{1}{\omega\mu\varepsilon} \frac{\mu I_0 l}{4\pi} \cos\theta \left[\frac{2}{r^3} + \frac{2jk}{r^2} \right] e^{-jkr}$$

$$= -j\frac{2}{\omega\varepsilon} \frac{I_0 l}{4\pi} \cos\theta \left[\frac{jk}{r^2} + \frac{1}{r^3} \right] e^{-jkr}$$

$$= -j\frac{2}{\omega\varepsilon} \frac{I_0 l}{4\pi} \cos\theta (jk) \left[\frac{1}{r^2} + \frac{1}{jkr^3} \right] e^{-jkr} \tag{6.162}$$

$$= \frac{2k}{\omega\varepsilon} \frac{I_0 l}{4\pi} \cos\theta \left[\frac{1}{r^2} + \frac{1}{jkr^3} \right] e^{-jkr}$$

Let's introduce a different notation (to conform to the formulas in Paul, pp. 422–423):

$$k = \beta_0 \tag{6.163a}$$

$$l = dl \tag{6.163b}$$

$$\frac{1}{\omega\varepsilon} = \frac{\eta}{\beta} \tag{6.163c}$$

Then Eq. (6.162) can be written as

$$
\begin{aligned}
E_r &= 2\beta \frac{\eta}{\beta} \frac{I_0 dl}{4\pi} \cos\theta \left[\frac{1}{r^2} + \frac{1}{j\beta r^3} \right] e^{-j\beta r} \\
&= 2\frac{I_0 dl}{4\pi} \eta\beta^2 \cos\theta \left[\frac{1}{\beta^2 r^2} + \frac{1}{j\beta^3 r^3} \right] e^{-j\beta r}
\end{aligned}
\tag{6.164}
$$

Or

$$E_r = 2\frac{I_0 dl}{4\pi} \eta\beta^2 \cos\theta \left[\frac{1}{\beta^2 r^2} - j\frac{1}{\beta^3 r^3} \right] e^{-j\beta r} \tag{6.165}$$

Now, the θ-component

$$
\begin{aligned}
E_\theta &= -j\omega \left(-\frac{\mu I_0 l e^{-jkr}}{4\pi r} \sin\theta \right) - j\frac{1}{\omega\mu\varepsilon} \frac{\mu I_0 l}{4\pi} \left(\frac{1}{r^3} + \frac{jk}{r^2} \right) e^{-jkr} \sin\theta \\
&= j\omega \frac{\mu I_0 l e^{-jkr}}{4\pi r} \sin\theta - j\frac{1}{\omega\varepsilon} \frac{I_0 l}{4\pi} \left(\frac{1}{r^3} + \frac{jk}{r^2} \right) e^{-jkr} \sin\theta
\end{aligned}
\tag{6.166}
$$

In Part III of this book, we will define the intrinsic impedance of a medium as

$$\eta = \sqrt{\frac{\mu}{\varepsilon}} = \sqrt{\frac{\mu\mu}{\varepsilon\mu}} = \frac{\mu}{\sqrt{\mu\varepsilon}} \quad \Rightarrow \quad \mu = \eta\sqrt{\mu\varepsilon} \tag{6.167a}$$

and express the angular frequency as

$$\omega = \frac{\beta}{\sqrt{\mu\varepsilon}} \tag{6.167b}$$

then, the following holds

$$\omega\mu = \frac{\beta}{\sqrt{\mu\varepsilon}} \eta\sqrt{\mu\varepsilon} = \beta\eta \tag{6.167c}$$

$$\frac{1}{\omega\varepsilon} = \frac{\sqrt{\mu\varepsilon}}{\beta} \frac{\eta}{\sqrt{\mu\varepsilon}} = \frac{\eta}{\beta} \tag{6.167d}$$

thus,

$$
\begin{aligned}
E_\theta &= j\omega \frac{\mu I_0 l e^{-jkr}}{4\pi r} \sin\theta - j\frac{1}{\omega\varepsilon}\frac{I_0 l}{4\pi}\left(\frac{1}{r^3}+\frac{jk}{r^2}\right)e^{-jkr}\sin\theta \\
&= j\frac{\beta\eta I_0 l e^{-j\beta r}}{4\pi r}\sin\theta + \frac{\eta}{\beta}\frac{I_0 l}{4\pi}\left(-\frac{j}{r^3}+\frac{\beta}{r^2}\right)e^{-j\beta r}\sin\theta \\
&= \frac{I_0 dl}{4\pi}\eta\sin\theta\left(\frac{j\beta}{r}-j\frac{1}{\beta r^3}+\frac{1}{r^2}\right)e^{-j\beta r} \\
&= \frac{I_0 dl}{4\pi}\eta\beta^2\sin\theta\left(\frac{j}{\beta r}-j\frac{1}{\beta^3 r^3}+\frac{1}{\beta^2 r^2}\right)e^{-j\beta r}
\end{aligned}
\tag{6.168}
$$

or

$$
E_\theta = \frac{I_0 dl}{4\pi}\eta\beta^2\sin\theta\left(\frac{j}{\beta r}+\frac{1}{\beta^2 r^2}-j\frac{1}{\beta^3 r^3}\right)e^{-j\beta r}
\tag{6.169}
$$

Equations (6.165) and (6.169) correspond to Equations (7.1d) and (7.1e) in Paul (2006, p. 422–423).

6.7.5 Electric Dipole Antenna Radiated Power

Electric dipole radiated power can be computed from (Paul, 2006, p. 425)

$$
\hat{S}(\mathbf{r}) = \frac{1}{2}\left\{\hat{E}(\mathbf{r})\times\hat{H}^*(\mathbf{r})\right\}
\tag{6.170}
$$

where the electric and magnetic fields were derived earlier as

$$
\hat{E} = \left(\hat{E}_r, \hat{E}_\theta, 0\right)
\tag{6.171a}
$$

$$
\hat{H} = \left(0, 0, \hat{H}_\phi\right)
\tag{6.171b}
$$

where

$$
\begin{aligned}
\hat{E}_r &= \frac{1}{\omega\varepsilon}\frac{kI_0 l\cos\theta}{2\pi r^2}\left[1+\frac{1}{jkr}\right]e^{-jkr} \\
&= \eta\frac{kI_0 l\cos\theta}{2\pi r^2}\left[1+\frac{1}{jkr}\right]e^{-jkr}
\end{aligned}
\tag{6.172a}
$$

$$
\begin{aligned}
\hat{E}_\theta &= \frac{1}{\omega\varepsilon}\frac{kI_0 l\sin\theta}{4\pi r}\left[1+\frac{1}{jkr}-\frac{1}{(kr)^2}\right]jke^{-jkr} \\
&= j\eta\frac{kI_0 l\sin\theta}{4\pi r}\left[1+\frac{1}{jkr}-\frac{1}{(kr)^2}\right]ke^{-jkr}
\end{aligned}
\tag{6.172b}
$$

$$\hat{H}_\varphi = j\frac{kI_0l\sin\theta}{4\pi r}\left[1+\frac{1}{jkr}\right]e^{-jkr} \tag{6.172c}$$

Substitution of Eq. (6.172) into Eq. (6.170) produces

$$\hat{S}(\mathbf{r}) = \frac{1}{2}\left(\hat{E}_\theta\hat{H}_\varphi^*\mathbf{a}_r - \hat{E}_r\hat{H}_\varphi^*\mathbf{a}_\theta\right) = \left(\hat{S}_r, \hat{S}_\theta, 0\right) \tag{6.173}$$

Let's calculate the r component of the radiated power.

$$\hat{S}_r = \frac{1}{2}\hat{E}_\theta\hat{H}_\varphi^*$$

$$= \frac{1}{2}\left\{j\eta\frac{kI_0l\sin\theta}{4\pi r}\left[1+\frac{1}{jkr}-\frac{1}{(kr)^2}\right]e^{-jkr}\right\}\left\{j\frac{kI_0l\sin\theta}{4\pi r}\left[1+\frac{1}{jkr}\right]e^{-jkr}\right\}^*$$

$$= \frac{1}{2}\left(\frac{kI_0l\sin\theta}{4\pi r}\right)^2\left\{j\eta\left[1+\frac{1}{jkr}-\frac{1}{(kr)^2}\right]\right\}\left\{j\left[1-j\frac{1}{kr}\right]\right\}^* \tag{6.174a}$$

$$= \frac{1}{2}\left(\frac{kI_0l\sin\theta}{4\pi r}\right)^2\left\{j\eta\left[1+\frac{1}{jkr}-\frac{1}{(kr)^2}\right]\right\}\left\{-j\left[1+j\frac{1}{kr}\right]\right\}$$

$$= \frac{1}{2}\eta\left(\frac{kI_0l\sin\theta}{4\pi r}\right)^2\left\{\left[1+\frac{1}{jkr}-\frac{1}{(kr)^2}\right]\left[1+\frac{j}{kr}\right]\right\}$$

or

$$\hat{S}_r = \frac{1}{2}\eta\left(\frac{kI_0l\sin\theta}{4\pi r}\right)^2\left\{\left[1+\frac{j}{kr}+\frac{1}{jkr}+\frac{1}{jkr}\frac{j}{kr}-\frac{1}{(kr)^2}-\frac{1}{(kr)^2}\frac{j}{kr}\right]\right\}$$

$$= \frac{1}{2}\eta\left(\frac{kI_0l\sin\theta}{4\pi r}\right)^2\left\{\left[1+\frac{j}{kr}-j\frac{1}{kr}+\frac{1}{(kr)^2}-\frac{1}{(kr)^2}-\frac{j}{(kr)^3}\right]\right\} \tag{6.174b}$$

$$= \frac{\eta}{2}\left(\frac{kI_0l\sin\theta}{4\pi r}\right)^2\left\{\left[1-\frac{j}{(kr)^3}\right]\right\} = \frac{\eta}{2}\left(\frac{\frac{2\pi}{\lambda}I_0l\sin\theta}{4\pi r}\right)^2\left\{\left[1-\frac{j}{(kr)^3}\right]\right\}, \quad k=\frac{2\pi}{\lambda}$$

leading to

$$\hat{S}_r = \frac{\eta}{2}\left(\frac{\frac{1}{\lambda}I_0l\sin\theta}{2r}\right)^2\left[1-\frac{j}{(kr)^3}\right] = \frac{\eta}{8}\left(\frac{I_0l}{\lambda}\right)^2\frac{\sin^2\theta}{r^2}\left[1-\frac{j}{(kr)^3}\right] \tag{6.174c}$$

$$= \frac{120\pi}{8}\left(\frac{I_0l}{\lambda}\right)^2\frac{\sin^2\theta}{r^2}\left[1-\frac{j}{(kr)^3}\right], \quad \eta=120\pi$$

or

$$\hat{S}_r = 15\pi \left(\frac{I_0 l}{\lambda}\right)^2 \frac{\sin^2\theta}{r^2}\left[1 - \frac{j}{(kr)^3}\right]\left(\frac{W}{m^2}\right)$$

(6.174d)

Equation (6.174d) describes the r component of the complex power density vector. To calculate the total average power radiated by the electric dipole antenna we evaluate the following surface integral

$$\hat{P} = \oint_S \hat{S}(\mathbf{r}) \cdot d\mathbf{S}$$

$$= \int_{\varphi=0}^{2\pi}\int_{\theta=0}^{\pi}\left(\hat{S}_r \mathbf{a}_r + \hat{S}_\theta \mathbf{a}_\theta\right)r^2\sin\theta d\theta d\varphi \mathbf{a}_r$$

$$= \int_{\varphi=0}^{2\pi}\int_{\theta=0}^{\pi}\hat{S}_r r^2\sin\theta d\theta d\varphi$$

(6.175a)

$$= \int_{\varphi=0}^{2\pi}\int_{\theta=0}^{\pi}\left\{15\pi\left(\frac{I_0 l}{\lambda}\right)^2 \frac{\sin^2\theta}{r^2}\left[1 - \frac{j}{(kr)^3}\right]\right\}r^2\sin\theta d\theta d\varphi$$

$$= \int_{\varphi=0}^{2\pi}\int_{\theta=0}^{\pi}\left\{15\pi\left(\frac{I_0 l}{\lambda}\right)^2\left[1 - \frac{j}{(kr)^3}\right]\right\}\sin^3\theta d\theta d\varphi$$

or

$$\hat{P} = 15\pi\left(\frac{I_0 l}{\lambda}\right)^2\left[1 - \frac{j}{(kr)^3}\right]\int_{\varphi=0}^{2\pi}\int_{\theta=0}^{\pi}\sin^3\theta d\theta d\phi$$

$$= 30\pi^2\left(\frac{I_0 l}{\lambda}\right)^2\left[1 - \frac{j}{(kr)^3}\right]\int_{\theta=0}^{\pi}\sin^3\theta d\theta d\varphi, \quad \int_{\theta=0}^{\pi}\sin^3\theta d\theta = \frac{4}{3}$$

(6.175b)

$$= \hat{P} = \frac{120\pi^2}{3}\left(\frac{I_0 l}{\lambda}\right)^2\left[1 - \frac{j}{(kr)^3}\right]$$

$$= 120\pi\frac{\pi}{3}\left(\frac{I_0 l}{\lambda}\right)^2\left[1 - \frac{j}{(kr)^3}\right]$$

and thus

$$\hat{P} = 120\pi\frac{\pi}{3}\left(\frac{I_0 l}{\lambda}\right)^2\left[1 - \frac{j}{(kr)^3}\right]$$

(6.175c)

The real radiated power is just the real part of the complex power in Eq. (6.175c).

$$P_{rad} = \eta\frac{\pi}{3}\left(\frac{I_0 l}{\lambda}\right)^2 = 80\pi^2\left(\frac{l}{\lambda}\right)^2\frac{I_0^2}{2}$$

(6.176)

References

Alexander, C.K. and Sadiku, N.O., *Fundamentals of Electric Circuits*, 4th ed., McGraw Hill, New York, 2009.

Balanis, C.A., *Antenna Theory Analysis and Design*, 3rd ed., Wiley Interscience, Hoboken. New Jersey, 2005.

Kreyszig, E., *Advanced Engineering Mathematics*, 8th ed., John Wiley and Sons, New York, 1999.

Paul, C.R., *Introduction to Electromagnetic Compatibility*, 2nd ed., John Wiley and Sons, New York, 2006.

Part II

Circuits Foundations of EMC

7

Basic Laws and Methods of Circuit Analysis

7.1 Fundamental Concepts

7.1.1 Current

The motion of electric charges constitutes an electric current, denoted by the letters i or I. As a matter of vocabulary, we say that a current flows along a path, from A to B, or through an element, as shown in Figure 7.1.

Note that a complete description of current requires both a value and a reference direction, as shown in Figure 7.1.

By definition, current is the time rate of change of charge, or

$$i = \frac{dq}{dt} \qquad \left[A = \frac{C}{s} \right] \tag{7.1}$$

We consider the network elements to be electrically neutral. That is, no net charge can accumulate in the element. Charges may not accumulate or be depleted at any point. Any charge entering the element must be accompanied by an equal charge leaving the element.

7.1.2 Voltage

Charges in a conductor may move in a random manner. To move the charges in a conductor in a particular direction requires some work or energy transfer.

We define the voltage v_{AB} between two points A and B in an electric circuit as the energy (work) needed to move a unit of charge from A to B.

Mathematically,

$$v_{AB} = \frac{dw}{dq} \qquad \left[V = \frac{J}{C} \right] \tag{7.2}$$

As a matter of vocabulary, we say that a voltage exists across an element, or between two points or nodes, as shown in Figure 7.2.

Note that a complete description of voltage requires both a value and a reference direction, as shown in Figure 7.2.

Foundations of Electromagnetic Compatibility with Practical Applications, First Edition. Bogdan Adamczyk.
© 2017 John Wiley & Sons Ltd. Published 2017 by John Wiley & Sons Ltd.

Figure 7.1 Current designation.

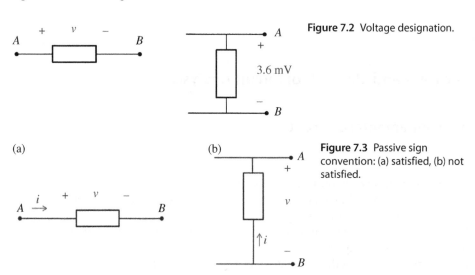

Figure 7.2 Voltage designation.

(a)

(b)

Figure 7.3 Passive sign convention: (a) satisfied, (b) not satisfied.

Passive Sign Convention   For any electrical element, if the current reference direction, upon entering the element, points into the positive voltage reference direction, the current and voltage so defined are said to satisfy the *passive sign convention* (Alexander and Sadiku, 2009, p.11).

In Figure 7.3(a) the passive sign convention is satisfied, while in Figure 7.3(b) it is not.

7.1.3   Power

Voltage and current are useful variables in the analysis and design of electrical circuits. The circuit specifications, in addition to voltage and current, often include the requirement on *power* that the circuit needs to deliver to a load.

In transferring charge through an element, work is being done, or energy is being transferred. We define power p, as the rate at which energy w is being transferred. Mathematically,

$$p = \frac{dw}{dt} \quad \left[W = \frac{J}{s} \right] \tag{7.3}$$

In the circuit analysis, it is more convenient to work with the circuit variables (voltage and current) than the field variables (energy or vector quantities). It is, therefore, preferred to express power in term of voltage and current.

Note that (7.3) can be written as

$$p = \frac{dw}{dt} = \frac{dw}{dq} \cdot \frac{dq}{dt} \tag{7.4}$$

and thus

$$p = vi \quad [\text{W} = \text{VA}] \tag{7.5}$$

When the passive sign convention is satisfied, the power calculated according to (7.5) is called the *power absorbed* or *dissipated by the element*.

If the current and voltage direction do not satisfy the passive sign convention, the power calculated using (7.5) is called the *power delivered by the element*.

7.1.4 Average Power in Sinusoidal Steady State

Average Power in the Time Domain When voltage and current in Eq. (7.5) are time varying, then the power that is obtained from

$$p(t) = v(t)i(t) \tag{7.6}$$

is often referred to as the *instantaneous power*.

Of special interest to us is the case when both voltage and current are sinusoidal function of time. Sinusoidal steady state analysis is of paramount importance in EMC engineering.

Let the voltage and current at the terminals of the circuit be

$$v(t) = V_m \cos(\omega t + \theta_v) \tag{7.7}$$

$$i(t) = I_m \cos(\omega t + \theta_i) \tag{7.8}$$

where V_m, I_m are the amplitudes, and θ_v and θ_i are the phase angles of the voltage and current, respectively.

Since we are operating in the sinusoidal steady state, we may choose any convenient reference for zero time. It is convenient to use a zero reference time corresponding to the instant the current is passing through a positive maximum.

This reference system requires a shift of both the voltage and current by θ_i. Thus, Eqs (7.7) and (7.8) become

$$v(t) = V_m \cos(\omega t + \theta_v - \theta_i) \tag{7.9}$$

$$i(t) = I_m \cos(\omega t) \tag{7.10}$$

When we substitute Eqs (7.9) and (7.10) into Eq. (7.6), the instantaneous power absorbed by the element is

$$p(t) = v(t)i(t) = V_m I_m \cos(\omega t + \theta_v - \theta_i)\cos(\omega t) \tag{7.11}$$

Let's make use of the trigonometric identity

$$\cos\alpha\cos\beta = \frac{1}{2}\big[\cos(\alpha - \beta) + \cos(\alpha + \beta)\big] \tag{7.12}$$

Then Eq. (7.11) can be expressed as

$$p(t) = \frac{V_m I_m}{2} \cos(\theta_v - \theta_i) + \frac{V_m I_m}{2} \cos(2\omega t + \theta_v - \theta_i) \tag{7.13}$$

Note that $p(t)$ is periodic with a period $T_0 = T/2$ since its frequency is twice that of voltage or current.

The instantaneous power may be positive, negative, or zero, depending on the time t at which it is evaluated, and thus does not convey much information about the element or system. Of much more use is the *average power* which, as we shall see, is not a function of time.

The average power is the average of the instantaneous power over one period. Thus, the average power is given by

$$P_{ave} = \frac{1}{T_0} \int_0^{T_0} p(t) dt \tag{7.14}$$

where T_0 is the period of $p(t)$, and the instantaneous power is expressed by (7.13).

We would get the same result is we performed the integration in (7.14) over the time interval of two periods, $T = 2T_0$; that is, the average power can also be computed from

$$P_{ave} = \frac{1}{T} \int_0^{T} p(t) dt \tag{7.15}$$

Substituting (7.13) into (7.15) results in

$$P_{ave} = \frac{1}{T} \int_0^{T} \frac{1}{2} V_m I_m \cos(\theta_v - \theta_i) dt + \frac{1}{T} \int_0^{T} \frac{1}{2} V_m I_m \cos(2\omega t + \theta_v - \theta_i) dt$$

$$= \frac{1}{2} V_m I_m \cos(\theta_v - \theta_i) + \frac{1}{2} V_m I_m \frac{1}{T} \int_0^{T} \cos(2\omega t + \theta_v - \theta_i) dt \tag{7.16}$$

Note that the second term contains the integral of the sinusoid over its period. This integral is zero because the area under the sinusoid during a positive half-cycle is cancelled by the area under it during the following negative half-cycle.

Thus, the second term in (7.16) vanishes and the *average power* becomes

$$P_{ave} = \frac{1}{2} V_m I_m \cos(\theta_v - \theta_i) \tag{7.17}$$

Average Power in the Phasor Form The phasor forms of $v(t)$ and $i(t)$ are

$$\hat{V} = V_m \angle \theta_v \tag{7.18a}$$

$$\hat{I} = I_m \angle \theta_i \tag{7.18b}$$

Let's evaluate the following expression

$$
\begin{aligned}
\frac{1}{2}\hat{V}\hat{I}^* &= \frac{1}{2}V_m I_m \angle(\theta_v - \theta_i) \\
&= \frac{1}{2}V_m I_m \left[\cos(\theta_v - \theta_i) + j\sin(\theta_v - \theta_i)\right]
\end{aligned}
\tag{7.19}
$$

Comparing (7.19) with (7.17) reveals that the average power can be computed from the phasor forms as

$$
P_{ave} = \mathrm{Re}\left\{\frac{1}{2}\hat{V}\hat{I}^*\right\}
\tag{7.20}
$$

Average Power Delivered to a Resistive Load   In the next section we will show that the voltage current relationship for a resistor in phasor domain is

$$
\hat{V} = R\hat{I}
\tag{7.21}
$$

Since

$$
\hat{I} = I_m \angle\theta_i
\tag{7.22}
$$

it follows that

$$
\hat{V} = R I_m \angle\theta_i
\tag{7.23}
$$

Now, utilizing Eq. (7.20), we obtain the average power delivered to a resistive load as

$$
\begin{aligned}
P_{ave} &= \mathrm{Re}\left\{\frac{1}{2}\hat{V}\hat{I}^*\right\} \\
&= \mathrm{Re}\left\{\frac{1}{2}(R I_m \angle\theta_i)(I_m \angle-\theta_i)\right\} \\
&= \mathrm{Re}\left\{\frac{1}{2}(R I_{mi})(I_m)\right\}
\end{aligned}
\tag{7.24}
$$

or

$$
P_{ave} = \frac{1}{2}R|\hat{I}|^2
\tag{7.25}
$$

7.2   Laplace Transform Basics

7.2.1   Definition of Laplace Transform

Consider a time function $f(t)$ defined for $t>0$. The Laplace transform operates on a time function and creates a new function that exists in a new domain, called the Laplace or s domain.

$$
F(s) = L\{f(t)\}
\tag{7.26}
$$

The formal definition of the Laplace transform is (Nilsson and Riedel, 2015, p. 428)

$$L\{f(t)\} = \int_{0^-}^{\infty} f(t)e^{-st}dt \qquad (7.27)$$

Is this useful in EMC? Extremely! As we shall see, the Laplace transform will lead to the concept of impedance, transfer function, and frequency transfer function, which will allow us to carry the sinusoidal steady state frequency domain analysis. It is the frequency domain analysis that is of utmost importance to an EMC engineer.

To get the feel for this definition let's calculate the Laplace transform of a constant and an exponential functions.

$$L\{A\} = \int_0^{\infty} Ae^{-st}dt = A\int_0^{\infty} e^{-st}dt = A\frac{e^{-st}}{-s}\Big|_{t=0}^{t=\infty} = \frac{A}{s} \qquad (7.28)$$

$$L\{e^{-at}\} = \int_0^{\infty} e^{-at}e^{-st}dt = \int_0^{\infty} e^{-(s+a)t}dt = -\frac{e^{-(s+a)t}}{s+a}\Big|_{t=0}^{t=\infty} = \frac{1}{s+a} \qquad (7.29)$$

Obtaining the Laplace transform of a time function using the definition is often time-consuming and cumbersome. In practice, we often use the tables of Laplace transform pairs, together with the properties of Laplace transforms to obtain the transform of a given function which might not be tabulated.

The most common transform pairs that we might encounter in EMC problems are

$$f(t) = 1 \quad \Leftrightarrow \quad F(s) = \frac{1}{s} \qquad (7.30a)$$

$$f(t) = t \quad \Leftrightarrow \quad F(s) = \frac{1}{s^2} \qquad (7.30b)$$

$$f(t) = e^{-at} \quad \Leftrightarrow \quad F(s) = \frac{1}{s+a} \qquad (7.30c)$$

$$f(t) = te^{-at} \quad \Leftrightarrow \quad F(s) = \frac{1}{(s+a)^2} \qquad (7.30d)$$

$$f(t) = \sin\omega t \quad \Leftrightarrow \quad F(s) = \frac{\omega}{s^2 + \omega^2} \qquad (7.30e)$$

$$f(t) = \cos\omega t \quad \Leftrightarrow \quad F(s) = \frac{s}{s^2 + \omega^2} \qquad (7.30f)$$

$$f(t) = e^{-at}\sin\omega t \quad \Leftrightarrow \quad F(s) = \frac{\omega}{(s+a)^2 + \omega^2} \qquad (7.30g)$$

$$f(t) = e^{-at}\cos\omega t \quad \Leftrightarrow \quad F(s) = \frac{s}{(s+a)^2 + \omega^2} \qquad (7.30h)$$

7.2.2 Properties of Laplace Transform

Next, we will present a few selected properties of Laplace transform that we will subsequently use in this book when discussing EMC applications.

Let

$$F(s) = \mathcal{L}\{f(t)\} \tag{7.31a}$$

$$G(s) = \mathcal{L}\{g(t)\} \tag{7.31b}$$

Then (*linearity property*)

$$\mathcal{L}\{af(t) + bg(t)\} = aF(s) + bG(s) \tag{7.32}$$

Before we present the next property, let's define the unit step and time-shifted unit step functions.

The unit step function is defined as

$$u(t) = \begin{cases} 1, & t \geq 0 \\ 0, & t < 0 \end{cases} \tag{7.33a}$$

Similarly, the time-shifted unit step function is defined as

$$u(t-a) = \begin{cases} 1, & t \geq a \\ 0, & t < a \end{cases} \tag{7.33b}$$

Both functions are shown in Figure 7.4.

The *time shift* property of the Laplace transform can be stated as

$$\mathcal{L}\{f(t-a)u(t-a)\} = e^{-as}F(s) \tag{7.34}$$

We will use this property when discussing the capacitive termination of a transmission line in the EMC application section of this chapter.

Figure 7.4 Unit step function.

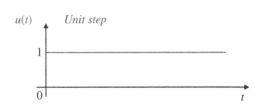

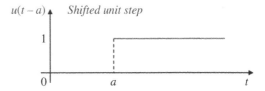

The final very important property is the *time differentiation*:

$$\mathcal{L}\left\{\frac{df(t)}{dt}\right\} = sF(s) - f(0) \tag{7.35}$$

$$\mathcal{L}\left\{\frac{d^2f(t)}{dt^2}\right\} = s^2 F(s) - sf(0) - \frac{df(0)}{dt} \tag{7.36}$$

When the initial value of the time function is zero, Eq. (7.33) becomes

$$\mathcal{L}\left\{\frac{df(t)}{dt}\right\} = sF(s) \tag{7.37}$$

We will utilize this property later in this chapter when introducing deriving the voltage–current relationships of inductors and capacitors in the s domain.

7.2.3 Inverse Laplace Transform

Often, we are given an expression for a function in the s domain, and we need to determine the corresponding time-domain function. This is accomplished by performing the inverse Laplace transform operation.

In order to obtain $f(t)$ in the time domain from $F(s)$ in the Laplace domain we may try to use the Laplace transform tables. The tables, however, have a limited number of transform pairs and many functions are not in them.

In many instances we can use an abbreviated version of Laplace transform pairs, together with the properties of Laplace transforms, and apply partial fraction expansion.

For linear, lumped-parameter circuits, the s domain expressions (functions) for the unknown voltages or currents are always rational functions of s. That is, $F(s)$ can be expressed as a ratio of two polynomials in s, such that the powers of s are non-negative integers.

Thus, in general, $F(s)$ has the form

$$F(s) = \frac{N(s)}{D(s)} = \frac{a_m s^m + a_{m-1} s^{m-1} + \cdots + a_1 s + a_0}{s^n + b_{n-1} s^{n-1} + \cdots + b_1 s + b_0} \tag{7.38}$$

where a and b are real constants, and m and n are positive integers. We will consider proper rational functions, where $m < n$.

A proper rational function can be expressed as the sum of partial fractions with constant coefficients. In order to apply partial fraction expansion, we express the denominator of $F(s)$ in the factor form:

$$F(s) = \frac{N(s)}{D(s)} = \frac{N(s)}{(s+p_1)(s+p_2)\ldots(s+p_k)^r} \tag{7.39}$$

Since the order of the denominator is n, it follows that the polynomial $D(s)$ will have n roots, or poles, which may be real or complex, and distinct or repeated. In the following, we will discuss distinct real roots.

When the denominator $D(s)$ has only distinct real roots then Eq. (7.39) can be written as

$$F(s) = \frac{N(s)}{D(s)} = \frac{N(s)}{(s+p_1)(s+p_2)\ldots(s+p_n)} \tag{7.40}$$

and $s = -p_1, -p_2, \ldots, -p_n$ are real distinct roots. Then, $F(s)$ can be expressed as

$$F(s) = \frac{N(s)}{D(s)} = \frac{A_1}{(s+p_1)} + \frac{A_2}{(s+p_2)} + \cdots + \frac{A_n}{(s+p_n)} \tag{7.41}$$

The coefficients A_k are known as the residues of $F(s)$. One way to obtain these residues is to apply *Heaviside's theorem* as follows.

To evaluate a typical coefficient A_k, multiply both sides of Eq. (7.41) by $(s+p_k)$. The result is

$$F(s)(s+p_k) = \frac{N(s)}{D(s)}(s+p_k) = \frac{N(s)(s+p_k)}{(s+p_1)(s+p_2)\ldots(s+p_k)\ldots(s+p_n)}$$
$$= A_1 \frac{(s+p_k)}{(s+p_1)} + A_2 \frac{(s+p_k)}{(s+p_2)} + \cdots + A_k + \cdots + A_n \frac{(s+p_k)}{(s+p_n)} \tag{7.42}$$

Equation (7.42) is valid for all values of s. We choose the values of s that lead to useful results. Thus, we let $= -p_k$. Then each term on the right-hand side of Eq. (7.42) vanishes, except A_k.

Thus,

$$A_k = \left[(s+p_k) \frac{N(s)}{D(s)} \right]_{s=-p_k} \tag{7.43}$$

Example 7.1 Inverse Laplace transform – distinct real roots

Determine the inverse Laplace transform of

$$F(s) = \frac{s+4}{s^3 + 4s^2 + 3s}$$

Solution:

$$F(s) = \frac{s+4}{s^3 + 4s^2 + 3s} = \frac{s+4}{s(s+1)(s+3)} = \frac{A_1}{s} + \frac{A_2}{s+1} + \frac{A_3}{s+3}$$

$$A_1 = \frac{s+4}{(s+1)(s+3)} \bigg|_{s=0} = \frac{4}{3}$$

$$A_2 = \frac{s+4}{s(s+3)} \bigg|_{s=-1} = -\frac{3}{2}$$

$$A_3 = \frac{s+4}{s(s+1)} \bigg|_{s=-3} = \frac{1}{6}$$

Therefore, using the table of transforms and the linearity property, the inverse Laplace transform is

$$f(t) = L^{-1}\{F(s)\} = L^{-1}\left\{\frac{4}{3}\frac{1}{s} - \frac{3}{2}\frac{1}{s+1} + \frac{1}{6}\frac{1}{s+3}\right\} = \left(\frac{4}{3} - \frac{3}{2}e^{-t} + \frac{1}{6}e^{-3t}\right), \quad t \geq 0$$

7.3 Fundamental Laws

We have, thus far, introduced the fundamental concepts such as current, voltage, and power in an electric circuit. To determine the values of these variables in a given circuit requires an understanding of some fundamental laws that govern electric circuits.

These laws, known as Ohm's law and Kirchhoff's laws, form the foundation upon which electric circuit analysis is built. We will present these laws in the time domain, as well as in the phasor and s domains.

7.3.1 Resistors and Ohm's Law

The physical property of material, the ability to resist the flow of current, is known as *resistance*. The circuit element used to model this behavior is the resistor.

The circuit symbol of a resistor and the voltage and current designations in the time domain are shown in Figure 7.5.

The resistor R, is a two-terminal device, connected between two nodes A and B, has a current i flowing through it, and a voltage v across it.

The relationship between current and voltage for a resistor is known as *Ohm's law*.

Ohm's law states that the voltage across a resistor is directly proportional to the current flowing through it. The constant of proportionality is the resistance value of the resistor.

When the passive sign convention is satisfied, the Ohm's law is expressed as

$$v = Ri \tag{7.44a}$$

or alternatively,

$$i = \frac{v}{R} \tag{7.44b}$$

Equations (7.44) constitute the *voltage–current relationship* for a resistor. We often refer to such a relationship as an *element constraint*.

R is measured in units called ohms that can be obtained from Eq. (7.44):

$$R = \frac{v}{i} \quad \left[\frac{V}{A} = \Omega\right] \tag{7.45}$$

When the passive sign convention is *not* satisfied, Ohm's law is expressed as

$$v = -Ri \tag{7.46a}$$

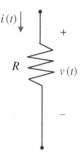

Figure 7.5 Resistor symbol and circuit variables.

or alternatively,

$$i = -\frac{v}{R} \qquad (7.46b)$$

Conductance Ohm's law can also be written as

$$i = \frac{1}{R}v = Gv \qquad (7.47)$$

where G denotes the *conductance* in siemens (S) and is the reciprocal of R.

$$G = \frac{1}{R} \qquad \left[S = \frac{1}{\Omega} = \frac{A}{V} \right] \qquad (7.48)$$

The concept of conductance not only simplifies circuit analysis (by avoiding the division by R) but also has a physical meaning.

Conductance is a measure of how well an element would conduct electric current.

Power Dissipated by a Resistor Using Ohm's law, the power dissipated by the resistor can be expressed as

$$p = vi = Ri^2 = \frac{v^2}{R} \qquad \left[W = VA = \Omega A^2 = \frac{V}{\Omega^2} \right] \qquad (7.49)$$

where v and i have been assumed to satisfy the passive sign convention.

Let's calculate the power dissipated by the resistor, when the passive sign convention is not satisfied. In this case, the power dissipated by the resistor is

$$p = -vi \qquad (7.50)$$

Now according to the Ohm's law

$$v = -Ri \qquad (7.51)$$

Substituting Eq. (7.51) into Eq. (7.50) produces

$$p = -vi = -(-Ri)i = Ri^2 \qquad (7.52)$$

Alternatively, substituting

$$i = -\frac{v}{R} \qquad (7.53)$$

into Eq. (7.50) results in

$$p = -vi = -v\left(-\frac{v}{R} \right) = \frac{v^2}{R} \qquad (7.54)$$

Results (7.52) and (7.54) agree with the result (7.49), and confirm the fact that the resistors always dissipate power.

Using the definition of conductance, the power dissipated by the resistor can alternatively be expressed as

$$p = vi = \frac{i^2}{G} = Gv^2 \qquad (7.55)$$

Open and Short Circuit The value of a resistance R can range from zero to infinity. Two important extreme cases arise when the resistance is zero or infinite.

When the resistance is zero, the resulting circuit, shown in Figure 7.6, is called a *short circuit*.

According to Ohm's law

$$v = Ri = 0i = 0 \qquad (7.56)$$

showing that the *voltage across a short circuit is zero*.

When the resistance of a resistor is infinite, the resulting circuit, shown in Figure 7.7, is called an *open circuit*.

According to Ohm's law

$$i = \frac{v}{R} = \frac{v}{\infty} = 0 \qquad (7.57)$$

showing that a *current through an open circuit is zero*.

Ideal Switch The ideal switch can be modeled as a combination of an open- and short-circuit elements. Figure 7.8 shows the circuit symbol of an ideal switch.

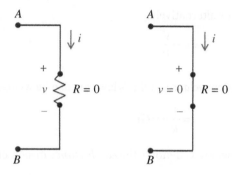

Figure 7.6 Short circuit.

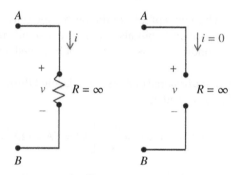

Figure 7.7 Open circuit.

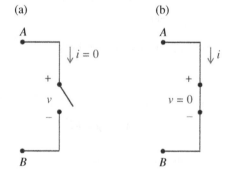

Figure 7.8 Ideal switch: (a) open (b) closed.

7.3.2 Inductors and Capacitors

Inductor The circuit symbol of an inductor and voltage–current designations in the time domain are shown in Figure 7.9.

When the passive sign convention is satisfied, the voltage–current relationships are

$$v(t) = L\frac{di(t)}{dt} \qquad (7.58a)$$

$$i(t) = \frac{1}{L}\int_{-\infty}^{t} v(t)dt = i(0) + \frac{1}{L}\int_{0}^{t} v(t)dt \qquad (7.58b)$$

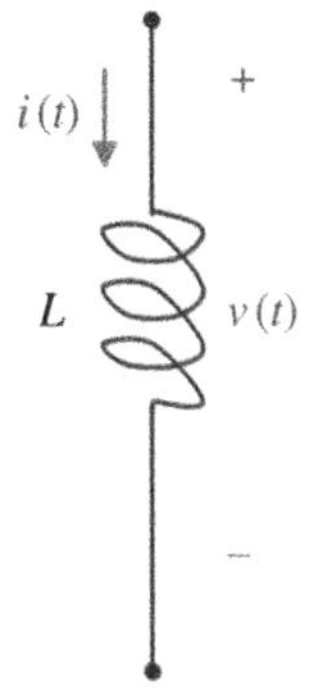

The unit of inductance is henry (H). Eq. (7.58a) can be used to determine its equivalence.

$$V = H\frac{A}{s} \quad \Rightarrow \quad H = \frac{Vs}{A} \qquad (7.59)$$

In many EMC problems, we are concerned about the parasitic inductances present in the circuit. These inductances are usually in the range of a few to a few tens of nH.

Energy stored in the magnetic field of an inductor can be calculated from

$$
\begin{aligned}
W_m &= \int_{-\infty}^{t} p(t)dt = \int_{-\infty}^{t} v(t)i(t)dt \\
&= \int_{-\infty}^{t} L\frac{di(t)}{dt}i(t)dt = L\int_{i(t-\infty)}^{t(t=t)} i(t)di \\
&= L\left[\frac{i^2(t)}{2}\right]_{i(-\infty)}^{i(t)} = \frac{1}{2}L\left[i^2(t) - 0\right]
\end{aligned}
\qquad (7.60)
$$

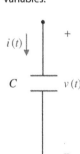

Figure 7.9 Inductor symbol and circuit variables.

or

$$W_m = \frac{1}{2}Li^2(t) \qquad (7.61)$$

Figure 7.10 Capacitor symbol and circuit variables.

Capacitors The circuit symbol of a capacitor and the voltage current designations in the time domain are shown in Figure 7.10.

When the passive sign convention is satisfied the voltage–current relationships are

$$i(t) = C\frac{dv(t)}{dt} \qquad (7.62a)$$

$$v(t) = \frac{1}{C}\int_{-\infty}^{t} i(t)dt = v(0) + \frac{1}{C}\int_{0}^{t} i(t)dt \qquad (7.62b)$$

The unit of capacitance is farad (F). Equation (7.62a) can be used to determine its equivalence.

$$A = F\frac{V}{s} \quad \Rightarrow \quad F = \frac{As}{V} \qquad (7.63)$$

In many EMC problems we are concerned about the parasitic capacitance present in the circuit. These capacitances are usually in the range of a few to a few tens of pF.

Energy stored in the electric field of a capacitor can be calculated from

$$
W_e = \int_{-\infty}^{t} p(t)\,dt = \int_{-\infty}^{t} v(t)i(t)\,dt
$$

$$
= \int_{-\infty}^{t} v(t)C\frac{dv(t)}{dt}\,dt = C \int_{v(t=-\infty)}^{v(t=t)} v(t)\,dv \tag{7.64}
$$

$$
= C\left[\frac{v^2(t)}{2}\right]_{v(-\infty)}^{v(t)} = \frac{1}{2}C\left[v^2(t) - 0\right]
$$

or

$$
W_e = \frac{1}{2}Cv^2(t) \tag{7.65}
$$

7.3.3 Phasor Relationships for Circuit Elements

Recall the time domain voltage–current relationships for resistors, inductors, and capacitors (when passive sign convention is satisfied):

$$
v_R(t) = Ri_R(t) \tag{7.66a}
$$

$$
v_L(t) = L\frac{di_L(t)}{dt} \tag{7.66b}
$$

$$
i_C(t) = C\frac{dv_C(t)}{dt} \tag{7.66c}
$$

We will transform the voltage–current relationship from the time domain to the phasor domain for each element. Since we are concerned with the sinusoidal steady state, all voltages and currents in the time domain are expressed as sinusoids.

Resistor   In the time domain, if the current through a resistor R is

$$
i = I_m \cos(\omega t + \theta) \tag{7.67}
$$

then, according to the Ohm's law, the voltage across it is

$$
v = Ri = RI_m \cos(\omega t + \theta) \tag{7.68}
$$

The phasor form of the current is

$$
\hat{I} = I_m \angle \theta = I_m e^{j\theta} \tag{7.69}
$$

while the phasor form of the voltage is

$$
\hat{V} = RI_m \angle \theta = RI_m e^{j\theta} \tag{7.70}
$$

Therefore,

$$
\hat{V} = R\hat{I} \tag{7.71}
$$

Figure 7.11 Resistor symbol and circuit variables in the (a) time domain, (b) phasor domain.

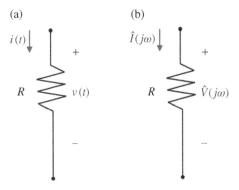

Thus, the voltage–current relationship for the resistor in the phasor domain continues to be Ohm's law, as in the time domain.

Figure 7.11 shows the resistor symbol and the circuit variables in both the time and phasor domains.

Inductor In the time domain, let the current through an inductor L be

$$i = I_m \cos(\omega t + \theta) \tag{7.72}$$

The phasor form of this current is

$$\hat{I} = I_m \angle \theta = I_m e^{j\theta} \tag{7.73}$$

The voltage across the inductor, in the time domain, is

$$v(t) = L\frac{di(t)}{dt} = L\frac{d}{dt}\left[I_m \cos(\omega t + \theta)\right]$$
$$= -\omega L I_m \sin(\omega t + \theta) = \omega L I_m \cos(\omega t + \theta + 90°) \tag{7.74}$$

and the phasor form of this voltage is

$$\hat{V} = \omega L I_m e^{j(\theta+90°)} = \omega L I_m e^{j\theta} e^{j90°} = j\omega L I_m e^{j\varphi} = j\omega L I_m \angle \theta \tag{7.75}$$

Since

$$e^{j90°} = \cos 90° + j\sin 90° = 0 + j1 = j \tag{7.76}$$

we have

$$\hat{V} = j\omega L I_m e^{j\varphi} = j\omega L I_m \angle \theta \tag{7.77}$$

Therefore, utilizing Eq. (7.73), we arrive at the voltage–current relationship for an inductor in the phasor domain as

$$\hat{V} = j\omega L \hat{I} \tag{7.78}$$

Figure 7.12 shows the inductor symbol and the circuit variables in both the time and phasor domains.

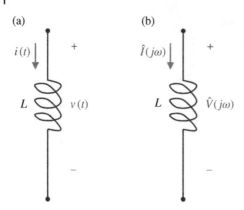

(a)   (b)

$i(t)$ $\hat{I}(j\omega)$

L $v(t)$ L $\hat{V}(j\omega)$

Figure 7.12 Inductor symbol and circuit variables in the (a) time domain, (b) phasor domain.

Capacitor   In the time domain, let the voltage across a capacitor C be

$$v = V_m \cos(\omega t + \theta)$$
(7.79)

The phasor form of this voltage is

$$\hat{V} = V_m \angle \theta = V_m e^{j\theta}$$
(7.80)

The current through the capacitor, in the time domain, is

$$i(t) = C\frac{dv(t)}{dt} = C\frac{d}{dt}\left[V_m \cos(\omega t + \theta)\right]$$
$$= -\omega C V_m \sin(\omega t + \theta) = \omega C V_m \cos(\omega t + \theta + 90°)$$
(7.81)

and the phasor form of this current is

$$\hat{I} = \omega C V_m e^{j(\theta + 90°)} = \omega C V_m e^{j\theta} e^{j90°}$$
$$= j\omega C V_m e^{j\theta} = j\omega C V_m \angle \theta$$
(7.82)

Therefore, utilizing Eq. (7.72), we arrive at the voltage–current relationship for a capacitor in the phasor domain as

$$\hat{I} = j\omega C \hat{V}$$
(7.83)

Figure 7.13 shows the capacitor symbol and circuit variables in both the time and phasor domains.

7.3.4 *s* Domain Relationships for Circuit Elements

Resistor   In the time domain, voltage and current are related by Ohm's law:

$$v(t) = Ri(t)$$
(7.84)

Taking the Laplace transform of both side of Eq. (7.84) we get

$$\mathcal{L}\{v(t)\} = \mathcal{L}\{Ri(t)\}$$
(7.85)

Figure 7.13 Capacitor symbol and circuit variables in the (a) time domain, (b) phasor domain.

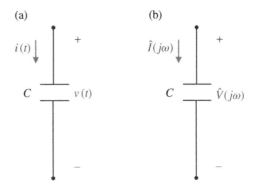

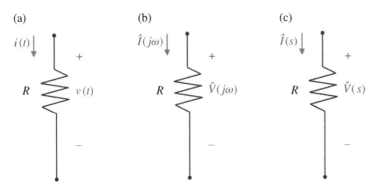

Figure 7.14 Resistor symbol and circuit variables in the (a) time domain, (b) phasor domain, (c) s domain.

or

$$\hat{V}(s) = R\hat{I}(s) \tag{7.86}$$

Thus, the voltage–current relationship for the resistor in the s domain continues to be Ohm's law, as in the time domain.

Figure 7.14 shows the resistor symbol and the circuit variables in the time domain, phasor domain, and s domain.

Inductor In the time domain, the current–voltage relationship is

$$v(t) = L\frac{di(t)}{dt} \tag{7.87}$$

Taking the Laplace transform of both side of Eq. (7.87) we get

$$\mathcal{L}\{v(t)\} = \mathcal{L}\left\{L\frac{di(t)}{dt}\right\} \tag{7.88}$$

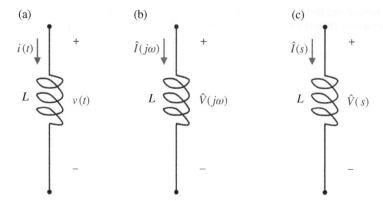

(a)   (b)   (c)

Figure 7.15 Inductor symbol and circuit variables in the (a) time domain, (b) phasor domain, (c) s domain.

When the initial inductor current is zero, we obtain

$$\hat{V}(s) = sL\hat{I}(s) \tag{7.89}$$

Figure 7.15 shows the inductor symbol and the circuit variables in the time domain, phasor domain, and s domain.

Capacitor   In the time domain, the current–voltage relationship is

$$i(t) = C\frac{dv(t)}{dt} \tag{7.90}$$

Taking the Laplace transform of both side of Eq. (7.90) we get

$$\mathcal{L}\{i(t)\} = \mathcal{L}\left\{C\frac{dv(t)}{dt}\right\} \tag{7.91}$$

When the initial inductor current is zero, we obtain

$$\hat{I}(s) = sC\hat{V}(s) \tag{7.92}$$

Figure 7.16 shows the inductor symbol and the circuit variables in the time domain, phasor domain and s domain.

7.3.5 Impedance in Phasor Domain

In the previous section, we obtained the voltage–current relations for the three passive elements as

$$\hat{V} = R\hat{I} \tag{7.93a}$$

$$\hat{V} = j\omega L\hat{I} \tag{7.93b}$$

$$\hat{I} = j\omega C\hat{V} \tag{7.93c}$$

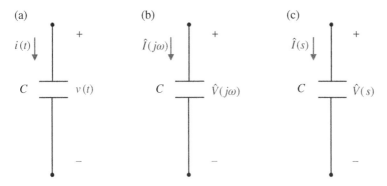

Figure 7.16 Capacitor symbol and circuit variables in the (a) time domain, (b) phasor domain, (c) s domain.

These equations may be written in terms of the *ratio of the phasor voltage to the phasor current* as

$$\frac{\hat{V}}{\hat{I}} = R \tag{7.94a}$$

$$\frac{\hat{V}}{\hat{I}} = j\omega L \tag{7.94b}$$

$$\frac{\hat{V}}{\hat{I}} = \frac{1}{j\omega C} \tag{7.94c}$$

From these three expressions, we obtain *Ohm's law in phasor form* for any type of element as

$$\frac{\hat{V}}{\hat{I}} = \hat{Z} \tag{7.95a}$$

or

$$\hat{V} = \hat{Z}\hat{I} \tag{7.95b}$$

$\hat{Z}$ is a frequency-dependent quantity known as impedance, measured in ohms.

The *impedance $\hat{Z}$ of a circuit element is the ratio of the phasor voltage $\hat{V}$, across it, to the phasor current $\hat{I}$, through it.*

The impedance represents the frequency-dependent opposition to the sinusoidal current flow.

Although the impedance is the ratio of two phasors, it is not a phasor, because it does not have a corresponding sinusoid in the time domain.

The impedances of a resistor, inductor, and capacitor are, respectively,

$$\hat{Z}_R = R \tag{7.96a}$$

$$\hat{Z}_L = j\omega L \tag{7.96b}$$

$$\hat{Z}_C = \frac{1}{j\omega C} = -\frac{j}{\omega C} \tag{7.96c}$$

Let's consider two extreme cases of frequency: dc circuits ($\omega = 0$) and high frequency circuits ($\omega \to \infty$).

Under dc conditions

$$\hat{Z}_L = 0 \tag{7.97a}$$

$$\hat{Z}_C = \infty \tag{7.97b}$$

confirming that, at dc, the inductor acts like a short circuit, while the capacitor acts like an open circuit.

At very high frequencies

$$\hat{Z}_L = \infty \tag{7.98a}$$

$$\hat{Z}_C = 0 \tag{7.98b}$$

verifying that, at very high frequencies, the inductor acts like an open circuit, while the capacitor acts like a short circuit.

Since the impedance is a complex quantity, it can be expressed as

$$\hat{Z} = R + jX \tag{7.99}$$

where $R = \mathrm{Re}\{\hat{Z}\}$ is the *resistance*, and $X = \mathrm{Im}\{\hat{Z}\}$ is the *reactance*. The impedance, resistance, and reactance are all measured in ohms.

For passive circuits, resistance R is always positive. The reactance X may be positive or negative. When the reactance X is positive, we say that the reactance is inductive. When the reactance X is negative, we say that the reactance is capacitive.

In some applications, it is convenient to work with the reciprocal of impedance, known as *admittance*.

$$\hat{Y} = \frac{1}{\hat{Z}} \tag{7.100}$$

The admittances of a resistor, inductor, and capacitor are, respectively,

$$\hat{Y}_R = \frac{1}{R} = G \tag{7.101a}$$

$$\hat{Y}_L = \frac{1}{j\omega L} \tag{7.101b}$$

$$\hat{Y}_C = j\omega C \tag{7.101c}$$

Since the admittance $\hat{Y}$ is a complex quantity, it can be expressed as

$$\hat{Y} = G + jB \tag{7.102}$$

where $G = \mathrm{Re}\{\hat{Y}\}$ is the *conductance*, and $B = \mathrm{Im}\{\hat{Y}\}$ is the *susceptance*. The admittance, conductance, and susceptance are all measured in siemens.

Often, when drawing circuit elements in phasor domain, we replace the component values with the corresponding impedance, as shown in Figure 7.17.

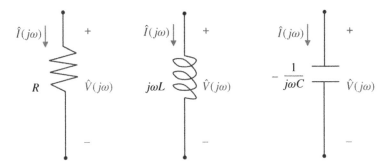

Figure 7.17 Circuit elements and their impedances.

7.3.6 Impedance in the s Domain

Recall the voltage–current relationships for the circuit elements under the zero initial conditions assumption:

$$\hat{V}(s) = R\hat{I}(s) \tag{7.103a}$$

$$\hat{V}(s) = sL\,\hat{I}(s) \tag{7.103b}$$

$$\hat{I}(s) = sC\hat{V}(s) \tag{7.103c}$$

These relationships may be written in terms of the ratio of *the Laplace transform of the voltage across the element* to *the Laplace transform of the current through the element*, as

$$\frac{\hat{V}(s)}{\hat{I}(s)} = R \tag{7.104a}$$

$$\frac{\hat{V}(s)}{\hat{I}(s)} = sL \tag{7.104b}$$

$$\frac{\hat{V}(s)}{\hat{I}(s)} = \frac{1}{sC} \tag{7.104c}$$

From these three expressions, we obtain *Ohm's law in the s domain* for any type of element as

$$\frac{\hat{V}(s)}{\hat{I}(s)} = \hat{Z} \tag{7.105a}$$

or

$$\hat{V}(s) = \hat{Z}\hat{I}(s) \tag{7.105b}$$

Where $\hat{Z}$ is a complex quantity known as impedance, measured in ohms.

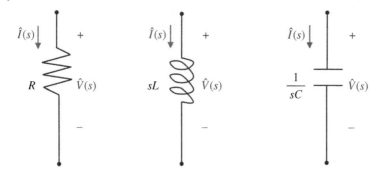

Figure 7.18 Circuit elements and their impedances.

The *impedance $\hat{Z}(s)$ of a circuit element in the s domain, is the ratio of the Laplace transform of the voltage $\hat{V}(s)$, across it, to the Laplace transform of the current $\hat{I}(s)$, through it.*

The impedances of a resistor, inductor, and capacitor are, respectively,

$$\hat{Z}_R = R \qquad\qquad (7.106a)$$

$$\hat{Z}_L = sL \qquad\qquad (7.106b)$$

$$\hat{Z}_C = \frac{1}{sC} \qquad\qquad (7.106c)$$

Often, when drawing circuit elements in the s domain, we replace the component values with the corresponding impedance, as shown in Figure 7.18.

7.3.7 Kirchhoff's Laws in the Time Domain

There are two fundamental laws governing electric circuit behavior: *Kirchhoff's current law (KCL)* and *Kirchhoff's voltage law (KVL)*.

Kirchhoff's laws can be regarded as the *connection constraints*, since they impose constraints the voltages and currents when different elements are connected to form an electric circuit.

Kirchhoff's laws, when coupled with Ohm's law, provide us with a set of tools for systematic analysis of a large variety of electric circuits. Virtually, all circuit laws to be discussed in the following chapters are based or can be derived from Kirchhoff's laws and the element constraints.

Kirchhoff's Current Law (KCL) *Kirchhoff's current law (KCL)* is based on the law of conservation of charge and is a consequence of the fact that charge cannot accumulate or be depleted at a node.

KCL states that the sum of the currents entering a node must be equal to the sum of the currents leaving the node.

To illustrate KCL, let's consider the node shown in Figure 7.19.

Currents i_3 and i_4 enter the node, while the currents i_1, i_2, and i_5 leave the node. Thus

$$i_3 + i_4 = i_1 + i_2 + i_5 \qquad\qquad (7.107a)$$

Note that KCL in Eq. (7.107) can also be written as

$$-i_1 - i_2 - i_5 + i_3 + i_4 = 0 \qquad (7.107b)$$

In general, KCL can stated as the algebraic sum

$$\sum_{k=1}^{N} i_k = 0 \qquad (7.108)$$

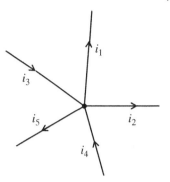

Figure 7.19 Illustration of the Kirchhoff's current law.

Where N is the number of branches (wires) connected to the node and i_k is the kth current entering or leaving the node.

The term *algebraic* implies the dependency on the current reference direction, that is, whether the current "enters" or "leaves" the node. We will adopt the following convention:

Current entering a node will have a "+1" multiplier preceding its value, whereas current leaving a node will have a "−1" multiplier preceding its value.

Note that Eq. (7.107b) conforms to this notation.

Example 7.2   KCL

Let the currents in Figure 7.19 have the values: $i_1 = -8A, i_2 = 3A, i_3 = 2A, i_5 = 4A$. Determine i_4.

Solution: Applying *KCL* at the node results in

$$2 + i_4 = -8 + 3 + 4$$

Thus $i_4 = -3A$.

∎

Example 7.3   Current sources in parallel

A simple application of *KCL* is combining current sources in parallel. For example, the current sources shown in Figure 7.20(a) can be combined as shown in Figure 7.20(b). Applying *KCL* at node A:

$$I_1 + I_3 = I_2 + I_T \quad \Rightarrow \quad I_T = I_1 - I_2 + I_3 \qquad (7.109)$$

When the current sources are connected in parallel, the equivalent current is the algebraic sum of the currents supplied by the individual sources.

∎

Note that a circuit cannot have two different current sources in series, unless their currents are equal, otherwise the law of conservation of charge is violated.

Kirchhoff's Voltage Law (KVL)   *Kirchhoff's voltage law* (*KVL*) states that the *algebraic* sum of all voltages around any loop in a circuit is identically zero at all times.

(a)

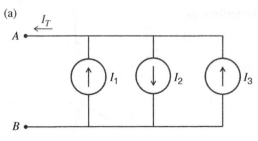

(b)

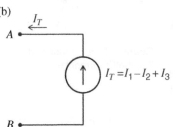

Figure 7.20 Combining current sources in parallel.

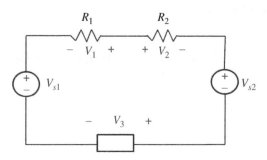

Figure 7.21 Illustration of the Kirchhoff's voltage law.

The term *algebraic* implies the dependency on the voltage polarity encountered as the closed path is traversed. We will adopt the following convention:

In traversing the element, when going from (+) to (−), we assign a "−" sign to the voltage, whereas, when going from (−) to (+), we assign a "+" sign to the voltage.

When traversing the loop, we can start with any element and go around the loop either clockwise or counterclockwise. To illustrate KVL, let us consider a circuit shown in Figure 7.21.

Suppose, we start with the voltage source v_{s1} and traverse the loop in the clockwise direction. Applying KVL yields

$$-v_{s1} - v_1 + v_2 + v_{s2} + v_3 = 0 \qquad (7.110)$$

In traversing the element, when going from + to −, we refer to the voltage as the voltage drop, whereas when going from − to +, we talk about a voltage rise.

In general, KVL can be expressed as the algebraic sum

$$\sum_{k=1}^{N} v_k = 0 \qquad (7.111)$$

where N is the number of voltages in the loop and v_k is the kth voltage.

Figure 7.22 Combining voltage sources in series.

(a)

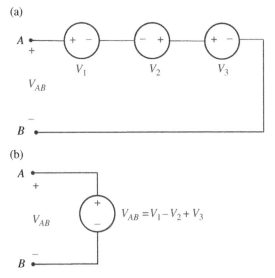

(b)

$V_{AB} = V_1 - V_2 + V_3$

Example 7.4 Voltage sources in series
A simple application of KVL is combining voltage sources in series. For example, the voltage sources shown in Figure 7.22(a) can be combined as shown in Figure 7.22(b).
The equivalent voltage source in Figure 7.22(b) is obtained by applying KVL:

$$-V_{AB} + V_1 - V_2 + V_3 = 0 \qquad (7.112a)$$

or equivalently

$$V_{AB} = V_1 - V_2 + V_3 \qquad (7.112b)$$

When the voltage sources are in series, the equivalent voltage is the algebraic sum of the voltages of the individual sources.

■

Note: A circuit cannot contain two different voltage sources in parallel, unless they are equal, otherwise the law of conservation of energy is violated.

7.3.8 Kirchhoff's Laws in the Phasor Domain

Kirchhoff's Current Law Let the KCL in the time domain be stated as

$$\sum_{k=1}^{N} i_k = 0 \qquad (7.113)$$

or, equivalently

$$i_1(t) + i_2(t) + \cdots + i_N(t) = 0 \qquad (7.114)$$

KCL holds for any time function of currents. Let each current in Eq. (7.113) be sinusoidal of the form

$$i_k(t) = I_{mk} \cos(\omega t + \theta_k) \qquad (7.115)$$

Then the corresponding phasor is

$$\hat{I}_k = I_{mk}\angle\theta_k = I_{mk}e^{j\theta_k} \tag{7.116}$$

It follows that Eq. (7.114) can be written in the phasor form as

$$\hat{I}_1 + \hat{I}_2 + \cdots \hat{I}_N = 0 \tag{7.117}$$

Eq. (7.117) represents KCL in the phasor domain.

Kirchhoff's Voltage Law Let the KVL in the time domain be stated as

$$\sum_{k=1}^{N} v_k(t) = 0 \tag{7.118}$$

or, equivalently

$$v_1(t) + v_2(t) + \cdots + v_N(t) = 0 \tag{7.119}$$

Let each voltage in Eq. (7.119) be sinusoidal of the form

$$v_k(t) = V_{mk}\cos(\omega t + \theta_k) \tag{7.120}$$

and each corresponding phasor be

$$\hat{V}_k = V_{mk}\angle\theta_k = V_{mk}e^{j\theta_k} \tag{7.121}$$

Then, the KVL in phasor domain holds, and has the form

$$\hat{V}_1 + \hat{V}_2 + \cdots + \hat{V}_N = 0 \tag{7.122}$$

7.3.9 Kirchhoff's Laws in the *s* Domain

Kirchhoff's Current Law Let the KCL in the time domain be stated as

$$i_1(t) + i_2(t) + \cdots + i_N(t) = 0 \tag{7.123}$$

Taking Laplace transform of Eq. (7.123) gives

$$\mathcal{L}\{i_1(t) + i_2(t) + \cdots + i_N(t)\} = \mathcal{L}\{0\} \tag{7.124}$$

By the linearity property, we have

$$\mathcal{L}\{i_1(t)\} + \mathcal{L}\{i_2(t)\} + \cdots + \mathcal{L}\{i_N(t)\} = 0 \tag{7.125}$$

or

$$\hat{I}_1(s) + \hat{I}_2(s) + \cdots + \hat{I}_N(s) = 0 \tag{7.126}$$

Equation (7.126) represents KCL in the *s* domain.

Kirchhoff's Voltage Law   Let the KVL in the time domain be stated as

$$v_1(t) + v_2(t) + \cdots + v_N(t) = 0 \tag{7.127}$$

Taking Laplace transform of Eq. (7.127) gives

$$\mathcal{L}\{v_1(t) + v_2(t) + \cdots + v_N(t)\} = 0 \tag{7.128}$$

By the linearity property, we have

$$\mathcal{L}\{v_1(t)\} + \mathcal{L}\{v_2(t)\} + \cdots + \mathcal{L}\{v_N(t)\} = 0 \tag{7.129}$$

or

$$\hat{V}_1(s) + \hat{V}_2(s) + \cdots + \hat{V}_N(s) = 0 \tag{7.130}$$

Equation (7.130) represents the KVL in the *s* domain.

7.3.10   Resistors in Series and the Voltage Divider

The analysis of an electric circuit can often be made easier by replacing a part of the circuit with one that is equivalent but simpler. This leads us to the definition of the equivalence.

Equivalence of Two Circuits   *Two circuits are said to be equivalent, with respect to the same two nodes, if they have identical i–v characteristics at these nodes.*

Consider the circuits shown in Figure 7.23(a) and 7.23(b).

In order for these two circuits to be equivalent with respect to nodes A and B, the *i–v* characteristics with respect to these two nodes must be the same. Thus V_{AB} and I in circuit 7.23(a) must be equal to V_{AB} and I in circuit 7.23(b).

Resistors in Series   Resistors R_1, R_2, and R_3 in Figure 7.23(a) are connected in *series* because each two resistors *exclusively* share a single node; resistors R_1 and R_2 share node C, whereas resistors R_2 and R_3 share node D.

This is the topological definition of a connection in series. Two or more elements are connected in series when each two exclusively share one common node.

Figure 7.23 Illustration of the circuit equivalence.

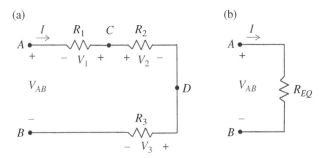

On the other hand, elements connected in series have the same current flowing through them. This is the circuit-variable definition of the connection in series.

Series Equivalent Resistance Looking at the two circuits in Figure 7.23 we may pose the following question:

Under what condition are the two circuits equivalent? That is, what is the relationship between R_1, R_2, R_3, and R_{EQ} which results in the voltage V_{AB} and the current I being the same in both circuits?

We will answer this question by applying basic circuit laws. Let's use the KVL (connection constraint) around the loop in the circuit shown in Figure 7.23(a):

$$-V_{AB} - V_1 + V_2 + V_3 = 0 \qquad (7.131)$$

Using Ohm's law for each resistor (element constraint) we get

$$V_1 = -R_1 I \qquad (7.132a)$$

$$V_2 = R_2 I \qquad (7.132b)$$

$$V_3 = R_3 I \qquad (7.132c)$$

Substituting Eq. (7.132) into Eq. (7.131) we get

$$-V_{AB} - (-R_1 I) + R_2 I + R_3 = 0 \qquad (7.133)$$

or

$$(R_1 + R_2 + R_3) I = V_{AB} \qquad (7.134)$$

On the other hand, writing Ohm's for the circuit shown in Figure 7.23(b) results in

$$R_{EQ} I = V_{AB} \qquad (7.135)$$

For the two circuits shown in Figure 7.23 to be equivalent, the following condition must be met:

$$R_{EQ} = R_1 + R_2 + R_3 \qquad (7.136)$$

In general, *the equivalent resistance R_{EQ} of a series of N resistors is*

$$R_{EQ} = R_1 + R_2 + \cdots + R_N = \sum_{k=1}^{N} R_k \qquad (7.137)$$

Voltage Divider Let's return to the circuit shown in Figure 7.23(a), redrawn as Figure 7.24.

From Eq. (7.126) we get

$$I = \frac{V_{AB}}{R_1 + R_2 + R_3} \qquad (7.138)$$

Substituting this result back into (7.132) produces

$$V_1 = -\frac{R_1}{R_1 + R_2 + R_3} V_{AB} \qquad (7.139a)$$

$$V_2 = \frac{R_2}{R_1 + R_2 + R_3} V_{AB} \qquad (7.139b)$$

$$V_3 = \frac{R_3}{R_1 + R_2 + R_3} V_{AB} \qquad (7.139c)$$

Thus the *magnitude* of the voltage appearing across each resistor connected in series is equal to the ratio of its resistance to the total resistance in the path formed, from the beginning of the string of the resistors (node A) to the end of the string of the resistors (node B) multiplied by the voltage between nodes A and B.

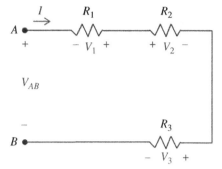

Figure 7.24 Illustration of the voltage divider.

In general, if we had N resistors connected in series, the voltage across the nth resistor would be

$$V_n = \pm \frac{R_n}{R_1 + R_2 + \cdots + R_N} V_{AB} \qquad (7.140)$$

This circuit demonstrates the principle of voltage division, and this rule is called a *voltage divider*.

The natural question arises: *when do we put a plus or a minus sign in this formula?*

We put the plus sign in the voltage divider formula when the following takes place:

When we traverse the loop moving through V_{AB}, we move through the minus to plus of its reference direction *and* when we move through a particular resistor, we encounter the plus of its reference direction first.

Any change to the reference directions results in a sign reversal. The following examples will test our understanding of the voltage divider.

Example 7.5   Voltage divider
Consider the circuit shown in Figure 7.25.
Use the voltage divider to express V_1 in terms of V_3.
Answer:

$$V_1 = \frac{R_1}{R_1 + R_5} V_3$$

∎

Example 7.6   Voltage divider
Consider the circuit shown in Figure 7.26.
Use the voltage divider to express V_7 in terms of V_3.
Answer:

$$V_7 = -\frac{R_7}{R_5 + R_7} V_3$$

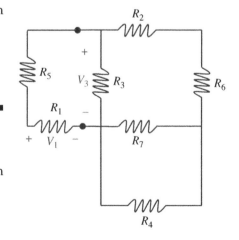

Figure 7.25 Circuit for Example 7.5.

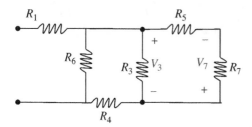

Figure 7.26 Circuit for Example 7.6.

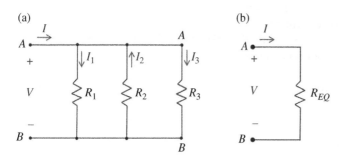

Figure 7.27 Resistors in parallel and circuit equivalence.

7.3.11 Resistors in Parallel and the Current Divider

Resistors in Parallel   Consider the circuit shown in Figure 7.27(a).

Resistors R_1, R_2, and R_3 in Figure 7.27(a) are connected in *parallel* because they all are connected between the same two nodes, A and B (any parts of the circuit connected with a wire, or a short circuit, electrically constitute the same node).

This is the topological definition of the connection in parallel. Two or more elements are connected in parallel when they are connected between the same two nodes.

On the other hand, if the elements are connected in parallel, then they have the same voltage across them. This is the circuit-variable definition of the connection in parallel.

Parallel Equivalent Resistance   Looking at the two circuits in Figure 7.27 we may pose the following question:

Under what condition are the two circuits equivalent? That is, what is the relationship between R_1, R_2, R_3 and R_{EQ} which results in the voltage V and the current I being the same in both circuits?

Let's use the KCL at node A in the circuit shown in Figure 7.27(a):

$$I + I_2 = I_1 + I_3 \tag{7.141}$$

Using Ohm's law for each resistor we get

$$I_1 = \frac{V}{R_1} \tag{7.142a}$$

$$I_2 = -\frac{V}{R_2} \tag{7.142b}$$

$$I_3 = \frac{V}{R_3} \tag{7.142c}$$

Substituting Eq. (7.142) into (7.141) we get

$$I + \frac{V}{R_1} = -\frac{V}{R_2} + \frac{V}{R_3} \tag{7.143}$$

or

$$\left(\frac{1}{R_1} + \frac{1}{R_2} + \frac{1}{R_3}\right)I = V \tag{7.144}$$

On the other hand, writing Ohm's for the circuit shown in Figure 7.27(b) results in

$$\frac{1}{R_{EQ}}I = V \tag{7.145}$$

For the two circuits shown in Figure 7.26 to be equivalent, the following condition must be met:

$$\frac{1}{R_{EQ}} = \frac{1}{R_1} + \frac{1}{R_2} + \frac{1}{R_3} \tag{7.146}$$

In general, *the equivalent resistance R_{EQ} of N resistors connected in parallel is*

$$\frac{1}{R_{EQ}} = \frac{1}{R_1} + \frac{1}{R_2} + \cdots + \frac{1}{R_N} = \sum_{k=1}^{N}\frac{1}{R_k} \tag{7.147}$$

In terms of conductance, Eq. (7.147) becomes

$$G_{EQ} = G_1 + G_2 + \cdots + G_N = \sum_{k=1}^{N}G_k \tag{7.148}$$

It is useful to derive the formula for two resistors connected in parallel, shown in Figure 7.28.

Figure 7.28 Equivalent resistance of two resistors in parallel.

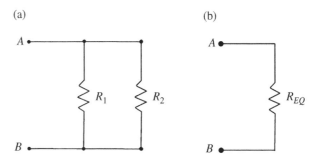

(a)   (b)

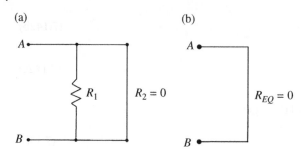

Figure 7.29 Equivalent resistance of a resistor in parallel with a short.

According to Eq. (7.147) we have

$$\frac{1}{R_{EQ}} = \frac{1}{R_1} + \frac{1}{R_2} \tag{7.149}$$

or

$$\frac{1}{R_{EQ}} = \frac{R_1 + R_2}{R_1 R_2} \tag{7.150}$$

thus

$$R_{EQ} = \frac{R_1 R_2}{R_1 + R_2} \tag{7.151}$$

We arrived at a very useful result: *the equivalent resistance of two resistors in parallel is equal to their product divided by their sum.*

Using this result we can easily determine the equivalent resistance of a resistor parallel to a short circuit ($R = 0$), shown in Figure 7.29.

According to Eq. (7.151) we have

$$R_{EQ} = \frac{R_1 R_2}{R_1 + R_2} = \frac{R_1 \times 0}{R_1 + 0} = 0 \tag{7.152}$$

Thus, the equivalent resistance of a resistor parallel to a short circuit is zero, the same as a short circuit itself. Therefore, when a resistor is bypassed by short circuit we can replace it by a short, as shown in Figure 7.29(b).

Current Divider   Let's return to the circuit shown in Figure 7.27(a), redrawn as Figure 7.30. Equation (7.144) repeated here

$$\left(\frac{1}{R_1} + \frac{1}{R_2} + \frac{1}{R_3} \right) I = V \tag{7.153}$$

leads to

$$V = \frac{I}{G_1 + G_2 + G_3} \tag{7.154}$$

Substituting this result back into Eq. (7.142) produces

$$I_1 = \frac{V}{R_1} = G_1 V = \frac{G_1}{G_1 + G_2 + G_3} I \qquad (7.155a)$$

$$I_2 = -\frac{V}{R_2} = -G_2 V = -\frac{G_2}{G_1 + G_2 + G_3} I \qquad (7.155b)$$

$$I_3 = \frac{V}{R_3} = G_3 V = \frac{G_3}{G_1 + G_2 + G_3} I \qquad (7.155c)$$

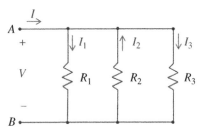

Figure 7.30 Illustration of the current divider.

Thus, the *magnitude* of the current flowing out of the node through each resistor connected in parallel is equal to the ratio of its conductance to the total conductance of the resistors in parallel multiplied by the current flowing into the node.

In general, if we had N resistors connected in parallel, the current through the nth resistor would be

$$I_k = \pm \frac{G_k}{G_1 + G_2 + \cdots + G_N} I \qquad (7.156)$$

This circuit demonstrates the principle of current division, and this rule is called a *current divider*.

The natural question arises: *when do we put a plus or a minus sign in this formula?*

We put the plus sign in the current divider formula when the following takes place:

The total current flows into the node and the individual current flows out of the node. Any change to the reference directions results in a sign reversal.

It is useful to derive the formula for the current divider when two resistors are connected in parallel, shown in Figure 7.31.

According to formula (7.156) we have

$$I_1 = \frac{G_1}{G_1 + G_2} I = \frac{\dfrac{1}{R_1}}{\dfrac{1}{R_1} + \dfrac{1}{R_2}} I \qquad (7.157)$$

or

$$I_1 = \frac{\dfrac{1}{R_1}}{\dfrac{R_1 + R_2}{R_1 R_2}} I \qquad (7.158)$$

resulting in

$$I_1 = \frac{R_2}{R_1 + R_2} I \qquad (7.159)$$

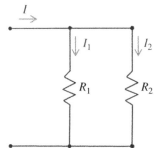

Figure 7.31 Current divider rule for two resistors in parallel.

Similarly,

$$I_2 = \frac{G_2}{G_1 + G_2} I = \frac{\dfrac{1}{R_2}}{\dfrac{1}{R_1} + \dfrac{1}{R_2}} I \tag{7.160}$$

or

$$I_2 = \frac{\dfrac{1}{R_2}}{\dfrac{R_1 + R_2}{R_1 R_2}} I \tag{7.161}$$

resulting in

$$I_2 = \frac{R_1}{R_1 + R_2} I \tag{7.162}$$

This is a very useful result: when we have two resistors in parallel, the current flowing in one path equals the resistance in the other path divided by the sum of both resistances in parallel, times the current flowing into the node.

The following examples will test our understanding of the current divider.

Example 7.7 Current divider
Consider the circuit shown in Figure 7.32.
Use the current divider to express I_7 in terms of I_2.

Solution:

$$I_7 = \frac{R_4}{R_4 + R_7} I_2$$

Example 7.8 Current divider
Consider the circuit shown in Figure 7.33.
Use the current divider to express I_7 in terms of I_2.

Solution:

$$I_7 = -\frac{R_5 + R_6}{R_5 + R_6 + R_7} I_2$$

7.3.12 Impedance Combinations and Divider Rules in Phasor Domain

Impedances in Series Consider the N impedances connected in series as shown in Figure 7.34.

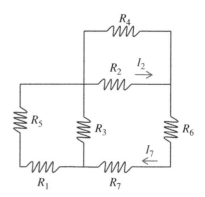

Figure 7.32 Circuit for Example 7.7.

Figure 7.33 Circuit for Example 7.8.

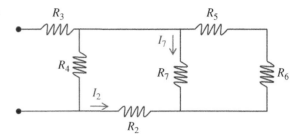

Figure 7.34 Impedances in series.

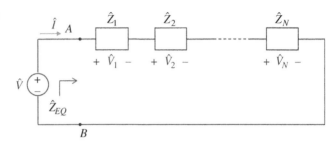

Applying KVL around the loop gives

$$\hat{V} = \hat{V}_1 + \hat{V}_2 + \cdots + \hat{V}_N \qquad (7.163)$$

Since the same current $\hat{I}$ flows through all impedances, we have

$$\hat{V} = \left(\hat{Z}_1 + \hat{Z}_2 + \cdots + \hat{Z}_N\right)\hat{I} \qquad (7.164)$$

The equivalent impedance at the input terminals is

$$\hat{Z}_{EQ} = \frac{\hat{V}}{\hat{I}} = \hat{Z}_1 + \hat{Z}_2 + \cdots + \hat{Z}_N \qquad (7.165)$$

leading to

$$\hat{Z}_{EQ} = \hat{Z}_1 + \hat{Z}_2 + \cdots + \hat{Z}_N \qquad (7.166)$$

Showing that *the equivalent impedance of series-connected impedances is the sum of the individual impedances.*

This relationship has the same mathematical form as that for the resistors in series.

Voltage Divider   Consider the circuit shown in Figure 7.35.

Combing the impedances in series and using Ohm's law we can express the current as

$$\hat{I} = \frac{\hat{V}}{\hat{Z}_1 + \hat{Z}_2} \qquad (7.167)$$

Now,

$$\hat{V}_1 = \hat{Z}_1\hat{I} \qquad (7.168a)$$

$$\hat{V}_2 = \hat{Z}_2\hat{I} \qquad (7.168b)$$

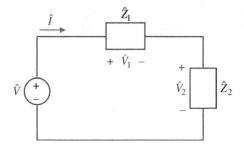

Figure 7.35 Voltage divider circuit.

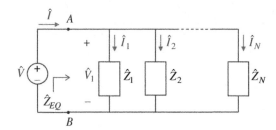

Figure 7.36 Impedances in parallel.

Using Eq. (7.167) in Eq. (7.168) produces the *voltage divider* relationships.

$$\hat{V}_1 = \frac{\hat{Z}_1}{\hat{Z}_1 + \hat{Z}_2} \hat{V} \tag{7.169a}$$

$$\hat{V}_2 = \frac{\hat{Z}_2}{\hat{Z}_1 + \hat{Z}_2} \hat{V} \tag{7.169b}$$

Impedances in Parallel   Consider the N impedances connected in parallel as shown in Figure 7.36.

Applying KCL at the upper node gives

$$\hat{I} = \hat{I}_1 + \hat{I}_2 + \cdots + \hat{I}_N \tag{7.170}$$

Since the voltage $\hat{V}$ across each impedance is the same, we have

$$\hat{I} = \left(\frac{1}{\hat{Z}_1} + \frac{1}{\hat{Z}_2} + \cdots + \frac{1}{\hat{Z}_N} \right) \hat{V} \tag{7.171}$$

The equivalent impedance at the input terminals is

$$\hat{I} = \frac{1}{\hat{Z}_{EQ}} \hat{V} \tag{7.172}$$

leading to

$$\frac{1}{\hat{Z}_{EQ}} = \frac{1}{\hat{Z}_1} + \frac{1}{\hat{Z}_2} + \cdots + \frac{1}{\hat{Z}_N} \tag{7.173}$$

Figure 7.37 Current divider circuit.

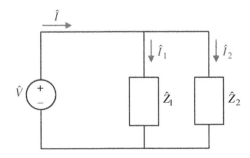

showing that *the reciprocal of the equivalent impedance of parallel-connected imped-*
ances is the sum of the reciprocals of the individual impedances.

This relationship has the same mathematical form as that for the resistors in parallel.
In terms of admittances, Eq. (7.173) can be written as

$$\hat{Y}_{EQ} = \hat{Y}_1 + \hat{Y}_2 + \cdots + \hat{Y}_N \tag{7.174}$$

Current Divider Consider the circuit shown in Figure 7.37.

The equivalent impedance of the two impedances in parallel is obtained as

$$\frac{1}{\hat{Z}_{EQ}} = \frac{1}{\hat{Z}_1} + \frac{1}{\hat{Z}_2} = \frac{\hat{Z}_1 + \hat{Z}_2}{\hat{Z}_1 \hat{Z}_2} \tag{7.175}$$

thus

$$\hat{Z}_{EQ} = \frac{\hat{Z}_1 \hat{Z}_2}{\hat{Z}_1 + \hat{Z}_2} \tag{7.176}$$

The voltage across each impedance is

$$\hat{V} = \hat{Z}_{EQ} \hat{I} = \frac{\hat{Z}_1 \hat{Z}_2}{\hat{Z}_1 + \hat{Z}_2} \tag{7.177}$$

Now,

$$\hat{I}_1 = \frac{\hat{V}}{\hat{Z}_1} \tag{7.178a}$$

$$\hat{I}_2 = \frac{\hat{V}}{\hat{Z}_2} \tag{7.178b}$$

Using Eq. (7.177) in Eq. (7.178) produces the *current divider* relationships.

$$\hat{I}_1 = \frac{\hat{Z}_2}{\hat{Z}_1 + \hat{Z}_2} \hat{I} \tag{7.179a}$$

$$\hat{I}_2 = \frac{\hat{Z}_1}{\hat{Z}_1 + \hat{Z}_2} \hat{I} \tag{7.179b}$$

Example 7.9 Impedance combinations

Determine the input impedance of the circuit shown in Figure 7.38. Assume that the circuit operates at $\omega = 1000\,\text{rad/s}$.

Solution: Combine the components in series, and redraw the circuit as shown in Figure 7.39.

Where

$$\hat{Z}_1 = \frac{1}{j\omega C_1} = -j\frac{1}{1000 \times 4 \times 10^{-6}} = -j250 \;\; [\Omega]$$

$$\hat{Z}_2 = \hat{Z}_{R_2} + \hat{Z}_{C_2} = R_2 + \frac{1}{j\omega C_2} = 5 - j\frac{1}{1000 \times 200 \times 10^{-6}} = 5 - j5 \;\; [\Omega]$$

$$\hat{Z}_3 = \hat{Z}_L + \hat{Z}_{R_1} = j\omega L + R_1 = j1000 \times 0.1 + 4 = 4 + j100 \;\; [\Omega]$$

The input impedance is thus

$$\hat{Z}_{in} = \hat{Z}_1 + \hat{Z}_2 \,\|\, \hat{Z}_3 = \hat{Z}_1 + \frac{\hat{Z}_2 \hat{Z}_3}{\hat{Z}_2 + \hat{Z}_3} = -j250 + \frac{(5 - j5)(4 + j100)}{5 - j5 + 4 + j100}$$

$$= -j250 + \frac{20 + j500 - j20 + 500}{9 + j95} = -j250 + \frac{520 + j480}{9 + j95}$$

$$= -j250 + \frac{520 + j480}{9 + j95} = 5.5216 - j254.95$$

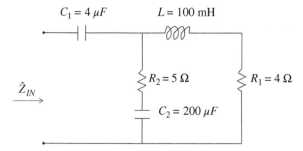

$C_1 = 4\,\mu F$ $L = 100\,\text{mH}$

$R_2 = 5\,\Omega$ $R_1 = 4\,\Omega$

$\hat{Z}_{IN}$

$C_2 = 200\,\mu F$

Figure 7.38 Circuit for Example 7.8.

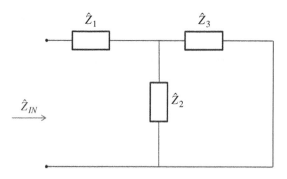

$\hat{Z}_1$ $\hat{Z}_3$

$\hat{Z}_2$

$\hat{Z}_{IN}$

Figure 7.39 Circuit for Example 7.9.

Example 7.10 Voltage divider

Consider the circuit shown in Figure 7.40. Let $V = 100V$.

Use the voltage divider to determine the voltage across the resistor.

Solution:

$$\hat{V}_R = \frac{3-j}{3-j+j5}(100) = \frac{3-j}{3+j4}(100) = \frac{(3-j)(3-j4)}{(3+j4)(3-j4)}(100)$$

$$= \frac{9-j12-j3-4}{9+16}(100) = \frac{5-j15}{25}(100) = 20-j60$$

Example 7.11 Current divider

Consider the circuit shown in Figure 7.41. Let

$$\hat{I} = 0.12 + j0.16A$$

Use current divider to determine the capacitor current

Solution:

$$\hat{I}_C = \frac{\hat{Z}_R}{\hat{Z}_R + \hat{Z}_C}\hat{I} = \frac{3000}{3000-j1000}(0.12 + j0.16) = 0.06 + j0.18 = 0.19\angle71.6° \text{ A}$$

Figure 7.40 Circuit for Example 7.10.

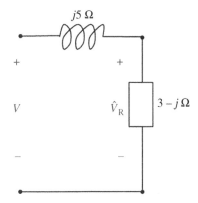

Figure 7.41 Circuit for Example 7.10.

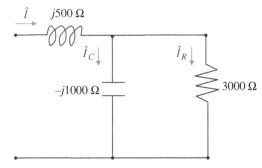

Example 7.12 s domain analysis

The initial conditions are zero in the circuit shown in Figure 7.42. Determine the equivalent impedance between nodes A and B. Use the voltage divider to determine $V_C(s)$. Use the current divider to determine $I_R(s)$.

Solution: Transform the circuit to the s domain (Figure 7.43).
The impedance of the parallel configuration of R and C is

$$R \| C = \frac{R \dfrac{1}{sC}}{R + \dfrac{1}{sC}} = \frac{R}{sRC + 1}$$

Thus, the input impedance is

$$Z_{in} = sL + \frac{R}{sRC + 1}$$

Using the voltage divider we get

$$V_C(s) = \frac{\dfrac{R}{sRC + 1}}{sL + \dfrac{R}{sRC + 1}} V(s)$$

Using the current divider we get

$$I_R(s) = \frac{\dfrac{1}{sC}}{R + \dfrac{1}{sC}} I(s)$$

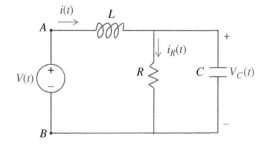

i(t) L

A

$i_R(t)$

V(t) R C $V_C(t)$

B

Figure 7.42 Circuit for Example 7.12.

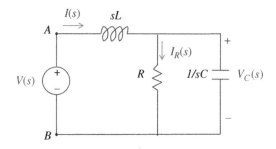

I(s) sL

A

$I_R(s)$

V(s) R 1/sC $V_C(s)$

B

Figure 7.43 s-domain circuit for Example 7.12.

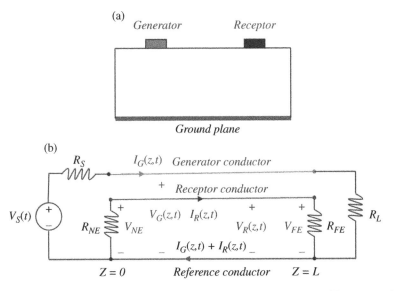

Figure 7.44 Three-conductor transmission line: (a) PCB arrangement; (b) circuit model.

7.4 EMC Applications

In this section we will present two EMC examples illustrating the applicability of the material covered in this chapter.

The first example will illustrate the use of a voltage and current divider as applied to a time-domain circuit model of crosstalk between PCB traces. The second example will show the applicability of the s domain analysis to describing the reflections on a transmission line terminated by a capacitive load.

7.4.1 Crosstalk between PCB Traces

Recall the crosstalk circuit model described in Section 1.10.1, and shown in Figure 7.44 (Adamczyk and Teune, 2009)

The current on the generator line, I_G, creates a magnetic field that results in a magnetic flux ψ_G crossing the loop of the receptor circuit, as shown in Figure 7.45(a).

If this flux is time varying, then according to Faraday's law, it induces a voltage V_R in the receptor circuit. The circuit model of this field phenomenon is represented by a mutual inductance and is shown in Figure 7.45(b). (We will describe this model in detail in Chapter 17.)

Using the current divider, we obtain the induced near- and far-end voltages as

$$V_{NE}(t) = \frac{R_{NE}}{R_{NE} + R_{FE}} L_m \frac{dI_G}{dt} \qquad (7.180a)$$

$$V_{FE}(t) = -\frac{R_{FE}}{R_{NE} + R_{FE}} L_m \frac{dI_G}{dt} \qquad (7.180b)$$

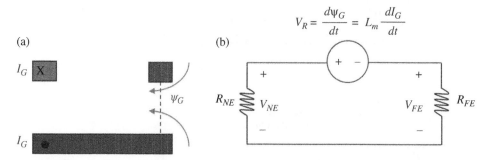

Figure 7.45 Inductive coupling between the circuits: (a) field model, (b) circuit model.

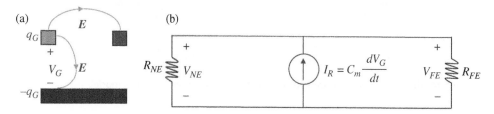

Figure 7.46 Capacitive coupling between the circuits: (a) field model, (b) circuit model.

Similarly, the voltage between the two conductors of the generator circuit, V_G, has associated with it a charge separation that creates the electric field lines, some of which terminate on the conductors of the receptor circuit as shown in Figure 7.46(a).

If this charge (voltage) varies with time, it induces a current in the receptor circuit. The circuit model of this field phenomenon is represented by a mutual capacitance and is shown in Figure 7.46(b).

Using the voltage divider, we obtain the induced near- and far-end voltages as

$$V_{NE}(t) = \frac{R_{NE}R_{FE}}{R_{NE} + R_{FE}} C_m \frac{dV_G}{dt} \tag{7.181a}$$

$$V_{FE}(t) = \frac{R_{FE}R_{NE}}{R_{NE} + R_{FE}} C_m \frac{dV_G}{dt} \tag{7.181b}$$

7.4.2 Capacitive Termination of a Transmission Line

In this section we will show the application of the s domain analysis to the transmission line terminated by a capacitive load. (We will discuss transmission lines in detail in Chapter 17.)

Consider the circuit shown in Figure 7.47.

A line of length d is terminated by a capacitor C with zero initial voltage. A constant voltage source with internal resistance equal to the characteristic impedance Z_C of the line is connected to the line at $t = 0$.

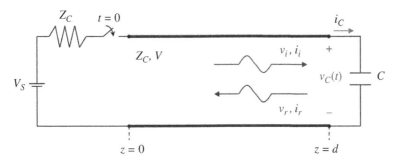

Figure 7.47 Transmission line terminated by a capacitive load.

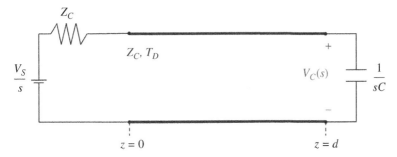

Figure 7.48 *s* domain circuit model.

As we will learn in Part III, the moment the switch closes at $t = 0$, the voltage and current waves (v_i and i_i) originate at $z = 0$ and travel down the line to reach the load end at time T.

Upon arriving at the load the reflected voltage and current waves (v_r and i_r) are created.

In order to determine the reflected waves, the circuit is transformed to the s domain, as shown in Figure 7.48.

The time-domain voltage at the capacitive load is given by (Paul, 2006, p. 235)

$$v_C(t) = (1 + \Gamma_L)\frac{1}{2}V_0u(t - T_D)$$ (7.182)

Taking Laplace transform of Eq. (7.182) we obtain

$$V_C(s) = (1 + \Gamma_L(s))\frac{1}{2}V_S(s)e^{-sT_D}$$ (7.183)

Where Γ_L is the load reflection coefficient given by

$$\Gamma_L = \frac{Z_L - Z_C}{Z_L + Z_C} = \frac{\dfrac{1}{sC} - Z_C}{\dfrac{1}{sC} + Z_C}$$ (7.184)

$$= \frac{1 - sZ_CC}{1 + sZ_CC} = \frac{1 - sT_C}{1 + sT_C} = \Gamma_L(s)$$

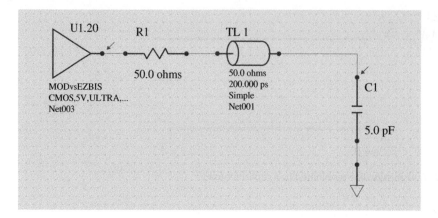

Figure 7.49 HyperLynx circuit model of a transmission line terminated by a capacitive load.

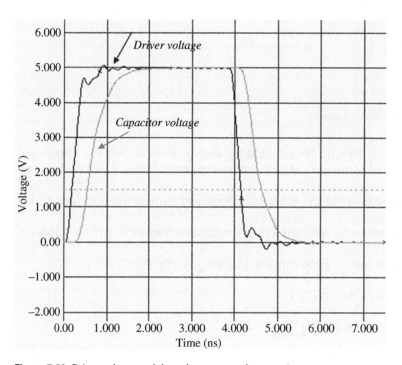

Figure 7.50 Driver voltage and the voltage across the capacitor.

Substituting Eq. (7.184) into Eq. (7.183) gives

$$V_C(s) = \left(1 + \Gamma_L(s)\right)\frac{1}{2}V_S(s)e^{-sT_D} = \left(1 + \frac{1 - sT_C}{1 + sT_C}\right)\frac{1}{2}\frac{V_0}{s}e^{-sT_D} \qquad (7.185)$$

Rearranging and using partial fraction expansion (see Section 7.2.3) we get

$$
V_C(s) = \left(1 + \frac{1 - sT_C}{1 + sT_C}\right) \frac{1}{2} \frac{V_0}{s} e^{-sT_D} = \frac{1 + sT_C + 1 - sT_C}{1 + sT_C} \frac{1}{2} \frac{V_0}{s} e^{-sT_D}
$$

$$
= \frac{1}{1 + sT_C} \frac{V_0}{s} e^{-sT_D} = \frac{\dfrac{1}{T_C}}{s + \dfrac{1}{T_C}} \frac{V_0}{s} e^{-sT_D} = \left[\frac{1}{s} - \frac{1}{s + \dfrac{1}{T_C}}\right] V_0 e^{-sT_D}
$$

(7.186)

Using inverse Laplace transform yields (Paul, 2006, p. 238)

$$
v_C(t) = V_0 u(t - T_D) - e^{-\frac{t - T_D}{T_C}} V_0 u(t - T_D)
$$

(7.187)

Figure 7.49 shows a circuit schematic of a transmission line driven by a 5 V CMOS and terminated in a capacitive load.

The driver voltage and the voltage across the capacitor are displayed in Figure 7.50.

References

Adamczyk, B. and Teune, J., "EMC Crosstalk vs. Circuit Topology", ASEE North Central Section Spring Conference, Grand Rapids, MI, 2009.

Alexander, C.K. and Sadiku, N.O., *Fundamentals of Electric Circuits*, 4th ed., McGraw Hill, New York, 2009.

Nilsson, J.W. and Riedel, S.A., *Electric Circuits*, 10th ed., Pearson, Upper Saddle River, NJ, 2015.

Paul, C.R., *Introduction to Electromagnetic Compatibility*, 2nd ed., John Wiley and Sons, New York, 2006.

8

Systematic Methods of Circuit Analysis

8.1 Node Voltage Analysis

8.1.1 Node Analysis for the Resistive Circuits

In Chapter 7 we discussed the basic circuit laws that impose constraints on voltages and currents in the circuit. To be more precise, the voltages and currents we referred to were the *element* voltages and *element* currents.

Using KCL, KVL, and Ohm's law to solve for the element voltages and currents can be quite cumbersome, except for the very simple circuits.

In this section we present a systematic method of circuit analysis in which node voltages are the circuit variables to be found. As we shall see, choosing node voltages instead of the element voltages as circuit variables allows us to develop a systematic method of circuit analysis that is applicable to more complex circuits.

We will use the circuit shown in Figure 8.1 to define node voltages and explain the node voltage analysis method.

To define a set of node voltages we first select one node in the network to be a reference node – node D, shown in Figure 8.2. By definition, the node voltage at the reference node is equal to zero (Nilsson and Riedel, 2015, p. 93).

$$V_D = 0 \tag{8.1}$$

The node voltages are then defined as the voltages between the remaining nodes and the selected reference node; voltages V_A, V_B, and V_C. By default, the node voltage polarities are such that the $(-)$ is always at the reference node.

Before proceeding with the node voltage analysis method, let's establish the relationship between the node voltages and the element voltage and element current.

Writing KVL for the loop containing resistors R_1, R_2, and R_3, we get

$$-V_A + V_2 + V_B = 0 \tag{8.2a}$$

or

$$V_2 = V_A - V_B \tag{8.2b}$$

Foundations of Electromagnetic Compatibility with Practical Applications, First Edition. Bogdan Adamczyk.
© 2017 John Wiley & Sons Ltd. Published 2017 by John Wiley & Sons Ltd.

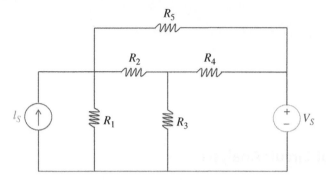

Figure 8.1 Node voltage analysis circuit.

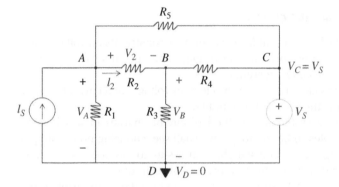

Figure 8.2 Node voltage assignments.

This example illustrates the following fundamental relationship between the node voltages and the element voltage:

If the two-terminal element is connected between two non-reference nodes, then the element voltage is equal to the difference of the two node voltages; we take the node voltage on the (+) side of the element voltage reference direction and subtract from it the node voltage on the (−) side.

If the two-terminal element is connected to the reference node, then the element voltage is equal to the node voltage (provided that the reference directions are the same).

According to the Ohm's law we have

$$I_2 = \frac{V_2}{R_2} \tag{8.3}$$

Now, utilizing Eq. (8.2) in Eq. (8.1) we obtain the relationship between the node voltages and the resistor current.

$$I_2 = \frac{V_A - V_B}{R_2} \tag{8.4}$$

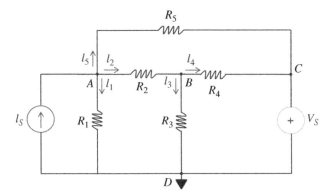

Figure 8.3 Current assignments.

Thus, *once the reference direction of the current is assigned, we take the node voltage at the node from which the current flows and subtract from it the node voltage at the node towards which the current flows; then we divide the result by the resistance value.*

We are now ready to proceed with the node voltage analysis method.

After having chosen the reference, the next step is to write KCL *at the nodes, where the node voltage is unknown.* Since at node C we have $V_C = V_S$, there is no need to write KCL at that node.

To write KCL at nodes A and B, we assign currents and their reference directions; one such assignment is shown in Figure 8.3

KCL at node A produces

$$I_S = I_1 + I_2 + I_5 \tag{8.5}$$

In terms of node voltages, Eq. (8.5) can be written as

$$I_S = \frac{V_A - V_D}{R_1} + \frac{V_A - V_B}{R_2} + \frac{V_A - V_C}{R_5} \tag{8.6}$$

Now, $V_D = 0$ and $V_C = V_S$ and Eq. (8.6) becomes

$$I_S = \frac{V_A}{R_1} + \frac{V_A - V_B}{R_2} + \frac{V_A - V_S}{R_5} \tag{8.7}$$

or in terms of conductances

$$I_S = G_1 V_A + G_2 (V_A - V_B) + G_5 (V_A - V_S) \tag{8.8}$$

Similarly, KCL at node B produces

$$I_2 = I_3 + I_4 \tag{8.9}$$

In terms of node voltages, Eq. (8.8) can be written as

$$\frac{V_A - V_B}{R_2} = \frac{V_B}{R_3} + \frac{V_B - V_S}{R_4} \tag{8.10}$$

or in terms of conductances

$$G_2(V_A - V_B) = G_3 V_B + G_4(V_B - V_S) \tag{8.11}$$

Rearranging Eqns. (8.7) and (8.10), we can rewrite them as

$$(G_1 + G_2 + G_5)V_A - G_2 V_B = I_S + G_5 V_S \tag{8.12a}$$

$$-G_2 V_A + (G_2 + G_3 + G_4)V_B = G_4 V_S \tag{8.12b}$$

Which in a matrix form become

$$\begin{bmatrix} G_1 + G_2 + G_5 & -G_2 \\ -G_2 & G_2 + G_3 + G_4 \end{bmatrix} \begin{bmatrix} V_A \\ V_B \end{bmatrix} = \begin{bmatrix} I_S + G_5 V_S \\ G_4 V_S \end{bmatrix} \tag{8.13}$$

We will obtain the solution using the Cramer's rule (see Section 1.8). The determinant of the conductance matrix is:

$$\Delta = \begin{vmatrix} G_1 + G_2 + G_5 & -G_2 \\ -G_2 & G_2 + G_3 + G_4 \end{vmatrix}$$
$$= (G_1 + G_2 + G_5)(G_2 + G_3 + G_4) - G_2^2 \tag{8.14}$$

and

$$\Delta_1 = \begin{vmatrix} I_S + G_5 V_S & -G_2 \\ G_4 V_S & G_2 + G_3 + G_4 \end{vmatrix}$$
$$= (I_S + G_5 V_S)(G_2 + G_3 + G_4) + G_2 G_4 V_S \tag{8.15}$$

$$\Delta_2 = \begin{vmatrix} G_1 + G_2 + G_5 & I_S + G_5 V_S \\ -G_2 & G_4 V_{in} \end{vmatrix}$$
$$= (G_1 + G_2 + G_5)(G_4 V_S) + G_2(I_S + G_5 V_S) \tag{8.16}$$

Now, using Cramer's rule, the node voltages are:

$$V_A = \frac{\Delta_1}{\Delta} = \frac{(I_S + G_5 V_S)(G_2 + G_3 + G_4) + G_2 G_4 V_S}{(G_1 + G_2 + G_5)(G_2 + G_3 + G_4) - G_2^2} \tag{8.17}$$

$$V_B = \frac{\Delta_2}{\Delta} = \frac{(G_1 + G_2 + G_5)(G_4 V_S) + G_2(I_S + G_5 V_S)}{(G_1 + G_2 + G_5)(G_2 + G_3 + G_4) - G_2^2} \tag{8.18}$$

8.2 Mesh Current Analysis

8.2.1 Mesh Analysis for the Resistive Circuits

Mesh analysis provides another systematic procedure for analyzing circuits, using so-called mesh currents as circuit variables (Alexander and Sadiku, 2009, p. 93).

Figure 8.4 Mesh current analysis circuit.

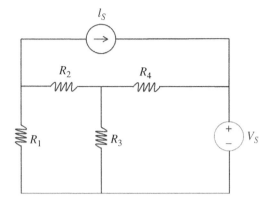

Figure 8.5 Mesh current assignments.

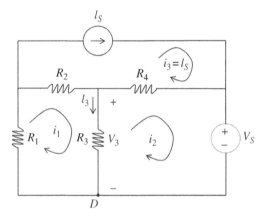

We will use the circuit shown in Figure 8.4 to define mesh currents and explain the method.

First let's define a mesh: *a mesh is a loop that does not contain any other loops within it.* Thus, a loop containing R_1, R_2, and R_3 is a mesh; so is the loop containing R_3, R_4, and V_S. But the loop obtained by combining the two meshes is not.

We define a mesh current as the current that flows through the elements constituting the mesh.

Three mesh currents i_1, i_2, and i_3 are shown in Figure 8.5.

Before proceeding with the mesh current analysis method, let's establish the relationship between the mesh currents and the element voltage and the element current.

Writing KCL at node D we get

$$I_3 + i_2 = i_1 \tag{8.19}$$

or

$$I_3 = i_1 - i_2 \tag{8.20}$$

This example illustrates the following fundamental relation between the mesh currents and the element current:

If the two-terminal element is connected between two meshes, then the element current is equal to the difference of the two mesh currents; we take the mesh current flowing in the same direction as the element current and subtract from it the mesh current flowing in the opposite direction.

If the two-terminal element is not being shared by two meshes, then the element current is equal to the mesh current (provided the reference directions are the same).

According to the Ohm's law we have

$$V_3 = R_3 I_3 \tag{8.21}$$

Now, utilizing Eq. (8.20) in Eq. (8.21) we obtain the relationship between the mesh currents and the resistor voltage.

$$V_3 = R_3 \left(i_1 - i_2 \right) \tag{8.22}$$

Thus, *once the reference direction of the voltage is assigned, we take the mesh current flowing from the (+) to the (−) and subtract from it the mesh current flowing in the opposite direction; then we multiply the result by the resistance value.*

We are now ready to proceed with the mesh current analysis method.

After having assigned the mesh current, the next step is to write KVL *around the meshes, where the mesh current is unknown.* Since in mesh 3 we have $i_3 = I_S$, there is no need to write KVL around that mesh.

To write KVL around meshes 1 and 2 we assign voltages and their reference directions; one such assignment is shown in Figure 8.6.

KVL around mesh 1 produces

$$-V_1 + V_2 + V_3 = 0 \tag{8.23}$$

In terms of mesh currents, Eq. (8.24) can be written as

$$-\left(-R_1 i_1 \right) + R_2 \left(i_1 - i_3 \right) + R_3 \left(i_1 - i_2 \right) = 0 \tag{8.24}$$

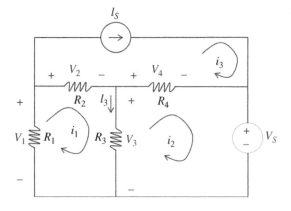

Figure 8.6 Current assignments.

Now, $i_3 = I_S$ and Eq. (8.24) becomes

$$R_1 i_1 + R_2 (i_1 - I_S) + R_3 (i_1 - i_2) = 0 \qquad (8.25)$$

Similarly, KVL around mesh 2 produces

$$-V_3 + V_4 + V_S = 0 \qquad (8.26)$$

In terms of mesh currents, Eq. (8.26) can be written as

$$-R_3 (i_1 - i_2) + R_4 (i_2 - I_S) + V_S = 0 \qquad (8.27)$$

Rearranging Eqs (8.25) and (8.27), we can rewrite them as

$$(R_1 + R_2 + R_3) i_1 - R_3 i_2 = R_2 I_S \qquad (8.28)$$
$$-R_3 i_1 + (R_3 + R_4) i_2 = R_4 I_S - V_S \qquad (8.29)$$

The above system of equations can be written in matrix form as

$$\begin{bmatrix} R_1 + R_2 + R_3 & -R_3 \\ -R_3 & R_3 + R_4 \end{bmatrix} \begin{bmatrix} i_1 \\ i_2 \end{bmatrix} = \begin{bmatrix} R_2 I_S \\ R_4 I_S - V_S \end{bmatrix} \qquad (8.30)$$

and it can easily be solved for the mesh currents using matrix algebra.

8.3 EMC Applications

8.3.1 Power Supply Filters – Common- and Differential-Mode Current Circuit Model

Virtually all electronic products need some form of internal power supply filter. A typical power supply filter topology is shown in Figure 8.7 (Paul, 2006, p. 389).

The common- and differential-mode currents at the output of the product (at the input to the filter) are denoted as $\hat{I}_C$ and $\hat{I}_D$, respectively. The common- and differential-mode currents at the input to the line impedance stabilization network (LISN), at the output of the filter, are denoted with primes as $\hat{I}'_C$ and $\hat{I}'_D$, respectively.

The object of the filter is to reduce the unprimed current levels to the primed current levels, which result in the LISN-measured voltages

$$\hat{V}_P = 50 \left(\hat{I}'_C + \hat{I}'_D \right) \qquad (8.31a)$$

$$\hat{V}_N = 50 \left(\hat{I}'_C - \hat{I}'_D \right) \qquad (8.31b)$$

that are below the allowable conducted emission limits over the required frequency range.

To study the effect of the filter components on the common- and differential-mode noise currents, we need to obtain a circuit model for the filter and LISN.

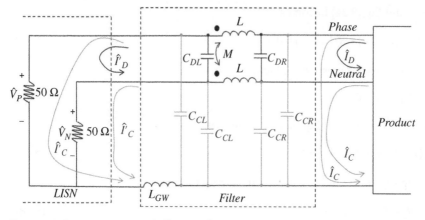

Figure 8.7 Generic power supply filter topology.

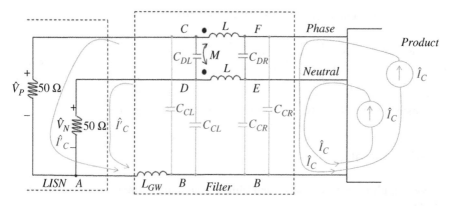

Figure 8.7 Generic power supply filter topology.

Common-Mode Currents Circuit Model Let's redraw the circuit shown in Figure 8.7, as shown in Figure 8.8. The common-mode currents are simulated with the current sources.

With the nodes labeled, it is easy to verify that the circuit shown in Figure 8.7 is equivalent to the circuit shown in Figure 8.8.

Now, we will write mesh equations for this circuit.

For mesh *D-A-C-D* (on the far left side) we have

$$\frac{1}{j\omega C_{DL}}\hat{I}_{DL} + 50\left(\hat{I}_{DL} + \hat{I}_2\right) + 50\left(\hat{I}_{DL} - \hat{I}_2\right) = 0 \tag{8.32}$$

or

$$\frac{1}{j\omega C_{DL}}\hat{I}_{DL} + 100\hat{I}_{DL} = 0 \tag{8.33}$$

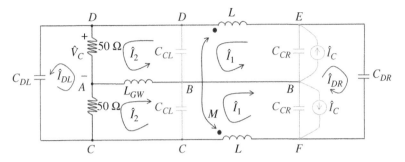

Figure 8.8 Equivalent circuit to that shown in Figure 8.7.

Figure 8.9 Equivalent mesh E-F-B-E.

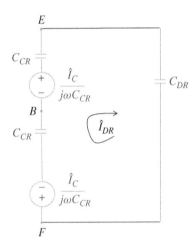

or

$$\left(\frac{1}{j\omega C_{DL}} + 100\right)\hat{I}_{DL} = 0 \tag{8.34}$$

showing that the differential-mode current $\hat{I}_{DL}$ is zero:

$$\check{I}_{DL} = 0 \tag{8.35}$$

Before writing the mesh equations for mesh E-F-B-E (on the far right side) let's apply source transformations to the current sources in parallel to C_{CR} resulting in the circuit shown in Figure 8.9.

$$\frac{1}{j\omega C_{DR}}\hat{I}_{DR} + \frac{\hat{I}_C}{j\omega C_{CR}} + \frac{1}{j\omega C_{CR}}\hat{I}_{DR} - \frac{\hat{I}_C}{j\omega C_{CR}} + \frac{1}{j\omega C_{CR}}\hat{I}_{DR} = 0 \tag{8.36}$$

or

$$\left(\frac{1}{j\omega C_{DR}} + \frac{1}{j\omega C_{CR}} + \frac{1}{j\omega C_{CR}}\right)\hat{I}_{DR} = 0 \tag{8.37}$$

showing that the differential-mode current $\hat{I}_{DR}$ is zero:

$$\hat{I}_{DR} = 0 \tag{8.38}$$

Thus, the line capacitors C_{DL} and C_{DR} have no effect on common-mode currents, since no current flows through them, and they are thus effectively acting as open circuits.

Let's write the mesh equation for the remaining meshes and see what conclusions can be drawn.

Mesh *A-B-C-A* or mesh *A-D-B-A*:

$$50\hat{I}_2 + j\omega L_{GW}\left(\hat{I}_2 + \hat{I}_2\right) + \frac{1}{j\omega C_{CL}}\left(\hat{I}_2 - \hat{I}_1\right) = 0 \tag{8.39}$$

or

$$50\hat{I}_2 + j\omega 2 L_{GW}\hat{I}_2 + \frac{1}{j\omega C_{CL}}\left(\hat{I}_2 - \hat{I}_1\right) = 0 \tag{8.40}$$

Mesh *B-C-F-B* or mesh *B-D-E-B*:

$$\frac{1}{j\omega C_{CR}}\left(\hat{I}_1 - \hat{I}_C\right) + j\omega L\hat{I}_1 + j\omega M\hat{I}_1 + \frac{1}{j\omega C_{CL}}\left(\hat{I}_1 - \hat{I}_2\right) = 0 \tag{8.41}$$

or

$$\frac{1}{j\omega C_{CR}}\left(\hat{I}_1 - \hat{I}_C\right) + j\omega(L + M)\hat{I}_1 + \frac{1}{j\omega C_{CL}}\left(\hat{I}_1 - \hat{I}_2\right) = 0 \tag{8.42}$$

Now, let's write mesh equations for the circuit shown in Figure 8.10. For the left-most mesh we have:

$$j\omega 2 L_{GW}\hat{I}_2 + 50\hat{I}_2 + \frac{1}{j\omega C_{CL}}\left(\hat{I}_2 - \hat{I}_1\right) = 0 \tag{8.43}$$

For the center mesh we write

$$\frac{1}{j\omega C_{CR}}\left(\hat{I}_1 - \hat{I}_C\right) + \frac{1}{j\omega C_{CL}}\left(\hat{I}_1 - \hat{I}_2\right) + j\omega(L + M)\hat{I}_1 = 0 \tag{8.44}$$

Compare Eq. (8.43) with Eq. (8.40), and Eq. (8.44) with Eq. (8.42). They are the same equations!

Thus, the circuit in Figure 8.10 is an equivalent circuit for common-mode currents for the filter and LISN for the phase-ground or neutral-ground configurations.

Differential-Mode Current Circuit Model   Let's redraw the circuit shown in Figure 8.7, as shown in Figure 8.11. The differential-mode currents are represented by a current source.

With the nodes labeled, it is easy to verify that the circuit shown in Figure 8.11 is equivalent to the circuit shown in Figure 8.12.

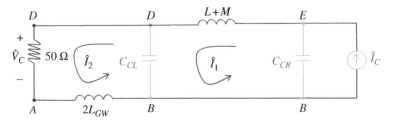

Figure 8.10 Equivalent circuit to that shown in Figure 8.8.

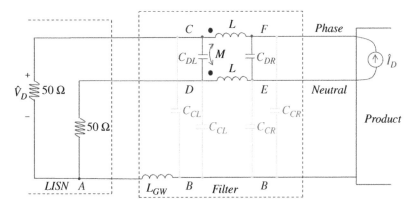

Figure 8.11 Power supply filter with the differential mode current.

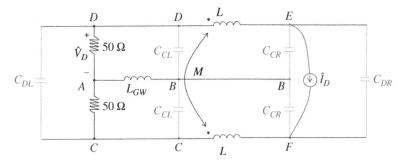

Figure 8.12 Equivalent circuit to that shown in Figure 8.11.

Furthermore, the circuit shown in Figure 8.13 is equivalent to that shown in Figure 8.12.

Let's write mesh equations for this circuit. For mesh D-A-C-D (on the far left side) we have

$$\frac{1}{j\omega C_{DL}}\hat{I}_3 + 50\left(\hat{I}_3 - \hat{I}_2\right) + 50\left(\hat{I}_3 - \hat{I}_2\right) = 0 \tag{8.45}$$

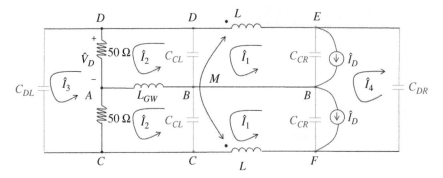

Figure 8.13 Equivalent circuit to that shown in Figure 8.12.

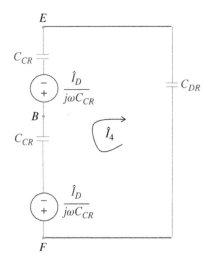

Figure 8.14 Equivalent mesh E-F-B-E.

or

$$\frac{1}{j\omega C_{DL}}\hat{I}_3 + 100\hat{I}_3 - 100\hat{I}_2 = 0 \tag{8.46}$$

or

$$\frac{1}{j\omega 2C_{DL}}\hat{I}_3 + 50\left(\hat{I}_3 - \hat{I}_2\right) = 0 \tag{8.47}$$

Let's apply source transformations to the current sources in parallel to C_{CR}. The resulting circuit is shown in Figure 8.14.

For mesh E-F-B-E we have

$$\frac{1}{j\omega C_{DR}}\hat{I}_4 + \frac{1}{j\omega C_{CR}}\left(\hat{I}_4 + \hat{I}_1\right) + \frac{1}{j\omega C_{CR}}\left(\hat{I}_4 + \hat{I}_1\right) = 0 \tag{8.48}$$

or

$$\frac{1}{j\omega C_{DR}}\hat{I}_4 + \frac{2}{j\omega C_{CR}}\left(\hat{I}_4 + \hat{I}_1\right) = 0 \tag{8.49}$$

and thus

$$\frac{1}{j\omega 2 C_{DR}}\hat{I}_4 + \frac{1}{j\omega C_{CR}}\left(\hat{I}_4 + \hat{I}_1\right) = 0 \tag{8.50}$$

Let's write mesh equations for the remaining meshes and see what conclusions can be drawn.

Mesh *A-B-C-A* or mesh *A-D-B-A*:

$$50\left(\hat{I}_2 - \hat{I}_3\right) + \frac{1}{j\omega C_{CL}}\left(\hat{I}_2 - \hat{I}_1\right) + j\omega L_{GW}\left(\hat{I}_2 - \hat{I}_2\right) = 0 \tag{8.51}$$

or

$$50\left(\hat{I}_2 - \hat{I}_3\right) + \frac{1}{j\omega C_{CL}}\left(\hat{I}_2 - \hat{I}_1\right) = 0 \tag{8.52}$$

Mesh *B-C-F-B* or mesh *B-D-E-B*:

$$\frac{1}{j\omega C_{CR}}\left(\hat{I}_1 + \hat{I}_4\right) + j\omega L\hat{I}_1 - j\omega M\hat{I}_1 + \frac{1}{j\omega C_{CL}}\left(\hat{I}_1 - \hat{I}_2\right) = 0 \tag{8.53}$$

or

$$\frac{1}{j\omega C_{CR}}\left(\hat{I}_1 + \hat{I}_4\right) + j\omega\left(L - M\right)\hat{I}_1 + \frac{1}{j\omega C_{CL}}\left(\hat{I}_1 - \hat{I}_2\right) = 0 \tag{8.54}$$

Now, let's write mesh equations for the circuit shown in Figure 8.15. For the left-most mesh we have:

$$50\left(\hat{I}_3 - \hat{I}_2\right) + \frac{1}{j\omega 2 C_{DL}}\hat{I}_3 = 0 \tag{8.55}$$

For the second mesh we write

$$50\left(\hat{I}_2 - \hat{I}_3\right) + \frac{1}{j\omega C_{CL}}\left(\hat{I}_2 - \hat{I}_1\right) = 0 \tag{8.56}$$

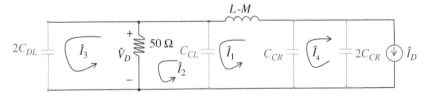

Figure 8.15 Equivalent circuit to that shown in Figure 8.13.

Mesh one produces

$$\frac{1}{j\omega C_{CR}}\left(\hat{I}_1 + \hat{I}_4\right) + j\omega\left(L - M\right)\hat{I}_1 + \frac{1}{j\omega C_{CL}}\left(\hat{I}_1 - \hat{I}_2\right) = 0 \tag{8.57}$$

and for the fourth mesh we write

$$\frac{1}{j\omega 2C_{DR}}\hat{I}_4 + \frac{1}{j\omega C_{CR}}\left(\hat{I}_4 + \hat{I}_1\right) = 0 \tag{8.58}$$

Compare Eq. (8.47) with Eq. (8.55); Eq. (8.50) with Eq. (8.58); Eq. (8.52) with Eq. (8.56); and Eq. (8.54) with Eq. (8.57). They are the same equations!

Thus, the circuit in Figure 8.15 is an equivalent circuit for differential-mode currents for the filter and LISN for the phase-ground or neutral-ground configurations.

References

Alexander, C.K. and Sadiku, N.O., *Fundamentals of Electric Circuits*, 4th ed., McGraw Hill, New York, 2009.

Nilsson, J.W. and Riedel, S.A., *Electric Circuits*, 10th ed., Pearson, Upper Saddle River, NJ, 2015.

Paul, C.R., *Introduction to Electromagnetic Compatibility*, 2nd ed., John Wiley and Sons, New York, 2006.

9

Circuit Theorems and Techniques

9.1 Superposition

Consider a linear circuit with several independent voltage or current sources, like the one shown in Figure 9.1.

Say, we want to calculate the voltage or current somewhere in the circuit; voltage V across R_2 in this case. We could, of course, solve this circuit using the node voltage or mesh current methods discussed previously in Chapter 7.

However, we could also solve this circuit using the *principle of superposition* (Nilsson and Riedel, 2015, p. 122) which states then whenever a linear circuit is driven by more than one independent source, the response of the circuit can be obtained as the sum of the individual responses due to each independent source acting alone.

We can think of each independent source as the input to the circuit, and the voltage or current somewhere in the circuit as the output. Then the principle of superposition can be illustrated in block diagram form as shown in Figure 9.2.

In Figure 9.2(a) the circuit is driven by several inputs u_1 to u_N. The output of the system is equal to y. In Figures 8.2(b)–(d), the circuit is driven by one input at time u_k, resulting in the corresponding output y_k.

According to the principle of superposition, the total output y, when all inputs are present, can be obtained by summing the individual outputs y_k due to each input acting alone. That is,

$$y = y_1 + y_2 + \ldots + y_N \tag{9.1}$$

When an individual source is acting alone, the other sources are deactivated, or suppressed: the voltage sources are replaced by a short circuit, while the current sources are replaced by an open circuit.

Let's apply the principle of superposition to the circuit shown in Figure 9.1. First, let's drive the circuit by a voltage source V_{S1}, as shown in Figure 9.3.

When the circuit is driven by a voltage source V_{S2}, we obtain the circuit shown in Figure 9.4.

And finally, when the circuit is driven by a current source, we have the configuration shown in Figure 9.5.

Foundations of Electromagnetic Compatibility with Practical Applications, First Edition. Bogdan Adamczyk.
© 2017 John Wiley & Sons Ltd. Published 2017 by John Wiley & Sons Ltd.

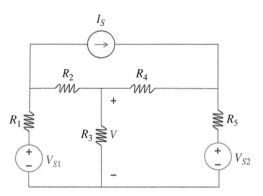

Figure 9.1 Linear circuit driven by several independent sources.

(a)

(b)

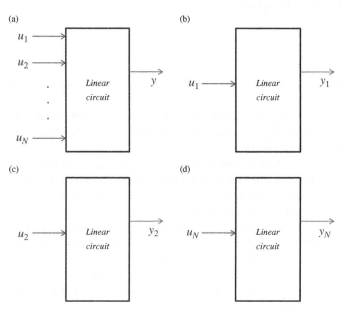

(c)

(d)

Figure 9.2 The principle of superposition.

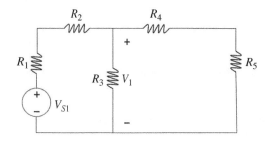

Figure 9.3 Circuit driven by the voltage source V_{S1}.

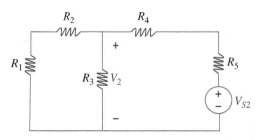

Figure 9.4 Circuit driven by the voltage source V_{S2}.

Figure 9.5 Circuit driven by the current source I$_S$.

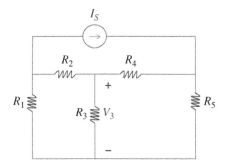

According to the principle of superposition we have

$$V = V_1 + V_2 + V_3 \tag{9.2}$$

Example 9.1 Superposition

Verify the principle of superposition for the circuit shown in Figure 9.6.

We will first solve for the output voltage V_0 using the node voltage method. Then, we will solve for it using the superposition approach.

Applying KCL at the upper node we have

$$\frac{V_S - V_o}{R_1} + I_S = \frac{V_o}{R_2} \tag{9.3}$$

Moving the inputs to the left side of this equation yields

$$\frac{V_S}{R_1} + I_S = \frac{V_o}{R_1} + \frac{V_o}{R_2} \tag{9.4}$$

or

$$\frac{V_S}{R_1} + I_S = \frac{R_1 + R_2}{R_1 R_2} V_o \tag{9.5}$$

Figure 9.6 Circuit for Example 9.1.

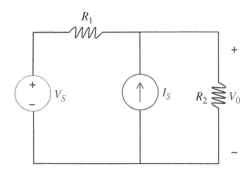

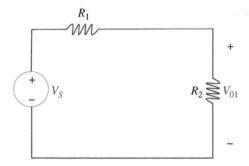

Figure 9.7 Circuit with the current source deactivated.

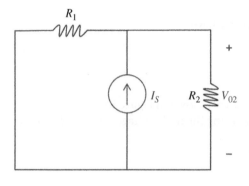

Figure 9.8 Circuit with the voltage source deactivated.

resulting in

$$V_o = \frac{R_2}{R_1 + R_2} V_S + \frac{R_1 R_2}{R_1 + R_2} I_S \qquad (9.6)$$

We will now find V_o using the superposition method. We will first deactivate the current source and replace it with an open circuit as shown in Figure 9.7.

Using the voltage divider we get

$$V_{01} = \frac{R_2}{R_1 + R_2} V_S \qquad (9.7)$$

Next, we deactivate the voltage source leaving the current source turned on, as shown in Figure 9.8.

Using Ohm's law and the current divider we obtain

$$V_{02} = R_2 \frac{R_1}{R_1 + R_2} I_S \qquad (9.8)$$

Applying the superposition principle, we find the response with both sources active by adding the two responses V_{01} and V_{02}

$$V_0 = V_{01} + V_{02} = \frac{R_2}{R_1 + R_2} V_S + \frac{R_1 R_2}{R_1 + R_2} I_S \qquad (9.9)$$

9.2   Source Transformation

The analysis of complex circuits can usually be accomplished by either the node voltage or the mesh current method. In both the node and the mesh methods, it is often desirable to have the sources of the same kind: current sources in the node voltage method, and voltage sources in the mesh current method.

If a circuit, however, has both current sources and voltage sources, it is desirable to make adjustments to the circuit so that all the sources are of the same type.

A source transformation (Alexander and Sadiku, 2009, p. 135), shown in Figure 9.9 allows a voltage source in series with a resistor to be replaced by a current source in parallel with a resistor, or vice versa.

The fundamental concept behind this technique is the concept of equivalence. We recall that two circuits are equivalent with respect to the same two nodes if they have the same $v–i$ characteristics at those nodes.

Let's determine the required relationships between the sources and the resistances, so that they are equivalent with respect to nodes A and B.

Since the two circuits are to be equivalent, we require that both circuits have the same $v–i$ characteristic for all values of an external resistor R connected between terminals A and B as shown in Figure 9.10.

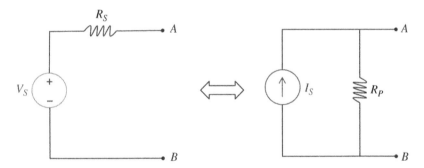

Figure 9.9 Source transformations.

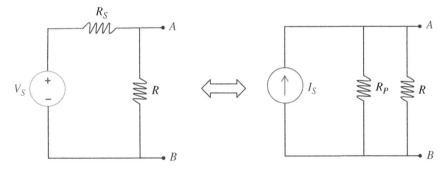

Figure 9.10 Equivalent circuits.

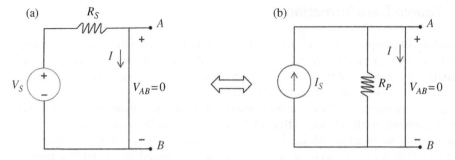

Figure 9.11 Equivalence for $R=0$.

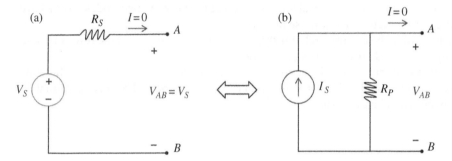

Figure 9.12 Equivalence for $R=\infty$.

First, let's try the extreme values of R first, namely $R=0$ and $R=\infty$. With $R=0$, we have a short circuit across the terminals A and B, as shown in Figure 9.11.

Equivalence requires that the currents in both circuits are the same. For the circuit in Figure 9.11(a) we have

$$I = \frac{V_S}{R_S} \tag{9.10}$$

While for the circuit in Figure 9.11(b) we write

$$I = I_S \tag{9.11}$$

Comparison of Eqs (9.10) and (9.11) leads to

$$I_S = \frac{V_S}{R_S} \tag{9.12}$$

With $R=\infty$, we have an open circuit across the terminals A and B, as shown in Figure 9.12.

Equivalence requires that the voltages in both circuits are the same. For the circuit in Figure 9.12(a) we have

$$V_{AB} = V_S \tag{9.13}$$

while for the circuit in Figure 9.12(b) we write

$$V_{AB} = R_P I_S \tag{9.14}$$

Comparison of Eqs (9.13) and (9.14) leads to

$$V_S = R_P I_S \tag{9.15}$$

At the same time, according to Eq. (9.12) we have

$$V_S = R_S I_S \tag{9.16}$$

Thus, for two circuits to be equivalent,

$$R_S = R_P \tag{9.17}$$

and

$$V_S = R_S I_S \tag{9.18}$$

We have shown the equivalence under the conditions (9.17) and (9.18) for the extreme values of $R = 0$ and $R = \infty$. Next, using the circuit shown in Figure 9.13, we will show that the equivalence under these conditions holds for any value of R.

For the circuit in Figure 9.13(a) we use KVL to obtain

$$V_S = R_S I + V_{AB} \tag{9.19}$$

or

$$\frac{V_S}{R_S} = I + \frac{V_{AB}}{R_S} \tag{9.20}$$

For the circuit 9.13(b) we use KCL to obtain

$$I_S = I + \frac{V_{AB}}{R_P} \tag{9.21}$$

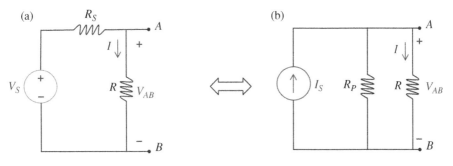

Figure 9.13 Equivalence for any *R*.

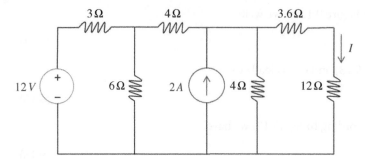

Figure 9.14 Circuit for Example 9.2.

Now, if

$$I_S = \frac{V_S}{R_S} \qquad\qquad (9.22)$$

and

$$R_S = R_P \qquad\qquad (9.23)$$

the two circuits are equivalent.

Example 9.2   Source transformations
In the circuit shown in Figure 9.14, use a series of source transformations to determine the current *I*.

Solution: Applying the source transformation to the 12 V voltage source and 3 Ω resistor produces the circuit shown in Figure 9.15.
The combination of 3 Ω in parallel with 6 Ω results in an equivalent resistance of 2 Ω, being in parallel with the 4 A current source. Applying source transformation to that configuration results in the circuit shown in Figure 9.16.

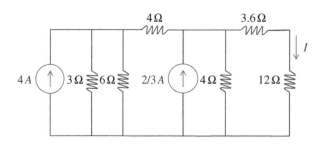

Figure 9.15 Source transformations – Example 9.2.

Figure 9.16 Source transformations – Example 9.2.

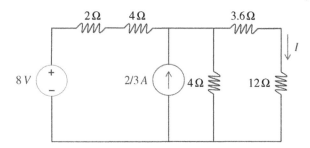

Figure 9.17 Source transformations – Example 9.2.

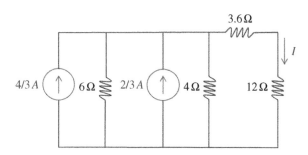

Next, we apply source transformation to the 8 V source in series with 6 Ω resistance, resulting in the circuit shown in Figure 9.17.

Combining current sources in parallel and resistor 6 Ω and 4 Ω in parallel results in the circuit shown in Figure 9.18.

Now applying the current divider rule we get

$$I = \frac{2.4}{2.4 + 3.6 + 12} \times 2 = 0.3 \quad [\text{A}]$$

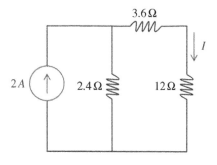

Figure 9.18 Source transformations – Example 9.2.

9.3 Thévenin Equivalent Circuit

Consider a linear circuit driving a load, as shown in Figure 9.19.

The driving circuit, with the load disconnected is shown in Figure 9.20.

According to the Thévenin theorem (Nilsson and Riedel, 2015, p. 113), the circuit shown in Figure 9.20(a) is equivalent (with respect to nodes A and B) to a circuit consisting of an independent voltage source in series with resistor, as shown in Figure 9.21(b).

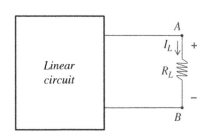

Figure 9.19 Linear circuit driving a load.

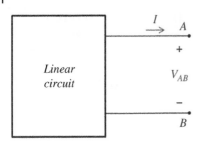

Figure 9.20 Driving circuit with the load disconnected.

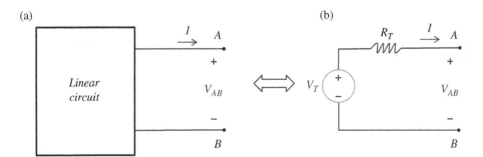

(a)   (b)

Figure 9.21 Thévenin equivalent circuit.

The equivalence, of course, means that the circuits shown in Figures 9.21(a) and (b) have the same i–v characteristics with respect to nodes A and B.

Since these are equivalent, we could replace the driving circuit in Figure 9.19 with its Thévenin equivalent, to obtain the circuit shown in Figure 9.22.

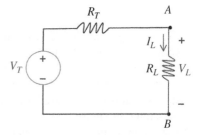

It should be obvious that calculating the voltage or current associated with the load is trivial for this circuit, whereas such calculations for the circuit shown in Figure 9.19 might be quite involved.

Figure 9.22 Thévenin equivalent circuit.

The question, of course, remains: *How do we determine the values of V_T and R_T?*

According to the Thévenin theorem, the value of V_T is just the value of the voltage between nodes A and B, V_{AB} when the load is not connected. This voltage is often referred to as an open-circuit voltage, V_{oc}, and is shown in Figure 9.23.

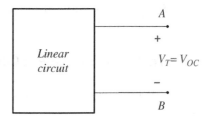

Figure 9.23 Thévenin voltage.

Thévenin resistance, R_T, can be obtained in a number of ways, depending on the circuit complexity and the types of sources present in the driving circuitry.

When the circuit consists of independent sources and resistors, we deactivate the sources and simply calculate the resistance between nodes A and B. The following example illustrates this approach.

Example 9.3 Thévenin equivalent circuit

Determine the Thévenin equivalent with respect to nodes A and B, for the circuit shown in Figure 9.24.

The Thévenin voltage is the voltage V_{oc} between nodes A and B when the load is disconnected (or not present), and is shown in Figure 9.25.

To calculate this voltage we can use any appropriate circuit analysis method. Let's use the principle of superposition, discussed earlier in this chapter. Let's suppress the current source first, as shown in Figure 9.26(a).

Since no current flows through the $2\,\Omega$ and $5\,\Omega$ resistors, the open circuit voltage V_{oc1} is the voltage across the $6\,\Omega$ resistor.

Using the voltage divider

$$V_{oc1} = \frac{6}{8+6+4}12 = 4 \; V$$

Figure 9.24 Circuit for Example 9.3.

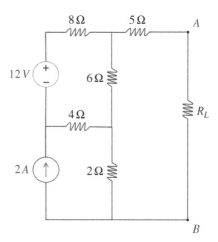

Figure 9.25 Open-circuit voltage.

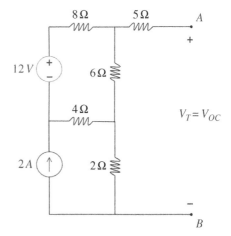

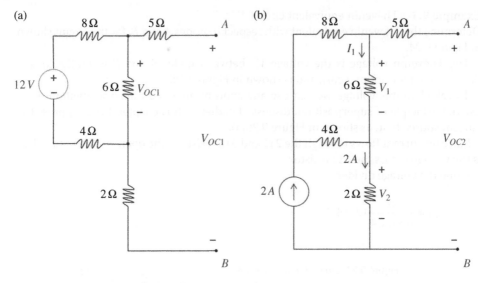

Figure 9.26 Circuits for calculation of open-circuit voltage.

Next, let's suppress the voltage source, as shown in Figure 9.26(b). Note that

$$V_{oc2} = V_1 + V_2$$

where

$$V_2 = (2)(2) = 4 \text{ V}$$

and V_1 can be obtained using the current divider and Ohm's law, as follows

$$V_1 = 6 \, I_1$$

I_1 can be obtained using current divider as

$$I_1 = \frac{4}{4+6+8} \cdot 2 = \frac{4}{9} \text{ A}$$

and thus

$$V_1 = 6 \, I_1 = \frac{24}{9} = 2.6667 \text{ V}$$

thus

$$V_{oc2} = V_1 + V_2 = 4 + 2.6667 = 6.6667 \text{ V}$$

Therefore, the Thévenin voltage is

$$V_T = V_{oc1} + V_{oc2} = 4 + 6.6667 = 10.6667 \text{ V}$$

Thévenin resistance is the resistance between nodes A and B when the load is disconnected and the independent sources are deactivated. The resulting circuit is shown in Figure 9.27.

Thus the Thévenin resistance is

$$R_T = 5 + \left[(8+4)6\right] + 2 = 11 \quad \Omega$$

The Thévenin equivalent circuit is shown in Figure 9.28.

When the circuit consists of independent and dependent sources and resistors, we use another approach to determine the Thévenin resistance. We deactivate the independent sources and drive the circuit with an external voltage or current source connected between nodes A and B. (We will use this approach later in this chapter when discussing the two-port networks.)

This approach is based on the following discussion. Consider a linear circuit with no independent sources (or the independent sources suppressed) and/or dependent sources, as shown in Figure 9.29(a).

Figure 9.29(b) shows its Thévenin equivalent resistance. The Thévenin resistance of this circuit can be obtained by applying an external voltage or current source as shown in Figures 9.30 (a) and (b).

Now consider the circuits shown in Figure 9.31.

Since the circuits to the left of nodes A and B are equivalent, it follows that in order to obtain the Thévenin resistance of an arbitrary linear circuit, we first deactivate the independent sources (if present). Then we drive the circuit with an arbitrary voltage V_S and calculate the resulting current I_s, or alternatively we drive the circuit with an arbitrary current I_S and calculate the resulting voltage V_S.

The following example illustrates this approach.

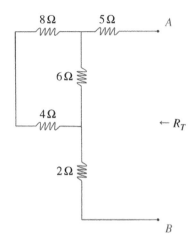

Figure 9.27 Circuit for calculation of Thévenin Resistance.

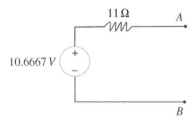

Figure 9.28 Thévenin equivalent for Example 9.3.

Figure 9.29 Thévenin equivalent resistance.

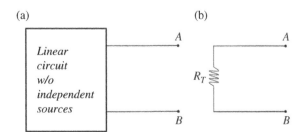

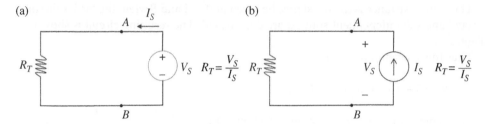

Figure 9.30 Calculation of Thévenin resistance.

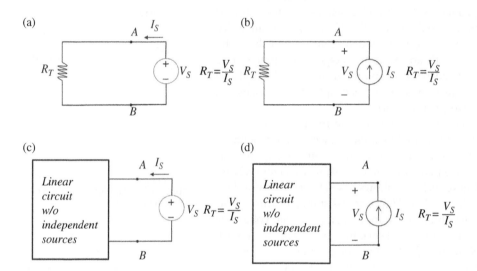

Figure 9.31 Calculation of Thévenin resistance.

Example 9.4 Calculation of Thévenin resistance

Determine the Thévenin resistance of the circuit shown in Figure 9.32 by energizing it with an external voltage source.

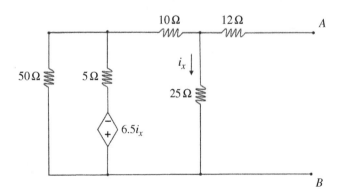

Figure 9.32 Calculation of Thévenin resistance.

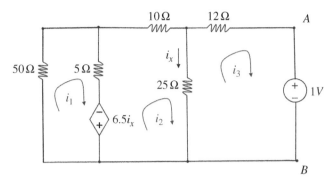

Figure 9.33 Calculation of Thévenin resistance.

Solution: Since the value of the external voltage source does not matter, we often energize the circuit with a 1 V source, as shown in Figure 9.33.

Next, we write the mesh equations

$$\text{Mesh } 1 : 50i_1 + 5(i_1 - i_2) - 6.5i_x = 0$$
$$i_x = i_2 - i_3$$
$$\text{Mesh } 2 : 6.5(i_2 - i_3) + 5(i_2 - i_1) + 10i_2 + 25(i_2 - i_3) = 0$$
$$\text{Mesh } 3 : 25(i_3 - i_2) + 12i_3 + 1 = 0$$

The solution to these equation is

$$i_1 = -0.0012$$
$$i_2 = -0.0340$$
$$i_3 = -0.05 \quad [\text{A}]$$

The Thévenin resistance is, therefore,

$$R_T = \frac{V_S}{I_S} = \frac{1}{-i_3} = \frac{1}{0.05} = 20 \quad \Omega$$

9.4 Norton Equivalent Circuit

Norton equivalent circuit provides an alternative to the Thévenin equivalent (Alexander and Sadiku, 2009, p. 145). The underlying concepts leading to that equivalent are the same.

Consider a linear two- terminal circuit driving a load and shown in Figure 9.34(a). The same driving circuit with the load disconnected is shown in Figure 9.34(b).

According to Norton's theorem, the circuit shown in Figure 9.34(b) is equivalent (with respect to nodes *A* and *B*) to a circuit consisting of an independent current source in parallel with resistor, as shown in Figure 9.35.

(a)   (b)

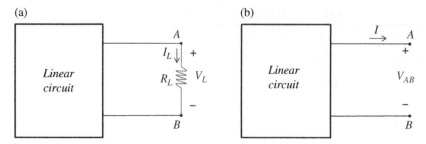

Figure 9.34 Driving circuit with and without the load.

(a)   (b)

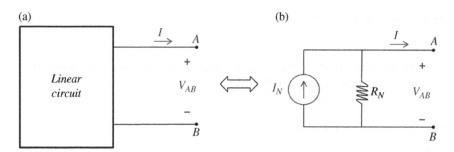

Figure 9.35 Norton equivalent circuit.

Figure 9.36 Short-circuit current.

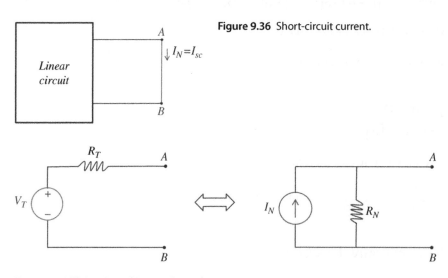

Figure 9.37 Thévenin and Norton Equivalence.

The Norton resistance, R_N, in Figure 9.35 (b), is the same as the Thévenin resistance. The Norton current, I_N, is obtained by placing a short circuit across nodes A and B, as shown in Figure 9.36, and calculating the so-called short-circuit current flowing from A to B.

Note that the Norton equivalent can of course be obtained from the Thévenin equivalent by a source transformation, as shown in Figure 9.37.

Therefore,

$$I_N = \frac{V_T}{R_T} \tag{9.24}$$

or

$$V_T = R_N I_N \tag{9.25}$$

It also follows that the Thévenin or Norton resistance can be obtained from

$$R_T = \frac{V_T}{I_N} = \frac{V_{oc}}{I_{sc}} \tag{9.26}$$

The following example illustrates the application of Norton's theorem.

Example 9.5 Norton equivalent circuit
Determine the Norton equivalent with respect to nodes A and B, for the circuit shown in Figure 9.38. (This is the same circuit as we used for the Thévenin equivalent in Example 9.3.)

Norton resistance is the same as the Thévenin resistance calculated in Example 9.3:

$$R_N = R_T = 11 \; \Omega$$

The Norton current is the current flowing through a short circuit from A to B when the load is disconnected (or not present), and is shown in Figure 9.39.

To calculate the short-circuit current we could use any appropriate method of circuit analysis. For this particular circuit, mesh analysis would be well suited.

Let's assign mesh currents, as shown in Figure 9.39. Note that

$$i_3 = 2 \, \text{A}$$

Figure 9.38 Circuit for Example 9.4.

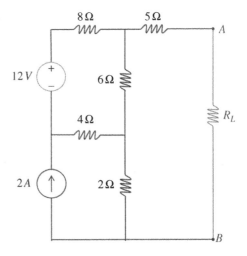

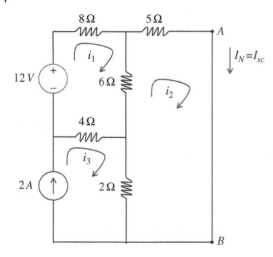

Figure 9.39 Short-circuit current.

and the Norton current is

$$I_N = i_2$$

Writing KVL around the first mesh results in

$$-12 + 8i_1 + 6(i_1 - i_2) + 4(i_1 - 2) = 0$$

Writing KVL around the second mesh results in

$$5i_2 + 2(i_2 - 2) - 6(i_1 - i_2) = 0$$

This system of equations yields:

$$i_1 = 1.4343 \ \text{A}$$
$$i_2 = 0.9697 \ \text{A}$$

and thus

$$I_N = 0.9697 \quad \text{A}$$

The Norton resistance can be now calculated from

$$R_N = \frac{V_T}{I_N} = \frac{10.6667}{0.9697} = 11 \ \Omega$$

which, of course agrees with the result of Example 9.3.

9.5   Maximum Power Transfer

9.5.1   Maximum Power Transfer – Resistive Circuits

When interfacing the driving circuitry to the load, it is important to consider the voltage, current, and power available at an interface between a fixed source and an adjustable load.

Figure 9.40 Thévenin equivalent of the driving circuitry.

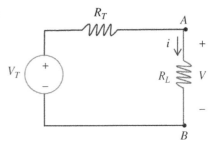

For simplicity we will consider the case in which both the source and the load are linear resistive circuits. The source can be represented by a Thévenin equivalent, and the load by an equivalent resistance R_L, as shown in Figure 9.40.

For a fixed source, the parameters V_T and R_T are given and the interface signal levels are functions of the load resistance R_L. By voltage division, the interface voltage is

$$v = \frac{R_L}{R_L + R_T} V_T \tag{9.27}$$

This relation can be rewritten as

$$v = \frac{1}{1 + \dfrac{R_T}{R_L}} V_T \tag{9.28}$$

For a fixed source (R_T = const), and a variable load R_L, the voltage v will be at maximum when R_L is made very large compared with R_T. Ideally, R_L should be made infinite (an open circuit), in which case

$$v_{MAX} = V_T = V_{oc} \tag{9.29}$$

Therefore, the maximum voltage available at the interface is the source open-circuit voltage V_{oc}.

The current delivered at the interface is

$$i = \frac{V_T}{R_L + R_T} \tag{9.30}$$

For a fixed source and a variable load, the current will be a maximum if R_L is made very small compared with R_T. Ideally, R_L should be zero (a short circuit), in which case

$$i_{MAX} = \frac{V_T}{R_T} = I_N = I_{sc} \tag{9.31}$$

Therefore, the maximum current available at the interface is the source short-circuit current I_{sc}.

The power delivered to the load is

$$p = v \cdot i \tag{9.32}$$

Using Eqs (9.27) and (9.30), the power delivered to the load is

$$p = \frac{R_L}{R_L + R_T} V_T \cdot \frac{V_T}{R_L + R_T} = \left(\frac{V_T}{R_L + R_T}\right)^2 R_L \tag{9.33}$$

For a given source, the parameters V_T and R_T are fixed, and the delivered power is a function of a single variable R_L. We wish to find the value of the load R_L such that the maximum power is delivered to it.

The condition for maximum voltage ($R_L = \infty$) and the condition for maximum current ($R_L = 0$) both produce zero power. The value of R_L that maximizes the power lies somewhere between these two extrema.

To find the value of R_L that maximizes the power, we differentiate Eq. (9.33) with respect to R_L and solve for the value of R_L for which the derivative $dp/dR_L = 0$.

$$\frac{dp}{dR_L} = V_T^2 \frac{d}{dR_L}\left(\frac{R_L}{(R_L + R_T)^2}\right) = V_T^2 \frac{1 \cdot (R_L + R_T)^2 - R_L 2(R_L + R_T) \cdot 1}{(R_L + R_T)^4} \tag{9.34}$$

Equating this derivative to zero gives

$$(R_L + R_T)^2 - 2R_L(R_L + R_T) = 0 \tag{9.35a}$$

or

$$(R_L + R_T)(R_L + R_T - 2R_L) = 0 \tag{9.35b}$$

Solving for R_L gives

$$R_L = R_T \tag{9.36}$$

This is the value of R_L at which the extremum of power p happens. To determine whether it is a maximum or minimum, the second derivative of p with respect to R_L needs to be evaluated at $R_L = R_T$. If the value of that derivative is negative, the extremum corresponds to the maximum.

One way to determine whether it is a maximum or minimum is to evaluate the second derivative of p (at $R_L = R_T$) with respect to R_L.

The maximum power delivered to R_L is then

$$p_{MAX} = p(R_L = R_T) = \left(\frac{V_T}{R_T + R_T}\right)^2 R_T = v_T^2 \frac{R_T}{4R_T^2} = \frac{v_T^2}{4R_T} \tag{9.37}$$

Since

$$V_T = R_T I_N \tag{9.38}$$

the above result can also be written as

$$p_{MAX} = \frac{R_T I_N^2}{4} \tag{9.39}$$

or

$$p_{MAX} = \frac{V_T I_N}{4} = \left(\frac{v_{oc}}{2}\right)\left(\frac{i_{sc}}{2}\right)$$ (9.40)

9.5.2 Maximum Power Transfer – Sinusoidal Steady State

To address the maximum power transfer in the sinusoidal steady state, we use the circuit shown in Figure 9.41.

The source is represented by a Thévenin equivalent with a phasor voltage $\hat{V}_T$ and the source impedance $\hat{Z}_T$, where

$$\hat{Z}_T = R_T + jX_T$$ (9.41)

The load circuit is represented by an equivalent impedance $\hat{Z}_L$, where

$$\hat{Z}_L = R_L + jX_L$$ (9.42)

In the maximum power transfer problem, the driving circuitry, $\hat{V}_T$, R_T, and X_T, is fixed, and the objective is to adjust the load impedance R_L and X_L so that the average power delivered to the load is at its maximum.

The average power delivered to the load is

$$P = \frac{1}{2}R_L\left|\hat{i}\right|^2$$ (9.43)

The magnitude of the load current is

$$\left|\hat{i}\right| = \left|\frac{\hat{V}_T}{\hat{Z}_T + \hat{Z}_L}\right| = \left|\frac{\hat{V}_T}{(R_T + R_L) + j(X_T + X_L)}\right|$$

$$= \frac{\left|\hat{V}_T\right|}{\sqrt{(R_T + R_L)^2 + (X_T + X_L)^2}}$$ (9.44)

Figure 9.41 Source-load interface in the sinusoidal steady state.

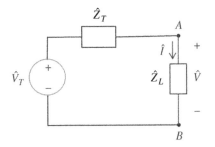

Substituting Eq. (9.44) into Eq. (9.43) produces

$$P = \frac{1}{2} \frac{R_L |\hat{V}_T|^2}{(R_T + R_L)^2 + (X_T + X_L)^2} \tag{9.45}$$

Since the driving circuitry is fixed, we can only adjust X_T and X_L to maximize the power delivered to the load.

Clearly, the power is maximized when

$$X_L = -X_T \tag{9.46}$$

Under this condition, the expression for average power reduces to

$$P = \frac{R_L |\hat{V}_T|^2}{2(R_T + R_L)^2} \tag{9.47}$$

This equation has the same form as Eq. (9.33) in the previous section. From the derivation in the previous section we know that the power is maximized when

$$R_L = R_T \tag{9.48}$$

The conditions in Eqs (9.46) and (9.48) can be combined as

$$Z_L = Z_T^* \tag{9.49}$$

Thus, the maximum power transfer occurs under a conjugate match condition.

Under the conjugate match condition, the maximum power available from the source equals

$$P = \left. \frac{R_L |\hat{V}_T|}{2(R_T + R_L)^2} \right|_{R_L = R_T} = \frac{R_T |\hat{V}_T|}{2(2R_T)^2} = \frac{|\hat{V}_T|}{8R_T} \tag{9.50}$$

9.6 Two-Port Networks

So far we have discussed several circuit analysis techniques including Kirchhoff's laws, node-voltage or mesh-current analysis, and Thévenin or Norton theorems.

Using Kirchhoff's laws or node-voltage/mesh current methods we can calculate voltages and current anywhere in the circuit. Thévenin or Norton theorems allow us to obtain an equivalent circuit model with respect to the specified pair of terminals (usually the output terminals, or the output port) of the network.

Another way of describing the circuit with respect to the two terminals is by treating the network as a two-port circuit. In many electrical circuits obtaining voltages and currents at the input and output ports, instead of any point in the circuit, is more convenient and practical.

Figure 9.42 Two-port network.

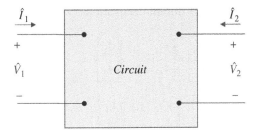

Thus, the fundamental principle underlying the two-port circuit analysis is that only the terminal variables (input voltage/current and output voltage/current) are of interest. We are not interested in calculating voltages and current inside the circuit.

The most general description of the two-port network is carried in the s domain (sinusoidal steady state is a special case of s domain analysis). Figure 9.42 shows the basic building block in terms of the s domain variables.

The voltage and current reference directions at each port are symmetric with respect to each other; that is, at each port the current flows into the upper terminal and the voltage at that terminal has a plus for its reference direction. This symmetry makes it easier to generalize the analysis of two-port networks.

Of the four terminal variables only two are independent. Thus, for any two-port network, once we specify two of the four variables, the other two can be obtained. It follows that the description of a two-port network requires only two simultaneous equations.

There are six different ways of writing the two equation involving the four variables:

$$\hat{V}_1 = z_{11}\hat{I}_1 + z_{12}\hat{I}_2$$
$$\hat{V}_2 = z_{21}\hat{I}_1 + z_{22}\hat{I}_2 \tag{9.51}$$

$$\hat{I}_1 = y_{11}\hat{V}_1 + y_{12}\hat{V}_2$$
$$\hat{I}_2 = y_{21}\hat{V}_1 + y_{22}\hat{V}_2 \tag{9.52}$$

$$\hat{V}_1 = a_{11}\hat{V}_2 - a_{12}\hat{I}_2$$
$$\hat{I}_1 = a_{21}\hat{V}_2 - a_{22}\hat{I}_2 \tag{9.53}$$

$$\hat{V}_2 = b_{11}\hat{V}_1 - b_{12}\hat{I}_1$$
$$\hat{I}_2 = b_{21}\hat{V}_1 - b_{22}\hat{I}_1 \tag{9.54}$$

$$\hat{V}_1 = h_{11}\hat{I}_1 + h_{12}\hat{V}_2$$
$$\hat{I}_2 = h_{21}\hat{I}_1 + h_{22}\hat{V}_2 \tag{9.55}$$

$$\hat{I}_1 = g_{11}\hat{V}_1 + g_{12}\hat{I}_2$$
$$\hat{V}_2 = g_{21}\hat{V}_1 + g_{22}\hat{I}_2 \tag{9.56}$$

In matrix notation, equations (9.51)–(9.56) may be written as

$$\begin{bmatrix} \hat{V}_1 \\ \hat{V}_2 \end{bmatrix} = \begin{bmatrix} z_{11} & z_{12} \\ z_{21} & z_{22} \end{bmatrix} \begin{bmatrix} \hat{I}_1 \\ \hat{I}_2 \end{bmatrix} \tag{9.57}$$

$$\begin{bmatrix} \hat{I}_1 \\ \hat{I}_2 \end{bmatrix} = \begin{bmatrix} y_{11} & y_{12} \\ y_{21} & y_{22} \end{bmatrix} \begin{bmatrix} \hat{V}_1 \\ \hat{V}_2 \end{bmatrix} \tag{9.58}$$

$$\begin{bmatrix} \hat{V}_1 \\ \hat{I}_1 \end{bmatrix} = \begin{bmatrix} a_{11} & -a_{12} \\ a_{21} & -a_{22} \end{bmatrix} \begin{bmatrix} \hat{V}_2 \\ \hat{I}_2 \end{bmatrix} \tag{9.59}$$

$$\begin{bmatrix} \hat{V}_2 \\ \hat{I}_2 \end{bmatrix} = \begin{bmatrix} b_{11} & -b_{12} \\ b_{21} & -b_{22} \end{bmatrix} \begin{bmatrix} \hat{V}_1 \\ \hat{I}_1 \end{bmatrix} \tag{9.60}$$

$$\begin{bmatrix} \hat{V}_1 \\ \hat{I}_2 \end{bmatrix} = \begin{bmatrix} h_{11} & h_{12} \\ h_{21} & h_{22} \end{bmatrix} \begin{bmatrix} \hat{I}_1 \\ \hat{V}_2 \end{bmatrix} \tag{9.61}$$

$$\begin{bmatrix} \hat{I}_1 \\ \hat{V}_2 \end{bmatrix} = \begin{bmatrix} g_{11} & g_{12} \\ g_{21} & g_{22} \end{bmatrix} \begin{bmatrix} \hat{V}_1 \\ \hat{I}_2 \end{bmatrix} \tag{9.62}$$

The coefficients in the square matrices in Eqs (9.57)–(9.62) are called the *parameters* of the two-port network. We refer to them as the *z* parameters, *y* parameters, *a* parameters, *b* parameters, *h* parameters, or *g* parameters of the network.

All parameter sets contain the same information about a network, and it is always possible to calculate any set in terms of any other set.

Consider Eq. (9.51), repeated here,

$$\hat{V}_1 = z_{11}\hat{I}_1 + z_{12}\hat{I}_2$$
$$\hat{V}_2 = z_{21}\hat{I}_1 + z_{22}\hat{I}_2 \tag{9.63}$$

The *z* parameters can be obtained as

$$z_{11} = \frac{\hat{V}_1}{\hat{I}_1}\bigg|_{\hat{I}_2=0} \quad (\Omega) \tag{9.64a}$$

$$z_{12} = \frac{\hat{V}_1}{\hat{I}_2}\bigg|_{\hat{I}_1=0} \quad (\Omega) \tag{9.64b}$$

$$z_{21} = \frac{\hat{V}_2}{\hat{I}_1}\bigg|_{\hat{I}_2=0} \quad (\Omega) \tag{9.64c}$$

$$z_{11} = \frac{\hat{V}_2}{\hat{I}_2}\bigg|_{\hat{I}_1=0} \quad (\Omega) \tag{9.64d}$$

Thus, the z parameters can be obtained from the voltage and current measurements when each port, one at a time, is open-circuited.

Example 9.6 Calculation of z parameters

Determine the z parameters of the circuit shown in Figure 9.43.

When port 2 is open, $I_2 = 0$, and we have a circuit shown in Figure 9.44. The z parameters for this circuit are obtained as

$$z_{11} = \frac{V_1}{I_1}\bigg|_{I_2=0} = \frac{\left[(6+8)\|4\right]I_1}{I_1} = \frac{14\times4}{14+4} = 3.1111 \ \ \Omega$$

$$z_{21} = \frac{V_2}{I_1}\bigg|_{I_2=0} = \frac{\dfrac{8}{8+6}V_1}{\dfrac{V_1}{(6+8)\|4}} = \frac{0.5714V_1}{\dfrac{V_1}{3.1111}} = 1.7778 \ \ \Omega$$

When port 1 is open, we have a circuit shown in Figure 9.45.

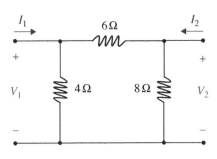

Figure 9.43 Resistive circuit for Example 9.6.

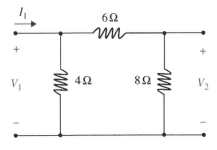

Figure 9.44 Port 2 open-circuited.

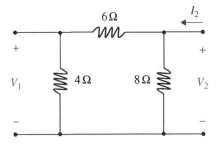

Figure 9.45 Port 1 open-circuited.

The z parameters for this circuit are obtained as

$$z_{22} = \frac{V_2}{I_2}\bigg|_{I_1=0} = \frac{[(6+4)\|8]I_2}{I_2} = \frac{10\times8}{10+8} = 4.4444 \quad \Omega$$

$$z_{12} = \frac{V_1}{I_2}\bigg|_{I_1=0} = \frac{\dfrac{4}{4+6}V_2}{\dfrac{V_2}{(6+4)\|8}} = \frac{0.4V_2}{\dfrac{V_2}{4.4444}} = 1.7778 \quad \Omega$$

∎

To determine the y parameters we reconsider Eq. (9.52), repeated here

$$\hat{I}_1 = y_{11}\hat{V}_1 + y_{12}\hat{V}_2$$
$$\hat{I}_2 = y_{21}\hat{V}_1 + y_{22}\hat{V}_2$$

(9.65)

The y parameters can be obtained as

$$y_{11} = \frac{\hat{I}_1}{\hat{V}_1}\bigg|_{\hat{V}_2=0} \quad (\text{S})$$

(9.66a)

$$y_{12} = \frac{\hat{I}_1}{\hat{V}_2}\bigg|_{\hat{V}_1=0} \quad (\text{S})$$

(9.66b)

$$y_{21} = \frac{\hat{I}_2}{\hat{V}_1}\bigg|_{\hat{V}_2=0} \quad (\text{S})$$

(9.66c)

$$y_{22} = \frac{\hat{I}_2}{\hat{V}_2}\bigg|_{\hat{V}_1=0} \quad (\text{S})$$

(9.66d)

Thus, the y parameters can be obtained from the voltage and current measurements when each port, one at a time, is short-circuited.

The remaining port parameters are obtained in a similar manner. For instance, since

$$\hat{V}_1 = a_{11}\hat{V}_2 - a_{12}\hat{I}_2$$
$$\hat{I}_1 = a_{21}\hat{V}_2 - a_{22}\hat{I}_2$$

(9.67)

we have

$$a_{11} = \frac{\hat{V}_1}{\hat{V}_2}\bigg|_{\hat{I}_2=0}$$

(9.68a)

$$a_{12} = -\frac{\hat{V}_1}{\hat{I}_2}\Bigg|_{\hat{V}_2=0} \quad (\Omega) \qquad\qquad (9.68\text{b})$$

$$a_{21} = \frac{\hat{I}_1}{\hat{V}_2}\Bigg|_{\hat{I}_2=0} \quad (S) \qquad\qquad (9.68\text{c})$$

$$a_{22} = -\frac{\hat{I}_1}{\hat{I}_2}\Bigg|_{\hat{V}_2=0} \qquad\qquad (9.68\text{d})$$

Thus, to obtain the a parameters, both the open-circuit and short-circuit measurements at port 2 are needed.

Example 9.7 Calculation of a parameters

The circuit operates in sinusoidal steady state. When the voltage of $v_1 = 160\cos 4000t$ is applied to port 1 of the two-port network, the following measurements are taken with port 2 open circuited:

$$i_1 = 10\cos(\omega t - 30°)$$
$$v_2 = 80\cos(\omega t + 20°)$$

With port 2 short circuited when the voltage of $v_1 = 60\cos 4000t$ is applied to port 1 the following measurements are taken:

$$i_1 = 6\cos(\omega t + 10°)$$
$$i_2 = 4\cos(\omega t - 40°)$$

Determine the a parameters that describe the sinusoidal steady state operation of the circuit.

Solution: The first set of measurements is described by

$$\hat{V}_1 = 160\angle 0° \quad \text{V}$$
$$\hat{I}_1 = 10\angle -30° \quad \text{A}$$
$$\hat{V}_2 = 80\angle 20° \quad \text{V}$$
$$\hat{I}_2 = 0 \quad \text{A}$$

From Eq. (9.68) we get

$$a_{11} = \frac{\hat{V}_1}{\hat{V}_2}\Bigg|_{\hat{I}_2=0} = \frac{160\angle 0°}{80\angle 20°} = 2\angle -20°$$

$$a_{21} = \frac{\hat{I}_1}{\hat{V}_2}\Bigg|_{\hat{I}_2=0} = \frac{10\angle -30°}{80\angle 20°} = 0.125\angle -50°$$

The second set of measurements is described by

$$\hat{V}_1 = 60\angle 0° \quad \text{V}$$

$$\hat{I}_1 = 3\angle 10° \quad \text{A}$$

$$\hat{V}_2 = 0 \quad \text{V}$$

$$\hat{I}_2 = 4\angle -40° \quad \text{A}$$

Thus,

$$a_{12} = -\frac{\hat{V}_1}{\hat{I}_2}\Bigg|_{\hat{V}_2=0} = -\frac{60\angle 0°}{4\angle -40°} = -15\angle 40° = 15\angle 220°$$

$$a_{22} = -\frac{\hat{I}_1}{\hat{I}_2}\Bigg|_{\hat{V}_2=0} = -\frac{3\angle 10°}{4\angle -40°} = -0.75\angle 50° = 0.75\angle 230°$$

∎

In the typical application of a two-port network, the circuit is driven at port 1 and terminated by a load at port 2, as shown in Figure 9.46.

In this case, we are usually interested in determining the port 2 voltage and current $\hat{V}_2, \hat{I}_2$ in terms of the two-port parameters and $\hat{V}_G, \hat{Z}_G$ and $\hat{Z}_L$. These terminal currents and voltages give rise to six characteristics describing this two-port network:

- input impedance $\hat{Z}_{in} = \hat{V}_1/\hat{I}_1$, or the input admittance $\hat{Y}_{in} = \hat{I}_1/\hat{V}_1$
- output current $\hat{I}_2$
- Thévenin voltage and impedance with respect to port 2, $\hat{V}_{TH}, \hat{Z}_{TH}$
- current gain $\hat{I}_2/\hat{I}_1$
- voltage gain $\hat{V}_2/\hat{V}_1$
- voltage gain $\hat{V}_2/\hat{V}_G$

To illustrate the approach we will use the z parameter set (Nielson and Riedel, 2015, p. 687). We begin with the two defining equations (9.51), repeated here,

$$\hat{V}_1 = z_{11}\hat{I}_1 + z_{12}\hat{I}_2 \tag{9.69}$$

$$\hat{V}_2 = z_{21}\hat{I}_1 + z_{22}\hat{I}_2 \tag{9.70}$$

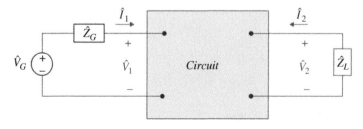

Figure 9.46 Typical two-port circuit.

The circuit shown in Figure 9.46 produces two additional equations:

$$\hat{V}_1 = \hat{V}_G - \hat{Z}_G \hat{I}_1 \tag{9.71}$$

$$\hat{V}_2 = -\hat{Z}_L \hat{I}_2 \tag{9.72}$$

Determining the input impedance $\hat{Z}_{in} = \hat{V}_1/\hat{I}_1$ Using Eq. (9.72) in Eq. (9.70) we obtain

$$-\hat{Z}_L \hat{I}_2 = z_{21} \hat{I}_1 + z_{22} \hat{I}_2 \tag{9.73}$$

and thus

$$\hat{I}_2 = -\frac{z_{21}}{z_{22} + \hat{Z}_L} \hat{I}_1 \tag{9.74}$$

Using Eq. (9.74) in Eq. (9.69) we get

$$\hat{V}_1 = z_{11} \hat{I}_1 - \frac{z_{12} z_{21}}{z_{22} + \hat{Z}_L} \hat{I}_1 \tag{9.75}$$

thus, the input impedance is

$$\boxed{\hat{Z}_{in} = z_{11} - \frac{z_{12} z_{21}}{z_{22} + \hat{Z}_L}} \tag{9.76}$$

Determining the current $\hat{I}_2$ From Eq. (9.69)

$$\hat{I}_1 = \frac{\hat{V}_1 - z_{12} \hat{I}_2}{z_{11}} \tag{9.77}$$

Using Eq. (9.71) in Eq. (9.77) produces

$$\hat{I}_1 = \frac{\hat{V}_G - \hat{Z}_G \hat{I}_1 - z_{12} \hat{I}_2}{z_{11}} \tag{9.78}$$

or

$$z_{11} \hat{I}_1 = \hat{V}_G - \hat{Z}_G \hat{I}_1 - z_{12} \hat{I}_2$$
$$\left(z_{11} + \hat{Z}_G \right) \hat{I}_1 = \hat{V}_G - z_{12} \hat{I}_2 \tag{9.79}$$

and

$$\hat{I}_1 = \frac{\hat{V}_G - z_{12} \hat{I}_2}{z_{11} + \hat{Z}_G} \tag{9.80}$$

Now, we substitute Eq. (9.80) into Eq. (9.74) to obtain

$$\hat{I}_2 = -\frac{z_{21}}{z_{22} + \hat{Z}_L} \frac{\hat{V}_G - z_{12} \hat{I}_2}{z_{11} + \hat{Z}_G} \tag{9.81}$$

or

$$\left(z_{22} + \hat{Z}_L\right)\left(z_{11} + \hat{Z}_G\right)\hat{I}_2 = -z_{21}\hat{V}_G + z_{21}z_{12}\hat{I}_2 \tag{9.82}$$

and therefore

$$\boxed{\hat{I}_2 = \frac{-z_{21}\hat{V}_G}{\left(z_{22} + \hat{Z}_L\right)\left(z_{11} + \hat{Z}_G\right) - z_{21}z_{12}}} \tag{9.83}$$

Determining the Thévenin voltage with respect to port 2   The Thévenin voltage with respect to port 2 is equal to $\hat{V}_2$ with the imposed condition of $\hat{I}_2 = 0$. Setting $\hat{I}_2 = 0$ in Eqs (9.69) and (9.70) produces

$$\hat{V}_1 = z_{11}\hat{I}_1 \quad \left(\hat{I}_2 = 0\right) \tag{9.84}$$

$$\hat{V}_2 = z_{21}\hat{I}_1 \quad \left(\hat{I}_2 = 0\right) \tag{9.85}$$

Utilizing Eq. (9.84) in Eq. (9.85) gives

$$\hat{V}_2 = \frac{z_{21}}{z_{11}}\hat{V}_1 \quad \left(\hat{I}_2 = 0\right) \tag{9.86}$$

Using Eq. (9.71) in Eq. (9.86) results in

$$\hat{V}_2 = \frac{z_{21}}{z_{11}}\left(\hat{V}_G - \hat{Z}_G\hat{I}_1\right) \quad \left(\hat{I}_2 = 0\right) \tag{9.87}$$

Setting $\hat{I}_2 = 0$ in Eqs (9.80) produces

$$\hat{I}_1 = \frac{\hat{V}_G}{z_{11} + \hat{Z}_G} \quad \left(\hat{I}_2 = 0\right) \tag{9.88}$$

Substituting Eq. (9.88) into Eq. (9.87) gives

$$\hat{V}_2 = \frac{z_{21}}{z_{11}}\left(\hat{V}_G - \hat{Z}_G\frac{\hat{V}_G}{z_{11} + \hat{Z}_G}\right) \quad \left(\hat{I}_2 = 0\right)$$

$$\hat{V}_2 = \frac{z_{21}}{z_{11}}\left[\frac{\left(z_{11} + \hat{Z}_G\right)\hat{V}_G - \hat{Z}_G\hat{V}_G}{z_{11} + \hat{Z}_G}\right] \tag{9.89}$$

$$\hat{V}_2 = \frac{z_{21}}{z_{11}}\left(\frac{z_{11}\hat{V}_G}{z_{11} + \hat{Z}_G}\right)$$

and thus the Thévenin voltage with respect to port 2 is given by

$$\boxed{\hat{V}_{TH} = \frac{z_{21}}{z_{11} + \hat{Z}_G}\hat{V}_G} \tag{9.90}$$

The Thévenin, or output impedance, can be obtained from

$$\hat{Z}_{TH} = \frac{\hat{V}_2}{\hat{I}_2}, \quad \left(\hat{V}_G = 0 \right) \tag{9.91}$$

When $\hat{V}_G = 0$, Eq, (9.71), repeated here

$$\hat{V}_1 = \hat{V}_G - \hat{Z}_G \hat{I}_1 \tag{9.92}$$

reduces to

$$\hat{V}_1 = -\hat{Z}_G \hat{I}_1 \tag{9.93}$$

Using Eq. (9.93) in Eq. (9.69), repeated here,

$$\hat{V}_1 = z_{11} \hat{I}_1 + z_{12} \hat{I}_2 \tag{9.94}$$

gives

$$-\hat{Z}_G \hat{I}_1 = z_{11} \hat{I}_1 + z_{12} \hat{I}_2 \tag{9.95}$$

or

$$\left(z_{11} + \hat{Z}_G \right) \hat{I}_1 = -z_{12} \hat{I}_2 \tag{9.96}$$

resulting in

$$\hat{I}_1 = \frac{-z_{12} \hat{I}_2}{z_{11} + \hat{Z}_G}, \quad \left(\hat{V}_G = 0 \right) \tag{9.97}$$

Using Eq. (9.97) in Eq. (9.70), repeated here

$$\hat{V}_2 = z_{21} \hat{I}_1 + z_{22} \hat{I}_2 \tag{9.98}$$

produces

$$\hat{V}_2 = z_{21} \left(\frac{-z_{12} \hat{I}_2}{z_{11} + \hat{Z}_G} \right) + z_{22} \hat{I}_2 \tag{9.99}$$

or

$$\hat{V}_2 = \left(z_{22} - \frac{z_{21} z_{12}}{z_{11} + \hat{Z}_G} \right) \hat{I}_2, \quad \left(\hat{V}_G = 0 \right) \tag{9.100}$$

resulting in Thévenin impedance of

$$\boxed{ \hat{Z}_{TH} = \frac{\hat{V}_2}{\hat{I}_2} = z_{22} - \frac{z_{21} z_{12}}{z_{11} + \hat{Z}_G} } \tag{9.101}$$

Determining the current gain $\hat{I}_2/\hat{I}_1$ The current gain can be obtained directly from Eq. (9.74), repeated here

$$\hat{I}_2 = -\frac{z_{21}}{z_{22} + \hat{Z}_L}\hat{I}_1 \tag{9.102}$$

Thus,

$$\boxed{\frac{\hat{I}_2}{\hat{I}_1} = -\frac{z_{21}}{z_{22} + \hat{Z}_L}} \tag{9.103}$$

Determining the voltage gain $\hat{V}_2/\hat{V}_1$ We start with Eq. (9.70) and (9.72), repeated here,

$$\hat{V}_2 = z_{21}\hat{I}_1 + z_{22}\hat{I}_2 \tag{9.104}$$

$$\hat{V}_2 = -\hat{Z}_L\hat{I}_2 \tag{9.105}$$

Solving Eq. (9.105) for $\hat{I}_2$ and substituting it in Eq. (9.104) produces

$$\hat{V}_2 = z_{21}\hat{I}_1 + z_{22}\left(\frac{-\hat{V}_2}{\hat{Z}_L}\right) \tag{9.106}$$

Now, solving Eq. (9.69), repeated here

$$\hat{V}_1 = z_{11}\hat{I}_1 + z_{12}\hat{I}_2 \tag{9.107}$$

for $\hat{I}_2$ gives

$$\hat{I}_1 = \frac{\hat{V}_1}{z_{11}} - \frac{z_{12}}{z_{11}}\hat{I}_2 \tag{9.108}$$

Solving Eq. (9.105) for $\hat{I}_2$ and substituting it in Eq. (9.108) produces

$$\hat{I}_1 = \frac{\hat{V}_1}{z_{11}} - \frac{z_{12}}{z_{11}}\left(-\frac{\hat{V}_2}{\hat{Z}_L}\right) \tag{9.109}$$

Substituting Eq. (9.109) into Eq. (9.106) results in

$$\hat{V}_2 = z_{21}\left[\frac{\hat{V}_1}{z_{11}} - \frac{z_{12}}{z_{11}}\left(-\frac{\hat{V}_2}{\hat{Z}_L}\right)\right] + z_{22}\left(\frac{-\hat{V}_2}{\hat{Z}_L}\right) \tag{9.110}$$

or

$$\hat{V}_2 - \frac{z_{12}z_{21}}{z_{11}}\left(\frac{\hat{V}_2}{\hat{Z}_L}\right) + z_{22}\left(\frac{\hat{V}_2}{\hat{Z}_L}\right) = \frac{z_{21}}{z_{11}}\hat{V}_1$$

$$\left(1 - \frac{z_{12}z_{21}}{z_{11}\hat{Z}_L} + \frac{z_{22}}{\hat{Z}_L}\right)\hat{V}_2 = \frac{z_{21}}{z_{11}}\hat{V}_1 \tag{9.111}$$

$$\left(\frac{z_{11}\hat{Z}_L - z_{12}z_{21} + z_{11}z_{22}}{z_{11}\hat{Z}_L}\right)\hat{V}_2 = \frac{z_{21}}{z_{11}}\hat{V}_1$$

and thus

$$\boxed{\dfrac{\hat{V}_2}{\hat{V}_1} = \dfrac{z_{21}\hat{Z}_L}{z_{11}\hat{Z}_L - z_{11}z_{22} - z_{12}z_{21}}}$$

(9.112)

Determining the voltage gain $\hat{V}_2/\hat{V}_G$ We start with Eqs (9.69), (9.71), and (9.72), repeated here,

$$\hat{V}_1 = z_{11}\hat{I}_1 + z_{12}\hat{I}_2$$

(9.113)

$$\hat{V}_1 = \hat{V}_G - \hat{Z}_G\hat{I}_1$$

(9.114)

$$\hat{V}_2 = -\hat{Z}_L\hat{I}_2$$

(9.115)

From Eq. (9.115) we obtain

$$\hat{I}_2 = -\dfrac{\hat{V}_2}{\hat{Z}_L}$$

(9.116)

Using Eqs (9.114) and (9.116) in Eq. (9.113) results in

$$\hat{V}_G - \hat{Z}_G\hat{I}_1 = z_{11}\hat{I}_1 + z_{12}\left(-\dfrac{\hat{V}_2}{\hat{Z}_L}\right)$$

(9.117)

or

$$z_{11}\hat{I}_1 + \hat{Z}_G\hat{I}_1 = \hat{V}_G + z_{12}\left(\dfrac{\hat{V}_2}{\hat{Z}_L}\right)$$
$$\left(z_{11} + \hat{Z}_G\right)\hat{I}_1 = \hat{V}_G + z_{12}\left(\dfrac{\hat{V}_2}{\hat{Z}_L}\right)$$

(9.118)

resulting in

$$\hat{I}_1 = \dfrac{\hat{V}_G}{z_{11} + \hat{Z}_G} + \dfrac{z_{12}\hat{V}_2}{\hat{Z}_L\left(z_{11} + \hat{Z}_G\right)}$$

(9.119)

We now use Eq. (9.119) together with Eqs (9.70) and (9.72), repeated here

$$\hat{V}_2 = z_{21}\hat{I}_1 + z_{22}\hat{I}_2$$

(9.120)

$$\hat{V}_2 = -\hat{Z}_L\hat{I}_2$$

(9.121)

to obtain,

$$\hat{V}_2 = z_{21}\left[\dfrac{\hat{V}_G}{z_{11} + \hat{Z}_G} + \dfrac{z_{12}\hat{V}_2}{\hat{Z}_L\left(z_{11} + \hat{Z}_G\right)}\right] + z_{22}\left(\dfrac{-\hat{V}_2}{\hat{Z}_L}\right)$$

(9.122)

or

$$\hat{V}_2 + z_{22}\left(\frac{\hat{V}_2}{\hat{Z}_L}\right) - \frac{z_{21}z_{12}\hat{V}_2}{\hat{Z}_L\left(z_{11} + \hat{Z}_G\right)} = \frac{z_{21}\hat{V}_G}{z_{11} + \hat{Z}_G}$$

$$\left[1 + \frac{z_{22}}{\hat{Z}_L} - \frac{z_{21}z_{12}}{\hat{Z}_L\left(z_{11} + \hat{Z}_G\right)}\right]\hat{V}_2 = \frac{z_{21}\hat{V}_G}{z_{11} + \hat{Z}_G} \tag{9.123}$$

$$\left[\frac{\hat{Z}_L\left(z_{11} + \hat{Z}_G\right) + z_{22}\left(z_{11} + \hat{Z}_G\right) - z_{21}z_{12}}{\hat{Z}_L\left(z_{11} + \hat{Z}_G\right)}\right]\hat{V}_2 = \frac{z_{21}\hat{V}_G}{z_{11} + \hat{Z}_G}$$

or

$$\left[\frac{\hat{Z}_L\left(z_{11} + \hat{Z}_G\right) + z_{22}\left(z_{11} + \hat{Z}_G\right) - z_{21}z_{12}}{\hat{Z}_L}\right]\hat{V}_2 = z_{21}\hat{V}_G \tag{9.124}$$

and thus

$$\boxed{\frac{\hat{V}_2}{\hat{V}_G} = \frac{z_{21}\hat{Z}_L}{\left(\hat{Z}_L + z_{22}\right)\left(z_{11} + \hat{Z}_G\right) - z_{21}z_{12}}} \tag{9.125}$$

9.7 EMC Applications

9.7.1 Fourier Series Representation of Signals

Fourier series representation of periodic signals is perhaps the greatest example of an application of the superposition principle in EMC. (We will devote the entire Chapter 12 to this important topic.)

A periodic signal $x(t)$ can be represented as an infinite series of the form

$$x(t) = \sum_{n=0}^{\infty} c_n \phi_n(t) = c_0\phi_0(t) + c_1\phi_1(t) + c_2\phi_2(t) + \cdots \tag{9.126}$$

where the c_n are called the expansion coefficients and the $\varphi_n(t)$ are called the basis functions – they are periodic with the same period as $x(t)$.

If we know the response of a linear system, $y_i(t)$, to each basis function, $\varphi_i(t)$, as illustrated in Figure 9.47, then the response of the system, $y(t)$, to the original signal $x(t)$, as shown in Figure 9.48, can be obtained as a weighted sum of the individual responses

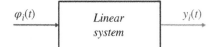

Figure 9.47 Response of a linear system to a basis function.

Figure 9.48 Response of a linear system to the signal $x(t)$.

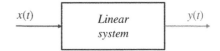

$$y(t) = \sum_{n=0}^{\infty} c_n y_n(t) = c_0 y_0(t) + c_1 y_1(t) + c_2 y_2(t) + \cdots \tag{9.127}$$

Not only are the individual responses easier to obtain or analyze, but they also give us an insight into the nature of the system.

There are infinitely many representations of the form in Eq. (9.126). One extremely useful form in EMC is the Fourier series representation (Kreyszig, 1999, p. 528) which, in the time domain, can be expressed as

$$x(t) = a_0 + \sum_{n=1}^{\infty} (a_n \cos n\omega_0 t + b_n \sin n\omega_0 t), \qquad t_1 < t < t_1 + T \tag{9.128}$$

where the Fourier coefficients are given by

$$a_0 = \frac{1}{T} \int_{t_1}^{t_1+T} x(t) dt \tag{9.129a}$$

$$a_n = \frac{2}{T} \int_{t_1}^{t_1+T} x(t) \cos n\omega_0 t dt \tag{9.129b}$$

$$b_n = \frac{2}{T} \int_{t_1}^{t_1+T} x(t) \sin n\omega_0 t dt \tag{9.129c}$$

In Chapter 12, we will show that the time-domain representation (9.128) is equivalent to the complex Fourier series representation

$$x(t) = c_0 + \sum_{n=1}^{\infty} 2c_n \cos(n\omega_0 t + \theta_{cn}) \tag{9.130a}$$

$$\hat{c}_n = c_n \angle \theta_{cn} = c_n e^{j\theta_{cn}} \tag{9.130b}$$

We will use this representation to analyze the spectrum of a digital signal, like the one shown in Figure 9.49.

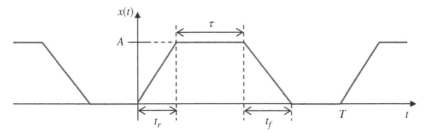

Figure 9.49 Trapezoidal clock signal.

That analysis will allow us to determine spectral bounds on clock signals and estimate the bandwidth of such signals.

9.7.2 Maximum Power Radiated by an Antenna

In Chapter 18 we will discuss the radiation mechanism from the typical EMC antennas. A physical model of an antenna in a transmitting mode is shown in Figure 9.50.

A circuit model of such an antenna is shown in Figure 9.51.

$\hat{Z}_{in}$ is the input impedance of the antenna, i.e. the impedance presented by the antenna to the generator circuit at the antenna's input terminals A-B. Figure 9.52 shows the details of the two impedances shown in Figure 9.51.

The antenna and generator impedances are given by

$$\hat{Z}_{in} = R_{in} + jX_{in} \tag{9.131}$$

$$R_{in} = R_{loss} + R_{rad} \tag{9.132}$$

$$\hat{Z}_g = R_g + jX_g \tag{9.133}$$

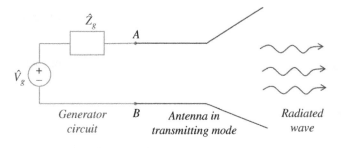

Figure 9.50 Physical model of an antenna in a transmitting mode.

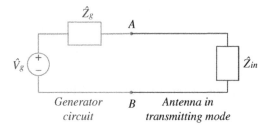

Figure 9.51 Circuit model of an antenna in a transmitting mode.

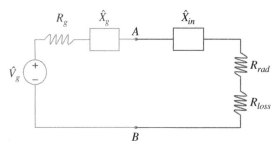

Figure 9.52 Detailed circuit model of an antenna in a transmitting mode.

Figure 9.53 Antenna current.

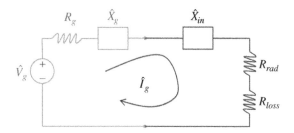

Here R_{in} is the antenna resistance at terminals *A-B*, X_{in} is the antenna reactance at terminals *A-B*, R_{loss} is the loss resistance of the antenna, R_{rad} is the radiation resistance of the antenna, R_g is the generator resistance, and X_g is the generator reactance.

In order to determine the maximum power transfer to the antenna, we first obtain an expression for the current flowing in the circuit, as shown in Figure 9.53.

The magnitude of the antenna current is

$$\left|\hat{I}_g\right| = \frac{\left|\hat{V}_g\right|}{\sqrt{\left(R_{rad} + R_{loss} + R_g\right)^2 + \left(X_{in} + X_g\right)^2}} \tag{9.134}$$

The power supplied by the generator is

$$P_s = \frac{1}{2}\hat{V}_g \hat{I}_g^* = \frac{1}{2}\hat{V}_g \left[\frac{\hat{V}_g}{\left(R_{rad} + R_{loss} + R_g\right) + j\left(X_{in} + X_g\right)}\right]^* \tag{9.135}$$

The power dissipated in the generator as heat is

$$P_g = \frac{1}{2}\left|\hat{I}_g\right|^2 R_g = \frac{\left|\hat{V}_g\right|^2}{2} \frac{R_g}{\left(R_{rad} + R_{loss} + R_g\right)^2 + \left(X_{in} + X_g\right)^2} \quad [\text{W}] \tag{9.136}$$

The power dissipated in the antenna as heat is

$$P_{loss} = \frac{1}{2}\left|\hat{I}_g\right|^2 R_{loss} = \frac{\left|\hat{V}_g\right|^2}{2} \frac{R_{loss}}{\left(R_{rad} + R_{loss} + R_g\right)^2 + \left(X_{in} + X_g\right)^2} \quad [\text{W}] \tag{9.137}$$

Finally, the power radiated by the antenna is

$$P_{rad} = \frac{1}{2}\left|\hat{I}_g\right|^2 R_{rad} = \frac{\left|\hat{V}_g\right|^2}{2} \frac{R_{rad}}{\left(R_{rad} + R_{loss} + R_g\right)^2 + \left(X_{in} + X_g\right)^2} \quad [\text{W}] \tag{9.138}$$

The maximum power delivered to the antenna for radiation occurs when

$$\hat{Z}_{in} = \hat{Z}_g^* \tag{9.139}$$

and thus

$$R_{rad} + R_{loss} = R_g \tag{9.140a}$$

$$X_{in} = -X_g \tag{9.140b}$$

Under the conditions in Eq. (9.140) the maximum power radiated by the antenna is

$$P_{rad} = \frac{\left|\hat{V}_g\right|^2}{2} \frac{R_{rad}}{\left[2\left(R_{rad} + R_{loss}\right)\right]^2} = \frac{\left|\hat{V}_g\right|^2}{8} \frac{R_{rad}}{\left(R_{rad} + R_{loss}\right)^2} \tag{9.141}$$

9.7.3 s Parameters

The two-port parameter sets described in this chapter require the input and output terminals of the network to be either open- or short-circuited. This can be hard to do at high frequencies where lead inductance and capacitance make short and open circuits difficult to obtain.

To characterize high-frequency circuits using *s* parameters, we use matched terminations instead of the open or short circuits.

Just like the other sets of parameters, *s* parameters completely describe the performance of a two-port network.

Unlike the other sets of parameters, *s* parameters do not make use of open-circuit or short-circuit measurements voltage or current measurements, but rather relate the traveling waves that are incident, reflected, and transmitted when a two-port network in inserted into a transmission line. This is depicted in Figure 9.54.

Travelling waves, unlike terminal voltages and currents, do not vary in magnitude at points along a lossless transmission line (waves and transmission lines will be discussed in Chapters 16 and 17, respectively). This means that the *s* parameters can be measured with the device at some distance from the measurement ports, provided that the line is a low-loss transmission line.

s Parameters are usually measured with the device embedded between a 50 Ω load and a 50 Ω source.

The incident waves (a_1, a_2) and reflected waves (b_1, b_2) used to define *s* parameters for a two-port network are shown in Figure 9.55.

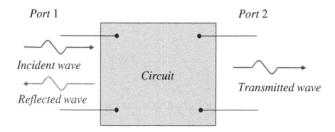

Figure 9.54 *s* Parameters are related to the traveling waves.

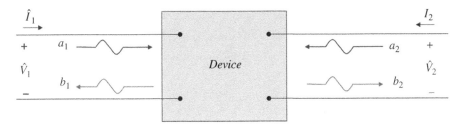

Figure 9.55 Incident and reflected waves.

The linear equations describing this two-port network in terms of the *s* parameters are

$$b_1 = s_{11}a_1 + s_{12}a_2$$
$$b_2 = s_{21}a_1 + s_{22}a_2$$
(9.142)

or in a matrix form

$$\begin{bmatrix} b_1 \\ b_2 \end{bmatrix} = \begin{bmatrix} s_{11} & s_{12} \\ s_{21} & s_{22} \end{bmatrix} \begin{bmatrix} a_1 \\ a_2 \end{bmatrix}$$
(9.143)

where *S* is the scattering matrix given by

$$S = \begin{bmatrix} s_{11} & s_{12} \\ s_{21} & s_{22} \end{bmatrix}$$
(9.144)

References

Alexander, C.K. and Sadiku, N.O., *Fundamentals of Electric Circuits*, 4th ed., McGraw Hill, New York, 2009.

Kreyszig, E., *Advanced Engineering Mathematics*, 8th ed., John Wiley and Sons, New York, 1999.

Nilsson, J.W. and Riedel, S.A., *Electric Circuits*, 10th ed., Pearson, Upper Saddle River, NJ, 2015.

Paul, C.R., *Introduction to Electromagnetic Compatibility*, 2nd ed., John Wiley and Sons, New York, 2006.

10

Magnetically Coupled Circuits

10.1 Self and Mutual Inductance

Consider the circuit shown in Figure 10.1.

Time-varying current $i_1(t)$ gives rise to the time-varying flux Φ_{11} that crosses the loop in which the current i_1 flows. According to Faraday's law this time-varying flux crossing the loop induces a voltage in the loop.

We model this by introducing the concept of the *self inductance* of circuit 1, defined as

$$L_1 = \frac{\Phi_{11}}{i_1} \tag{10.1}$$

We then augment the circuit of Figure 10.1 to that shown in Figure 10.2.

Note that the loop inductance is a property of the loop itself and does not depend on the voltage V_S or the current i (just like the resistance of a resistor does not depend on the voltage across it or the current through it).

That is, any closed loop will have its own self inductance, whether current flows through it or not. It can, therefore, be represented by the circuit model shown in Figure 10.3.

Now consider the situation where another circuit (with its own self inductance) is placed next to the original circuit, as shown in Figure 10.4.

The time-varying flux Φ_{12} created by the current i_1 flowing in loop 1 and intersecting loop 2 induces a voltage in loop 2.

We model this by introducing the concept of the *mutual inductance* between the circuits 1 and 2 as

$$M_{12} = \frac{\Phi_{12}}{i_1} \tag{10.2}$$

Using the concept of the mutual inductance we can now augment the circuit of Figure 10.4 to that shown in Figure 10.5.

Similar discussion leads to the scenario shown in Figure 10.6. Here circuit 2 is driven by a voltage source and circuit 1 is represented by a loop containing resistance and self inductance.

In many practical cases (Alexander and Sadiku, 2009, p. 558)

$$M_{12} = M_{12} = M \tag{10.3}$$

Foundations of Electromagnetic Compatibility with Practical Applications, First Edition. Bogdan Adamczyk.
© 2017 John Wiley & Sons Ltd. Published 2017 by John Wiley & Sons Ltd.

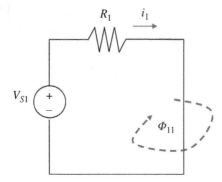

Figure 10.1 Magnetic flux produced by the current flowing in a loop.

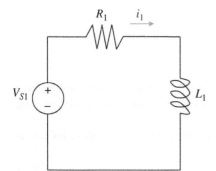

Figure 10.2 Loop self-inductance.

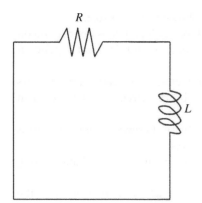

Figure 10.3 Circuit model of a loop.

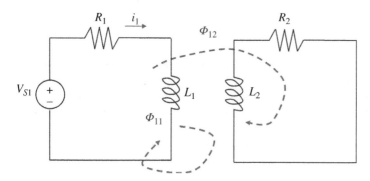

Figure 10.4 Flux caused by current in loop 1 intersects loop 2.

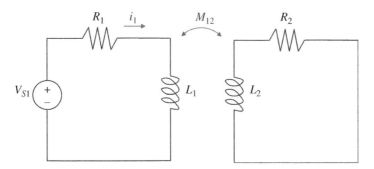

Figure 10.5 Mutual inductance between loops 1 and 2.

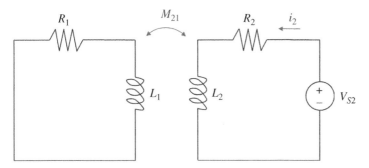

Figure 10.6 Mutual inductance between loops 2 and 1.

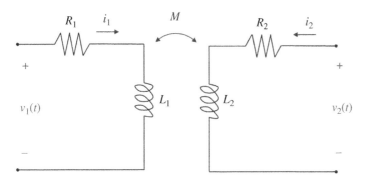

Figure 10.7 Mutual inductance between loops 2 and 1.

Let's combine the two scenarios shown in Figures 10.5 and 10.6, as shown in Figure 10.7.

When the time-varying current flows through the self inductance it gives rise to the voltage across it according to

$$v = L\frac{di}{dt}$$

(10.4a)

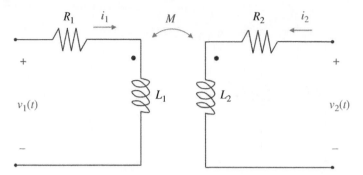

Figure 10.8 Mutual inductance with dot convention.

when the passive sign convention is satisfied, or according to

$$v = -L\frac{di}{dt} \tag{10.4b}$$

when the passive sign convention is not satisfied.

The polarity of the voltage due to the mutual inductance cannot be determined using the passive sign convention.

$$v = \pm M\frac{di}{dt} \tag{10.5}$$

The choice of the correct polarity for $M\,di/dt$ depends on the physical configuration of the circuit. This is indicated in the circuit by the dot marking, as shown in Figure 10.8.

The dot convention allows us to determine the polarity of the induced voltage according to the following rule (Nilsson and Riedel, 2015, p. 190).

When the reference direction of the current in one circuit enters the dotted terminal, the reference polarity of the induced voltage in another circuit is positive at the dotted terminal.

Figure 10.9 shows how to apply the dot convention.

Figure 10.10 shows the polarities of the induced voltages for the circuit shown in Figure 10.8.

KVL for the circuit on the left results in

$$v_1 = R_1 i_1 + L_1\frac{di_1}{dt} + M\frac{di_2}{dt} \tag{10.6a}$$

while for the circuit on the right we have

$$v_2 = R_2 i_2 + L_2\frac{di_2}{dt} + M\frac{di_1}{dt} = 0 \tag{10.6b}$$

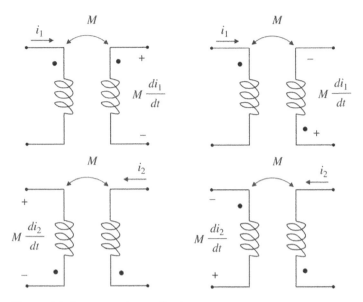

Figure 10.9 Dot convention application.

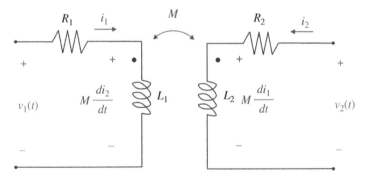

Figure 10.10 Polarities of the induced voltages.

Example 10.1 Coupled transmission lines

In Chapter 15 we will discuss the distributed-parameter transmission lines. The per-unit-length equivalent circuit of a three-conductor transmission line is shown in Figure 10.11 (Paul, 2006, p. 566).

KVL along the outside loop yields

$$-V_G(z,t)+l_G\Delta z\frac{\partial I_G(z,t)}{\partial t}+l_m\Delta z\frac{\partial I_R(z,t)}{\partial t}+V_G(z+\Delta z,t)=0 \tag{10.7a}$$

while the KVL along the lower loop in Figure 10.11 produces

$$-V_R(z,t)+l_R\Delta z\frac{\partial I_R(z,t)}{\partial t}+l_m\Delta z\frac{\partial I_G(z,t)}{\partial t}+V_R(z+\Delta z,t)=0 \tag{10.7b}$$

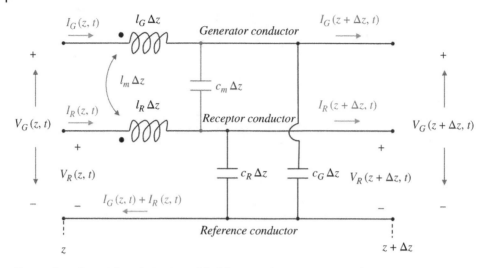

Figure 10.11 Per-unit length circuit model of three-conductor transmission line.

10.2 Energy in a Coupled Circuit

Recall that energy stored in an inductor is given by

$$w = \frac{1}{2}Li^2 \tag{10.8}$$

Let's determine the energy stored in magnetically coupled coils. Consider the circuit shown in Figure 10.12.

Let's assume that the initial current sin the coils are zero so there is no initial energy stored in the coils.

First we increase i_1 from zero to I_1, while keeping $i_2 = 0$. Since $i_2 = 0$ there is no mutual voltage induced in coil 1. The mutual voltage induced in coil 2 is $M\, di_1/dt$.

The energy stored in both coils is

$$w_1 = \int_0^t p_1(t)dt = \int_0^t \left(L_1 \frac{di_1}{dt} \right) i_1(t)dt + \int_0^t \left(M \frac{di_1}{dt} \right) i_2 dt$$

$$= L_1 \int_{i_1=0}^{i_1=I_1} i_1(t)di_1 + 0 = \frac{1}{2}L_1 I_1^2 \tag{10.9}$$

If we maintain $i_1 = I_1$ and increase i_2 from zero to I_2 the mutual voltage induced in coil 1 is $M\, di_2/dt$. The mutual voltage induced in coil 2 is zero since $i_1 = const$.

The energy stored in both coils now is

$$w_2 = \int_0^t p_2(t)dt = \int_0^t \left(L_2 \frac{di_2}{dt} \right) i_2(t)dt + \int_0^t \left(M \frac{di_2}{dt} \right) I_1 dt$$

$$= L_1 \int_{i_2=0}^{i_2=I_2} i_2(t)di_2 + MI_1 \int_{i_2=0}^{i_2=I_2} di_2 = \frac{1}{2}L_2 I_2^2 + MI_1 I_2 \tag{10.10}$$

Figure 10.12 Coupled coils.

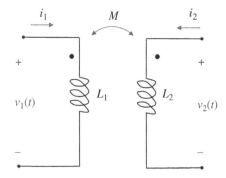

The total energy stored in the coils when both i_1 and i_2 have reached constant values is

$$w = w_1 + w_2 = \frac{1}{2}L_1I_1^2 + \frac{1}{2}L_2I_2^2 + MI_1I_2 \qquad (10.11)$$

Equation (10.11) was derived with the assumption that the coil currents both entered the dotted terminals. If one current enters the dotted terminal and the other does not, the mutual voltage is negative and the mutual energy MI_1I_2 is also negative.

The total energy in the system then is

$$w = w_1 + w_2 = \frac{1}{2}L_1I_1^2 + \frac{1}{2}L_2I_2^2 - MI_1I_2 \qquad (10.12)$$

Since I_1 and I_2 were arbitrarily chosen, we may replace them by any other values. Let's replace them by i_1 and i_2. Then we obtain a general expression for the instantaneous energy stored in the system:

$$w = \frac{1}{2}L_1i_1^2 + \frac{1}{2}L_2i_2^2 \pm Mi_1i_2 \qquad (10.13)$$

Equation (10.13) allows us to determine the upper limit for the value of the mutual inductance. Since the circuit consists of passive elements, the energy stored in it cannot be negative. Thus,

$$\frac{1}{2}L_1i_1^2 + \frac{1}{2}L_2i_2^2 - Mi_1i_2 \geq 0 \qquad (10.14)$$

Then, since

$$\frac{1}{2}\left(i_1\sqrt{L_1} - i_2\sqrt{L_2}\right)^2 = \frac{1}{2}L_1i_1^2 - i_1i_2\sqrt{L_1L_2} + \frac{1}{2}L_2i_2^2 \qquad (10.15)$$

inequality (10.14) can be written as

$$\frac{1}{2}\left(i_1\sqrt{L_1} - i_2\sqrt{L_2}\right)^2 + i_1i_2\sqrt{L_1L_2} - Mi_1i_2 \geq 0 \qquad (10.16)$$

The first term is never negative. This leads to the condition

$$i_1i_2\left(\sqrt{L_1L_2} - M\right) \geq 0 \qquad (10.17)$$

thus

$$\sqrt{L_1 L_2} - M \geq 0 \qquad (10.18)$$

or

$$M \leq \sqrt{L_1 L_2} \qquad (10.19)$$

Thus, the mutual inductance cannot be greater than the geometric mean of the self inductances of the coils.

Coefficient of coupling specifies the extent to which the mutual inductance approaches its limit.

$$k = \frac{M}{\sqrt{L_1 L_2}} \qquad (10.20)$$

or, equivalently,

$$M = k\sqrt{L_1 L_2} \qquad (10.21)$$

10.3 Linear (Air-Core) Transformers

Figure 10.13 shows a circuit model of a linear transformer.

The coil connected directly to the voltage source is called the *primary winding*. The coil connected to the load is the *secondary winding*. The resistors R_1 and R_2 account for the power losses. The transformer is said to be linear if the coils are wound on a magnetically linear material – a material with a constant magnetic permeability. Linear transformers are sometimes called *air-core transformers* (Alexander and Sadiku, 2009, p. 568).

Let's obtain the input impedance, $\hat{Z}_{in}$, as seen by the source. Applying KVL to the primary and the secondary coils gives

$$\hat{V} = R_1 \hat{I}_1 + j\omega L_1 \hat{I}_1 - j\omega M \hat{I}_2 \qquad (10.22a)$$

$$0 = R_2 \hat{I}_2 + j\omega L_2 \hat{I}_2 + \hat{Z}_L \hat{I}_2 - j\omega M \hat{I}_1 \qquad (10.22b)$$

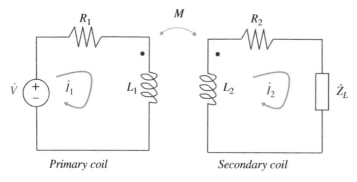

Figure 10.13 A linear transformer.

From Eq. (10.22b) we get

$$\hat{I}_2 = \frac{j\omega\, M\hat{I}_1}{R_2 + j\omega\, L_2 + \hat{Z}_L} \tag{10.23}$$

Substituting Eq. (10.23) into Eq. (10.22a) we obtain

$$\hat{V} = \left(R_1 + j\omega\, L_1\right)\hat{I}_1 - j\omega\, M\,\frac{j\omega\, M\hat{I}_1}{R_2 + j\omega\, L_2 + \hat{Z}_L} \tag{10.24}$$

or

$$\hat{V} = \left(R_1 + j\omega\, L_1\right)\hat{I}_1 + \frac{j\omega^2\, M^2}{R_2 + j\omega\, L_2 + \hat{Z}_L}\,\hat{I}_1 \tag{10.25}$$

Thus the input impedance seen by the source is

$$\hat{Z}_{in} = \frac{\hat{V}}{\hat{I}_1} = R_1 + j\omega\, L_1 + \frac{j\omega^2\, M^2}{R_2 + j\omega\, L_2 + \hat{Z}_L} \tag{10.26}$$

The first term on the right-hand side of Eq. (10.26), $R_1 + j\omega L_1$, is the primary impedance. The second term is the result of the coupling between the primary and secondary windings. It is known as the *reflected impedance* $\hat{Z}_R$,

$$\hat{Z}_R = \frac{j\omega^2\, M^2}{R_2 + j\omega\, L_2 + \hat{Z}_L} \tag{10.27}$$

10.4 Ideal (Iron-Core) Transformers

An ideal transformer consists of two or more coils with a large number of turns wound on a common core of high permeability (with no losses or magnetic flux leakage). Iron-core transformers are close approximations to ideal transformers (Alexander and Sadiku, 2019, p. 574).

The circuit symbol of an ideal transformer is shown in Figure 10.14.

Figure 10.14 Circuit symbol of an ideal transformer.

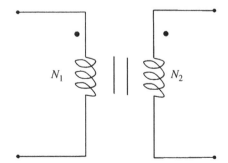

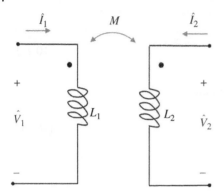

$\hat{I}_1$ M $\hat{I}_2$ **Figure 10.15** Coupled coils in frequency domain.

The vertical lines between the coils indicate an iron core, as distinct from the air core used in linear transformers. In an ideal transformer, the primary and secondary coils are lossless ($R_1 = 0$, $R_2 = 0$).

Consider the circuit shown in Figure 10.15.

The circuit is governed by the following equations

$$\hat{V}_1 = j\omega L_1 \hat{I}_1 + j\omega M \hat{I}_2 \tag{10.28a}$$

$$\hat{V}_2 = j\omega M \hat{I}_1 + j\omega L_2 \hat{I}_2 \tag{10.28b}$$

From Eq. (10.28a) we get

$$\hat{I}_1 = \frac{\hat{V}_1 - j\omega M \hat{I}_2}{j\omega L_1} \tag{10.29}$$

Substituting Eq. (10.29) into Eq. (10.28b) gives

$$\hat{V}_2 = j\omega M \frac{\hat{V}_1 - j\omega M \hat{I}_2}{j\omega L_1} + j\omega L_2 \hat{I}_2 \tag{10.30}$$

or

$$\hat{V}_2 = \frac{M \hat{V}_1}{L_1} + \frac{\omega^2 M^2 \hat{I}_2}{j\omega L_1} + j\omega L_2 \hat{I}_2 \tag{10.31}$$

leading to

$$\hat{V}_2 = j\omega L_2 \hat{I}_2 + \frac{M \hat{V}_1}{L_1} - \frac{j\omega M^2 \hat{I}_2}{L_1} \tag{10.32}$$

The ideal transformer is characterized by perfect coupling, that is, $k = 1$, and therefore

$$M = \sqrt{L_1 L_2} \tag{10.33}$$

Substituting Eq. (10.33) into Eq. (10.32) we get

$$\hat{V}_2 = j\omega L_2 \hat{I}_2 + \frac{\sqrt{L_1 L_2} \hat{V}_1}{L_1} - \frac{j\omega L_1 L_2 \hat{I}_2}{L_1} \tag{10.34}$$

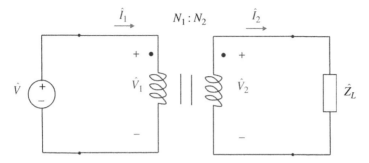

Figure 10.16 Voltages and currents in an ideal transformer.

or

$$\hat{V}_2 = \sqrt{\frac{L_2}{L_1}}\hat{V}_1 = n\hat{V}_1 \tag{10.35}$$

where

$$n = \sqrt{\frac{L_2}{L_1}} \tag{10.36}$$

is called the *turns ratio.*

When a sinusoidal voltage is applied to the primary winding, as shown in Figure 10.16, the same magnetic flux Φ flows through both windings.

According to Faraday's law, the voltages across the primary and the secondary windings, are, respectively,

$$v_1 = N_1 \frac{d\Phi}{dt} \tag{10.37a}$$

$$v_2 = N_2 \frac{d\Phi}{dt} \tag{10.37b}$$

Dividing Eq. (10.37b) by Eq. (10.37a) gives

$$\frac{v_2}{v_1} = \frac{N_2}{N_1} = n \tag{10.38}$$

where n is, again, the *turns ratio.* When $n = 1$, the transformer is usually called an *isolation transformer.*

Since there are no losses in an ideal transformer, we have

$$p_1 = p_2 \tag{10.39}$$

or

$$v_1 i_1 = v_2 i_2 \tag{10.40}$$

and thus

$$\frac{i_2}{i_1} = \frac{v_1}{v_2} = \frac{N_1}{N_2} = \frac{1}{n} \tag{10.41}$$

In phasor form, we have

$$\frac{\hat{V}_2}{\hat{V}_1} = \frac{N_2}{N_1} = n \tag{10.42a}$$

$$\frac{\hat{I}_2}{\hat{I}_1} = \frac{N_1}{N_2} = \frac{1}{n} \tag{10.42b}$$

From Eq. (10.42) we get

$$\hat{V}_1 = \frac{\hat{V}_2}{n} \tag{10.43a}$$

$$\hat{I}_1 = n\hat{I}_2 \tag{10.43b}$$

The input impedance as seen by the source in Figure 10.14 is

$$\hat{Z}_{in} = \frac{\hat{V}_1}{\hat{I}_1} \tag{10.44}$$

Using Eq. (10.43) in Eq. (10.44) we get

$$\hat{Z}_{in} = \frac{1}{n^2} \frac{\hat{V}_2}{\hat{I}_2} \tag{10.45}$$

From Figure 10.14, it is evident that

$$\frac{\hat{V}_2}{\hat{I}_2} = \hat{Z}_L \tag{10.46}$$

and thus the input impedance seen by the source is

$$\hat{Z}_{in} = \frac{\hat{Z}_L}{n^2} \tag{10.47}$$

The input impedance is also called the *reflected impedance*, since it appears as the load impedance reflected to the primary side.

As we shall see, the reflected impedance is used in *impedance matching* for maximum power transfer, as explained next.

Consider the circuit shown in Figure 10.17.

Recall that for maximum power transfer the load must be matched to the source resistance, i.e. $R_L = R_S$. In most cases, however, these two resistance are fixed and not equal.

An iron-core transformer can be used to match the load resistance to the source resistance. The ideal transformer reflects its load back to the primary with a scaling factor of n^2. To match this reflected load with the source resistance we set them equal

$$R_S = \frac{R_L}{n^2} \tag{10.48}$$

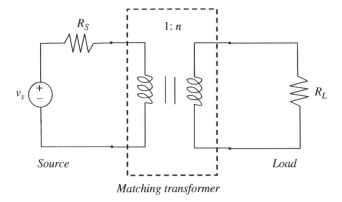

Figure 10.17 Circuit with a matching transformer.

10.5 EMC Applications

10.5.1 Common-Mode Choke

A common-mode choke, shown in Figure 10.18, consists of a pair of wires carrying currents $\hat{I}_1$ and $\hat{I}_2$ wound around a ferromagnetic core (Paul, 2006, p. 350).

As we shall see, the common-mode choke blocks the common-mode (CM) currents and has no effect on the differential-mode (DM) currents.

The currents shown in Figure 10.18 and the total current flowing in each wire, shown in Figure 10.19, are related by

$$\hat{I}_1 = \hat{I}_{CM} + \hat{I}_{DM} \tag{10.49a}$$

$$\hat{I}_1 = \hat{I}_{CM} - \hat{I}_{DM} \tag{10.49b}$$

Equivalently, the CM and DM currents can be expressed as

$$\hat{I}_C = \frac{1}{2}\left(\hat{I}_1 + \hat{I}_2\right) \tag{10.50a}$$

$$\hat{I}_D = \frac{1}{2}\left(\hat{I}_1 - \hat{I}_2\right) \tag{10.50b}$$

Let's investigate the effect of the choke on the CM and DM mode currents. The circuit model of the choke is shown in Figure 10.20.

Using the model in Figure 10.20, we calculate the impedance of each winding as

$$\hat{Z}_1 = \frac{\hat{V}_1}{\hat{I}_1} = \frac{j\omega L\hat{I}_1 + j\omega M\hat{I}_2}{\hat{I}_1} \tag{10.51a}$$

$$\hat{Z}_2 = \frac{\hat{V}_2}{\hat{I}_2} = \frac{j\omega L\hat{I}_2 + j\omega M\hat{I}_1}{\hat{I}_2} \tag{10.51b}$$

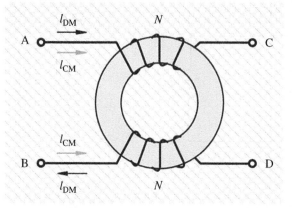

Figure 10.18 Common-mode choke.

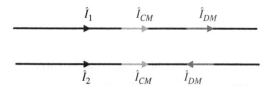

Figure 10.19 Total currents flowing in each wire.

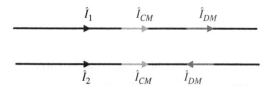

Figure 10.20 Circuit model of the CM choke.

To determine the effect of the choke on the DM currents let

$$\hat{I}_1 = \hat{I}_D \tag{10.52a}$$

$$\hat{I}_2 = -\hat{I}_D \tag{10.52b}$$

Using Eq. (10.52) in Eq. (10.51a) we get

$$\hat{Z}_{DM} = \frac{\hat{V}_1}{\hat{I}_1} = \frac{j\omega L\hat{I}_1 + j\omega M\hat{I}_2}{\hat{I}_1}$$

$$= \frac{j\omega L\hat{I}_D - j\omega M\hat{I}_D}{\hat{I}_D} = j\omega(L - M) \tag{10.53a}$$

Similarly, using Eq. (10.52) in Eq. (10.51b) we get

$$\hat{Z}_{DM} = \frac{\hat{V}_2}{\hat{I}_2} = \frac{j\omega\, L\hat{I}_2 + j\omega\, M\hat{I}_1}{\hat{I}_2}$$

$$= \frac{-j\omega\, L\hat{I}_D + j\omega\, M\hat{I}_D}{-\hat{I}_D} = j\omega\left(L - M\right)$$

(10.53b)

Thus, the impedance seen by the DM current in each winding is

$$\hat{Z}_{DM} = j\omega\left(L - M\right)$$

(10.54)

In the ideal case, where $L = M$, we have

$$\hat{Z}_{DM} = 0$$

(10.55)

Thus, the (ideal) CM choke is transparent to the DM currents, i.e. it does not affect them at all.

Now let's determine the effect of the choke on the CM currents. To this end, let

$$\hat{I}_1 = \hat{I}_C$$

(10.56a)

$$\hat{I}_2 = \hat{I}_C$$

(10.56b)

Using Eq. (10.56) in Eq. (10.51a) we get

$$\hat{Z}_{CM} = \frac{\hat{V}_1}{\hat{I}_1} = \frac{j\omega\, L\hat{I}_1 + j\omega\, M\hat{I}_2}{\hat{I}_1}$$

$$= \frac{j\omega\, L\hat{I}_C + j\omega\, M\hat{I}_C}{\hat{I}_C} = j\omega\left(L + M\right)$$

(10.57a)

Similarly, using Eq. (10.56) in Eq. (10.51b) we get

$$\hat{Z}_{CM} = \frac{\hat{V}_2}{\hat{I}_2} = \frac{j\omega L\hat{I}_2 + j\omega M\hat{I}_1}{\hat{I}_2}$$

$$= \frac{j\omega\, L\hat{I}_C + j\omega\, M\hat{I}_C}{\hat{I}_C} = j\omega\left(L + M\right)$$

(10.57b)

Thus, the impedance seen by the CM current in each winding is

$$\hat{Z}_{CM} = j\omega\left(L + M\right)$$

(10.58)

Thus, the CM choke inserts an inductance $L + M$ in each winding, and consequently it tends to block CM currents.

References

Alexander, C.K. and Sadiku, N.O., *Fundamentals of Electric Circuits*, 4th ed., McGraw Hill, New York, 2009.

Nilsson, J.W. and Riedel, S.A., *Electric Circuits*, 10th ed., Pearson, Upper Saddle River, NJ, 2015.

Paul, C.R., *Introduction to Electromagnetic Compatibility*, 2nd ed., John Wiley and Sons, New York, 2006.

11

Frequency-Domain Analysis

In Chapter 7, we defined the Laplace transformation and used it to obtain the s domain expression for a given time-domain function, and conversely, we obtained the time-domain expression from a given s domain function using inverse Laplace transform.

The real power of Laplace transformation in engineering applications emerges when we transform the electrical circuit itself from time domain to s domain and analyze it directly in the s domain. The s domain analysis leads to a definition of a transfer function, and subsequently to the concept of the frequency transfer function, some of the most important concepts in circuit analysis.

In EMC, we are predominantly interested in the sinusoidal steady state and therefore we will focus on the frequency transfer function techniques. We begin by defining the concept a transfer function.

11.1   Transfer Function

The concept of a transfer function is perhaps the most important concept in frequency domain analysis.

A *transfer function* is defined as the ratio of a Laplace transform of the output $Y(s)$ to the Laplace transform of the input $X(s)$, under the assumptions of zero initial conditions in the circuit (Nilsson and Riedel, 2015, p. 482).

Thus,

$$\hat{H}(s) = \frac{\hat{Y}(s)}{\hat{X}(s)}\bigg|_{IC's=0} \tag{11.1}$$

Figure 11.1 shows a typical representation of a circuit in the s domain, used to define the voltage transfer function.

The transfer function depends on what variables we define as the input and the output. Perhaps the voltage transfer function is the most important one; it is used to define several frequency-domain concepts (e.g. frequency transfer function and electrical filters).

$$\hat{H}(s) = \frac{\hat{V}_{OUT}(s)}{\hat{V}_{IN}(s)} \tag{11.2}$$

Foundations of Electromagnetic Compatibility with Practical Applications, First Edition. Bogdan Adamczyk.
© 2017 John Wiley & Sons Ltd. Published 2017 by John Wiley & Sons Ltd.

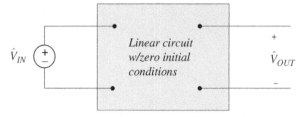

Figure 11.1 Circuit used to define the voltage transfer function.

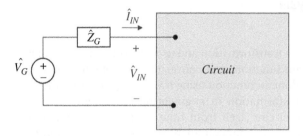

Figure 11.2 Circuit used to define input impedance.

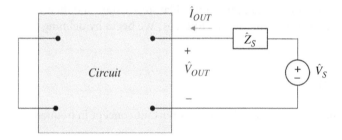

Figure 11.3 Circuit used to define output impedance.

Two additional very useful transfer (or network) functions can be defined using the circuits shown in Figures 11.2 and 11.3.

Using the circuit shown Figure 11.2 we define the input (driving point) impedance (Alexander and Sadiku, 2009, p. 852) as

$$\hat{Z}_{IN}(s) = \frac{\hat{V}_{IN}(s)}{\hat{I}_{IN}(s)} \bigg| \qquad (11.3)$$

while using the circuit shown Figure 11.3, we define the output impedance as

$$\hat{Z}_{OUT}(s) = \frac{\hat{V}_{OUT}(s)}{\hat{I}_{OUT}(s)} \bigg| \qquad (11.4)$$

Once the transfer function of the system, $\hat{H}(s)$, is obtained, then the output of the system, $\hat{Y}(s)$, due to an input $\hat{X}(s)$ can be obtained as

$$\hat{Y}(s) = \hat{H}(s)\hat{X}(s) \qquad (11.5)$$

The following example illustrates the above defined concepts.

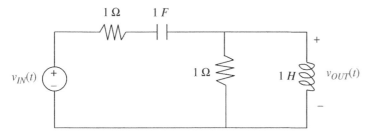

Figure 11.4 Circuit for Example 11.1.

Example 11.1 System network functions

Consider the circuit shown in Figure 11.4.

Determine:

1) the circuit's transfer function
2) the input impedance
3) the output impedance
4) the output $\hat{V}_{OUT}(s)$ when the input is $v_{IN}(t) = 1, t \geq 0$
5) the output $v_{OUT}(t)$ when the input is $v_{IN}(t) = 3cos2t, t \geq 0$
6) the steady state output $v_{OUT}(t)$ for the input $v_{IN}(t) = 3cos2t, t \geq 0$

Solution:

1. Transfer function

To determine the system's transfer function let's transform the circuit into the Laplace domain under the zero initial conditions assumption, as shown in Figure 11.5.

The desired transfer function is defined as:

$$\hat{H}(s) = \frac{\hat{V}_{OUT}(s)}{\hat{V}_{IN}(s)}$$

First, combine impedances in series:

$$1 + \frac{1}{s} = \frac{s+1}{s}$$

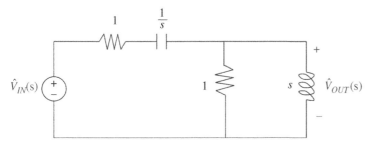

Figure 11.5 Circuit in the *s* domain.

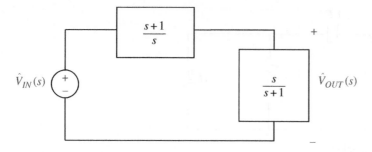

Figure 11.6 Simplified circuit.

Next, combine impedances in parallel:

$$1\|s = \frac{s}{s+1}$$

The resulting circuit is shown in Figure 11.6.
Using a voltage divider yields:

$$\hat{H}(s) = \frac{\hat{V}_{OUT}(s)}{\hat{V}_{IN}(s)} = \frac{\dfrac{s}{s+1}}{\dfrac{s+1}{s} + \dfrac{s}{s+1}} = \frac{\dfrac{s}{s+1}}{\dfrac{(s+1)^2 + s^2}{s(s+1)}} = \left(\frac{s}{s+1}\right)\frac{s(s+1)}{2s^2 + 2s + 1}$$

and thus the circuit's transfer function is

$$\hat{H}(s) = \frac{s^2}{2s^2 + 2s + 1}$$

2. Input impedance
The input impedance is obtained from the circuit shown in Figure 11.7.

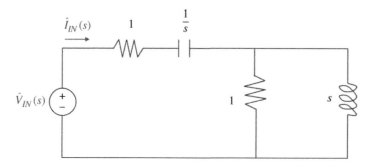

Figure 11.7 Circuit for obtaining the input impedance.

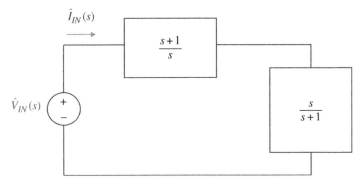

Figure 11.8 Equivalent circuit for obtaining the input impedance.

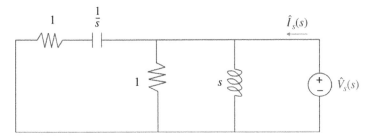

Figure 11.9 Circuit for obtaining the output impedance.

This circuit is, of course, equivalent to the one shown in Figure 11.8, from which the input impedance is obtained as

$$\hat{Z}_{IN}(s) = \frac{\hat{V}_{IN}(s)}{\hat{I}_{IN}(s)} = \frac{s+1}{s} + \frac{s}{s+1} = \frac{(s+1)^2 + s^2}{s(s+1)} = \frac{2s^2 + 2s + 1}{s^2 + s}$$

3. Output impedance
The input impedance is obtained from the circuit shown in Figure 11.9.
This circuit is equivalent to the one shown in Figure 11.10, from which the output impedance is obtained as

$$\hat{Z}_{OUT}(s) = \frac{\hat{V}_S(s)}{\hat{I}_S(s)} = \left(\frac{s+1}{s}\right)\left(\frac{s}{s+1}\right) = \frac{\left(\dfrac{s+1}{s}\right)\left(\dfrac{s}{s+1}\right)}{\left(\dfrac{s+1}{s}\right) + \left(\dfrac{s}{s+1}\right)}$$

$$= \frac{1}{\dfrac{s+1}{s} + \dfrac{s}{s+1}} = \frac{1}{\dfrac{(s+1)^2 + s^2}{s(s+1)}} = \frac{1}{\dfrac{2s^2 + 2s + 1}{s^2 + s}} = \frac{s^2 + s}{2s^2 + 2s + 1}$$

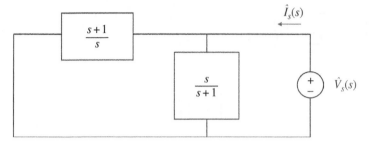

Figure 11.10 Equivalent circuit for obtaining the output impedance.

4. *Output* $\hat{V}_{OUT}(s)$ *when the input is* $v_{IN}(t)=1, t \geq 0$

The output can be obtained by utilizing the derived transfer function:

$$\hat{V}_{OUT}(s) = \hat{H}(s)\hat{V}_{IN}(s)$$

Since

$$v_{IN}(t)=1$$

we have

$$\hat{V}_{in}(s) = \frac{1}{s}$$

and the output in the s domain is

$$\hat{V}_{OUT}(s) = \frac{s^2}{\left(2s^2+2s+1\right)s}\frac{1}{s} = \frac{s}{s^2+s+\dfrac{1}{2}}$$

5. *Output* $v_{OUT}(t)$ *when the input is* $v_{IN}(t) = 3\cos 2t, t \geq 0$

Again, the output in the s domain can be obtained by utilizing the derived transfer function.

$$\hat{V}_{OUT}(s) = \hat{H}(s)\hat{V}_{IN}(s)$$

Since

$$v_{IN}(t) = 3\cos 2t$$

we have

$$\hat{V}_{in}(s) = \frac{3s}{s^2+4}$$

and the output in the s domain is

$$\hat{V}_{OUT}(s) = \frac{s^2}{\left(2s^2+2s+1\right)}\frac{3s}{s^2+4}$$

Now, partial fraction expansion yields

$$\hat{V}_{OUT}(s) = \frac{s^2}{2s^2 + 2s + 1} \frac{3s}{s^2 + 4} = \frac{As + B}{2s^2 + 2s + 1} + \frac{Cs + D}{s^2 + 4}$$

thus

$$3s^3 = (As + B)(s^2 + 4) + (Cs + D)(2s^2 + 2s + 1)$$

or

$$3s^3 = As^3 + 4As + Bs^2 + 4B + 2Cs^3 + 2Cs^2 + Cs + 2Ds^2 + 2Ds + D$$

or

$$3s^3 = (A + 2C)s^3 + (B + 2C + 2D)s^2 + (4A + C + 2D)s + (4B + D)$$

therefore,

$$A + 2C = 3$$
$$B + 2C + 2D = 0$$
$$4A + C + 2D = 0$$
$$4B + D = 0$$

leading to

$$A = 0.4154$$
$$B = 0.3692$$
$$C = 1.2923$$
$$D = -1.4769$$

thus

$$\hat{V}_{OUT}(s) = \frac{0.4154s + 0.3692}{2s^2 + 2s + 1} + \frac{1.2923s - 1.4769}{s^2 + 4}$$

Let's obtain the time domain function represented by the above expression. We will use it to obtain the steady state solution for part (5) of this example.

Let's complete the square for the first term.

$$\frac{0.4154s + 0.3692}{2s^2 + 2s + 1} = \frac{0.2077s + 0.1846}{s^2 + s + \frac{1}{2}} = \frac{0.2077s + 0.1846}{\left(s + \frac{1}{2}\right)^2 + \frac{1}{4}}$$

Then the result leads to the damped cosine and sine functions, as follows:

$$\frac{0.2077s+0.1846}{\left(s+\dfrac{1}{2}\right)^2+\dfrac{1}{4}}=0.2077\frac{s+0.88878}{\left(s+\dfrac{1}{2}\right)^2+\dfrac{1}{4}}$$

$$=0.2077\frac{s+0.5+0.38878}{\left(s+\dfrac{1}{2}\right)^2+\dfrac{1}{4}}=0.2077\left[\frac{s+0.5}{\left(s+\dfrac{1}{2}\right)^2+\dfrac{1}{4}}+\frac{0.38878}{\left(s+\dfrac{1}{2}\right)^2+\dfrac{1}{4}}\right]$$

$$=0.2077\frac{s+0.5}{\left(s+\dfrac{1}{2}\right)^2+\dfrac{1}{4}}+\frac{0.08075}{\left(s+\dfrac{1}{2}\right)^2+\dfrac{1}{4}}=0.2077\frac{s+0.5}{\left(s+\dfrac{1}{2}\right)^2+\dfrac{1}{4}}+\frac{0.08075}{0.5}\frac{0.5}{\left(s+\dfrac{1}{2}\right)^2+\dfrac{1}{4}}$$

$$=0.2077\frac{s+0.5}{\left(s+\dfrac{1}{2}\right)^2+\dfrac{1}{4}}+0.1615\frac{0.5}{\left(s+\dfrac{1}{2}\right)^2+\dfrac{1}{4}}$$

The second term in

$$\hat{V}_{OUT}(s)=\frac{0.4154s+0.3692}{2s^2+2s+1}+\frac{1.2923s-1.4769}{s^2+4}$$

can be written as

$$\frac{1.2923s-1.4769}{s^2+4}=\frac{1.2923s}{s^2+4}-\frac{1.4769}{s^2+4}$$

$$=1.2923\frac{s}{s^2+4}-\frac{1.4769}{2}\frac{2}{s^2+4}$$

$$=1.2923\frac{s}{s^2+4}-0.7384\frac{2}{s^2+4}$$

Thus, the output voltage can be expressed as

$$\hat{V}_{OUT}(s)=\frac{0.4154s+0.3692}{2s^2+2s+1}+\frac{1.2923s-1.4769}{s^2+4}$$

$$=0.2077\frac{s+0.5}{\left(s+\dfrac{1}{2}\right)^2+\dfrac{1}{4}}+0.1615\frac{0.5}{\left(s+\dfrac{1}{2}\right)^2+\dfrac{1}{4}}+1.2923\frac{s}{s^2+4}-0.7384\frac{2}{s^2+4}$$

Using the tables of Laplace transform pairs we obtain the time domain output as

$$v_{OUT}(t)=0.2077e^{-0.5t}\sin 2t+0.1615e^{-0.5t}\cos 2t+1.2923\cos 2t-0.7384\sin 2t$$

5. Steady-state output $v_{OUT}(t)$ for the input $v_{IN}(t)=3\cos 2t, t\geq 0$
The sinusoidal steady state output is obtained directly from the solution of part (5) as

$$v_{OUT,\,ss}(t)=1.2923\cos 2t-0.7384\sin 2t$$

This output can also be expressed as

$$v_{OUT, SS}(t) = 1.4884\cos(2t + 29.74°)$$

In the next section we will show an alternative (and much easier) method of obtaining the steady state response due to a sinusoidal input.

11.2 Frequency-Transfer Function

When the linear system is driven by a sinusoid, the response at steady state is sinusoidal at the same frequency as the input. The steady state output differs from the input only in the amplitude and phase angle.

The output amplitude and phase angle at each frequency can be determined by using the sinusoidal or frequency transfer function.

The frequency transfer function is defined as the transfer function H(s) in which s is replaced by jω.

$$\hat{H}(j\omega) = \hat{H}(s)\Big|_{s=j\omega} \tag{11.6}$$

Example 11.2 Sinusoidal transfer function
Given the system's transfer function

$$\hat{H}(s) = \frac{s+2}{s^2 + 3s + 4}$$

Obtain the system's frequency transfer function

Solution: To obtain the frequency transfer function from the system's transfer function, we simply replace s by jω in the system's transfer function.

$$\hat{H}(j\omega) = \hat{H}(s)\Big|_{s=j\omega} = \frac{s+2}{s^2 + 3s + 4}\Big|_{s=j\omega} = \frac{j\omega + 2}{(j\omega)^2 + 3j\omega + 4} = \frac{2 + j\omega}{4 - \omega^2 + j3\omega}$$

■

Obviously, the frequency transfer function is a complex function. As such it has a magnitude (which is a function of the frequency ω) and a phase (which is also a function of the frequency ω).

Let's calculate these for the frequency transfer function obtained in the above example.

Example 11.3 Magnitude and phase of the sinusoidal transfer function
Determine the magnitude and phase of the frequency transfer function

$$\hat{H}(j\omega) = \frac{2 + j\omega}{4 - \omega^2 + j3\omega}$$

Solution: First, let's determine the magnitude.

$$\left|\hat{H}(j\omega)\right| = \left|\frac{2+j\omega}{4-\omega^2+j3\omega}\right| = \frac{|2+j\omega|}{|4-\omega^2+j3\omega|} = \frac{\sqrt{4+\omega^2}}{\sqrt{\left(4-\omega^2\right)^2+9\omega^2}} = \sqrt{\frac{4+\omega^2}{\omega^4+\omega^2+9}}$$

The phase of the frequency transfer function is

$$\angle\hat{H}(j\omega) = \angle\left(\frac{2+j\omega}{4-\omega^2+3j\omega}\right) = \frac{\angle(2+j\omega)}{\angle\left(4-\omega^2+3j\omega\right)} = \tan^{-1}\frac{\omega}{2} - \tan^{-1}\frac{3\omega}{4-\omega^2}$$

11.2.1 Sinusoidal Steady-State Output

Frequency response characteristics of a system can be obtained directly from the sinusoidal transfer function. Consider the linear time invariant (LTI) system shown in Figure 11.11.

The input $x(t)$ is a sinusoid given by

$$x(t) = A\sin\omega t \tag{11.7}$$

Let the transfer function be expressed as a ratio of two polynomials (Nilsson and Riedel, 2015, p. 442)

$$\hat{H}(s) = \frac{\hat{N}(s)}{\hat{D}(s)} = \frac{\hat{N}(s)}{(s+p_1)(s+p_2)\cdots(s+p_n)} \tag{11.8}$$

The output in Laplace domain can be obtained as

$$\hat{Y}(s) = \hat{H}(s)\hat{X}(s) = \frac{\hat{N}(s)}{\hat{D}(s)}\hat{X}(s) \tag{11.9}$$

(a)

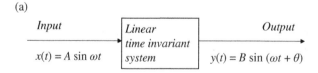

(b)

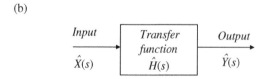

Figure 11.11 LTI system driven by a sinusoid (a) time domain representation, (b) frequency domain representation.

Let's consider the systems that are stable. For such systems the real parts of the roots in the denominator of $\hat{D}(s)$ are negative.

Also, since the steady state response of a stable LTI system to a sinusoidal input does not depend upon the initial conditions (they give rise to transient terms), we can assume that the initial conditions are zero.

In the following discussion, we will consider several cases of roots of the output's denominator $\hat{D}(s)$.

Case 1: $\hat{D}(s)$ has only distinct poles (real or complex) In this case, the output $\hat{Y}(s)$ can be expressed as

$$
\hat{Y}(s) = \frac{\hat{N}(s)}{\hat{D}(s)}\hat{X}(s) = \frac{\hat{N}(s)}{\hat{D}(s)}\frac{A\omega}{s^2 + \omega^2}
$$

$$
= \frac{\hat{N}(s)}{(s + p_1)(s + p_2)\cdots(s + p_n)}\frac{A\omega}{s^2 + \omega^2} \tag{11.10}
$$

$$
= \frac{\hat{a}}{s + j\omega} + \frac{\hat{a}^*}{s - j\omega} + \frac{b_1}{s + p_1} + \frac{b_2}{s + p_2} + \cdots + \frac{b_n}{s + p_n}
$$

where $\hat{a}$ is a complex constant and b_i $(i = 1,\ldots,n)$ are real or complex constants. The inverse Laplace transform gives

$$
y(t) = \hat{a}e^{-j\omega t} + \hat{a}^* e^{+j\omega t} + b_1 e^{-p_1 t} + b_2 e^{-p_2 t} + \cdots + b_n e^{-p_n t} \tag{11.11}
$$

For a stable system the roots, $-p_i$ have negative real parts. Thus, as $t \to \infty, e^{-pt} \to 0$. Therefore, at steady state, all the terms, except for the first two vanish, and the output is

$$
y_{ss}(t) = \hat{a}e^{-j\omega t} + \hat{a}^* e^{+j\omega t} \tag{11.12}
$$

Case 2: $\hat{D}(s)$ has repeated poles (real or complex) If $\hat{D}(s)$ has multiple poles p_j of multiplicity m_j, then $y(t)$ will be of the form

$$
y(t) = \hat{a}e^{-j\omega t} + \hat{a}^* e^{+j\omega t} + b_1 e^{-p_1 t} + b_2 e^{-p_2 t} + \cdots + t^{h_j} e^{-p_j t},
$$
$$
(h_j = 0,1,2,\ldots,m_j - 1) \tag{11.13}
$$

Since the real parts of the $-p_j$ are negative for a stable system, in steady state the terms $t^{h_j} e^{-p_j t}$ will also vanish, and again the *steady state response* becomes

$$
y_{ss}(t) = \hat{a}e^{-j\omega t} + \hat{a}^* e^{+j\omega t} \tag{11.14}
$$

The constant $\hat{a}$ can be evaluated as follows:

$$
\hat{a} = \hat{H}(s)\hat{X}(s)\Big|_{s=-j\omega} = \hat{H}(s)\frac{A\omega}{s^2 + \omega^2}(s + j\omega)\Big|_{s=-j\omega}
$$
$$
= \hat{H}(s)\frac{A\omega}{s - j\omega}\Big|_{s=-j\omega} = -\frac{A\hat{H}(-j\omega)}{2j} \tag{11.15}
$$

and

$$\hat{a}^* = \hat{H}(s)\hat{X}(s)\Big|_{s=+j\omega} = \hat{H}(s)\frac{A\omega}{s^2+\omega^2}(s-j\omega)\Big|_{s=+j\omega}$$

$$= \hat{H}(s)\frac{A\omega}{s+j\omega}\Big|_{s=+j\omega} = \frac{A\hat{H}(j\omega)}{2j}$$

(11.16)

$\hat{H}(j\omega)$ can be expressed in an exponential form as

$$\hat{H}(j\omega) = \left|\hat{H}(j\omega)\right|e^{j\theta}$$

(11.17)

Similarly,

$$\hat{H}(-j\omega) = \left|\hat{H}(-j\omega)\right|e^{-j\theta} = \left|\hat{H}(j\omega)\right|e^{-j\theta}$$

(11.18)

Now, Eq. (11.14) can be written as

$$y_{ss}(t) = \hat{a}e^{-j\omega t} + \hat{a}^* e^{+j\omega t} = -\frac{A}{2j}\hat{H}(-j\omega)e^{-j\omega t} + \frac{A}{2j}\hat{H}(j\omega)e^{j\omega t}$$

$$= -\frac{A}{2j}\left|\hat{H}(-j\omega)\right|e^{-j\theta}e^{-j\omega t} + \frac{A}{2j}\left|\hat{H}(j\omega)\right|e^{j\theta}e^{j\omega t}$$

(11.19)

$$= A\left|\hat{H}(j\omega)\right|\frac{e^{j(\omega t+\theta)} - e^{-j(\omega t+\theta)}}{2j} = A\left|\hat{H}(j\omega)\right|\sin\left(\omega t + \theta\right)$$

Therefore, the steady state output due to

$$x(t) = A\sin\omega t$$

(11.20)

is

$$y_{ss}(t) = A\left|\hat{H}(j\omega)\right|\sin\left(\omega t + \angle\hat{H}(j\omega)\right)$$

(11.21)

More generally, if the input is of the form

$$x(t) = A\sin\left(\omega t + \phi\right)$$

(11.23)

then the steady state output is

$$y_{ss}(t) = A\left|\hat{H}(j\omega)\right|\sin\left(\omega t + \phi + \angle\hat{H}(j\omega)\right)$$

(11.24)

Similarly, if the input is a cosine function of the form

$$x(t) = A\cos\left(\omega t + \phi\right)$$

(11.25)

then the steady state output is

$$y_{ss}(t) = A\left|\hat{H}(j\omega)\right|\cos\left(\omega t + \phi + \angle\hat{H}(j\omega)\right)$$

(11.26)

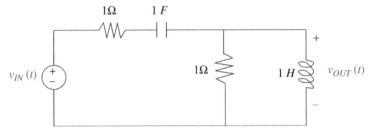

Figure 11.12 Circuit for Example 11.4.

This is an extremely important and useful result! The output in steady state, is a sinusoid of the same frequency as the input, and its amplitude and phase are determined from the frequency transfer function and the input's amplitude and phase.

Example 11.4 Sinusoidal steady-state
Let's use the circuit analyzed in Example 11.1 and redrawn in Figure 11.12.

Determine:

1) the system's transfer function
2) the steady-state solution when the input to the system is $v_{in}(t) = 3\cos 2t, t \geq 0$.

Solution: In Example 11.1 we found the transfer function of this circuit to be

$$\hat{H}(s) = \frac{s^2}{2s^2 + 2s + 1}$$

Thus the frequency transfer function is

$$\hat{H}(j\omega) = \frac{-\omega^2}{-2\omega^2 + 2j\omega + 1} = \frac{-\omega^2}{1 - 2\omega^2 + j2\omega}$$

Since $\omega = 2$, we have

$$\hat{H}(j\omega) = \hat{H}(j2) = \frac{-4}{-7 + j4}$$

The magnitude of the frequency transfer function at the frequency of $\omega = 2$ is

$$\left| \hat{H}(j\omega) \right| = \left| \hat{H}(j2) \right| = \frac{4}{\sqrt{49 + 16}} = 0.4961$$

The angle of the transfer function is

$$\angle \hat{H}(j\omega) = \angle \hat{H}(j2) = \angle \left(\frac{-4}{-7 + j4} \right) = \angle(-4) - \angle(-7 + j4)$$

$$= 180° - \tan^{-1} \frac{4}{(-7)} = 180° - (-29.74° + 180°) = 29.74°$$

Since $v_{in}(t) = 3\cos 2t, t \geq 0$, we have

$$v_{OUT,SS}(t) = (3)(0.4961)\cos(2t + 29.74°)$$

which, of course, agrees with the steady state solution obtained in Example 11.1.

11.3 Bode Plots

On numerous occasions in EMC we plot the output of the system using a dB scale for magnitude, and a logarithmic scale for frequency. We often refer to such plots as *Bode plots* (Alexander and Sadiku, 2009, p. 619). The exact Bode plots can be obtained using many available software packages. In many cases, it is more convenient and expedient to sketch an approximate magnitude plot using straight-line approximations.

Let's consider a transfer function with real, first-order poles and zeros:

$$\hat{H}(s) = \frac{K(s + z_1)}{s(s + p_1)} \tag{11.27}$$

The first step in creating Bode diagrams is transforming Eq. (11.27) into a *standard form*:

$$\hat{H}(s) = \frac{Kz_1\left(1 + \dfrac{s}{z_1}\right)}{p_1 s\left(1 + \dfrac{s}{p_1}\right)} \tag{11.28}$$

The corresponding frequency transfer function is

$$\hat{H}(j\omega) = \frac{Kz_1\left(1 + \dfrac{j\omega}{z_1}\right)}{p_1(j\omega)\left(1 + \dfrac{j\omega}{p_1}\right)} \tag{11.29}$$

If we let

$$K_0 = \frac{Kz_1}{p_1} \tag{11.30}$$

then Eq. (11.29) becomes

$$\hat{H}(j\omega) = \frac{K_0\left(1 + \dfrac{j\omega}{z_1}\right)}{(j\omega)\left(1 + \dfrac{j\omega}{p_1}\right)} \tag{11.31}$$

Example 11.5 Standard form of a frequency transfer function

Let

$$\hat{H}(s) = \frac{150s}{(s+10)(s+100)}$$

Expressing this transfer function in a standard form we get

$$\hat{H}(s) = \frac{150s}{(10)(100)\left(1+\dfrac{s}{10}\right)\left(1+\dfrac{s}{100}\right)} = \frac{0.15s}{\left(1+\dfrac{s}{10}\right)\left(1+\dfrac{s}{100}\right)}$$

It follows that the frequency transfer function in a standard form is

$$\hat{H}(j\omega) = \frac{0.15\,j\omega}{\left(1+j\dfrac{\omega}{10}\right)\left(1+j\dfrac{\omega}{100}\right)}$$

∎

Let's return to the frequency transfer function in a standard form given in Eq. (11.31). Expressing this frequency transfer function in polar form gives

$$\hat{H}(j\omega) = \frac{K_0\left|1+\dfrac{j\omega}{z_1}\right|\angle\alpha_1}{|\omega|\angle 90°\left|1+\dfrac{j\omega}{p_1}\right|\angle\beta_1} = \frac{K_0\left|1+\dfrac{j\omega}{z_1}\right|}{|\omega|\left|1+\dfrac{j\omega}{p_1}\right|}\angle(\alpha_1-90°-\beta_1) \qquad (11.32)$$

where

$$\alpha_1 = \tan^{-1}\frac{\omega}{z_1} \qquad (11.33a)$$

$$\beta_1 = \tan^{-1}\frac{\omega}{p_1} \qquad (11.33b)$$

and thus, the magnitude and the phase of the transfer function are given by

$$\left|\hat{H}(j\omega)\right| = \frac{K_0\left|1+\dfrac{j\omega}{z_1}\right|}{|\omega|\left|1+\dfrac{j\omega}{p_1}\right|} \qquad (11.34)$$

$$\angle\hat{H}(j\omega) = \alpha_1 - 90° - \beta_1 \qquad (11.35)$$

Let's focus on the magnitude. Expressing Eq. (11.34) in dB gives

$$\left|\hat{H}(j\omega)\right|_{dB} = 20\log_{10} \frac{K_0\left|1+\dfrac{j\omega}{z_1}\right|}{|\omega|\left|1+\dfrac{j\omega}{p_1}\right|} \tag{11.36}$$

$$= 20\log_{10} K_0 + 20\log_{10}\left|1+\dfrac{j\omega}{z_1}\right| - 20\log_{10}\omega - 20\log_{10}\left|1+\dfrac{j\omega}{p_1}\right|$$

Thus the magnitude of the transfer function in dB can be obtained by plotting each term in the equation separately and then combining the separate plots graphically. The individual factors are easy to plot because they can be approximated in all cases by straight lines, as discussed next.

The plot of $20\log_{10}K_0$ is a horizontal straight line because K_0 is not a function of ω. The value of this term is:

$$\begin{aligned} K_0 > 1 &\implies K_{0,dB} > 0 \\ K_0 = 1 &\implies K_{0,dB} = 0 \\ 0 < K_0 < 1 &\implies K_{0,dB} < 0 \end{aligned} \tag{11.37}$$

The plot of this term is shown in Figure 11.13.
Next, let's look at the term of the form

$$\left|\hat{H}(j\omega)\right|_{dB} = 20\log_{10}\left|1+\dfrac{j\omega}{z_1}\right| \tag{11.38}$$

For small values of ω, we obtain

$$20\log_{10}\left|1+\dfrac{j\omega}{z_1}\right| \cong 20\log_{10}1 = 0, \quad \omega \to 0 \tag{11.39}$$

On a log scale, this is a horizontal line at a dB = 0 value. For the large values of ω, we have

$$20\log_{10}\left|1+\dfrac{j\omega}{z_1}\right| \cong 20\log_{10}\left(\dfrac{\omega}{z_1}\right), \quad \omega \to \infty \tag{11.40}$$

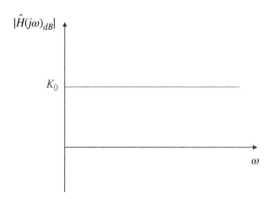

Figure 11.13 Magnitude plot for the factor in Eq. (11.37).

On a log scale, this is a straight line with a slope of $+20\,\text{dB/decade}$. This straight line intersects the $0\,\text{dB}$ axis at $\omega = z_1$, since

$$20\log_{10}\left(\omega = z_1 \middle/ z_1\right) = 20\log_{10}(1) = 0\,\text{dB} \tag{11.41}$$

This value of ω is called the *corner frequency*. Figure 11.14 shows the Bode plot for the factor in Eq. (11.38)

When $z_1 = 0$, i.e,

$$\hat{H}(s) = (s + z_1) = s \tag{11.42}$$

and subsequently

$$\hat{H}(j\omega) = j\omega \tag{11.43}$$

the magnitude plot takes on the form shown in Figure 11.15.

Next, let's look at the term of the form

$$\left|\hat{H}(j\omega)\right|_{dB} = -20\log_{10}\left|1 + \frac{j\omega}{p_1}\right| \tag{11.44}$$

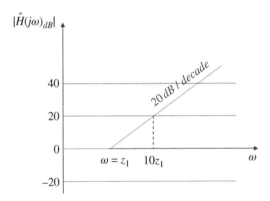

Figure 11.14 Magnitude plot for the factor in Eq. (11.38).

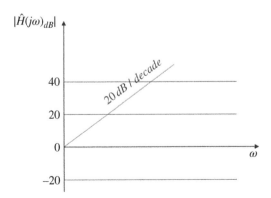

Figure 11.15 Magnitude plot for the transfer function in Eq. (11.43).

For small values of ω, we obtain

$$-20\log_{10}\left|1+\frac{j\omega}{p_1}\right| \cong -20\log_{10}1 = 0, \quad \omega \to 0 \tag{11.45}$$

For the large values of ω, we have

$$-20\log_{10}\left|1+\frac{j\omega}{p_1}\right| \cong -20\log_{10}\left(\frac{\omega}{p_1}\right), \quad \omega \to \infty \tag{11.46}$$

On a log scale, this is a straight line with a slope of $-20\,\mathrm{dB/decade}$. This straight line intersects the 0 dB axis at $\omega = p_1$, since

$$-20\log_{10}\left(\frac{\omega = p_1}{p_1}\right) = -20\log_{10}(1) = 0dB \tag{11.47}$$

Figure 11.16 shows the Bode plot for the factor in Eq. (11.44). When $p_1 = 0$, i.e,

$$\hat{H}(s) = \frac{1}{(s+p_1)} = \frac{1}{s} \tag{11.48}$$

and subsequently

$$\hat{H}(j\omega) = \frac{1}{j\omega} \tag{11.49}$$

the magnitude plot takes on the form shown in Figure 11.17.

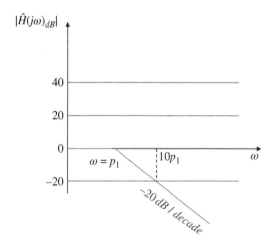

Figure 11.16 Magnitude plot for the factor in Eq. (11.44).

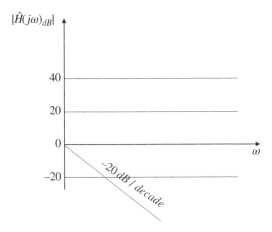

Figure 11.17 Magnitude plot for the transfer function in Eq. (11.49).

11.4 Passive Filters

To study the behavior of electrical filters we utilize the frequency transfer function discussed in the previous section.

Recall: the output of the circuit in steady state is given by

$$y_{ss}(t) = A\left|\hat{H}(j\omega)\right|\sin\left(\omega t + \angle\hat{H}(j\omega)\right)$$ (11.50)

In our discussion of passive filters (filters consisting of resistors, inductors, and capacitors) we will use the transfer function of the form

$$\hat{H}(s) = \frac{\hat{V}_{out}(s)}{\hat{V}_{in}(s)}$$ (11.51)

That is, we will study the behavior of the electrical filter where both the input and the output signals are voltages. In our study we vary the frequency over the frequency range of interest and determine the magnitude and phase of $\hat{H}(j\omega)$ in that frequency range.

Based on the magnitude response plots, the passive filters fall into four major categories, as shown in Figure 11.18.

11.4.1 RL and RC Low-Pass Filters

In this section we will examine two of the most basic low-pass RL and RC filters.

RL low-pass filter An RL low-pass filter is shown in Figure 11.19.

In order to analyze this filter (and all the remaining passive filters) we need to transform this circuit to the s domain. This is shown in Figure 11.20.

The voltage transfer function for this circuit can be obtained using the voltage divider

$$\hat{H}(s) = \frac{\hat{V}_{OUT}(s)}{\hat{V}_{IN}(s)} = \frac{R}{sL + R} = \frac{\dfrac{R}{L}}{s + \dfrac{R}{L}}$$ (11.52)

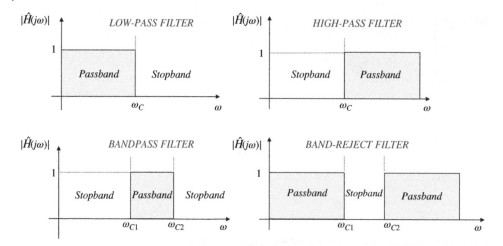

Figure 11.18 Frequency response of the four types of ideal filters.

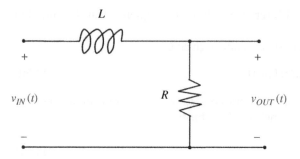

Figure 11.19 *RL* low-pass filter.

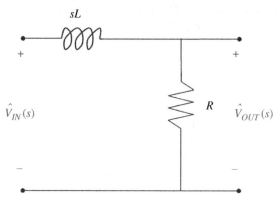

Figure 11.20 *RL* low-pass filter.

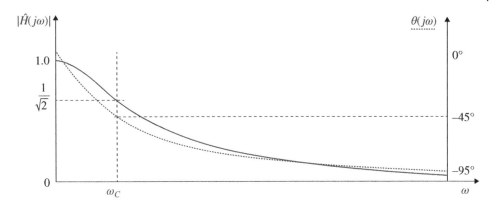

Figure 11.21 *RL* low-pass filter frequency response.

The corresponding frequency transfer function is

$$\hat{H}(j\omega) = \frac{\dfrac{R}{L}}{j\omega + \dfrac{R}{L}} \tag{11.53}$$

Figure 11.21 shows the frequency response of this RL filter.
The corner frequency of this filter is

$$\omega_C = \frac{R}{L} \tag{11.54}$$

Note that, at the corner frequency, the angle of the transfer function is $-45°$.
The RL low-pass filter transfer function can be now be expressed in terms of the corner frequency as

$$\hat{H}(s) = \frac{\omega_C}{s + \omega_C} \tag{11.55}$$

RC low-pass filter An RC circuit shown in Figure 11.22 is also a low-pass filter.

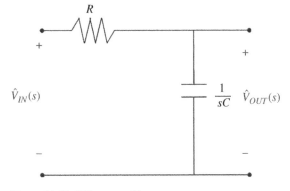

Figure 11.22 RC low-pass filter.

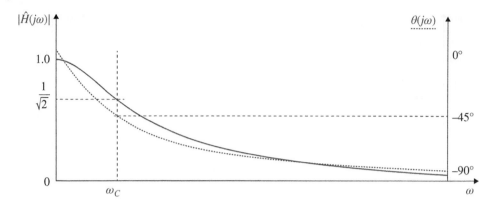

Figure 11.23 *RC* low-pass filter frequency response.

The voltage transfer function for this circuit can be obtained using the voltage divider as

$$\hat{H}(s) = \frac{\dfrac{1}{sC}}{R + \dfrac{1}{sC}} = \frac{1}{RsC + 1} = \frac{\dfrac{1}{RC}}{s + \dfrac{1}{RC}} \tag{11.56}$$

The corresponding frequency transfer function is

$$\hat{H}(j\omega) = \frac{\dfrac{1}{RC}}{j\omega + \dfrac{1}{RC}} \tag{11.57}$$

Figure 11.23 shows the frequency response of this RC filter.

Note that this filter has the same shape of frequency response as that for the low-pass RL filter. The corner frequency of this filter is

$$\omega_C = \frac{1}{RC} \tag{11.58}$$

The RC low-pass filter transfer function can be now expressed in terms of the corner frequency as

$$\hat{H}(s) = \frac{\omega_C}{s + \omega_C} \tag{11.59}$$

which has the same form as the RL low-pass filter transfer function in Eq. (11.55).

The Bode plot of the RL and RC low pass filter is shown in Figure 11.24.

11.4.2 RL and RC High-Pass Filters

RL high-pass filter   An RL high-pass filter is shown in Figure 11.25.

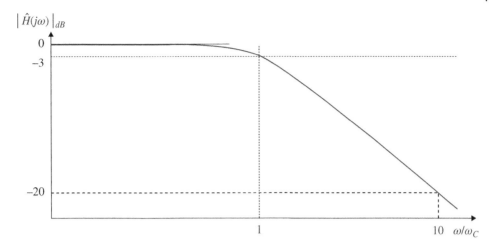

Figure 11.24 Bode magnitude plot of the *RL* and *RC* low-pass filter.

Figure 11.25 *RL* high-pass filter.

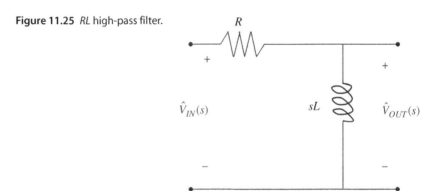

The voltage transfer function for this filter is

$$\hat{H}(s) = \frac{\hat{V}_{OUT}(s)}{\hat{V}_{IN}(s)} = \frac{sL}{sL + R} = \frac{s}{s + \dfrac{R}{L}}$$

(11.60)

The corresponding frequency transfer function is

$$\hat{H}(j\omega) = \frac{j\omega}{j\omega + \dfrac{R}{L}}$$

(11.61)

Figure 11.26 shows the frequency response of this RL filter.
The corner frequency of this filter is

$$\omega_C = \frac{R}{L}$$

(11.62)

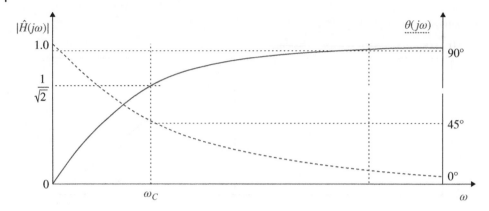

Figure 11.26 *RL high-pass filter frequency response.*

Note that at the corner frequency the angle of the transfer function is 45°.

The RL high-pass filter transfer function can be now expressed in terms of the corner frequency as

$$\hat{H}(s) = \frac{s}{s + \omega_C} \tag{11.63}$$

RC high-pass filter   An RC high-pass filter is shown in Figure 11.27.

The voltage transfer function for this circuit is

$$\hat{H}(s) = \frac{R}{R + \dfrac{1}{sC}} = \frac{RCs}{RCs + 1} = \frac{s}{s + \dfrac{1}{RC}} \tag{11.64}$$

The corresponding frequency transfer function is

$$\hat{H}(j\omega) = \frac{j\omega}{j\omega + \dfrac{1}{RC}} \tag{11.65}$$

Figure 11.27 *RC high-pass filter.*

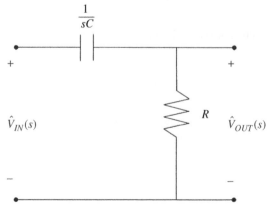

Figure 11.28 shows the frequency response of this RC filter.

Note that this filter has the same shape of frequency response as that for the high-pass RL filter. The corner frequency of this filter is

$$\omega_C = \frac{1}{RC} \tag{11.66}$$

Note that this is the same value as that obtained for the low-pass RC filter. The RC high-pass filter transfer function can be now expressed in terms of the corner frequency as

$$\hat{H}(s) = \frac{s}{s + \omega_C} \tag{11.67}$$

which has the same form as the RL high-pass filter transfer function.

The Bode plot of the RC and RL high-pass filter is shown in Figure 11.29.

The summary of the first order low-pass and high-pass filters is presented in Table 11.1.

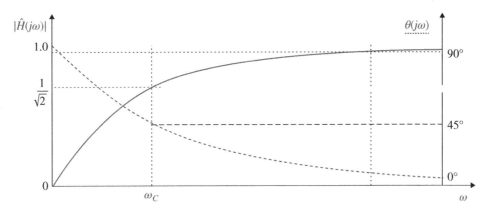

Figure 11.28 *RC* high-pass filter frequency response.

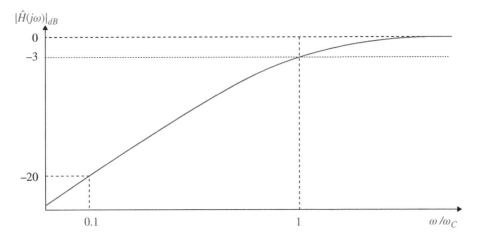

Figure 11.29 Bode magnitude plot of *RL* and *RC* high-pass filters.

Table 11.1 First-order filter descriptions.

	Low-pass RL	Low-pass RC	High-pass RL	High-pass RC
Corner frequency	$\omega_C = \dfrac{R}{L}$	$\omega_C = \dfrac{1}{RC}$	$\omega_C = \dfrac{R}{L}$	$\omega_C = \dfrac{1}{RC}$
Transfer function	$H(s) = \dfrac{\omega_C}{s + \omega_C}$	$H(s) = \dfrac{\omega_C}{s + \omega_C}$	$H(s) = \dfrac{s}{s + \omega_C}$	$H(s) = \dfrac{s}{s + \omega_C}$

11.4.3 Series and Parallel RLC Bandpass Filters

In this section we will examine two fundamental RLC filter configurations: the series RLC filter and the parallel RLC filter. Both configurations can be implemented as either bandpass or band-reject filters.

Understanding of these filters facilitates the discussion of the very important topic of resonance presented in the following section. We begin with the bandpass configurations.

Series RLC bandpass filter   A *series RLC bandpass filter* is shown in Figure 11.30.

The transfer function of this circuit can be obtained from the voltage divider as

$$\hat{H}(s) = \frac{R}{R + sL + \dfrac{1}{sC}} = \frac{RCs}{RCs + s^2 LC + 1} = \frac{\dfrac{R}{L}s}{s^2 + \dfrac{R}{L}s + \dfrac{1}{LC}} \qquad (11.68)$$

The frequency transfer function is

$$\hat{H}(j\omega) = \frac{j\dfrac{R}{L}\omega}{\dfrac{1}{LC} - \omega^2 + j\dfrac{R}{L}\omega} \qquad (11.69)$$

The frequency plot of this filter is shown in Figure 11.31. The Bode magnitude plot is shown in Figure 11.32.

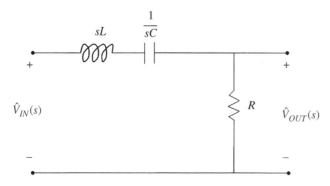

Figure 11.30 Series RLC bandpass filter.

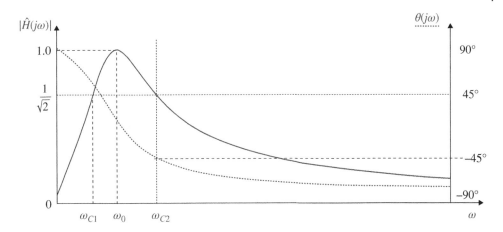

Figure 11.31 Series RLC bandpass filter frequency response.

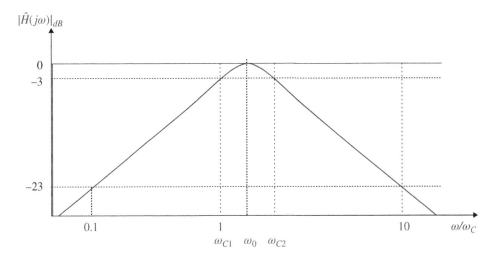

Figure 11.32 Series RLC bandpass filter – Bode magnitude plot.

The maximum magnitude occurs at

$$\omega_M = \frac{1}{\sqrt{LC}} \tag{11.70}$$

This frequency is often called the center frequency. For this series RLC circuit, this frequency is equal to the undamped natural frequency of a pure LC tank, which is denoted as

$$\omega_0 = \frac{1}{\sqrt{LC}} \tag{11.71}$$

We will see in the next section that it is also the resonant frequency of this circuit.

The filter has two corner frequencies that can be calculated by setting the magnitude in Eq. (11.69) equal to:

$$\left|\hat{H}(j\omega_c)\right| = \frac{1}{\sqrt{2}} \tag{11.72}$$

The result is (Nilsson and Riedel, 2015, p. 537)

$$\omega_{C1} = -\frac{R}{2L} + \sqrt{\left(\frac{R}{2L}\right)^2 + \frac{1}{LC}} \tag{11.73a}$$

$$\omega_{c2} = \frac{R}{2L} + \sqrt{\left(\frac{R}{2L}\right)^2 + \frac{1}{LC}} \tag{11.73b}$$

The *bandwidth* β of the bandpass filter is defined as the difference of the corner frequencies:

$$\beta = \omega_{C2} - \omega_{C1} = \frac{R}{L} \tag{11.74}$$

In terms of the bandwidth and the center frequency, the corner frequencies can be expressed as

$$\omega_{c1} = -\frac{\beta}{2} + \sqrt{\left(\frac{\beta}{2}\right)^2 + \omega_0^2} \tag{11.75a}$$

$$\omega_{c2} = \frac{\beta}{2} + \sqrt{\left(\frac{\beta}{2}\right)^2 + \omega_0^2} \tag{11.75b}$$

Additionally, for this filter, the center frequency is the geometric mean of the corner frequencies.

$$\omega_0 = \sqrt{\omega_{c1}\omega_{c2}} \tag{11.76}$$

The *quality factor* Q of the bandpass filter is defined as the ratio of its center frequency to its bandwidth:

$$Q = \frac{\omega_0}{\beta} = \frac{1}{R}\sqrt{\frac{L}{C}} \tag{11.77}$$

The corner frequencies can be expressed in terms of the center frequency and the quality factor as

$$\omega_{c1} = \omega_0 \left[-\frac{1}{2Q} + \sqrt{1 + \left(\frac{1}{2Q}\right)^2} \right] \tag{11.78a}$$

$$\omega_{c2} = \omega_0 \left[\frac{1}{2Q} + \sqrt{1 + \left(\frac{1}{2Q}\right)^2} \right] \tag{11.78b}$$

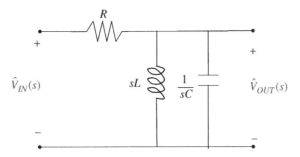

Figure 11.33 Parallel RLC bandpass filter.

Finally, the transfer function of the series RLC circuit can be expressed in terms of the bandwidth and the center frequency as

$$\hat{H}(s) = \frac{\dfrac{R}{L}s}{s^2 + \dfrac{R}{L}s + \dfrac{1}{LC}} = \frac{\beta s}{s^2 + \beta s + \omega_0^2} \tag{11.79}$$

Parallel RLC bandpass filter   A *parallel RLC bandpass filter* is shown in Figure 11.33. The impedance of the parallel configuration of L and C is

$$sL \| \frac{1}{sC} = \frac{(sL)\left(\dfrac{1}{sC}\right)}{sL + \dfrac{1}{sC}} = \frac{\left(\dfrac{L}{C}\right)(sC)}{s^2 LC + 1} = \frac{sL}{s^2 LC + 1} \tag{11.80}$$

Using the voltage divider, we obtain the transfer function of this filter as

$$\hat{H}(s) = \frac{\dfrac{sL}{s^2 LC + 1}}{R + \dfrac{sL}{s^2 LC + 1}} = \frac{sL}{s^2 RLC + sL + R} = \frac{\dfrac{1}{RC}s}{s^2 + \dfrac{1}{RC}s + \dfrac{1}{LC}} \tag{11.81}$$

The corresponding frequency transfer function is

$$\hat{H}(j\omega) = \frac{j\dfrac{1}{RC}\omega}{\dfrac{1}{LC} - \omega^2 + j\dfrac{1}{RC}\omega} \tag{11.82}$$

The frequency plots for the parallel RLC filter have the same general shape as those for the series RLC filter shown in Figures (11.31) and (11.32). The maximum magnitude of the transfer function is

$$\left|\hat{H}(j\omega)\right|_{\max} = 1 \tag{11.83}$$

when

$$\omega_0 = \frac{1}{\sqrt{LC}} \qquad (11.84)$$

The two corner frequencies are:

$$\omega_{C1} = -\frac{1}{2RC} + \sqrt{\left(\frac{1}{2RC}\right)^2 + \frac{1}{LC}} \qquad (11.85a)$$

$$\omega_{C2} = \frac{1}{2RC} + \sqrt{\left(\frac{1}{2RC}\right)^2 + \frac{1}{LC}} \qquad (11.85b)$$

The *bandwidth β* of the parallel RLC bandpass filter is

$$\beta = \omega_{c2} - \omega_{c1} = \frac{1}{RC} \qquad (11.86)$$

In terms of the bandwidth and the center frequency, the corner frequencies can be expressed as

$$\omega_{c1} = -\frac{\beta}{2} + \sqrt{\left(\frac{\beta}{2}\right)^2 + \omega_0^2} \qquad (11.87a)$$

$$\omega_{c2} = \frac{\beta}{2} + \sqrt{\left(\frac{\beta}{2}\right)^2 + \omega_0^2} \qquad (11.87b)$$

Additionally, the center frequency is the geometric mean of the corner frequencies.

$$\omega_0 = \sqrt{\omega_{c1}\omega_{c2}} \qquad (11.88)$$

The *quality factor Q* of the bandpass filter is

$$Q = \frac{\omega_0}{\beta} = \omega_0 RC = \sqrt{\frac{R^2C}{L}} = R\sqrt{\frac{C}{L}} \qquad (11.89)$$

The corner frequencies can be expressed in terms of the center frequency and quality factor as

$$\omega_{c1} = \omega_0 \left[-\frac{1}{2Q} + \sqrt{1 + \left(\frac{1}{2Q}\right)^2} \right] \qquad (11.90a)$$

$$\omega_{c2} = \omega_0 \left[\frac{1}{2Q} + \sqrt{1 + \left(\frac{1}{2Q}\right)^2} \right] \qquad (11.90b)$$

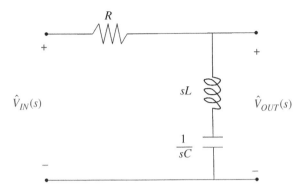

Figure 11.34 Series RLC band reject filter.

Finally, the transfer function of the parallel RLC circuit can be expressed in terms of the bandwidth and the center frequency as

$$\hat{H}(s) = \frac{\dfrac{s}{RC}}{s^2 + \dfrac{s}{RC} + \dfrac{1}{LC}} = \frac{\beta s}{s^2 + \beta s + \omega_0^2} \qquad (11.91)$$

11.4.4 Series and Parallel RLC Band-Reject Filters

Series RLC band reject filter   A *series RLC band reject filter* is shown in Figure 11.34. The transfer function of this circuit can be obtained from the voltage divider as

$$\hat{H}(s) = \frac{sL + \dfrac{1}{sC}}{R + sL + \dfrac{1}{sC}} = \frac{s^2 LC + 1}{s^2 LC + sRC + 1} = \frac{s^2 + \dfrac{1}{LC}}{s^2 + \dfrac{R}{L}s + \dfrac{1}{LC}} \qquad (11.92)$$

The frequency transfer function is

$$\hat{H}(j\omega) = \frac{\dfrac{1}{LC} - \omega^2}{\dfrac{1}{LC} - \omega^2 + j\dfrac{R}{L}\omega} \qquad (11.93)$$

The frequency plot of this filter is shown in Figure 11.35.
Bode magnitude plot is shown in Figure 11.36
The magnitude minimum is zero

$$\left| \hat{H}(j\omega) \right|_{min} = 0 \qquad (11.94)$$

when

$$\omega = \omega_0 = \frac{1}{\sqrt{LC}} \qquad (11.95)$$

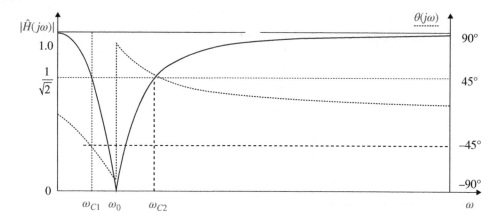

Figure 11.35 Series RLC band reject filter frequency response.

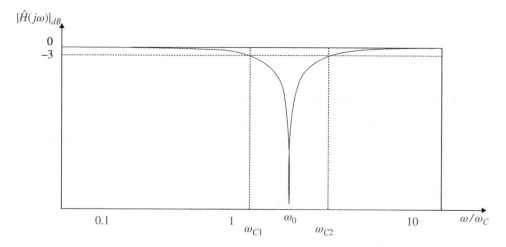

Figure 11.36 Series RLC band reject filter – Bode magnitude plot.

The two corner frequencies are

$$\omega_{C1} = -\frac{R}{2L} + \sqrt{\left(\frac{R}{2L}\right)^2 + \frac{1}{LC}} \qquad (11.96a)$$

$$\omega_{c2} = \frac{R}{2L} + \sqrt{\left(\frac{R}{2L}\right)^2 + \frac{1}{LC}} \qquad (11.96b)$$

The *bandwidth β* of the series band reject bandpass filter is

$$\beta = \frac{R}{L} \qquad (11.97)$$

In terms of the bandwidth and the center frequency, the corner frequencies can be expressed as

$$\omega_{c1} = -\frac{\beta}{2} + \sqrt{\left(\frac{\beta}{2}\right)^2 + \omega_0^2} \tag{11.98a}$$

$$\omega_{c2} = \frac{\beta}{2} + \sqrt{\left(\frac{\beta}{2}\right)^2 + \omega_0^2} \tag{11.98b}$$

Additionally, the center frequency is the geometric mean of the corner frequencies.

$$\omega_0 = \sqrt{\omega_{c1}\omega_{c2}} \tag{11.99}$$

The *quality factor* Q is

$$Q = \frac{\omega_0}{\beta} = \frac{1}{R}\sqrt{\frac{L}{C}} \tag{11.100}$$

The corner frequencies can be expressed in terms of the center frequency and quality factor as

$$\omega_{c1} = \omega_0\left[-\frac{1}{2Q} + \sqrt{1 + \left(\frac{1}{2Q}\right)^2}\right] \tag{11.101a}$$

$$\omega_{c2} = \omega_0\left[\frac{1}{2Q} + \sqrt{1 + \left(\frac{1}{2Q}\right)^2}\right] \tag{11.101b}$$

Finally, the transfer function of the series RLC band reject filter can be expressed in terms of the bandwidth and the center frequency as

$$\hat{H}(s) = \frac{s^2 + \dfrac{1}{LC}}{s^2 + \dfrac{R}{L}s + \dfrac{1}{LC}} = \frac{s^2 + \omega_0^2}{s^2 + \beta s + \omega_0^2} \tag{11.102}$$

Parallel RLC Band Reject Filter   A *parallel RLC band reject filter* is shown in Figure 11.37. The impedance of the parallel configuration of L and C is

$$sL \| \frac{1}{sC} = \frac{(sL)\left(\dfrac{1}{sC}\right)}{sL + \dfrac{1}{sC}} = \frac{\left(\dfrac{L}{C}\right)(sC)}{s^2LC + 1} = \frac{sL}{s^2LC + 1} \tag{11.103}$$

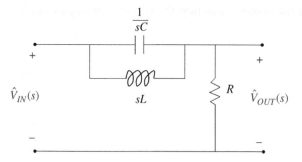

Figure 11.37 Parallel RLC band reject filter.

Using the voltage divider we have

$$\hat{H}(s) = \frac{R}{R + \frac{sL}{s^2LC + 1}} = \frac{s^2RLC + R}{s^2RLC + sL + R} = \frac{s^2 + \frac{1}{LC}}{s^2 + \frac{1}{RC}s + \frac{1}{LC}} \tag{11.104}$$

The frequency transfer function is

$$\hat{H}(j\omega) = \frac{\frac{1}{LC} - \omega^2}{\frac{1}{LC} - \omega^2 + j\frac{R}{L}\omega} \tag{11.105}$$

The frequency plots for the parallel RLC band-reject filter have the same general shape as those for the series RLC filter shown in Figures (11.35) and (11.36). Following the same steps as those for the series RLC band reject filter, it's easy to show that the two corner frequencies are:

$$\omega_{C1} = -\frac{1}{2RC} + \sqrt{\left(\frac{1}{2RC}\right)^2 + \frac{1}{LC}} \tag{11.106a}$$

$$\omega_{C2} = \frac{1}{2RC} + \sqrt{\left(\frac{1}{2RC}\right)^2 + \frac{1}{LC}} \tag{11.106b}$$

The *bandwidth* β of the parallel RLC bandpass filter is

$$\beta = \frac{1}{RC} \tag{11.107}$$

In terms of the bandwidth and the center frequency, the corner frequencies can be expressed as

$$\omega_{c1} = -\frac{\beta}{2} + \sqrt{\left(\frac{\beta}{2}\right)^2 + \omega_0^2} \tag{11.108a}$$

$$\omega_{c2} = \frac{\beta}{2} + \sqrt{\left(\frac{\beta}{2}\right)^2 + \omega_0^2} \tag{11.108b}$$

Additionally, the center frequency is the geometric mean of the corner frequencies.

$$\omega_0 = \sqrt{\omega_{c1}\omega_{c2}} \tag{11.109}$$

The *quality factor* Q of this filter is

$$Q = \frac{\omega_0}{\beta} = \omega_0 RC = \sqrt{\frac{R^2 C}{L}} = R\sqrt{\frac{C}{L}} \tag{11.110}$$

The corner frequencies can be expressed in terms of the center frequency and quality factor as

$$\omega_{c1} = \omega_0 \left[-\frac{1}{2Q} + \sqrt{1 + \left(\frac{1}{2Q}\right)^2} \right] \tag{11.111a}$$

$$\omega_{c2} = \omega_0 \left[\frac{1}{2Q} + \sqrt{1 + \left(\frac{1}{2Q}\right)^2} \right] \tag{11.111b}$$

Finally, the transfer function of the parallel band reject filter can be expressed in terms of the bandwidth and the center frequency as

$$\hat{H}(s) = \frac{s^2 + \dfrac{1}{LC}}{s^2 + \dfrac{1}{RC}s + \dfrac{1}{LC}} = \frac{s^2 + \omega_0^2}{s^2 + \beta s + \omega_0^2} \tag{11.112}$$

The summary of the second-order filters is presented in Table 11.2.

Table 11.2 Second-order filter descriptions.

	Series RLC Bandpass	Parallel RLC Bandpass	Series RLC Band Reject	Parallel RLC Band Reject
Transfer function	$H(s) = \dfrac{\beta s}{s^2 + \beta s + \omega_0^2}$	$H(s) = \dfrac{\beta s}{s^2 + \beta s + \omega_0^2}$	$H(s) = \dfrac{s^2 + \omega_0^2}{s^2 + \beta s + \omega_0^2}$	$H(s) = \dfrac{s^2 + \omega_0^2}{s^2 + \beta s + \omega_0^2}$
Bandwith	$\beta = \dfrac{R}{L}$	$\beta = \dfrac{1}{RC}$	$\beta = \dfrac{R}{L}$	$\beta = \dfrac{1}{RC}$
Center frequency		$\omega_0 = \dfrac{1}{\sqrt{LC}}$		
Quality factor	$Q = \dfrac{1}{R}\sqrt{\dfrac{L}{C}}$	$Q = R\sqrt{\dfrac{C}{L}}$	$Q = \dfrac{1}{R}\sqrt{\dfrac{L}{C}}$	$Q = R\sqrt{\dfrac{C}{L}}$
Corner frequencies	$\omega_{c1} = -\dfrac{\beta}{2} + \sqrt{\left(\dfrac{\beta}{2}\right)^2 + \omega_0^2}$ $\omega_{c2} = -\dfrac{\beta}{2} + \sqrt{\left(\dfrac{\beta}{2}\right)^2 + \omega_0^2}$		$\omega_{c1} = \omega_0\left[-\dfrac{1}{2Q} + \sqrt{1 + \left(\dfrac{1}{2Q}\right)^2}\right]$ $\omega_{c2} = \omega_0\left[-\dfrac{1}{2Q} + \sqrt{1 + \left(\dfrac{1}{2Q}\right)^2}\right]$	

11.5 Resonance in RLC Circuits

11.5.1 Resonance in Series RLC Bandpass Filter

Let's consider a series RLC bandpass filter analyzed in Section 11.4.3 and shown in Figure 11.38.

Let's calculate the input impedance for this filter.

$$\hat{Z}_{in}(j\omega) = \frac{V_{in}(s)}{I_{in}(s)}\bigg|_{s=j\omega} = R + j\omega L + \frac{1}{j\omega C} \tag{11.113}$$

or

$$\hat{Z}_{in}(j\omega) = R + j\left(\omega L - \frac{1}{\omega C}\right) \tag{11.114}$$

Let's determine the frequency when the input impedance is purely real; this happens when

$$\omega L - \frac{1}{\omega C} = 0 \tag{11.115}$$

or

$$\omega = \omega_0 = \frac{1}{\sqrt{LC}} \tag{11.116}$$

So, what does this mean? It means that at the frequency of ω_0 the input impedance is purely real, and thus the voltage phasor and the current phasor are in phase (with respect to the input terminals).

$$Z_{in} = R \tag{11.117}$$

The frequency at which the voltage and current phasors (with respect to the two terminals of the circuit) are in phase is called the *resonant frequency*, ω_r.

As we shall see, when the circuit has a resonant frequency (not every circuit does), several very interesting and important consequences may follow.

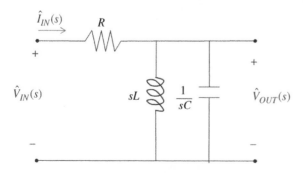

Figure 11.38 Series RLC bandpass filter.

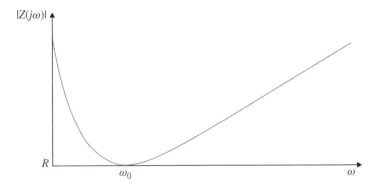

Figure 11.39 Magnitude of the input impedance.

The magnitude of the input impedance in Eq. (11.114) is

$$\left|\hat{Z}_{in}\right| = \sqrt{R^2 + \left(\omega L - \frac{1}{\omega C}\right)^2} \tag{11.118}$$

Note that the magnitude of the input impedance is minimum at the resonant frequency and is given by Eq. (11.117); it is infinite at $\omega = 0$ and $\omega = \infty$.

The magnitude plot of the input impedance using a linear scale is shown in Figure 11.39. Let's investigate the input admittance of this circuit

$$\hat{Y}_{in}(j\omega) = \frac{1}{\hat{Z}_{in}(j\omega)} = \frac{1}{R + j\left(\omega L - \frac{1}{\omega C}\right)} \tag{11.119}$$

Its magnitude is

$$\left|\hat{Y}(j\omega)\right| = \frac{1}{\sqrt{R^2 + \left(\omega L - \frac{1}{\omega C}\right)^2}} \tag{11.120}$$

This magnitude is maximum at the resonant frequency

$$\left|\hat{Y}(j\omega)\right|_{max} = \frac{1}{R} \tag{11.121}$$

and goes to zero at very low and very high frequencies, as shown in Figure 11.40.

Note that the shape of the magnitude of the input admittance is of the same form as that of the series RLC bandpass filter voltage transfer function.

We note that, at resonance, the input impedance of the circuit is minimum while the input admittance is maximum.

It is very instructive to look at the current and component voltages as functions of frequency, especially at resonance. Let's start with the magnitude plot of the current, shown in Figure 11.41.

$|Y(j\omega)|$

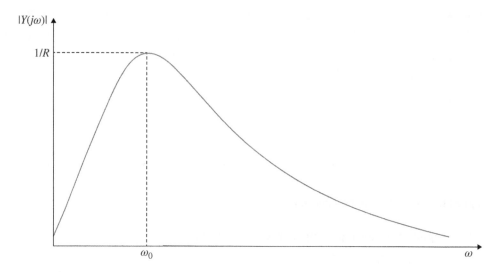

Figure 11.40 Magnitude of the input admittance.

$|I(j\omega)|$

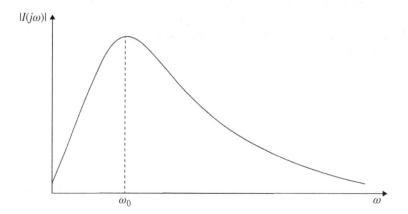

Figure 11.41 Magnitude of the current.

We observe that the maximum current occurs at resonant frequency, which is consistent with the input impedance plot.

Now, let's reveal something extremely interesting. Let's plot the voltage across the circuit elements; to focus on the phenomena occurring here let's use the prototype circuit with all circuit element values equal to one, as shown in Figure 11.42. The circuit is driven by a 1 V amplitude ac source.

The magnitudes of the voltages across the circuit element are plotted in Figure 11.43.

As expected, the voltage magnitude across the resistor is maximum at the resonant frequency, and is equal to one.

But notice that the magnitudes of the voltages across the capacitor and inductor are *greater than one*, even though the circuit was driven with a 1 V magnitude source!

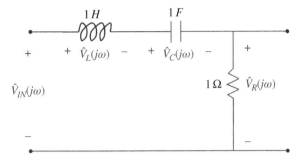

Figure 11.42 Voltages across the circuit elements.

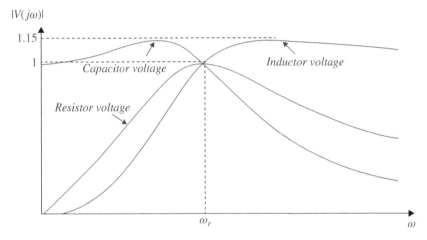

Figure 11.43 Magnitudes of the voltages across the circuit elements.

Let's look closer at both the magnitudes and the phases of the voltages across the capacitor and inductor, shown in Figure 11.44.

At resonant frequency ω_r we have

$$\hat{V}_R = V_{R,max}\angle 0° \tag{11.122}$$

$$\hat{V}_L = V_L\angle +90°, \quad V_L = V_C \tag{11.123}$$

$$\hat{V}_C = V_C\angle -90°, \quad V_L = V_C \tag{11.124}$$

and

$$\hat{V}_L + \hat{V}_C = V_L\angle +90° + V_L\angle -90° = 0 \tag{11.125}$$

We observe that the maximum magnitude of the capacitor voltage ($V_{C,max} = 1.15\,\text{V}$) is larger than the magnitude of the input voltage ($V_{in} = 1\,\text{V}$), and occurs at frequency $\omega_C < \omega_r$.

The maximum magnitude of the inductor voltage ($V_{L,max} = 1.15\,\text{V}$) is larger than magnitude of the input voltage ($V_{in} = 1\,\text{V}$) and occurs at frequency $\omega_L > \omega_r$.

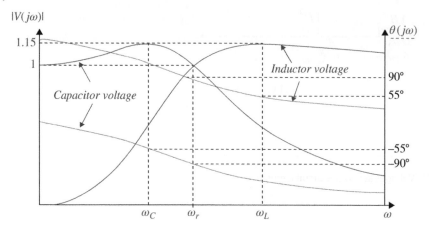

Figure 11.44 Voltages across the capacitor and inductor.

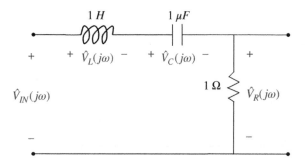

Figure 11.45 Circuit with a new capacitor value.

Additionally we have, $V_{C,max} = V_{L,max}$ and the phase angle of the capacitor voltage $\theta_C = -55°$ is of the opposite polarity to that of the inductor voltage $\theta_L = 55°$.

Recall: the quality factor of this circuit is given by Eq. (11.77), repeated here,

$$Q = \frac{\omega_0}{\beta} = \frac{1}{R}\sqrt{\frac{L}{C}} \tag{11.126}$$

which for the prototype circuit shown is equal to one.

For the illustration purposes, let's change the capacitor value to $C = 100\,\mu F$, as shown in Figure 11.45.

The quality factor now is

$$Q = \frac{1}{1}\sqrt{\frac{10^{-6}}{100}} = 100 \tag{11.127}$$

First let's look at the voltage across the resistor, shown in Figure 11.46.

Its maximum occurs at resonance and it is still equal to one. Now let's look at the voltages across the capacitor and inductor, shown in Figure 11.47.

$|V(j\omega)|$

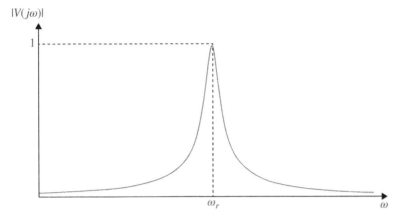

Figure 11.46 Voltages across the resistor Q=100.

$|V(j\omega)|$

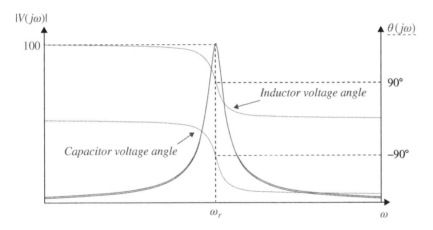

Inductor voltage angle

Capacitor voltage angle

Figure 11.47 Voltages across the capacitor and inductor Q=100.

At resonance, the magnitude of the capacitor and inductor voltage is 100 V for a 1 V input signal!

To be more specific

$$V_L = QV_m \tag{11.128a}$$

$$V_C = QV_m \tag{11.128b}$$

where V_m is the amplitude of the input voltage.

Let's prove it. Recall the definition of the quality factor

$$Q = \frac{\omega_0}{\beta} \tag{11.129}$$

Since

$$\beta = \frac{R}{L} \tag{11.130}$$

$$\omega_0 = \frac{1}{\sqrt{LC}} \tag{11.131}$$

we can rewrite Eq. (11.129) as

$$Q = \frac{\omega_0 L}{R} \tag{11.132}$$

or

$$Q = \frac{\omega_0 L}{R} = \frac{\omega_0^2 L}{\omega_0 R} = \frac{\frac{1}{LC}L}{\omega_0 R} = \frac{1}{\omega_0 RC} \tag{11.133}$$

and thus

$$\left| \hat{V}_L \right| = Z_L I = (\omega_0 L)\left(\frac{V_m}{R} \right) = \frac{\omega_0 L}{R} V_m = QV_m \tag{11.134a}$$

$$\left| \hat{V}_C \right| = Z_C I = \left(\frac{1}{\omega_0 C} \right)\left(\frac{V_m}{R} \right) = \frac{1}{\omega_0 RC} V_m = QV_m \tag{11.134b}$$

At resonance, the voltage across the capacitor and the voltage across the inductor can be many times larger than the input voltage.

This is especially dangerous in high Q circuits where this voltage might be destructive to the capacitor (inductors handle high voltages much better).

Note that at resonance

$$\hat{V}_L = V_{L,max} \angle + 90°, \quad V_{L,max} = V_{C,max} \tag{11.135a}$$

$$\hat{V}_C = V_{C,max} \angle - 90°, \quad V_{L,max} = V_{C,max} \tag{11.135b}$$

And again, the total voltage across the LC configuration is zero.

$$\hat{V}_L + \hat{V}_C = V_{L,max} \angle + 90° + V_{L,max} \angle - 90° = 0 \tag{11.136}$$

11.5.2 Resonance in Parallel RLC Bandpass Filter

Let's now consider a parallel RLC bandpass filter analyzed in Section 11.4.3, and shown in Figure 11.48.

Let's apply the source transformation to the input voltage source and the resistor. The resulting circuit is shown in Figure 11.49 and is known as a parallel RLC circuit.

Note that the circuit to the left of the inductor, consisting of a current source in parallel to a resistor, is equivalent to the one in Figure 11.48, consisting of a voltage source in series with a resistor. This means that the results obtained for the inductor and capacitor currents from the circuit in Figure 11.49 also apply to the circuit in Figure 11.48.

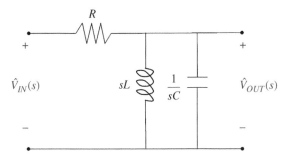

Figure 11.48 Parallel RLC bandpass filter.

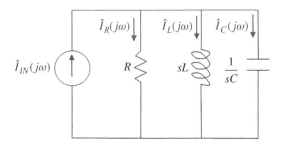

Figure 11.49 Parallel RLC circuit.

The input admittance of this circuit is

$$\hat{Y}_{in}(j\omega) = \frac{1}{R} + \frac{1}{j\omega L} + j\omega C = \frac{1}{R} + j\left(\omega C - \frac{1}{\omega L}\right) \tag{11.137}$$

At

$$\omega = \omega_0 = \frac{1}{\sqrt{LC}} \tag{11.138}$$

the input admittance is purely real. Thus the frequency in Eq. (11.138) is the resonant frequency of the parallel RLC circuit. Let's look at the magnitudes of the element currents, shown in Figure 11.50.

Let's look closer at both the magnitudes and the phases of the currents through the capacitor and inductor, shown in Figure 11.51.

At resonant frequency ω_r we have

$$\hat{I}_R = I_{R,max}\angle 0° \tag{11.139}$$

$$\hat{I}_L = I_L\angle -90°, \quad I_L = I_C \tag{11.140}$$

$$\hat{I}_C = I_C\angle +90°, \quad I_L = I_C \tag{11.141}$$

$|I(j\omega)|$

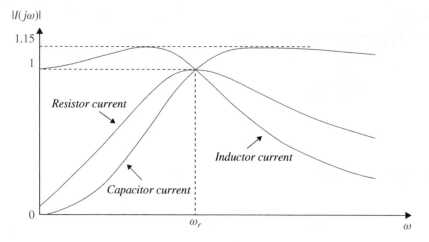

Figure 11.50 Magnitudes of the currents through the circuit elements.

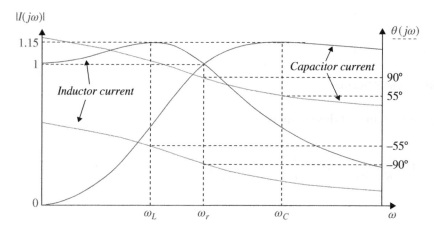

Figure 11.51 Currents through the capacitor and inductor.

and

$$\hat{I}_L + \hat{I}_C = I_L\angle - 90^\circ + I_L\angle - 90^\circ = 0 \qquad (11.142)$$

We observe that the maximum magnitude of the inductor current ($I_{L,max} = 1.15$ A) is larger than magnitude of the input current ($I_{in} = 1$ A) and occurs at frequency $\omega_L < \omega_r$.

The maximum magnitude of the capacitor current ($I_{C,max} = 1.15$ A) is larger than magnitude of the input current ($I_{in} = 1$ A) and occurs at frequency $\omega_L < \omega_r$.

Additionally, we have $I_{L,max} = I_{C,max}$ and the phase angle of the inductor current $\theta_L = -55^\circ$ is of the opposite polarity to that of the capacitor current $\theta_C = 55^\circ$.

The quality factor of this circuit is given by

$$Q = \frac{\omega_0}{\beta} = R\sqrt{\frac{C}{L}} \qquad (11.143)$$

$|I_L(j\omega)|$

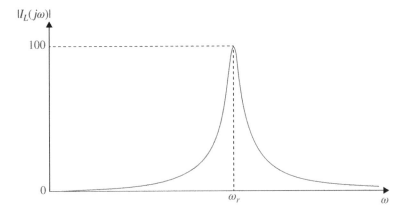

Figure 11.52 Magnitude of the capacitor and inductor currents, $Q = 100$.

which for the prototype circuit ($R = 1\,\Omega$, $L = 1\,H$, $C = 1\,F$) is equal to one. For the illustration purposes let's change the resistor value to $R = 100\,\Omega$.

The quality factor now is

$$Q = 100\sqrt{\frac{1}{1}} = 100 \tag{11.144}$$

Now, let's look at the capacitor and inductor currents shown in Figure 11.52

At resonance, the magnitude of the capacitor and inductor current is 100A for a 1 A input signal!

At resonance the current through the capacitor and the current through the inductor can be many times larger than the input current.

This is especially dangerous in high Q circuits where this current might be destructive to the inductor (capacitors handle high currents much better).

To be more specific

$$I_L = QI_m \tag{11.145a}$$

$$I_C = QI_m \tag{11.145b}$$

where I_m is the amplitude of the input current. Let's prove it. Recall

$$Q = \frac{\omega_0}{\beta} = \frac{\omega_0}{1/RC} = \omega_0 RC = \frac{R}{\omega_0 L} \tag{11.146}$$

The impedance of the parallel combination of R and C is

$$\hat{Z}_{RC} = \frac{R\dfrac{1}{j\omega_0 C}}{R + \dfrac{1}{j\omega_0 C}} = \frac{R}{1 + j\omega_0 RC} \tag{11.147}$$

Using the current divider, we get the inductor current as

$$\hat{I}_L = \frac{\dfrac{R}{1+j\omega_0 RC}}{j\omega_0 L + \dfrac{R}{1+j\omega_0 RC}} \hat{I} = \frac{R}{R + j\omega_0 L(1+j\omega_0 RC)} \hat{I}$$

$$= \frac{R}{R - \omega_0^2 RLC + j\omega_0 L} \hat{I} = \frac{R}{R - \dfrac{1}{LC}RLC + j\omega_0 L} \hat{I} = \frac{R}{j\omega_0 L} \hat{I} \tag{11.148}$$

The magnitude of this current is

$$I_L = \frac{R}{\omega_0 L} I_m = Q I_m \tag{11.149}$$

The capacitor current is

$$\hat{I}_C = \frac{\dfrac{Rj\omega_0 L}{R+j\omega_0 L}}{\dfrac{Rj\omega_0 L}{R+j\omega_0 L} + \dfrac{1}{j\omega_0 C}} \hat{I} = \frac{j\omega_0 RL}{j\omega_0 RL + \dfrac{R+j\omega_0 L}{j\omega_0 C}} \hat{I}$$

$$= \frac{(j\omega_0 RL)(j\omega_0 C)}{(j\omega_0 RL)(j\omega_0 C) + R + j\omega_0 L} \hat{I} = \frac{-\omega_0^2 RLC}{-\omega_0^2 RLC + R + j\omega_0 L} \hat{I} \tag{11.150}$$

$$= \frac{-R}{-R + R + j\omega_0 L} \hat{I} = \frac{-R}{j\omega_0 L} \hat{I} = j\frac{R}{\omega_0 L} \hat{I} = jQ\hat{I}$$

The magnitude of this current is

$$I_C = \frac{R}{\omega_0 L} I_m = Q I_m \tag{11.151}$$

11.5.3 Resonance in Other RLC Circuits

In series and parallel RLC circuits discussed so far, we encounter a "pure" series and parallel connection of L and C, as shown in Figure 11.53.

For these pure configurations, the resonant frequency was equal to

$$\omega_r = \omega_0 = \frac{1}{\sqrt{LC}} \tag{11.152}$$

In general, when L and C are not purely in series or parallel, this is not the case. Let's consider the parallel LC circuit shown in Figure 11.54.

Even though this is not a pure parallel LC configuration, it is often called such. Let's determine the input admittance and the resonant frequency of this circuit.

$$\hat{Y}(j\omega) = \frac{1}{-\dfrac{j}{\omega C}} + \frac{1}{R + j\omega L} = j\omega C + \frac{1}{R + j\omega L}$$

$$= j\omega C + \frac{R - j\omega L}{R^2 + (\omega L)^2} = \frac{R}{R^2 + (\omega L)^2} + j\left(\omega C - \frac{\omega L}{R^2 + (\omega L)^2}\right) \tag{11.153}$$

(a)

(b)

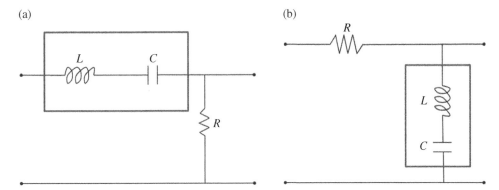

Figure 11.53 Pure (a) series and (b) parallel LC configurations.

Figure 11.54 Another parallel LC configuration.

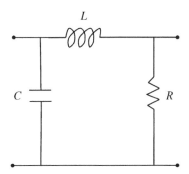

For resonance to occur, the admittance needs to be real. This occurs when

$$\omega_r C - \frac{\omega_r L}{R^2 + (\omega_r L)^2} = 0 \tag{11.154}$$

or

$$\omega_r C \left[R^2 + (\omega_r L)^2 \right] - \omega_r L = 0 \tag{11.155}$$

Dividing by ω_r and rearranging produces

$$\omega_r^2 = \frac{L - RC^2}{L^2} \tag{11.156}$$

and therefore the resonant frequency of the circuit is expressed as

$$\omega_r = \sqrt{\frac{1}{LC} - \frac{R^2}{L^2}} \tag{11.157}$$

Substituting this value into Eq. (11.153) gives the value of the input admittance at resonance.

$$\hat{Y}(j\omega_r) = \left\{ \frac{R}{R^2 + (\omega L)^2} + j\left(\omega C - \frac{\omega L}{R^2 + (\omega L)^2}\right) \right\}_{\omega=\omega_r}$$

$$= \left\{ \frac{R}{R^2 + (\omega L)^2} \right\}_{\omega=\omega_r} = \frac{R}{R^2 + (\omega_r L)^2} = \frac{R}{R^2 + \left(\dfrac{1}{LC} - \dfrac{R^2}{L^2}\right)L^2} \qquad (11.158)$$

$$= \frac{R}{R^2 + \dfrac{L}{C} - R^2} = \frac{RC}{L}$$

The input impedance at resonance is therefore

$$Z(j\omega_r) = \frac{L}{RC} \qquad (11.159)$$

Returning to Eq. (11.157) we observe that if

$$\frac{1}{LC} - \frac{R^2}{L^2} < 0 \quad \Leftrightarrow \quad \frac{1}{LC} < \frac{R^2}{L^2} \quad \Leftrightarrow \quad \frac{CR^2}{L} > 1 \qquad (11.160)$$

then the resonant frequency is complex and thus there is no real solution for the resonant frequency; this means that the source voltage and the source current cannot be in phase at any frequency.

If

$$\frac{1}{LC} - \frac{R^2}{L^2} > 0 \quad \Leftrightarrow \quad \frac{1}{LC} > \frac{R^2}{L^2} \quad \Leftrightarrow \quad \frac{CR^2}{L} < 1 \qquad (11.161)$$

then there is a unique non-zero resonant frequency.

For completeness, let's consider the series LC circuit shown in Figure 11.55.

Even though this is not a pure series LC configuration, it is often called such. Let's determine the input impedance and the resonant frequency of this circuit.

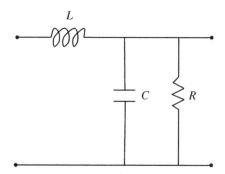

Figure 11.55 Another series LC configuration.

$$\hat{Z}(j\omega) = j\omega L + \frac{R\dfrac{1}{j\omega C}}{R + \dfrac{1}{j\omega C}} = j\omega L + \frac{R}{1 + j\omega RC}$$

(11.162)

$$= j\omega L + \frac{R(1 - j\omega RC)}{1 + (\omega RC)^2} = \frac{R}{1 + (\omega RC)^2} + j\left[\omega L - \frac{\omega R^2 C}{1 + (\omega RC)^2}\right]$$

For resonance to occur the impedance needs to be real. This occurs when

$$\omega_r L - \frac{\omega_r R^2 C}{1 + (\omega_r RC)^2} = 0$$

(11.163)

or

$$\omega_r L\left[1 + (\omega_r RC)^2\right] - \omega_r R^2 C = 0$$

(11.164)

Dividing by ω_r and rearranging produces

$$\omega_r^2 = \frac{R^2 C - L}{R^2 C^2 L} = \frac{1}{LC} - \frac{1}{R^2 C^2}$$

(11.165)

and therefore the resonant frequency of the circuit is

$$\omega_r = \sqrt{\frac{1}{LC} - \frac{1}{R^2 C^2}}$$

(11.166)

Substituting this value into Eq. (11.162) gives the value of the input impedance at resonance.

$$\hat{Z}(j\omega_r) = \left\{\frac{R}{1 + (\omega RC)^2} + j\left[\omega L - \frac{\omega R^2 C}{1 + (\omega RC)^2}\right]\right\}_{\omega = \omega_r}$$

$$= \left\{\frac{R}{1 + (\omega RC)^2}\right\}_{\omega = \omega_r} = \frac{R}{1 + R^2 C^2\left(\dfrac{1}{LC} - \dfrac{1}{R^2 C^2}\right)}$$

(11.167)

$$= \frac{R}{1 + \dfrac{R^2 C}{L} - 1} = \frac{L}{RC}$$

The input impedance at resonance is therefore

$$Z(j\omega_r) = \frac{L}{RC}$$

(11.168)

Returning to Eq. (11.166) we observe that if

$$\frac{1}{LC} - \frac{1}{R^2 C^2} < 0 \quad \Leftrightarrow \quad \frac{1}{LC} < \frac{1}{R^2 C^2} \quad \Leftrightarrow \quad \frac{L}{CR^2} > 1$$

(11.169)

then the resonant frequency is complex and thus there is no real solution for the resonant frequency; this means that the source voltage and the source current cannot be in phase at any frequency.

If

$$\frac{L}{CR^2} < 1 \tag{11.170}$$

then there is a unique non-zero resonant frequency.

11.6 EMC Applications

11.6.1 Non-Ideal Behavior of Capacitors and Inductors

Capacitors The impedance of an *ideal* capacitor is equal to

$$\hat{Z}(j\omega) = \frac{1}{j\omega C} \tag{11.171}$$

The magnitude of this impedance decreases linearly with frequency or at a rate of $-20\,\text{dB/decade}$, as shown in Figure 11.56.

The equivalent circuit model of a physical capacitor is shown in Figure 11.57.
The impedance of this RLC circuit is

$$H(j\omega) = j\omega + R + \frac{1}{j\omega C} = \frac{-\omega^2 LC + j\omega CR + 1}{j\omega C}$$

$$= \frac{-\omega^2 L + j\omega R + \dfrac{1}{C}}{j\omega} = L\,\frac{-\omega^2 + j\dfrac{\omega R}{L} + \dfrac{1}{LC}}{j\omega} \tag{11.172}$$

$$= L\,\frac{\dfrac{1}{LC} - \omega^2 + j\dfrac{\omega R}{L}}{j\omega}$$

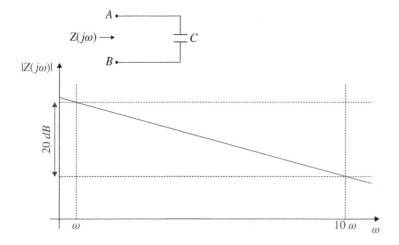

Figure 11.56 Impedance magnitude of the ideal capacitor.

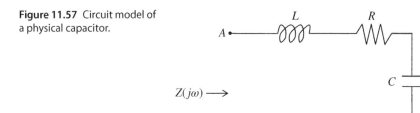

Figure 11.57 Circuit model of a physical capacitor.

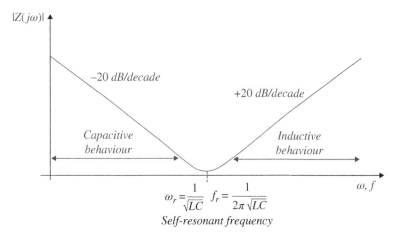

Figure 11.58 Impedance magnitude of a physical capacitor.

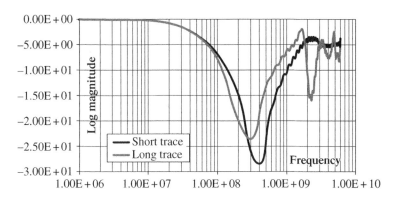

Figure 11.59 The effect of the connection leads on the impedance of a capacitor.

The magnitude of this impedance is shown in Figure 11.58.

One of the most important factors that affects the behavior of a capacitor (and an inductor) in a practical circuit is the length of its leads. The longer the leads the larger the inductance. Figure 11.59 shows the impedance measurements of a 120 pF capacitor with the short and long connection leads, respectively.

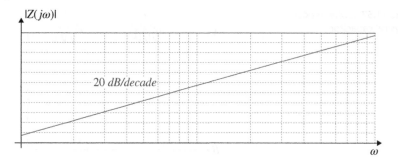

Figure 11.60 Impedance magnitude of the ideal inductor.

Note that increasing the length of the connection leads moves the self-resonant frequency of a capacitor to the left. This is consistent with the model we have used. Increasing the connection leads increases the inductance and thus the resonant frequency becomes smaller.

Inductors The impedance of an *ideal* inductor is equal to

$$\hat{Z}(j\omega) = j\omega L \tag{11.173}$$

The magnitude of this impedance increases linearly with frequency or at a rate of 20 dB/decade as shown in Figure 11.60.

Figure 11.61 shows the impedance measurements of a 56 nH inductor with short and long connection leads, respectively.

Again, we note that increasing the connection leads increases the inductance and thus the resonant frequency becomes smaller.

11.6.2 Decoupling Capacitors

One of the most interesting EMC examples of resonance involves the use of decoupling capacitors. Figure 11.62 shows the circuit model and the current flow for two cascaded CMOS inverters with the adjacent decoupling capacitors.

When a decoupling capacitor is placed adjacent to an IC to supply the transient switching current, an RLC circuit is created. The parasitic inductance comes from several sources (Ott, 2009, p. 432):

- the capacitor itself
- the interconnecting PCB traces and vias
- the lead frame of the IC

This inductance is shown in Figure 11.63.

Effectively, this RLC circuit will be resonant! Let's look at this resonance for several different decoupling schemes.

First, let's consider three different capacitors in series with 15 nH of parasitic inductance, as shown in Figure 11.64.

Figure 11.65 shows the plot of the magnitude of the impedance of these LC networks vs frequency.

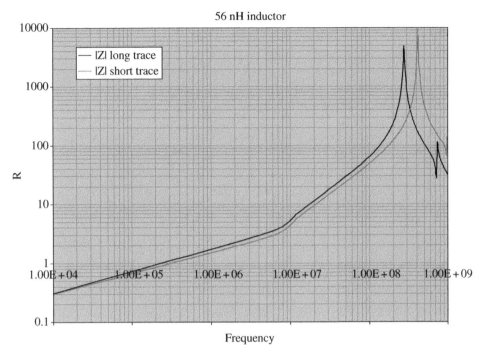

Figure 11.61 The effect of the connection leads on the impedance of an inductor.

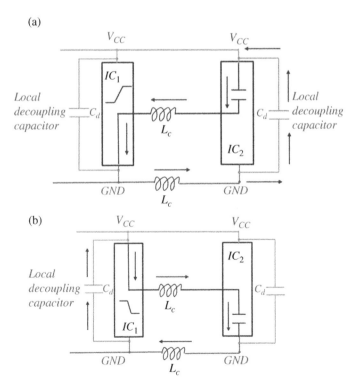

Figure 11.62 Local decoupling capacitor in a CMOS circuitry (a) low-to-high transition, (b) high-to-low transition.

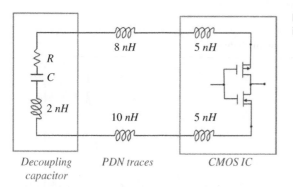

Figure 11.63 Parasitic circuit inductance.

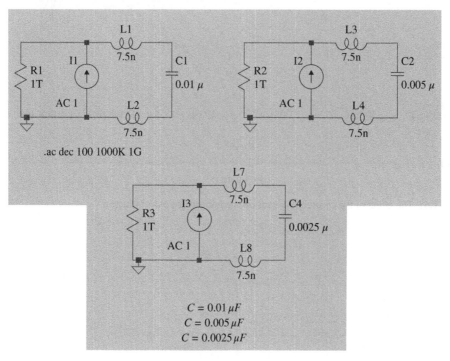

Figure 11.64 Circuit model – three different capacitors in series with 15 nH of parasitic inductance.

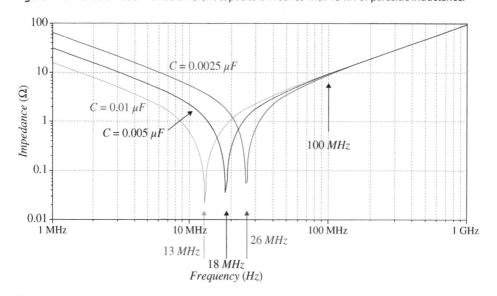

Figure 11.65 Impedance plot – three different capacitors in series with 15 nH of parasitic inductance.

Note that above 100 MHz, the impedance of the decoupling network is dominated by the 15 nH of inductance, regardless of what value capacitor is used.

Figure 11.66 shows multiple capacitors of the same value.

The resulting impedance plots are shown in Figure 11.67.

Figure 11.68 shows a single capacitor vs multiple capacitor configurations, where the total capacitance in each circuit is the same.

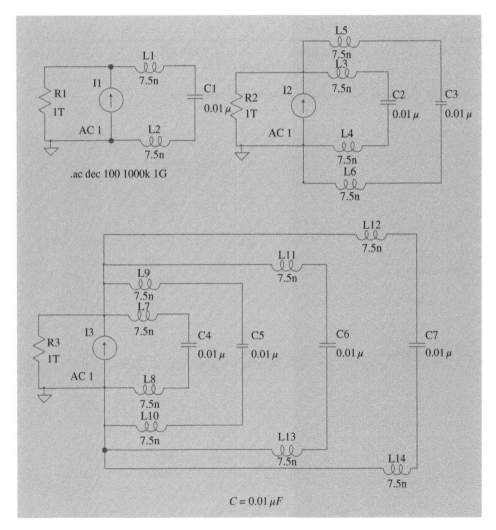

Figure 11.66 Circuit model – multiple capacitors of the same value.

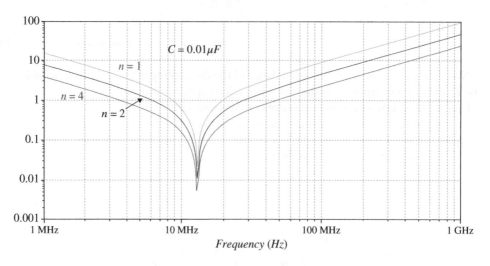

Figure 11.67 Impedance plot – multiple capacitors of the same value.

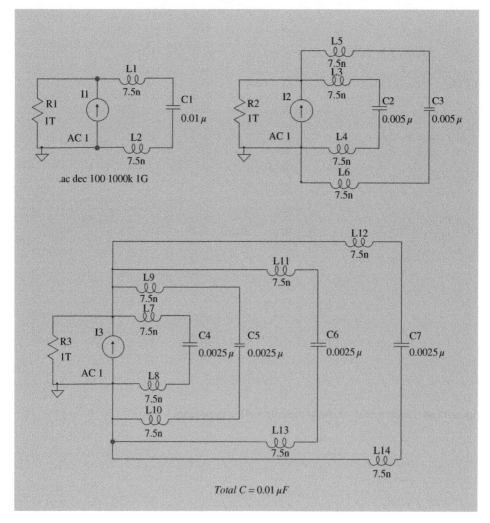

Figure 11.68 Circuit model – single cap vs multiple capacitors when the total capacitance is the same.

The resulting impedance plots are shown in Figure 11.69.
Figure 11.70 shows two capacitors one decade apart in values.
The resulting impedance plots are shown in Figure 11.71.
Figure 11.72 shows three capacitors one decade apart in values.

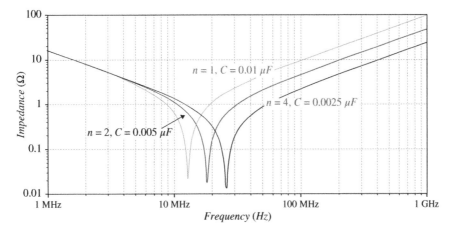

Figure 11.69 Impedance plot – multiple capacitors of the same value.

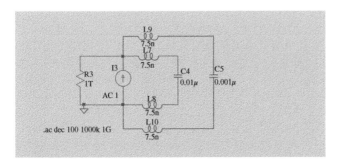

Figure 11.70 Circuit model – two capacitors one decade apart.

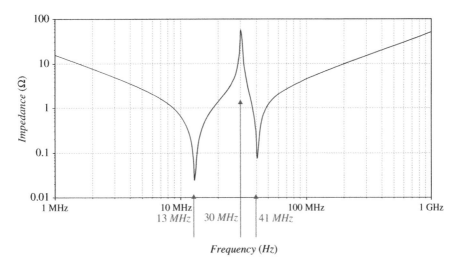

Figure 11.71 Impedance plot – two capacitors one decade apart.

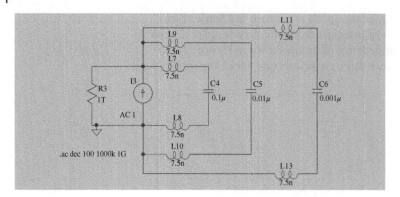

Figure 11.72 Circuit model – three capacitors one decade apart.

The resulting impedance plots are shown in Figure 11.73.

Finally, it is interesting to compare the case of three capacitors decades apart vs three capacitors of the same value. This is shown in Figure 11.74.

The resulting impedance plots are shown in Figure 11.75.

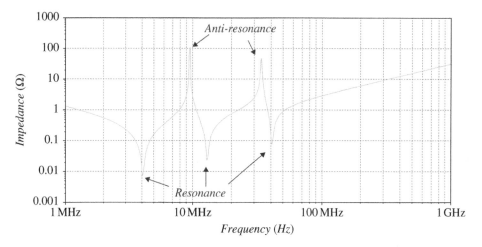

Figure 11.73 Impedance plot – three capacitors one decade apart.

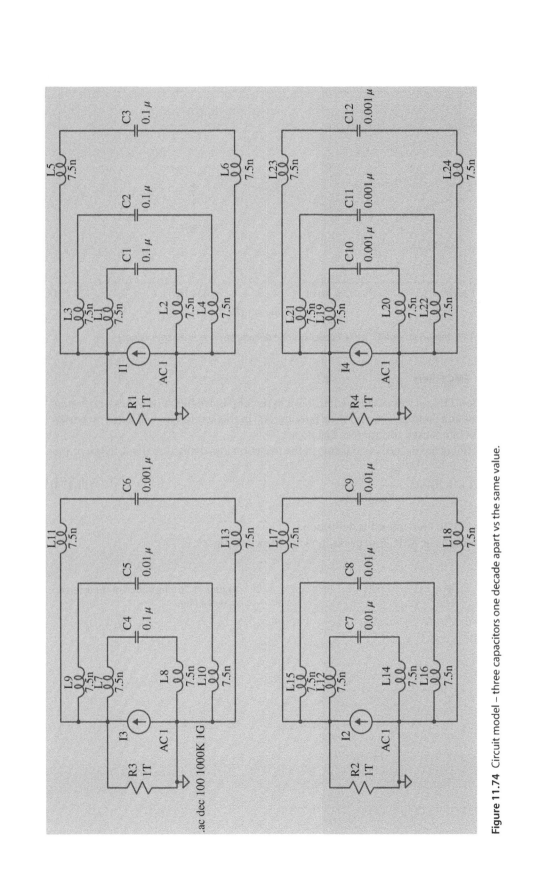

Figure 11.74 Circuit model – three capacitors one decade apart vs the same value.

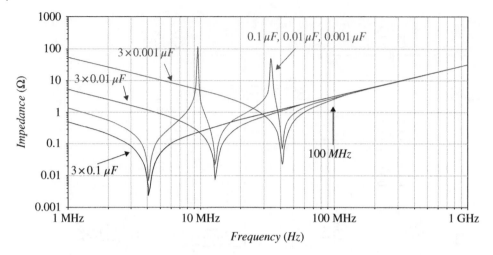

Figure 11.75 Impedance plot – three capacitors one decade apart vs. the same value.

11.6.3 EMC Filters

In Section 11.4 we discussed the four basic types of passive filters (low-pass, high-pass, bandpass and band-reject) and their parameters. In this section we will focus on passive low-pass filters used to suppress EM noise.

EMC filters are described in terms of the insertion loss defined as (Paul, 2006, p. 386)

$$IL_{dB} = 20\log_{10}\frac{V_{L,without\ filter}}{V_{L,with\ filter}} \tag{11.173}$$

Figure 11.76 illustrates this definition.

The most basic EMC low-pass filters are shown in Figure 11.77.

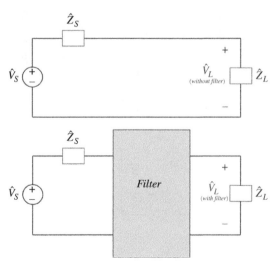

Figure 11.76 Illustration of the insertion loss of a filter.

Figure 11.77 First-order low-pass filters.

A typical higher-order EMC low-pass filter consists of a series inductance and shunt capacitance. Figure 11.78 shows two different LC configurations.

The filters shown in Figure 11.78 can be cascaded to produce higher-order filters. This is shown in Figures 11.79 and 11.80.

Third-order π and T filters are shown in Figure 11.81.

(a)   (b)

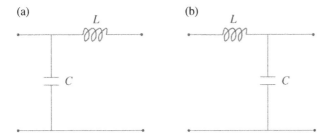

Figure 11.78 LC low-pass filters: (a) configuration 1, (b) configuration 2.

Figure 11.79 Cascaded LC filters – configuration 1.

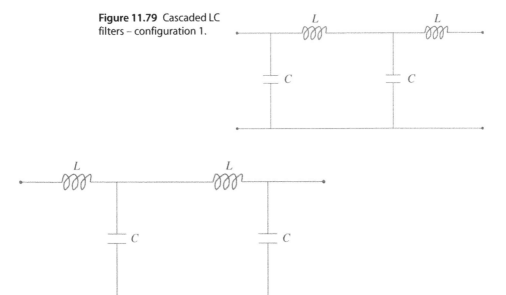

Figure 11.80 Cascaded LC filters – configuration 2.

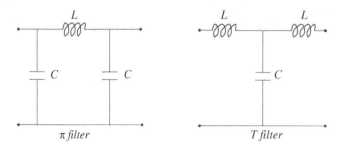

π *filter*   T *filter*

Figure 11.81 π and T low-pass filters.

The higher the order of the filter the sharper the transition from the pass band to the rejection region. Note that for each order of the filter we have two different configurations. Which one will perform better? That depends on the impedance of the source and the load.

The general rule is that the inductor should be on the low-impedance side and the capacitor should be on the high-impedance side.

Figure 11.82 shows the appropriate configurations when both the source and the load impedances are low.

Figure 11.83 shows the appropriate configurations when both the source and the load impedances are high.

Figure 11.84 shows the appropriate configurations when the source impedance is low and the load impedance is high.

Finally, Figure 11.85 shows the appropriate configurations when the source impedance is high and the load impedance is low.

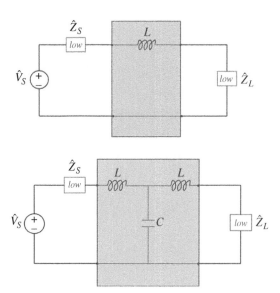

Figure 11.82 Filter configurations when both the source and the load impedances are low.

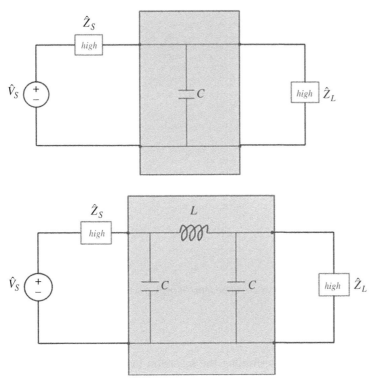

Figure 11.83 Filter configurations when both the source and the load impedances are high.

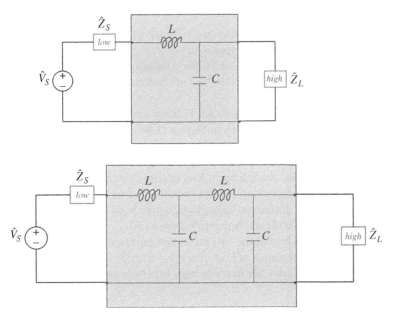

Figure 11.84 Filter configurations when the source impedance is low and the load impedance is high.

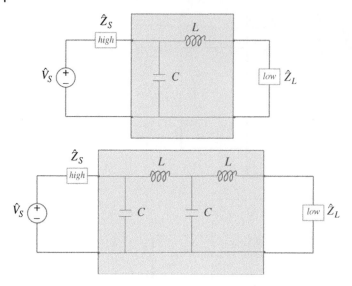

Figure 11.85 Filter configurations when the source impedance is high and the load impedance is low.

Let's verify the above claims by determining the insertion loss of second-order LC filters. First, let's investigate the configurations shown in Figure 11.86, where the source impedance is low and the load impedance is high.

Figure 11.87 shows the insertion loss of the two filter configurations in Figure 11.86.

As can be seen from Figure 11.87, the insertion loss of configuration 1 is about 25 dB higher than that of configuration 2.

Next, let's investigate the configurations shown in Figure 11.88, where the source impedance is high and the load impedance is low.

Figure 11.89 shows the insertion loss of the two filter configurations in Figure 11.88.

As can be seen from Figure 11.89, the insertion loss of configuration 1 is again about 25 dB higher than that of configuration 2.

Next, let's compare the performance of the π and T filters as shown in Figure 11.83. First, let's investigate the configurations shown in Figure 11.90, where both the source impedance and the load impedance are low.

Figure 11.91 shows the insertion loss of the two filter configurations in Figure 11.90.

As can be seen from Figure 11.91, the insertion loss of the T configuration is about 50 dB higher than that of the π configuration (except for the low frequency region).

Next, let's investigate the π and T filters configurations, where both the source impedance and the load impedance are 100 Ω, as shown in Figure 11.92.

Figure 11.93 shows the insertion loss of the two filter configurations in Figure 11.92.

What we notice is that the performance of both filters is virtually the same. The reason is that the 100 Ω impedance is neither low nor high for these configurations (and filter component values).

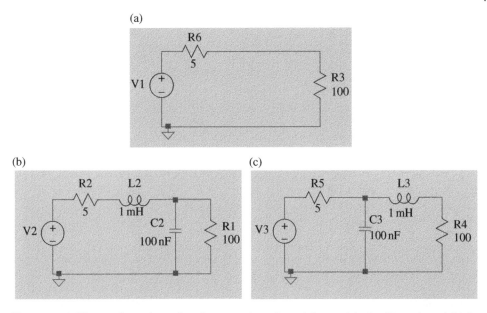

Figure 11.86 Filter configurations when the source impedance is low and the load impedance is high: (a) no filter, (b) inductor on the low impedance side (configuration 1), (c) inductor on the high impedance side (configuration 2).

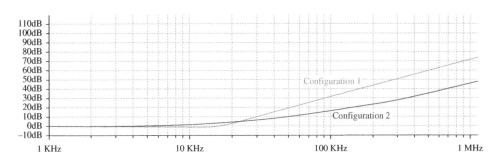

Figure 11.87 Insertion loss of the two configurations shown in Figure 11.86.

Finally, let's investigate the π and T filters configurations, where both the source impedance and the load impedance are high, as shown in Figure 11.94.

Figure 11.95 shows the insertion loss of the two filter configurations in Figure 11.94.

As can be seen from Figure 11.95, this time the insertion loss of the π configuration is about 25 dB higher than that of the T configuration (except for the low frequency region). Increasing the source and load impedances would increase the insertion loss.

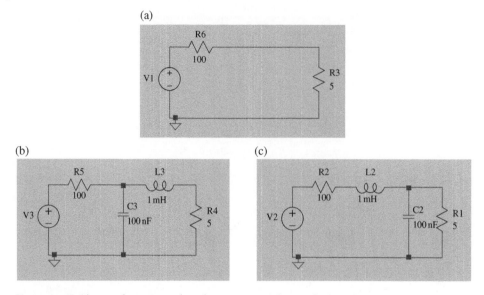

Figure 11.88 Filter configurations when the source impedance is high and the load impedance is low: (a) no filter, (b) inductor on the low impedance side (Configuration 1), (c) inductor on the high impedance side (Configuration 2).

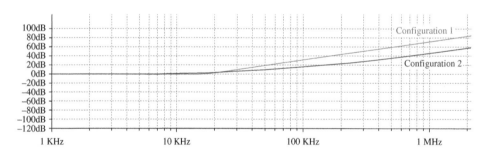

Figure 11.89 Insertion loss of the two configurations shown in Figure 11.86.

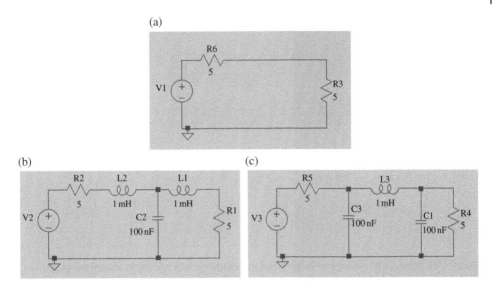

Figure 11.90 Filter configurations when both the source impedance and the load impedance are low: (a) no filter, (b) inductors on the low impedance sides (π configuration), (c) capacitors on the low impedance sides (T configuration).

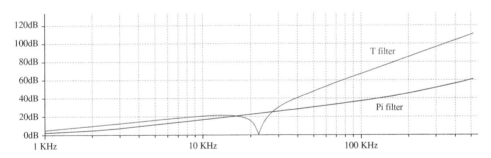

Figure 11.91 Insertion loss of the two configurations shown in Figure 11.90.

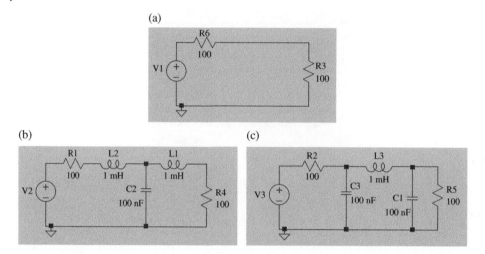

Figure 11.92 Filter configurations when both the source impedance and the load impedance are 100 Ω: (a) no filter, (b) T filter configuration, (c) π filter configuration.

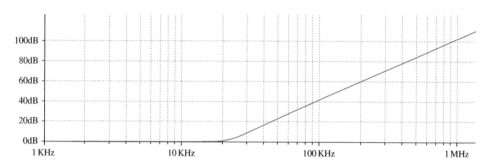

Figure 11.93 Insertion loss of the two configurations shown in Figure 11.90.

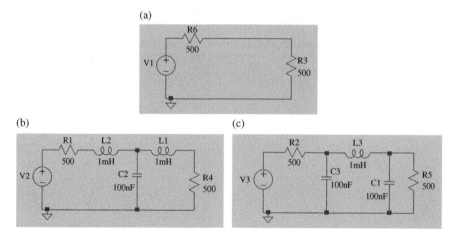

Figure 11.94 Filter configurations when both the source impedance and the load impedance are high: (a) no filter, (b) T filter configuration, (c) π filter configuration.

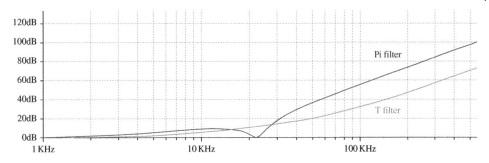

Figure 11.95 Insertion loss of the two configurations shown in Figure 11.92.

References

Alexander, C.K. and Sadiku, N.O., *Fundamentals of Electric Circuits*, 4th ed., McGraw Hill, New York, 2009.

Nilsson, J.W. and Riedel, S.A., *Electric Circuits*, 10th ed., Pearson, Upper Saddle River, NJ, 2015.

Paul, C.R., *Introduction to Electromagnetic Compatibility*, 2nd ed., John Wiley and Sons, New York, 2006.

Ott, H.W., *Electromagnetic Compatibility Engineering*, John Wiley and Sons, Hoboken, New Jersey, 2009.

12

Frequency Content of Digital Signals

12.1 Fourier Series and Frequency Content of Signals

12.1.1 Trigonometric Fourier Series

Any periodic function can be represented as an infinite sum of sinusoids:

$$x(t) = a_0 + \sum_{n=1}^{\infty} \left(a_n \cos 2\pi n f_0 t + b_n \sin 2\pi n f_0 t \right)$$

$$= a_0 + \sum_{n=1}^{\infty} \left(a_n \cos n\omega_0 t + b_n \sin n\omega_0 t \right), \qquad t_1 < t < t_1 + T \tag{12.1}$$

An expansion of this type is known as a *Fourier Series Expansion* (Kreyszig, 1999, p. 530).

Note that each sinusoidal component has a frequency that is a multiple of the fundamental frequency, $f_0 = 1/T$, and the radian fundamental frequency is $\omega = 2\pi f_0 = 2\pi / T$.

There are two forms of the Fourier series: trigonometric and exponential. The form in Eq. (12.1) is called the trigonometric form.

The multiples of the fundamental frequency, f_0, are called *harmonics* of that fundamental frequency. The coefficients a_0, a_n, b_n are called the *Fourier coefficients*.

The Fourier coefficients are determined from the following formulas:

$$a_0 = \frac{1}{T} \int_{t_1}^{t_1+T} x(t) dt \tag{12.2}$$

Note that a_0 is the average value of $x(t)$ over $t_1 < t < t_1 + T$.

$$a_n = \frac{2}{T} \int_{t_1}^{t_1+T} x(t) \cos n\omega_0 t dt \tag{12.3}$$

$$b_n = \frac{2}{T} \int_{t_1}^{t_1+T} x(t) \sin n\omega_0 t dt \tag{12.4}$$

Example 12.1   Trigonometric Fourier series – triangular wave
Determine the Fourier series for the periodic voltage waveform shown in Figure 12.1.
The expression for $v(t)$ over one period is

$$v(t) = \frac{V_m}{T} t$$

Now, we are ready to calculate the Fourier coefficients.

$$a_0 = \frac{1}{T} \int_{t_1}^{t_1+T} v(t)\,dt = \frac{1}{T} \int_0^T \frac{V_m}{T} t\,dt = \frac{V_m}{T^2} \int_0^T t\,dt = \frac{V_m}{T^2} \left(\frac{T^2}{2}\Big|_{t=0}^{t=T}\right) = \frac{V_m}{2}$$

That is,

$$a_0 = \frac{V_m}{2}$$

This is clearly the average value of this waveform.

$$a_n = \frac{2}{T} \int_{t_1}^{t_1+T} v(t)\cos n\omega_0 t\,dt = \frac{2}{T} \int_0^T \frac{V_m}{T} t \cos n\omega_0 t\,dt = \frac{2V_m}{T^2} \int_0^T t \cos n\omega_0 t\,dt$$

From the integral tables we find

$$\int x(\cos ax)\,dx = \frac{1}{a^2}\cos ax + \frac{x}{a}\sin ax$$

Thus,

$$\int_0^T t \cos n\omega_0 t\,dt = \left[\frac{1}{(n\omega_0)^2}\cos n\omega_0 t + \frac{t}{n\omega_0}\sin n\omega_0 t\right]_{t=0}^{T}$$

$$= \left[\frac{1}{(n\omega_0)^2}\cos n\omega_0 T + \frac{T}{n\omega_0}\sin n\omega_0 T\right] - \left[\frac{1}{(n\omega_0)^2}\cos 0 + \frac{t}{n\omega_0}\sin 0\right]$$

$$= \left[\frac{1}{(n\omega_0)^2}\cos n\frac{2\pi}{T}T + \frac{T}{n\omega_0}\sin n\frac{2\pi}{T}T\right] - \left[\frac{1}{(n\omega_0)^2} + 0\right]$$

$$= \left[\frac{1}{(n\omega_0)^2}\cos n2\pi + \frac{T}{n\omega_0}\sin n2\pi\right] - \frac{1}{(n\omega_0)^2}$$

$$= \frac{1}{(n\omega_0)^2} + 0 - \frac{1}{(n\omega_0)^2} = 0 \quad \text{for all } n$$

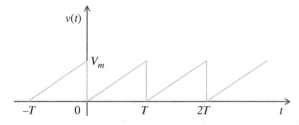

Figure 12.1 Periodic waveform for Example 12.1.

Thus,

$$a_n = 0$$

$$b_n = \frac{2}{T}\int_{t_1}^{t_1+T} v(t)\sin n\omega_0 t\,dt = \frac{2}{T}\int_0^T \frac{V_m}{T}t\sin n\omega_0 t\,dt = \frac{2V_m}{T^2}\int_0^T t\sin n\omega_0 t\,dt$$

From the integral tables we find

$$\int x(\sin ax)\,dx = \frac{1}{a^2}\sin ax - \frac{x}{a}\cos ax$$

Thus,

$$\int_0^T t\sin n\omega_0 t\,dt = \left[\frac{1}{(n\omega_0)^2}\sin n\omega_0 t - \frac{t}{n\omega_0}\cos n\omega_0 t\right]_{t=0}^T$$

$$= \left[\frac{1}{(n\omega_0)^2}\sin n\omega_0 T - \frac{T}{n\omega_0}\cos n\omega_0 T\right] - \left[\frac{1}{(n\omega_0)^2}\sin 0 - 0\right]$$

$$= \frac{1}{(n\omega_0)^2}\sin n\frac{2\pi}{T}T - \frac{T}{n\omega_0}\cos n\frac{2\pi}{T}T$$

$$= \frac{1}{(n\omega_0)^2}\sin n2\pi - \frac{T}{n\omega_0}\cos n2\pi$$

$$= 0 - \frac{T}{n\omega_0}$$

Thus

$$b_n = \frac{2V_m}{T^2}\int_0^T t\sin n\omega_0 t\,dt = \left(\frac{2V_m}{T^2}\right)\left(-\frac{T}{n\omega_0}\right) = \left(\frac{2V_m}{T^2}\right)\left(-\frac{T}{n\frac{2\pi}{T}}\right) = -\frac{V_m}{n\pi}$$

That is

$$b_n = -\frac{V_m}{n\pi}$$

And according to Fourier series expansion, $v(t)$ can be expressed as

$$x(t) = a_0 + \sum_{n=1}^{\infty}(a_n\cos n\omega_0 t + b_n\sin n\omega_0 t), \qquad t_1 < t < t_1 + T$$

$$= \frac{V_m}{2} + \sum_{n=1}^{\infty}(b_n\sin n\omega_0 t) = \frac{V_m}{2} + \sum_{n=1}^{\infty}\left[\left(-\frac{V_m}{n\pi}\right)\sin n\omega_0 t\right]$$

$$= \frac{V_m}{2} - \frac{V_m}{\pi}\sin \omega_0 t - \frac{V_m}{2\pi}\sin 2\omega_0 t - \frac{V_m}{3\pi}\sin 3\omega_0 t - \cdots$$

■

Example 12.2   Trigonometric Fourier series – square wave
Determine the Fourier coefficients for the square wave shown in Figure 12.2.

Let the duty cycle of this square wave be 50%, that is, $=T/2$. An expression for a square wave on the interval $0 < t < T$ is

$$x(t) = \begin{cases} A & 0 < t < \dfrac{T}{2} \\ -A & \dfrac{T}{2} < t < T \end{cases}$$

The Fourier coefficients can be calculated as:

$$a_0 = \frac{1}{T} \int_{t_1}^{t_1+T} x(t)\,dt = 0$$

$$a_n = \frac{2}{T} \int_{t_1}^{t_1+T} x(t)\cos n\omega_0 t\,dt = 0$$

$$b_n = \frac{2}{T} \int_{t_1}^{t_1+T} v(t)\sin n\omega_0 t\,dt = \frac{2A}{n\pi}\big[1-\cos(n\pi)\big]$$

The term $\dfrac{2A}{n\pi}\big[1-\cos(n\pi)\big] = 2$ if n is odd and zero if n is even. Hence, b_n can be written as

$$b_n = \begin{cases} \dfrac{4A}{n\pi} & n \ \ \text{odd} \\ 0 & n \ \ \text{even} \end{cases}$$

The Fourier series of this square wave is

$$x(t) = \frac{4A}{\pi}\left[\sin 2\pi(f_0)t + \frac{1}{3}\sin 2\pi(3f_0)t + \frac{1}{5}\sin 2\pi(5f_0)t + \cdots\right].$$

Note that this series contains only odd harmonic terms. Let's recreate the square wave using Fourier expansion.

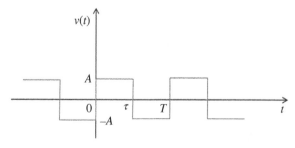

Figure 12.2 Periodic waveform for Example 12.2.

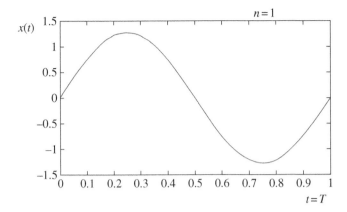

Figure 12.3 Waveform for Example 12.3, $n = 1$.

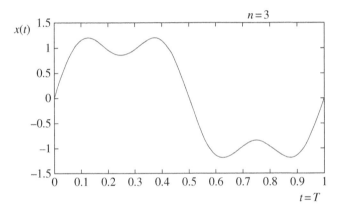

Figure 12.4 Waveform for Example 12.3, $n = 3$.

For $n = 1$ we get

$$x(t) = \frac{4A}{\pi} \Big[\sin 2\pi (f_0) t \Big]$$

This waveform is shown in Figure 12.3.
For $n = 3$ we have

$$x(t) = \frac{4A}{\pi} \left[\sin 2\pi (f_0) t + \frac{1}{3} \sin 2\pi (3 f_0) t \right]$$

The resulting waveform is shown in Figure 12.4.
For $n = 7$; the waveform is shown in Figure 12.5.
Two more waveforms: for $n = 19$ the waveform is shown in Figure 12.6.
And finally, for $n = 101$ we obtain the waveform shown in Figure 12.7.

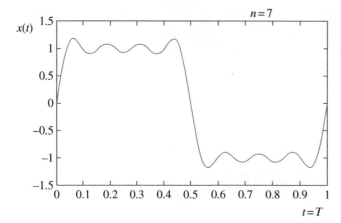

Figure 12.5 Waveform for Example 12.3, $n=7$.

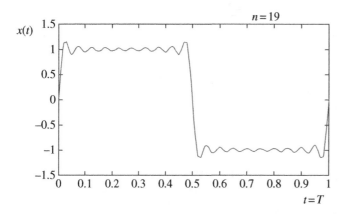

Figure 12.6 Waveform for Example 12.3, $n=19$.

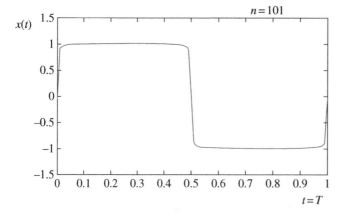

Figure 12.7 Waveform for Example 12.3, $n=101$.

12.1.2 Exponential Fourier Series

The Fourier series of Eq. (12.1) can be put into a much simpler and more elegant form with the use of complex exponentials. According to Euler's identity

$$e^{jn\omega_0 t} = \cos(n\omega_0 t) + j\sin(n\omega_0 t), \quad n = -\infty, \ldots, -1, 0 1, \ldots, \infty \tag{12.5}$$

From Eq. (12.5) we obtain two useful expressions

$$\cos(n\omega_0 t) = \frac{e^{jn\omega_0 t} + e^{-jn\omega_0 t}}{2} \tag{12.6}$$

$$\sin(n\omega_0 t) = \frac{e^{jn\omega_0 t} - e^{-jn\omega_0 t}}{2j} \tag{12.7}$$

Substituting Eq. (12.6) and Eq. (12.7) into Eq. (12.1) we obtain the complex-exponential form of Fourier series as

$$
\begin{aligned}
x(t) &= a_0 + \sum_{n=1}^{\infty} (a_n \cos n\omega_0 t + b_n \sin n\omega_0 t) \\
&= a_0 + \sum_{n=1}^{\infty} \left[a_n \left(\frac{e^{jn\omega_0 t} + e^{-jn\omega_0 t}}{2} \right) + b_n \left(\frac{e^{jn\omega_0 t} - e^{-jn\omega_0 t}}{2j} \right) \right] \\
&= a_0 + \sum_{n=1}^{\infty} \left[\left(\frac{a_n}{2} + \frac{b_n}{2j} \right) e^{jn\omega_0 t} + \left(\frac{a_n}{2} - \frac{b_n}{2j} \right) e^{-jn\omega_0 t} \right] \\
&= a_0 + \sum_{n=1}^{\infty} \left[\left(\frac{a_n}{2} - j\frac{b_n}{2} \right) e^{jn\omega_0 t} + \left(\frac{a_n}{2} + j\frac{b_n}{2} \right) e^{-jn\omega_0 t} \right] \\
&= c_0 + \sum_{n=1}^{\infty} \left[\hat{c}_n e^{jn\omega_0 t} + \hat{c}_n^* e^{-jn\omega_0 t} \right] = \sum_{n=-\infty}^{\infty} \hat{c}_n e^{jn\omega_0 t}
\end{aligned}
\tag{12.8}
$$

or

$$x(t) = \sum_{n=-\infty}^{\infty} \hat{c}_n e^{jn\omega_0 t} \tag{12.9}$$

where

$$\hat{c}_n = \frac{1}{T} \int_{t_1}^{t_1+T} x(t) e^{-jn\omega_0 t} \, dt \tag{12.10}$$

The Fourier series as expressed in Eq. (12.9) is called the *two-sided spectrum*, since it contains both the positive and the negative frequencies.

The complex exponential form is more useful and more easily computed than the trigonometric form. Note that the summation in Eq. (12.10) extends from $-\infty$ to $+\infty$.

Each expansion coefficient, $\hat{c}_n$, will be, in general, a complex number that can be expressed in polar or exponential form as

$$\hat{c}_n = c_n \angle \theta_{cn} = c_n e^{j\theta_{cn}} \tag{12.11}$$

Note, that for $n = 0$, the expansion coefficient becomes

$$c_0 = \left[\frac{1}{T} \int_{t_1}^{t_1+T} x(t)e^{-jn\omega_0 t} dt \right]_{n=0} = \frac{1}{T} \int_{t_1}^{t_1+T} x(t) dt \tag{12.12}$$

which is a real number and is the average value of $x(t)$.

Note that the complex exponential form of the Fourier series contains both the positive-valued harmonic frequencies $\omega_0, 2\omega_0, 3\omega_0, \ldots$ and the negative-valued harmonic frequencies $-\omega_0, -2\omega_0, -3\omega_0, \ldots$.

Thus, for each positive value of n (and harmonic frequency $n\omega_0$) there is a corresponding negative value of n (and harmonic frequency $-n\omega_0$). The coefficients corresponding to these values of n and $-n$ are

$$c_n = \frac{1}{T} \int_{t_1}^{t_1+T} x(t)e^{-jn\omega_0 t} dt \tag{12.13a}$$

$$c_{-n} = \frac{1}{T} \int_{t_1}^{t_1+T} x(t)e^{-j(-n)\omega_0 t} dt = \frac{1}{T} \int_{t_1}^{t_1+T} x(t)e^{jn\omega_0 t} dt \tag{12.13b}$$

Note that these coefficients are the complex conjugates of each other. That is,

$$\hat{c}_{-n} = \hat{c}_n^* \tag{12.14}$$

Thus

$$\hat{c}_n^* = c_n e^{-j\theta_{cn}} \tag{12.15}$$

Note that the complex exponential Fourier series in Eq. (12.9) can be written as

$$x(t) = \sum_{n=-\infty}^{\infty} \hat{c}_n e^{jn\omega_0 t} = \sum_{n=-\infty}^{-1} \hat{c}_n e^{jn\omega_0 t} + c_0 + \sum_{n=1}^{\infty} \hat{c}_n e^{jn\omega_0 t} \tag{12.16}$$

Since

$$\sum_{n=-\infty}^{-1} \hat{c}_n e^{jn\omega_0 t} = \sum_{n=1}^{\infty} \hat{c}_{-n} e^{-jn\omega_0 t} = \sum_{n=1}^{\infty} \hat{c}_n^* e^{-jn\omega_0 t} \tag{12.17}$$

we rewrite Eq. (12.16) as

$$x(t) = \sum_{n=-\infty}^{\infty} \hat{c}_n e^{jn\omega_0 t} = c_0 + \sum_{n=1}^{\infty} \hat{c}_n e^{jn\omega_0 t} + \sum_{n=1}^{\infty} \hat{c}_n^* e^{-jn\omega_0 t} \tag{12.18}$$

Using Eq. (12.11) and Eq. (12.15) in Eq. (12.18) produces

$$x(t) = c_0 + \sum_{n=1}^{\infty}\left(c_n e^{j\theta_{cn}}\right)e^{jn\omega_0 t} + \sum_{n=1}^{\infty}\left(c_n e^{-j\theta_{cn}}\right)e^{-jn\omega_0 t}$$

$$= c_0 + \sum_{n=1}^{\infty} c_n e^{j(n\omega_0 t + \theta_{cn})} + \sum_{n=1}^{\infty} c_n e^{-j(n\omega_0 t + \theta_{cn})} \tag{12.19}$$

$$= c_0 + \sum_{n=1}^{\infty} c_n \left[e^{j(n\omega_0 t + \theta_{cn})} + e^{-j(n\omega_0 t + \theta_{cn})} \right]$$

Since

$$\frac{e^{j(n\omega_0 t + \theta_{cn})} + e^{-j(n\omega_0 t + \theta_{cn})}}{2} = \cos(n\omega_0 t + \theta_{cn}) \tag{12.20}$$

the complex exponential Fourier series can be expressed as

$$x(t) = c_0 + \sum_{n=1}^{\infty} 2c_n \cos(n\omega_0 t + \theta_{cn}) \tag{12.21}$$

Note that c_n and θ_{cn} in Eq. (12.21) are real values, since

$$\hat{c}_n = c_n \angle \theta_{cn} = c_n e^{j\theta_{cn}} \tag{12.22}$$

The Fourier series, as expressed in Eq. (12.21), is called the *one-sided spectrum* since it contains only the positive frequencies.

Note that in order to obtain the expansion coefficients for the one-sided spectrum, the magnitudes of the expansion coefficients for the two-sided spectrum need to be doubled, while the dc component c_0 remains unchanged (Paul, 2006, p.97).

12.1.3 Spectrum of the Digital Clock Signals

Clock waveforms can be represented as periodic trains of trapezoid-shaped pulses as shown in Figure 12.8.

Each pulse is described by the key parameters: period T (and thus the fundamental frequency $f_0 = 1/T$), amplitude A, rise time t_r, fall time t_f, and the on-time or pulse width τ. We will investigate the effect of these pulse parameters on the spectrum of the clock waveform.

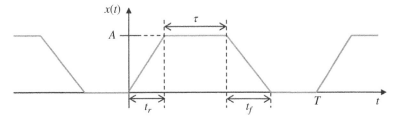

Figure 12.8 Trapezoidal clock signal.

To obtain the complex-exponential Fourier series of this waveform, we will use the computational techniques described in Paul, 2009, Section 3.1.3.

The expansion coefficients for a periodic function $x(t)$ are related to the expansion coefficients of its derivative in the following manner.

Let $x(t)$ be represented by its Fourier series as

$$x(t) = \sum_{n=-\infty}^{\infty} \hat{c}_n e^{jn\omega_0 t} \tag{12.23}$$

and let its kth derivative be represented as

$$\frac{d^k x(t)}{dt^k} = \sum_{n=-\infty}^{\infty} \hat{c}_n^{(k)} e^{jn\omega_0 t} \tag{12.24}$$

Then the expansion coefficients are related by (Paul, 2009, p. 115)

$$\hat{c}_n = \frac{1}{(jn\omega_0)^k} \hat{c}_n^{(k)}, \quad n \neq 0 \tag{12.25}$$

Thus, the coefficients of the waveform that is the second derivative of the original waveform are related to the coefficient of the original waveform by

$$\hat{c}_n = \frac{1}{(jn\omega_0)^2} \hat{c}_n^{(2)} = -\frac{\hat{c}_n^{(2)}}{(n\omega_0)^2} \tag{12.26}$$

Figure 12.9(a) shows the original clock waveform and Figure 12.9(c) shows its second derivative.

The second derivative waveform consists of four impulses, repeating themselves every period T:

$$\frac{A}{t_r}\delta(0), \quad -\frac{A}{t_r}\delta(t_r), \quad -\frac{A}{t_f}\delta\left(\tau + \frac{t_r - t_f}{2}\right), \quad \frac{A}{t_f}\delta\left(\tau + \frac{t_r + t_f}{2}\right) \tag{12.27}$$

where unit impulse $\delta(t)$ is defined by

$$\delta(t) = \begin{cases} 0 & \text{for} \quad t < 0 \\ 0 & \text{for} \quad t > 0 \\ \int_{0^-}^{0^+} \delta(t)dt = 1 \end{cases} \tag{12.28}$$

Each train of pulses has its own Fourier representation as

$$x_1(t) = \frac{A}{t_r}\delta(0 \pm kT) = \sum_{n=-\infty}^{\infty} \hat{c}_{n1} e^{jn\omega_0 t} \tag{12.29a}$$

$$x_2(t) = -\frac{A}{t_r}\delta(t_r \pm kT) = \sum_{n=-\infty}^{\infty} \hat{c}_{n2} e^{jn\omega_0 t} \tag{12.29b}$$

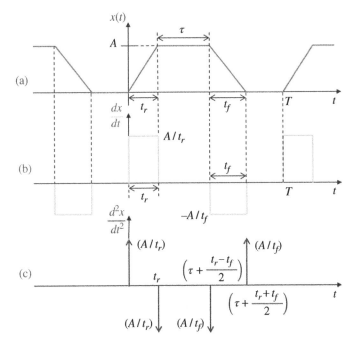

Figure 12.9 Trapezoidal clock signal and its derivative waveforms.

$$x_3(t) = -\frac{A}{t_f}\delta\left(\tau + \frac{t_r - t_f}{2}\right) = \sum_{n=-\infty}^{\infty} \hat{c}_{n3}e^{jn\omega_o t} \tag{12.29c}$$

$$x_4(t) = \frac{A}{t_f}\delta\left(\tau + \frac{t_r + t_f}{2}\right) = \sum_{n=-\infty}^{\infty} \hat{c}_{n4}e^{jn\omega_o t} \tag{12.29d}$$

By the property of linearity, the expansion coefficient for these four trains of pulses is equal to the sum of the expansion coefficients for each individual train of pulses. That is,

$$\hat{c}_n^{(2)} = \hat{c}_{n1} + \hat{c}_{n2} + \hat{c}_{n3} + \hat{c}_{n4} \tag{12.30}$$

The expansion coefficient for the train of pulses occurring at $t = 0$ (and repeating itself every T) is

$$
\begin{aligned}
\hat{c}_{n1} &= \frac{1}{T}\int_{t_1}^{t_1+T} x_1(t)e^{-jn\omega_o t}\,dt = \frac{1}{T}\int_0^T \frac{A}{t_r}\delta(t)e^{-jn\omega_o t}\,dt \\
&= \frac{A}{Tt_r}\int_{0^-}^{0^+}\delta(t)e^{-jn\omega_o t}\,dt = \frac{A}{Tt_r}\int_{0^-}^{0^+}\delta(t)\,dt = \frac{A}{Tt_r}
\end{aligned}
\tag{12.31}
$$

The expansion coefficients for a given waveform $x(t)$ are related to the expansion coefficients of the shifted version of it $x(t - \alpha)$ as follows.

Let $x(t)$ be represented by its Fourier series as

$$x(t) = \sum_{n=-\infty}^{\infty} \hat{c}_n e^{jn\omega_0 t} \tag{12.32}$$

and its shifted version $x(t-\alpha)$ be represented by its Fourier series as

$$x(t-\alpha) = \sum_{n=-\infty}^{\infty} \hat{c}_n' e^{jn\omega_0 t} \tag{12.33}$$

Then the coefficients of the shifted version are related to the coefficients of the unshifted one by

$$c_n' = c_n e^{-jn\omega_0 \alpha} \tag{12.34}$$

Thus, the expansion coefficients for the remaining three trains of pulses are

$$\hat{c}_{n2} = \frac{1}{T}\left(-\frac{A}{t_r}\right)e^{-jn\omega_0 t_r} \tag{12.35}$$

and

$$\hat{c}_{n3} = \frac{1}{T}\left(-\frac{A}{t_f}\right)e^{-jn\omega_0\left(\tau+\frac{t_r-t_f}{2}\right)} \tag{12.36}$$

and

$$\hat{c}_{n4} = \frac{1}{T}\left(\frac{A}{t_f}\right)e^{-jn\omega_0\left(\tau+\frac{t_r+t_f}{2}\right)} \tag{12.37}$$

And thus

$$\hat{c}_n^{(2)} = \frac{A}{Tt_r} + \frac{1}{T}\left(-\frac{A}{t_r}\right)e^{-jn\omega_0 t_r} + \frac{1}{T}\left(-\frac{A}{t_f}\right)e^{-jn\omega_0\left[\tau+(t_r-t_f)/2\right]} + \frac{1}{T}\left(\frac{A}{t_f}\right)e^{-jn\omega_0\left[\tau+(t_r+t_f)/2\right]}$$

$$= \frac{A}{Tt_r}\left(1-e^{-jn\omega_0 t_r}\right) - \frac{A}{Tt_f}e^{-jn\omega_0 \tau}\left[e^{-jn\omega_0(t_r-t_f)/2} - e^{-jn\omega_0(t_r+t_f)/2}\right]$$

$$= \frac{A}{Tt_r}\left(1-e^{-jn\omega_0 t_r}\right) - \frac{A}{Tt_f}e^{-jn\omega_0 \tau}\left[e^{-jn\omega_0 t_r/2}e^{jn\omega_0 t_f/2} - e^{-jn\omega_0 t_r/2}e^{-jn\omega_0 t_f/2}\right] \tag{12.38}$$

Or

$$\hat{c}_n^{(2)} = \frac{A}{T}\left[\frac{1}{t_r}e^{-jn\omega_0 t_r/2}\left(e^{jn\omega_0 t_r/2} - e^{-jn\omega_0 t_r/2}\right) - \frac{1}{t_f}e^{-jn\omega_0 \tau}e^{-jn\omega_0 t_r/2}\left(e^{jn\omega_0 t_f/2} - e^{-jn\omega_0 t_f/2}\right)\right]$$

$$= \frac{A}{T}e^{-jn\omega_0 t_r/2}\left[\frac{1}{t_r}\left(e^{jn\omega_0 t_r/2} - e^{-jn\omega_0 t_r/2}\right) - \frac{1}{t_f}e^{-jn\omega_0 \tau}\left(e^{jn\omega_0 t_f/2} - e^{-jn\omega_0 t_f/2}\right)\right]$$

$$= \frac{A}{T}e^{-jn\omega_0 t_r/2}e^{-jn\omega_0 \tau/2}$$

$$\left[\frac{1}{t_r}\left(e^{jn\omega_0 t_r/2} - e^{-jn\omega_0 t_r/2}\right)e^{jn\omega_0 \tau/2} - \frac{1}{t_f}\left(e^{jn\omega_0 t_f/2} - e^{-jn\omega_0 t_f/2}\right)e^{-jn\omega_0 \tau/2}\right] \tag{12.39}$$

Now, let's utilize the identity

$$\sin\left(\frac{n\omega_0 t_r}{2}\right) = \frac{e^{jn\omega_0 t_r/2} - e^{-jn\omega_0 t_r/2}}{2j} \tag{12.40}$$

or

$$e^{jn\omega_0 t_r/2} - e^{-jn\omega_0 t_r/2} = j2\sin\left(\frac{n\omega_0 t_r}{2}\right) \tag{12.41a}$$

similarly

$$e^{jn\omega_0 t_f/2} - e^{-jn\omega_0 t_f/2} = j2\sin\left(\frac{n\omega_0 t_f}{2}\right) \tag{12.41b}$$

Substituting Eqs (12.41) into Eq. (12.39) leads to

$$\hat{c}_n^{(2)} = \frac{A}{T} e^{-jn\omega_0 t_r/2} e^{-jn\omega_0 \tau/2} \left[\frac{1}{t_r} j2\sin\left(\frac{n\omega_0 t_r}{2}\right) e^{jn\omega_0 \tau/2} - \frac{1}{t_f} j2\sin\left(\frac{n\omega_0 t_f}{2}\right) e^{-jn\omega_0 \tau/2}\right]$$

$$= j\frac{A}{T} e^{-jn\omega_0(\tau+t_r)/2} \left[\frac{\sin\left(\frac{n\omega_0 t_r}{2}\right)}{\frac{1}{2}t_r} e^{jn\omega_0 \tau/2} - \frac{\sin\left(\frac{n\omega_0 t_f}{2}\right)}{\frac{1}{2}t_f} e^{-jn\omega_0 \tau/2}\right]$$

$$= j\frac{A}{T} n\omega_0 e^{-jn\omega_0(\tau+t_r)/2} \left[\frac{\sin\left(\frac{1}{2}n\omega_0 t_r\right)}{\frac{1}{2}n\omega_0 t_r} e^{jn\omega_0 \tau/2} - \frac{\sin\left(\frac{1}{2}n\omega_0 t_f\right)}{\frac{1}{2}n\omega_0 t_f} e^{-jn\omega_0 \tau/2}\right] \tag{12.42}$$

That is,

$$\hat{c}_n^{(2)} = j\frac{A}{T} n\omega_0 e^{-jn\omega_0(\tau+t_r)/2} \left[\frac{\sin\left(\frac{1}{2}n\omega_0 t_r\right)}{\frac{1}{2}n\omega_0 t_r} e^{jn\omega_0 \tau/2} - \frac{\sin\left(\frac{1}{2}n\omega_0 t_f\right)}{\frac{1}{2}n\omega_0 t_f} e^{-jn\omega_0 \tau/2}\right] \tag{12.43}$$

According to Eq. (12.26), the expansion coefficient for the original trapezoidal waveform is related to the coefficient in Eq. (12.43) by

$$\hat{c}_n = -\frac{\hat{c}_n^{(2)}}{\left(n\omega_0\right)^2} \tag{12.44}$$

Thus

$$\hat{c}_n = -j\frac{A}{Tn\omega_0} e^{-jn\omega_0(\tau+t_r)/2} \left[\frac{\sin\left(\frac{1}{2}n\omega_0 t_r\right)}{\frac{1}{2}n\omega_0 t_r} e^{jn\omega_0 \tau/2} - \frac{\sin\left(\frac{1}{2}n\omega_0 t_f\right)}{\frac{1}{2}n\omega_0 t_f} e^{-jn\omega_0 \tau/2}\right] \tag{12.45}$$

Since

$$\omega_0 = \frac{2\pi}{T} \quad \Leftrightarrow \quad T = \frac{2\pi}{\omega_0} \tag{12.46}$$

We finally obtain

$$\hat{c}_n = -j\frac{A}{2\pi n} e^{-jn\omega_0(\tau+\tau_r)/2} \left[\frac{\sin\left(\frac{1}{2}n\omega_0 t_r\right)}{\frac{1}{2}n\omega_0 t_r} e^{jn\omega_0\tau/2} - \frac{\sin\left(\frac{1}{2}n\omega_0 t_f\right)}{\frac{1}{2}n\omega_0 t_f} e^{-jn\omega_0\tau/2} \right] \tag{12.47}$$

Now, if the pulse rise time equals the fall time, $t_r = t_f$, we obtain a very useful result that leads to very important conclusions. Letting $t_r = t_f$ in Eq. (12.47) gives

$$\hat{c}_n = -j\frac{A}{2\pi n} e^{-jn\omega_0(\tau+\tau_r)/2} \left[\frac{\sin\left(\frac{1}{2}n\omega_0 t_r\right)}{\frac{1}{2}n\omega_0 t_r} e^{jn\omega_0\tau/2} - \frac{\sin\left(\frac{1}{2}n\omega_0 t_r\right)}{\frac{1}{2}n\omega_0 t_r} e^{-jn\omega_0\tau/2} \right]$$

$$= -j\frac{A}{2\pi n} \frac{\sin\left(\frac{1}{2}n\omega_0 t_r\right)}{\frac{1}{2}n\omega_0 t_r} \left(e^{jn\omega_0\tau/2} - e^{-jn\omega_0\tau/2} \right) e^{-jn\omega_0(\tau+\tau_r)/2} \tag{12.48}$$

$$= -j\frac{A}{2\pi n} \frac{\sin\left(\frac{1}{2}n\omega_0 t_r\right)}{\frac{1}{2}n\omega_0 t_r} j2\sin\left(\frac{1}{2}n\omega_0\tau\right) e^{-jn\omega_0(\tau+\tau_r)/2}$$

or

$$\hat{c}_n = \frac{A}{\pi n} \frac{\sin\left(\frac{1}{2}n\omega_0 t_r\right)}{\frac{1}{2}n\omega_0 t_r} \sin\left(\frac{1}{2}n\omega_0\tau\right) e^{-jn\omega_0(\tau+\tau_r)/2} \tag{12.49}$$

Since

$$\omega_0 = \frac{2\pi}{T} \quad \Leftrightarrow \quad \pi = \frac{\omega_0 T}{2} \tag{12.50}$$

we rewrite Eq. (12.49) as

$$\hat{c}_n = A\frac{\tau}{T} \frac{\sin\left(\frac{1}{2}n\omega_0\tau\right)}{\frac{1}{2}n\omega_0\tau} \frac{\sin\left(\frac{1}{2}n\omega_0 t_r\right)}{\frac{1}{2}n\omega_0 t_r} e^{-jn\omega_0(\tau+\tau_r)/2} \tag{12.51}$$

or, substituting for ω_0 from Eq. (12.50) as

$$\hat{c}_n = A\frac{\tau}{T}\frac{\sin(n\pi\tau/T)}{n\pi\tau/T}\frac{\sin(n\pi\tau_r/T)}{n\pi\tau_r/T}e^{-jn\omega_0(\tau+\tau_r)/2} \qquad (12.52)$$

Thus, with the *rise and fall times being equal*, we obtain the *one-sided* Fourier spectrum of the trapezoidal clock signal as

$$x(t) = c_0 + \sum_{n=1}^{\infty} 2c_n \cos(n\omega_0 t + \theta_{cn}) \qquad (12.53)$$

where

$$2c_n = 2A\frac{\tau}{T}\left|\frac{\sin\left(n\pi\tau/T\right)}{n\pi\tau/T}\right|\left|\frac{\sin\left(n\pi\tau_r/T\right)}{n\pi\tau_r/T}\right| \quad for\, n \neq 0 \qquad (12.54)$$

and

$$c_0 = A\frac{\tau}{T} \qquad (12.55)$$

The angle of the Fourier coefficient is

$$\theta_{cn} = \angle\hat{c}_n = -n\omega_0(\tau+\tau_r)/2 \pm \pi \qquad (12.56)$$

The $\pm\pi$ term in Eq. (12.56) appears when the product of the two $\sin x/x$ terms in Eq. (12.52) is a negative real number (and thus a complex number with an angle of $\pm\pi$).

A very interesting and useful result is obtained when we consider a 50% duty cycle signal. That is, when

$$\frac{\tau}{T} = \frac{1}{2} \qquad (12.57)$$

Under this condition the first sine term in Eq. (12.54) becomes

$$\frac{\left|\sin\left(\dfrac{n\pi\tau}{T}\right)\right|}{\left|\dfrac{n\pi\tau}{T}\right|} = \frac{\left|\sin\dfrac{1}{2}n\pi\right|}{\left|\dfrac{1}{2}n\pi\right|} \qquad (12.58)$$

which is zero for even n. Thus, there are no even harmonics when the duty cycle is 50%.

Figure 12.10 shows the frequency spectrum of a 1 V trapezoidal pulse, with a fundamental frequency of 10 MHz, and 5 ns risetime and two different duty cycles.

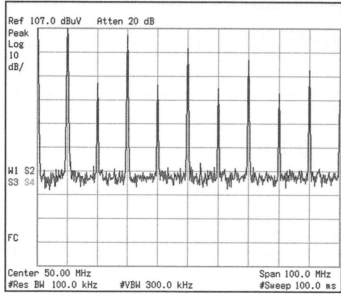

49% Duty cycle

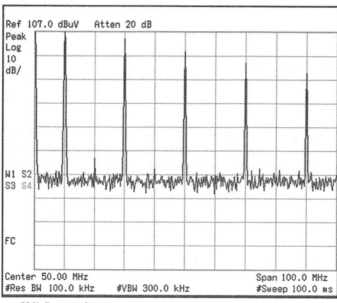

50% Duty cycle

Figure 12.10 Frequency spectrum of a clock signal with 49% and 50 % duty cycle.

12.1.4   Spectral Bounds on Digital Clock Signals

Recall: with the *rise and fall times being equal,* the *one-sided* Fourier spectrum of the trapezoidal clock signal is given by

$$x(t) = c_0 + \sum_{n=1}^{\infty} 2c_n \cos(n\omega_0 t + \theta_{cn}) \tag{12.59}$$

Where

$$2c_n = 2A\frac{\tau}{T} \left| \frac{\sin\left(n\pi\tau/T\right)}{n\pi\tau/T} \right| \left| \frac{\sin\left(n\pi\tau_r/T\right)}{n\pi\tau_r/T} \right| \quad for\, n \neq 0 \tag{12.60}$$

These coefficients (spectral components) exist only at the discrete frequencies $f = n/T$. The continuous envelope of these spectral components is obtained by replacing $n/T = f$ in Eq. (12.60).

$$\text{Envelope} = 2A\frac{\tau}{T} \left| \frac{\sin\left(\pi\tau f\right)}{\pi\tau f} \right| \left| \frac{\sin\left(\pi\tau_r f\right)}{\pi\tau_r f} \right|, \quad f = \frac{n}{T} \tag{12.61}$$

or in dB,

$$20\log_{10}\left(\text{Envelope}\right) = 20\log_{10}\left(2A\frac{\tau}{T}\right) + 20\log_{10}\left| \frac{\sin\left(\pi\tau f\right)}{\pi\tau f} \right|$$
$$+ 20\log_{10}\left| \frac{\sin\left(\pi\tau_r f\right)}{\pi\tau_r f} \right| \tag{12.62}$$

These bounds are shown in Figure 12.11.

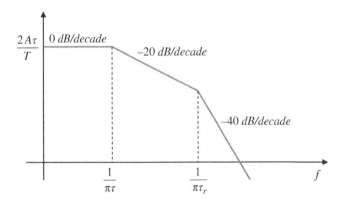

Figure 12.11 Bounds on the one-sided magnitude spectrum of a trapezoidal clock signal.

Trapezoidal pulse, 1V, 10 MHz, 50% duty cycle

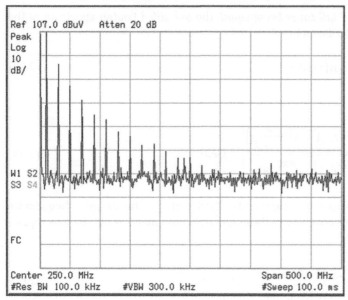

20 ns rise time

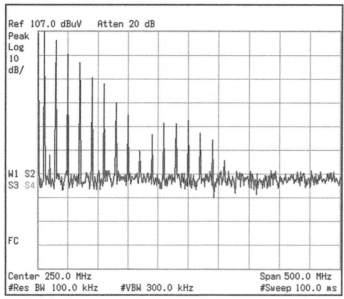

5 ns rise time

Figure 12.12 Frequency spectrum of a clock signal with 20 ns vs 5 ns risetime.

There is one extremely important observation we can make from the plots in Figure 12.11. Note that above the frequency $f = 1/\pi\tau_r$ the amplitudes of the spectral components are attenuated at a rate of 40 dB/decade.

It seems reasonable, therefore, to postulate that somewhere beyond this frequency these amplitudes are negligible (compared to the magnitudes of the components at lower frequencies) and can be neglected in the Fourier series expansion.

A reasonable choice for that frequency is (Paul, 2009, p. 133)

$$f_{max} = 3 \times \frac{1}{\pi\tau_r} \cong \pi \times \frac{1}{\pi\tau_r} = \frac{1}{\tau_r} \tag{12.63}$$

With the above choice, the bandwidth (BW) of a trapezoidal signal is

$$BW = \frac{1}{\tau_r} \tag{12.64}$$

Returning to Figure 12.10, we make another important observation: the pulses having short rise/fall times have larger high-frequency content than do pulses with long rise/fall times.

This is illustrated in Figure 12.12.

12.2 EMC Applications

12.2.1 Effect of the Signal Amplitude, Fundamental Frequency, and Duty Cycle on the Frequency Content of Trapezoidal Signals

The effect of the signal amplitude on the frequency content of a trapezoidal signal is shown in Figure 12.13.

As can be seen, reducing the signal amplitude reduces the frequency content over the entire frequency range. This is verified by the measurement shown in Figure 12.14.

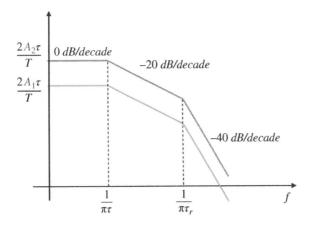

Figure 12.13 Effect of the signal amplitude.

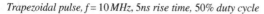

Trapezoidal pulse, f = 10 MHz, 5ns rise time, 50% duty cycle

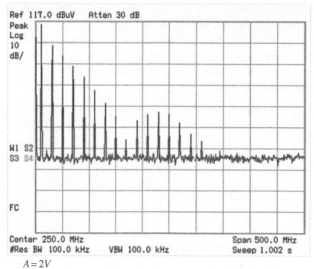

$A = 2V$

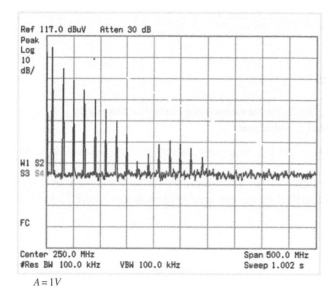

$A = 1V$

Figure 12.14 Effect of the amplitude reduction.

The effect of reducing the fundamental frequency while maintaining the same duty cycle on the frequency content of signal is shown in Figure 12.15.

Reducing the fundamental frequency (while maintain the duty cycle) reduces the high-frequency spectral content of the waveform, but does not affect the low-frequency content. This is shown in Figure 12.16.

The effect of reducing the duty cycle while maintaining the fundamental frequency is shown in Figure 12.17.

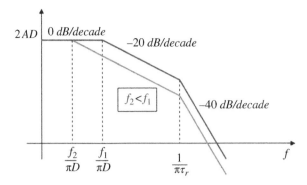

Figure 12.15 Effect of the fundamental frequency while maintaining the duty cycle.

Trapezoidal pulse, 1V, 5 ns rise time, 50% duty cycle

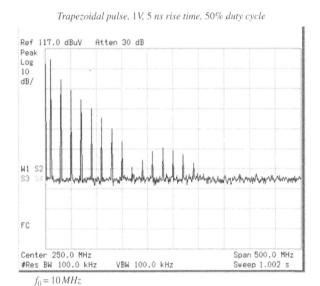

$f_0 = 10 \, MHz$

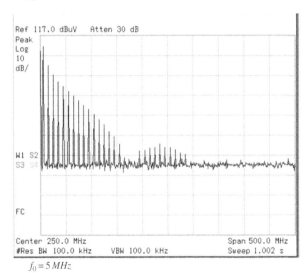

$f_0 = 5 \, MHz$

Figure 12.16 Effect of the fundamental frequency while maintaining the duty cycle.

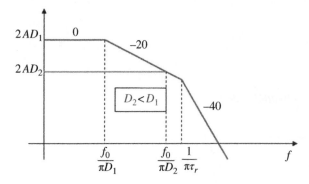

Figure 12.17 Effect of the duty cycle while maintaining the fundamental frequency.

Trapezoidal pulse, 1V, 5 ns rise time, 10 MHz

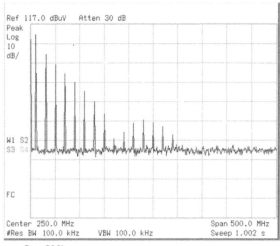

$D = 50\%$

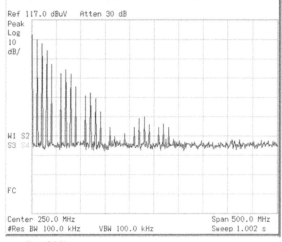

$D = 20\%$

Figure 12.18 Effect of the duty cycle while maintaining the fundamental frequency.

Reducing the duty cycle (the pulsewidth) reduces the low-frequency spectral content of the waveform, but does not affect the high-frequency content. This is shown in Figure 12.18.

References

Kreyszig, E., *Advanced Engineering Mathematics*, 8th ed., John Wiley and Sons, New York, 1999.

Paul, C.R., *Introduction to Electromagnetic Compatibility*, 2nd ed., John Wiley and Sons, New York, 2006.

Part III

Electromagnetics Foundations of EMC

13

Static and Quasi-Static Electric Fields

Modern theory of electromagnetics is based on a set of four fundamental relations known as Maxwell's equations. These equations hold in any material, at any spatial location, and involve the time-varying, coupled electric and magnetic fields.

When the fields are time-invariant (static) Maxwell's four equations separate into two uncoupled pairs, one for the electric field and one for the magnetic field. This allows us to study the electrostatics and magnetostatics separately.

13.1 Charge Distributions

The concept of electric charge is the basis for the study of electromagnetics. The electric charge can be either positive or negative, and exists in integer multiples of a charge of an electron (negative charge).

We often use the idealized model of an electric charge, called the point charge, where we assume that the charge is dimensionless (the charge is on a body whose dimensions are much smaller than other relevant dimensions).

In addition to a single point charge or to the discrete distribution of point charges, we will discuss continuous charge distributions: line, surface, and volume charge distributions. These distributions are shown in Figure 13.1.

If the charge is distributed along a line we characterize the distribution by the line charge density.

$$\rho_l = \frac{dq}{dl} \quad \left[\frac{C}{m}\right] \tag{13.1}$$

The total charge contained along a given length l is then obtained from

$$Q = \int_l \rho_l dl \quad [C] \tag{13.2}$$

If the charge is distributed across a surface we characterize the distribution by the *surface charge density*,

$$\rho_s = \frac{dq}{ds} \quad \left[\frac{C}{m^2}\right] \tag{13.3}$$

Foundations of Electromagnetic Compatibility with Practical Applications, First Edition. Bogdan Adamczyk.
© 2017 John Wiley & Sons Ltd. Published 2017 by John Wiley & Sons Ltd.

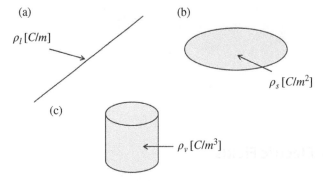

Figure 13.1 Charge distributions: (a) line, (b) surface, (c) volume.

The total charge contained in a given surface S is then obtained from

$$Q = \int_S \rho_s ds \quad [\text{C}] \tag{13.4}$$

Finally, if the electric charge is distributed over a volume in space we define *volume charge density*,

$$\rho_v = \frac{dq}{dv} \quad \left[\frac{\text{C}}{\text{m}^3}\right] \tag{13.5}$$

The total charge contained in a given volume v is then obtained from

$$Q = \int_v \rho_v dv \quad [\text{C}] \tag{13.6}$$

13.2 Coulomb's Law

There are two fundamental laws governing electrostatic fields:

1) Coulomb's law – applicable in finding the electric field due to any charge configuration
2) Gauss's law – practical to use when charge distribution is symmetrical

Coulomb's law describes the force that a point charge exerts on another point charge. The magnitude of that force is given by (Sadiku, 2010, p. 108)

$$F = \frac{1}{4\pi\varepsilon} \frac{Q_1 Q_2}{R^2} \quad [\text{N}] \tag{13.7}$$

where Q_1 and Q_2 are the magnitudes of the point charges, R is the distance between them, and ε is the permittivity of the surrounding medium. Often the surrounding medium is air and we use for it the permittivity of free space

$$\varepsilon_0 \approx \frac{10^{-9}}{36\pi} \quad \left[\frac{\text{F}}{\text{m}}\right] \tag{13.8}$$

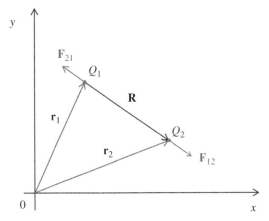

Figure 13.2 Forces between two point charges.

Consider that point charges Q_1 and Q_2 are located at points having position vectors r_1 and r_2, as shown in Figure 13.2.

The force F_{12} on Q_2 due to Q_1, is given by

$$F_{12} = \frac{1}{4\pi\varepsilon_0} \frac{Q_1 Q_2}{R^2} \mathbf{a}_R \qquad (13.9)$$

where R is the vector along the line connecting the charges and pointing from charge Q_1 to charge Q_2

$$\mathbf{R} = \mathbf{r}_2 - \mathbf{r}_1 \qquad (13.10)$$

and $\mathbf{a}_R$ is its unit vector

$$\mathbf{a}_R = \frac{\mathbf{R}}{R} \qquad (13.11)$$

The force, F_{21}, on charge Q_1 due to charge Q_2 (the order of the subscripts is source-destination) is given by

$$\mathbf{F}_{21} = -\mathbf{F}_{12} \qquad (13.12)$$

If we have more than two point charges, we use the *principle of superposition* to determine the force on a particular charge due to all the other charges.

13.3 Electric Field Intensity

Consider a positive electric point charge Q placed in space and shown in Figure 13.3.

If another positive test charge q is introduced into the vicinity of Q, then according to Coulomb's law, an electric force will be exerted on it by the charge Q.

Thus we may associate *an electric field around the point charge Q where electric forces act*. This concept leads to the first of the four fundamental vectors describing electromagnetic fields: the *electric field intensity vector* **E**.

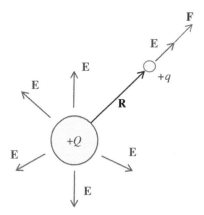

Figure 13.3 Electric field around a point charge.

The electric field intensity vector is defined as the force per unit test charge that is exerted on that charge:

$$E = \frac{F}{q} \quad \left[\frac{V}{m}\right] \tag{13.13}$$

Note that the electric field intensity vector for a positive point charge is directed radially away from that charge. If the test charge is also positive, then the force acting on it is in the direction of vector **E**.

If the charge Q is located at the origin of the coordinate system and the point charge q at a distance R from it, then using Coulomb's law, we obtain the electric field intensity as

$$E = \frac{\dfrac{1}{4\pi\varepsilon_0}\dfrac{Qq}{R^2}\mathbf{a}_R}{q} \tag{13.14}$$

or

$$E = \frac{1}{4\pi\varepsilon_0}\frac{Q}{R^2}\mathbf{a}_R \tag{13.15}$$

Note that the electric field intensity vector for a point charge decays inversely proportional to the square of the distance.

For N charges $Q_1, Q_2, \ldots, Q_N$ located respectively at points with position vectors $r_1, r_2, \ldots, r_N$, the electric field intensity at point r is obtained using superposition.

13.4   Electric Field Due to Charge Distributions

We now extend the results of the previous section for the discrete charge distribution to the case of the continuous charge distributions shown in Figure 13.1.

The electric field intensity due to each of the charge distributions, ρ_l, ρ_S, and ρ_v, is obtained as the superposition of the fields contributed by the numerous point charges making up the charge distribution.

Mathematically, the point charges are expressed as differential charges dQ and the field due to each such charge is

$$dE = \frac{1}{4\pi\varepsilon_0} \frac{dQ}{R^2} a_R \quad \left[\frac{V}{m}\right] \tag{13.16}$$

Mathematically, the superposition of the fields due to all differential charges dQ, corresponds to an integral of over the location where the charge is distributed.

Replacing Q in Eq. (13.15) with Q in Eqs (13.2), (13.3), and (13.4), respectively, results in expression for electric field intensity due to the line, surface, and volume charge density as

$$E = \int_l \frac{\rho_l dl}{4\pi\varepsilon_0 R^2} a_R \tag{13.17}$$

$$E = \int_S \frac{\rho_s dS}{4\pi\varepsilon_0 R^2} a_R \tag{13.18}$$

$$E = \int_v \frac{\rho_v dv}{4\pi\varepsilon_0 R^2} a_R \tag{13.19}$$

When computing the electric fields due using the above integrals, we usually do not determine the fields anywhere in space about the charge distributions, but only at certain locations where we can utilize symmetry to simplify the calculations.

13.5 Electric Flux Density

Recall from calculus: given a vector A, the *flux* of A is defined as a surface integral of A:

$$\Psi = \int_S A \cdot dS \tag{13.20}$$

The vector A is then called the *flux density* vector. Adhering to this definition, we could define the electric flux as

$$\Psi = \int_S E \cdot dS \tag{13.21}$$

Since the electrostatic field intensity E is a function of ε, the permittivity of the medium, it follows that the flux defined by Eq. (13.21) is dependent on the medium in which the charge is placed. This leads to a different definition of the electric flux that is independent of the medium, as explained next.

Suppose, we define a new vector D in free space as

$$D = \varepsilon_0 E \tag{13.22}$$

It is apparent, that all the formulas for E, derived from Coulomb's law can be used for calculating D. All that needs to be done is to multiply those formulas by ε_0.

In electrostatics, the electric flux is defined in terms of the vector D as

$$\Psi = \int_S D \cdot dS \quad [C] \tag{13.23}$$

The vector D is called the *electric flux density* and is measured in $[C/m^2]$.

13.6 Gauss's Law for the Electric Field

Gauss's law gives us a very powerful tool in calculations of the electric filed intensity, or the electric flux density due to the various charge distributions. It not only greatly simplifies the calculations, but also gives us an insight into the $\mathbf{E}$ and $\mathbf{D}$ fields surrounding the distributions.

According to Gauss's law states the total electric flux Ψ through any *closed* surface is equal to the total charge Q enclosed by that surface (Rao, 2004, p. 107):

$$Q = \oint_S \mathbf{D} \cdot d\mathbf{S} \tag{13.24}$$

When the enclosed charge distributed over a volume (within the closed surface S), we can express Gauss's law as

$$\oint_S \mathbf{D} \cdot d\mathbf{S} = \int_v \rho_v \, dv \tag{13.25}$$

where ρ_v is the volume charge density. Now, according to the divergence theorem we have

$$\oint_S \mathbf{D} \cdot d\mathbf{S} = \int_v \nabla \cdot \mathbf{D} \, dv \tag{13.26}$$

Equating the right-hand sides of Eqs (13.25) and (13.26), we obtain

$$\int_v \nabla \cdot \mathbf{D} \, dv = \int_v \rho_v \, dv \tag{13.27}$$

resulting in

$$\nabla \cdot \mathbf{D} = \rho_v \tag{13.28}$$

Equation (13.28) is referred to as Gauss's law in differential form, while Eq. (13.24) is referred to as Gauss's law in integral form. Each of these equations is one of the four Maxwell's equations (either in differential or integral form), which we will discuss in detail in Chapter 15.

13.7 Applications of Gauss's Law

Gauss's law is most useful when the charge distribution is symmetric. When the charge distribution is not symmetric, to determine $\mathbf{E}$ or $\mathbf{D}$ we resort to Coulomb's law.

In evaluating the surface integral in Eq. (13.24) we are free to choose any closed surface encompassing the charge. When symmetry in charge distribution exists, we choose the surface that mirrors the symmetry exhibited by the charge distribution.

On such a surface, $\mathbf{E}$ and $\mathbf{D}$ vectors are either tangential to it or normal to it while constant in magnitude. Such a surface is called a *Gaussian surface*. Next we apply Gauss's law to several symmetric charge distributions.

Example 13.1 Point charge
Consider a single point charge Q located at the origin, as shown in Figure 13.4.

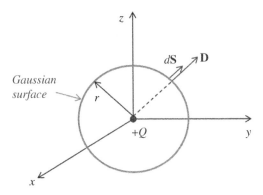

Figure 13.4 Determination of an electric field of a single point.

To determine **D** at any point, we choose a spherical surface with a center at the origin. Note that **D** is everywhere normal to this surface and constant in magnitude on it. That is,

$$\mathbf{D} = D_r \mathbf{a}_r \tag{13.29}$$

Applying Gauss's law we have

$$Q = \oint_S \mathbf{D} \cdot d\mathbf{S} = D_r \oint_S dS = D_r \, 4\pi r^2 \tag{13.30}$$

Solving this gives

$$\mathbf{D} = \frac{Q}{4\pi r^2} \mathbf{a}_r \tag{13.31}$$

Substituting $\mathbf{D} = \varepsilon \mathbf{E}$ gives the electric field intensity

$$\mathbf{E} = \frac{Q}{4\pi \varepsilon r^2} \mathbf{a}_r \tag{13.32}$$

Example 13.2   Sphere with a uniform charge
Consider a sphere centered with a uniform surface charge density ρ_S centered at the origin, as shown in Figure 13.5.

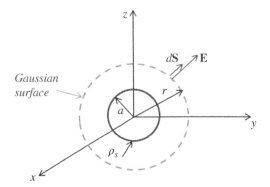

Figure 13.5 Determination of an electric field of a sphere of charge.

As a Gaussian surface, we choose a spherical surface with a center at the origin that encompasses the charge distribution. Note that $\mathbf{D}$ is everywhere normal to this surface and constant in magnitude on it. That is,

$$\mathbf{D} = D_r \mathbf{a}_r \tag{13.33}$$

Applying Gauss's law we have

$$Q = \oint_S \mathbf{D} \cdot d\mathbf{S} = D_r \oint dS = D_r \, 4\pi r^2 \tag{13.34}$$

or

$$Q = D_r \, 4\pi r^2 \tag{13.35}$$

On the other hand the total charge Q distributed over the surface of radius a is equal to

$$Q = \int_S \rho_S dS = \rho_S \int_S dS = \rho_S \, 4\pi a^2 \tag{13.36}$$

Equating the right-hand sides of Eqs (13.35) and (13.36) gives

$$D_r \, 4\pi r^2 = \rho_S \, 4\pi a^2 \tag{13.37}$$

Resulting in

$$D_r = \rho_S \frac{a^2}{r^2} \tag{13.38}$$

or

$$\begin{aligned} \mathbf{D} &= \rho_S \frac{a^2}{r^2} \mathbf{a}_r \quad r > a \\ \mathbf{D} &= 0, \quad r < a \end{aligned} \tag{13.39}$$

Substituting $\mathbf{D} = \varepsilon \mathbf{E}$ gives the electric field intensity

$$\begin{aligned} \mathbf{E} &= \rho_S \frac{a^2}{\varepsilon r^2} \mathbf{a}_r, \quad r > a \\ \mathbf{E} &= 0, \quad r < a \end{aligned} \tag{13.40}$$

Example 13.3 Infinite Plane of Charge
Determine the electric field of an infinite plane of charge with a uniform surface charge density ρ_S [C/m^2], shown in Figure 13.6.

Solution: Let's consider an infinite plane of charge lying on the $z = 0$ plane. Due to the infinite extent of the plane and the uniform charge distribution, the electric field will be perpendicular to its surface.

To determine $\mathbf{D}$ above the surface, we choose a cylindrical Gaussian surface that is cut symmetrically by the sheet of charge and has two of its faces parallel to the sheet, as shown in Figure 13.6. The electric field is perpendicular to the top and bottom surfaces, and is tangential to the sides of the cylinder.

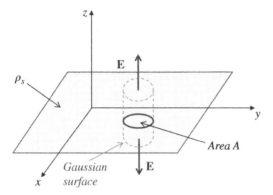

Figure 13.6 Determination of an electric field of plane of charge.

At the top surface we have

$$\mathbf{D} \cdot d\mathbf{S} = (D_z \mathbf{a}_z) \cdot (dS\mathbf{a}_z) = D_z dS \tag{13.41}$$

while at the bottom surface we have

$$\mathbf{D} \cdot d\mathbf{S} = \left[D_z (-\mathbf{a}_z) \right] \cdot \left[dS(-\mathbf{a}_z) \right] = D_z dS \tag{13.42}$$

Applying Gauss's law gives

$$Q = \oint_S \mathbf{D} \cdot d\mathbf{S} = \underbrace{\int \mathbf{D} \cdot d\mathbf{S}}_{side} + \underbrace{\int \mathbf{D} \cdot d\mathbf{S}}_{top} + \underbrace{\int \mathbf{D} \cdot d\mathbf{S}}_{bottom}$$

$$= \underbrace{\int D_z (\pm \mathbf{a}_z) \cdot dS\mathbf{a}_\rho}_{side} + \underbrace{\int D_z \mathbf{a}_z \cdot dS\mathbf{a}_z}_{top} + \underbrace{\int D_z (-\mathbf{a}_z) \cdot dS(-\mathbf{a}_z)}_{bottom} \tag{13.43}$$

$$= 0 + D_z \underbrace{\int dS}_{top} + D_z \underbrace{\int dS}_{top} = D_Z A + D_Z A = 2D_Z A$$

or

$$Q = 2D_Z A \tag{13.44}$$

where A is the area of the top and bottom surfaces of the Gaussian cylinder. The total charge enclosed by the Gaussian surface is

$$Q = \rho_S \int_S dS = \rho_S A \tag{13.45}$$

Combining Eqs (13.44) and (13.45) produces

$$2D_Z A = \rho_S A \tag{13.46}$$

and thus

$$D_Z = \frac{\rho_S}{2} \tag{13.47a}$$

$$\mathbf{D} = \frac{\rho_S}{2} \mathbf{a}_z, \quad z > 0 \tag{13.47b}$$

Similarly, the field below the surface $z = 0$ is given by

$$D = -\frac{\rho_S}{2} \mathbf{a}_z, \quad z < 0 \tag{13.48}$$

Therefore, the electric field **E** due to the plane of charge is given by

$$E = \begin{cases} \dfrac{\rho_S}{2\varepsilon} \mathbf{a}_z, & z > 0 \\[2mm] -\dfrac{\rho_S}{2\varepsilon} \mathbf{a}_z, & z < 0 \end{cases} \tag{13.49}$$

Example 13.4   Infinite line charge
Next we determine the electric field due to the infinite line of uniform charge ρ_l [C/m] shown in Figure 13.7.

By symmetry, the electric flux density **D** and the electric field intensity **E** are directed radially away from the line. In order to take advantage of this symmetry, we choose a Gaussian surface as a cylinder of radius ρ.

Let's consider a length l of the cylinder. Over the top and bottom surfaces of the cylinder, vector $d\mathbf{S}$ is perpendicular to the vector **D** and the dot product of the two is zero.

On the side of the cylinder **D** is constant and pointing in the same direction as $d\mathbf{S}$. Thus, Gauss's law produces

$$Q = \oint_S \mathbf{D} \cdot d\mathbf{S} = \oint_S D_\rho \mathbf{a}_\rho \cdot d S \mathbf{a}_\rho = D_\rho \oint_S dS = D_\rho 2\pi\rho l \tag{13.50}$$

The total charge enclosed by the cylinder of length l is

$$Q = \rho_l l \tag{13.51}$$

Combining Eqs (13.50) and (13.51) results in

$$D_\rho 2\pi\rho l = \rho_L l \tag{13.52}$$

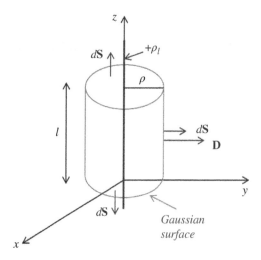

Figure 13.7 Determination of an electric field of an infinite line of charge.

and thus

$$D_\rho = \frac{\rho_L}{2\pi\rho} \tag{13.53a}$$

and

$$D = \frac{\rho_L}{2\pi\rho} \mathbf{a}_\rho \tag{13.53b}$$

Also

$$E = \frac{\rho_L}{2\pi\varepsilon\rho} \mathbf{a}_\rho \tag{13.54}$$

■

Example 13.5 Infinite cylinder with surface charge density
Determine the electric outside and within an infinitely long cylinder of radius a with a uniform surface charge density ρ_S [C/m^2], as shown in Figure 13.8.

Solution: The natural choice of the Gaussian surface is a cylinder of radius ρ, as shown in Figure 13.8. Because of the charge symmetry, the electric field lines are directed radially away from the cylinder.

The electric field is normal to the side of this Gaussian surface and parallel to its ends.

$$\mathbf{D} = D_\rho \mathbf{a}_\rho \tag{13.55}$$

Gauss's law produces

$$\begin{aligned}
Q = \oint_S \mathbf{D} \cdot d\mathbf{S} &= \int_{side} \mathbf{D} \cdot d\mathbf{S} + \int_{top} \mathbf{D} \cdot d\mathbf{S} + \int_{bottom} \mathbf{D} \cdot d\mathbf{S} \\
&= \int_{side} D_\rho \mathbf{a}_\rho \cdot dS \mathbf{a}_\rho + \int_{top} D_\rho \mathbf{a}_\rho \cdot dS \mathbf{a}_z + \int_{bottom} D_\rho \mathbf{a}_\rho \cdot dS (-\mathbf{a}_z) \\
&= \int_{side} D_\rho dS + 0 + 0 = D_\rho \oint dS = D_\rho 2\pi\rho l
\end{aligned} \tag{13.56}$$

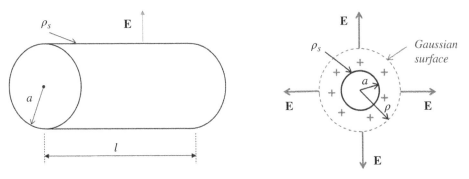

Figure 13.8 Determination of an electric field of a cylinder of charge.

or

$$Q = D_\rho 2\pi\rho l \qquad (13.57)$$

The total charge enclosed by the cylinder of length l is

$$Q = \rho_S 2\pi a l \qquad (13.58)$$

Combining Eqs (13.57) and (13.58) results in

$$D_\rho 2\pi\rho l = \rho_S 2\pi a l \qquad (13.59)$$

and thus

$$D_\rho = \frac{\rho_S a}{\rho} \qquad (13.60a)$$

$$\mathbf{D} = \frac{\rho_S a}{\rho}\mathbf{a}_\rho, \quad \rho > a \qquad (13.60b)$$

Also

$$\mathbf{E} = \frac{\rho_S a}{\varepsilon\rho}\mathbf{a}_\rho, \quad \rho > a \qquad (13.61a)$$

Since there is no charge interior to the cylinder of radius a, the electric field inside the cylinder is zero,

$$\mathbf{E} = 0, \quad \rho < a \qquad (13.61b)$$

Example 13.6 Coaxial transmission line
A coaxial transmission line is shown in Figure 13.9. The inner cylinder has a radius a and the outer cylinder has a radius b. The inner cylinder has a surface charge density of

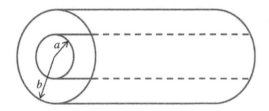

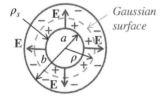

Figure 13.9 Determination of the electric field of coaxial transmission line.

ρ_S [C/m^2] distributed uniformly along its length and around its periphery. The outer cylinder has the same total charge as the inner cylinder distributed over its inner surface and of the opposite polarity.

Determine the electric field between the two cylinders.

Solution: Because of the uniform charge distribution and the infinite length of the cylinders, the electric field will be radially directed away from the inner cylinder toward the outer cylinder.

To determine the electric field distribution we choose a cylindrical Gaussian surface of radius ρ, as shown in Figure 13.9. The electric field is perpendicular to the side of this surface and parallel to the end surfaces. The Gaussian surface is the same as in the previous example. Thus, in the space between the two cylinders the electric field is given by

$$\mathbf{E} = \frac{\rho_S a}{\varepsilon\rho}\mathbf{a}_\rho, \quad a<\rho<b \tag{13.62}$$

Again, the electric field inside the inner cylinder is zero,

$$\mathbf{E} = 0, \quad \rho<a \tag{13.63}$$

The electric field outside the outer cylinder is also zero since the total charge enclosed by a cylindrical Gaussian surface surrounding both cylinders is zero.

$$\mathbf{E} = 0, \quad \rho>b \tag{13.64}$$

This is a very important observation. Since there is no **E** field outside the (ideal) coaxial cable, it is often referred to as a "shielded cable".

13.8   Electric Scalar Potential and Voltage

The concept of electric potential leads to the definition of voltage, and serves as a bridge between the field theory and circuit theory.

The electric scalar potential is defined through the work done by the electric field in moving a point charge. Suppose we wish to move a positive charge Q from point A to point B, in the presence of an electrostatic field **E**, as shown in Figure 13.10.

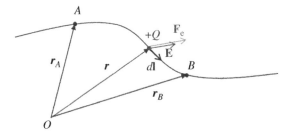

Figure 13.10 Determination of the work required to move a charge.

The force exerted on Q by the field is

$$\mathbf{F} = Q\mathbf{E} \tag{13.65}$$

and the work done by the field in moving the charge Q by a differential distance $d\mathbf{l}$ is

$$dW_f = \mathbf{F} \cdot d\mathbf{l} = Q\mathbf{E} \cdot d\mathbf{l} \tag{13.66}$$

The force that an external agent would have to apply to move the charge at constant velocity (i.e. with no acceleration) would have to counteract the force exerted by the field, so that the total net force on the charge is zero.

$$\mathbf{F}_{ext} = -\mathbf{F} = -Q\mathbf{E} \tag{13.67}$$

The work done, or energy expended in moving the charge Q by a differential distance $d\mathbf{l}$ under the influence of an external force is

$$dW = \mathbf{F}_{ext} \cdot d\mathbf{l} = -Q\mathbf{E} \cdot d\mathbf{l} \tag{13.68}$$

The total work done (by an external force), or the potential energy required, in moving Q from A to B is

$$W = -Q\int_A^B \mathbf{E} \cdot d\mathbf{l} \tag{13.69}$$

Dividing both sides by Q gives

$$\frac{W}{Q} = -\int_A^B \mathbf{E} \cdot d\mathbf{l} \qquad \left[\frac{J}{C} = V \right] \tag{13.70}$$

This quantity is known as the *voltage* or the *potential difference* between points A and B

$$V_{AB} = -\int_A^B \mathbf{E} \cdot d\mathbf{l} \qquad \left[\frac{J}{C} = V \right] \tag{13.71}$$

When evaluating this integral it is assume that point A is at a lower potential than point B. Thus,

$$V_{AB} = - \int_{\substack{lower \\ potential}}^{\substack{higher \\ potential}} \mathbf{E} \cdot d\mathbf{l} \qquad \left[\frac{J}{C} = V \right] \tag{13.72}$$

Notice that the potential difference between two points does not depend on the charge being moved between them.

To illustrate the application of Eq. (13.71), let's consider a positive point charge Q at the origin of the coordinate system. This is the charge that generates the electric field $\mathbf{E}$:

$$\mathbf{E} = \frac{Q}{4\pi\varepsilon_0 r^2}\mathbf{a}_r \tag{13.73}$$

Thus

$$V_{AB} = -\int_A^B \mathbf{E} \cdot d\mathbf{l} = -\int_{r_A}^{r_B} \frac{Q}{4\pi\varepsilon_0 r^2} \mathbf{a}_r \cdot d\mathbf{l} \tag{13.74}$$

Let's decompose the vector differential element $d\mathbf{l}$ into a vector $d\mathbf{r}$ along the $\mathbf{a}_r$ direction and vector $d\mathbf{t}$ perpendicular to it,

$$d\mathbf{l} = d\mathbf{r} + d\mathbf{t} \tag{13.75}$$

then

$$\mathbf{a}_r \cdot d\mathbf{l} = \mathbf{a}_r \cdot (d\mathbf{r} + d\mathbf{t}) = \mathbf{a}_r \cdot d\mathbf{r} a_r = dr \tag{13.76}$$

and

$$V_{AB} = -\int_{r_A}^{r_B} \frac{Q}{4\pi\varepsilon_0 r^2} dr \tag{13.77}$$

or

$$V_{AB} = \frac{Q}{4\pi\varepsilon_0} \left(\frac{1}{r_B} - \frac{1}{r_A} \right) \tag{13.78}$$

This is a very important result that we will refer to often.

Absolute potential It is often convenient to determine the potential or absolute potential, at a point, rather than the potential difference between two points. The potential at any point is defined as the potential difference between that point and a chosen point at which potential is zero.

Perhaps the most universal reference point in practical applications is "ground", by which we mean a reference point or surface where the potential is zero. Another widely used reference point with zero potential is infinity. This is very convenient in theoretical problems.

If we choose the reference point at infinity, then the voltage between this point and infinity is referred to as the *absolute potential*, or just the *potential*, at a point, and is defined as

$$V = -\int_\infty^r \mathbf{E} \cdot d\mathbf{l} \tag{13.79}$$

Thus if $V_A = 0$ as $r_A \to \infty$, the potential at any point ($r_B \to r$) due to a point charge Q located at the origin is

$$V = \frac{Q}{4\pi\varepsilon_0 r} \tag{13.80}$$

13.9   Voltage Calculations due to Charge Distributions

In this section we will calculate the voltage between two points in space due to the various charge distributions considered earlier in this chapter.

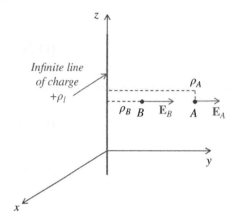

Figure 13.11 Two points away from a line of charge.

Example 13.7   Voltage between two points away from a line of charge

Let an infinite line of charge have a uniform line charge density ρ_l [C/m]. We want to determine the voltage between two points at distances ρ_A and ρ_B away ($\rho_A > \rho_B$) from the line shown in Figure 13.11.

The electric field at a distance ρ away from the line was previously calculated as

$$E = \frac{\rho_L}{2\pi\varepsilon\rho}\mathbf{a}_\rho \tag{13.81}$$

The voltage between two points is calculated from

$$V_{AB} = -\int_{\substack{lower \\ potential}}^{\substack{higher \\ potential}} \mathbf{E}\cdot d\mathbf{l} = -\int_{\rho_A}^{\rho_B} \frac{\rho_L}{2\pi\varepsilon\rho}\mathbf{a}_\rho \cdot d\rho\mathbf{a}_\rho \tag{13.82}$$

$$= -\frac{\rho_L}{2\pi\varepsilon}\int_{\rho_A}^{\rho_B}\frac{d\rho}{\rho} = -\frac{\rho_L}{2\pi\varepsilon}\left(\ln\rho_B - \ln\rho_A\right) = \frac{\rho_L}{2\pi\varepsilon}\left(\ln\rho_A - \ln\rho_B\right)$$

or

$$V_{AB} = \frac{\rho_L}{2\pi\varepsilon}\ln\left(\frac{\rho_A}{\rho_B}\right) \tag{13.83}$$

Again, this is a very important result that we will encounter on several occasions.

Example 13.8   Voltage between two points away from a plane of charge

Let an infinite plane of charge have a uniform surface charge density ρ_S [C/m²]. We want to determine the voltage between two points at z_A and z_B away ($z_A > z_B$) from the plane, shown in Figure 13.12.

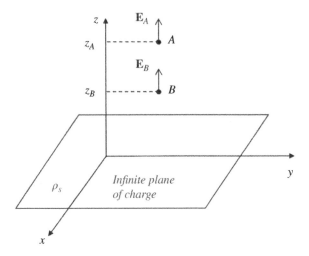

Figure 13.12 Two points away from a plane of charge.

The electric field was previously calculated as

$$
\mathbf{E} =
\begin{cases}
\dfrac{\rho_S}{2\varepsilon}\mathbf{a}_z, & z>0 \\[2mm]
-\dfrac{\rho_S}{2\varepsilon}\mathbf{a}_z, & z<0
\end{cases}
\tag{13.84}
$$

The voltage between two points is calculated from

$$
V_{AB} = -\int_{\substack{lower \\ potential}}^{\substack{higher \\ potential}} \mathbf{E}\cdot d\mathbf{l} = -\int_{z_A}^{z_B}\frac{\rho_S}{2\varepsilon}\mathbf{a}_z \cdot dz\mathbf{a}_z
\tag{13.85}
$$

$$
= -\frac{\rho_S}{2\varepsilon}\left(z_B - z_A\right)
$$

or

$$
V_{AB} = \frac{\rho_S}{2\varepsilon}\left(z_A - z_B\right)
\tag{13.86}
$$

Example 13.9 Voltage between the inner and outer cylinders of a coaxial cable
Next, let's determine the voltage between the inner and outer cylinders of a coaxial cable shown in Figure 13.13.

The inner cylinder has a surface charge density of ρ_S [C/m²] distributed uniformly along its length and around its periphery. The outer cylinder has the same total charge as the inner cylinder distributed over its inner surface and of the opposite polarity.

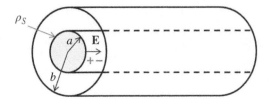

Figure 13.13 Voltage between two concentric cylinders.

The electric field was previously calculated as

$$E = \frac{\rho_s a}{\varepsilon \rho} \mathbf{a}_\rho, \quad a < \rho < b \tag{13.87}$$

The voltage between two points is calculated from

$$V_{AB} = - \int_{\substack{lower \\ potential}}^{\substack{higher \\ potential}} \mathbf{E} \cdot d\mathbf{l} = - \int_b^a \frac{\rho_s a}{\varepsilon \rho} \mathbf{a}_\rho \cdot d\rho \mathbf{a}_\rho \tag{13.88}$$

$$= - \frac{\rho_s a}{\varepsilon} \int_b^a \frac{d\rho}{\rho} = - \frac{\rho_s a}{\varepsilon} \big[\ln(a) - \ln(b) \big]$$

or

$$V_{AB} = \frac{\rho_s a}{\varepsilon} \ln\left(\frac{b}{a}\right) \tag{13.89}$$

Example 13.10   Voltage between two concentric spheres
Determine the voltage between two concentric spheres, shown in Figure 13.14.

The inner sphere of radius a has a uniformly distributed surface charge density of ρ_s [C/m²]. The outer sphere of radius b has the same total charge as the inner sphere distributed over its surface and of the opposite polarity.

The electric field between the spheres was previously calculated as

$$E = \frac{\rho_s a^2}{\varepsilon r^2} \mathbf{a}_r, \quad a < r < b \tag{13.90}$$

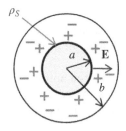

Figure 13.14 Voltage between two concentric spheres.

The voltage between two points is calculated from

$$V_{AB} = -\int\limits_{\substack{lower \\ potential}}^{\substack{higher \\ potential}} \mathbf{E} \cdot d\mathbf{l} = -\int_b^a \frac{\rho_S a^2}{\varepsilon r^2} \mathbf{a}_r \cdot d r \mathbf{a}_r$$ (13.91)

$$= -\frac{\rho_S a^2}{\varepsilon} \int_b^a \frac{dr}{r^2} = -\frac{\rho_S a^2}{\varepsilon} \left(-\frac{1}{r} \right)\Big|_b^a$$

or

$$V_{AB} = \frac{\rho_S a^2}{\varepsilon} \left(\frac{1}{a} - \frac{1}{b} \right)$$ (13.92)

Since the surface charge density can be expressed in terms of the total charge as

$$\rho_S = \frac{Q}{4\pi a^2}$$ (13.93)

The above result can be written as

$$V_{AB} = \frac{Q}{4\pi\varepsilon} \left(\frac{1}{a} - \frac{1}{b} \right)$$ (13.94)

13.10 Electric Flux Lines and Equipotential Surfaces

The concept of electric flux lines was introduced by Michael Faraday as a way of visualizing the electric field. An electric flux line is an imaginary path or line drawn in such a way that its direction at any point is the same as the direction of the electric field at that point.

Thus the electric flux lines are the lines to which the electric field intensity **E** or the electric flux density **D** is tangential at every point. These lines do not intersect and always start at positive charges and terminate at negative charges (or infinity).

Let's now define an equipotential surface (or line) as any surface (line) on which the potential is constant. Figure 13.15 shows the equipotential surfaces around a point charge.

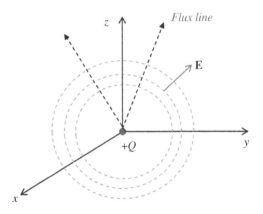

Figure 13.15 Equipotential surfaces around a point charge.

Since the potential is constant along the equipotential line or on the equipotential surface, the work done in moving a charge along such line or surface is zero,

$$V_A - V_B = 0 = \int \mathbf{E} \cdot d\mathbf{l} \tag{13.95}$$

Analyzing the above equation we note that vectors **E** and $d\mathbf{l}$, in general, are not zero, and therefore we conclude that the flux lines are always normal to the equipotential line or surface.

13.11 Maxwell's Equations for Static Electric Field

Recall from Chapter 4 that the line integral $\int_C \mathbf{F} \cdot d\mathbf{l}$ is independent of the path of integration if the function **F** is a gradient of some scalar function f, i.e. $\mathbf{F} = \nabla f$. It can be shown that for static electric fields, the electric field intensity **E** is related to the scalar potential V by

$$\mathbf{E} = -\nabla V \tag{13.96}$$

It follows that the integral

$$V_{AB} = V_B - V_A = -\int \mathbf{E} \cdot d\mathbf{l} \tag{13.97}$$

is independent of the integration path and therefore

$$\oint_C \mathbf{E} \cdot d\mathbf{l} = \left(-\int_A^B \mathbf{E} \cdot d\mathbf{l} \right) + \left(-\int_B^A \mathbf{E} \cdot d\mathbf{l} \right) = V_B - V_A + V_A - V_B = 0 \tag{13.98}$$

or

$$\oint_C \mathbf{E} \cdot d\mathbf{l} = 0 \tag{13.99}$$

According to the Stokes theorem we have

$$\oint_C \mathbf{E} \cdot d\mathbf{l} = \int_S (\nabla \times \mathbf{E}) \cdot d\mathbf{S} = 0 \tag{13.100}$$

and therefore

$$\nabla \times \mathbf{E} = 0 \tag{13.101}$$

Equations (13.99) and (13.110) are referred to as the Maxwell equations for static electric field.

13.12 Capacitance Calculations of Structures

13.12.1 Definition of Capacitance

When separated by an insulating (dielectric medium), any two conducting bodies, regardless of their shapes and sizes, form a *capacitor*.

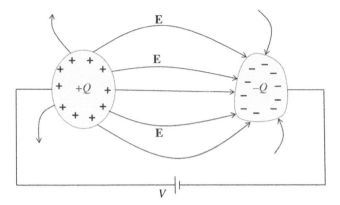

Figure 13.16 Capacitive structure.

If a dc voltage source is connected to the two conductors, charge of equal and opposite polarity is transferred to the conductor's surfaces. The surface of the conductor connected to the positive side of the source will accumulate charge $+Q$, and charge $-Q$ will accumulate on the surface of the other conductor, as shown in Figure 13.16.

Capacitance of a two-conductor structure is defined as

$$C = \frac{Q}{V} \tag{13.102}$$

where V is the voltage between the conducting surfaces and Q is the magnitude of the charge on either surface.

The presence of free charges on the conductors' surfaces gives rise to an electric field **E**. The field lines (flux lines) originate on the positive charges and terminate on the negative charges. Since a conductor's surface constitutes an equipotential surface, **E** is always perpendicular to the conducting surfaces.

The normal component of **E** at any point on the surface of either conductor is given by (see Section 13.7),

$$E_n = E_n a_n = \frac{\rho_s}{\varepsilon} \tag{13.103}$$

Charge Q distributed over the surface of either conductor is

$$Q = \int_S \rho_s dS \tag{13.104}$$

and according to Gauss's law can be calculated from

$$Q = \oint_S \varepsilon \mathbf{E} \cdot d\mathbf{S} \tag{13.105}$$

The voltage V is related to **E** by

$$V = -\int_l \mathbf{E} \cdot d\mathbf{l} \tag{13.106}$$

where the path of integration is from the conductor at the lower potential to the conductor at the higher potential.

Substituting Eqs (13.105) and (13.106) into Eq. (13.102) produces a general formula for calculating capacitance:

$$C = \frac{Q}{V} = \frac{\oint_S \varepsilon \mathbf{E} \cdot d\mathbf{S}}{-\int_l \mathbf{E} \cdot d\mathbf{l}} \tag{13.107}$$

Next we will calculate capacitance of the typical structures encountered in EMC:

- parallel-plate capacitor
- two-wire transmission line
- coaxial cable
- spherical capacitor

In our calculations we will follow these steps:

1) Assume $+Q$ charge on one conductor and $-Q$ charge on the other
2) Calculate **E** from Gauss's law

$$Q = \oint_s \varepsilon \mathbf{E} \cdot d\mathbf{S} \tag{13.108}$$

3) Calculate V from

$$V = -\int_l \mathbf{E} \cdot d\mathbf{l} \tag{13.109}$$

4) Determine the capacitance from

$$C = \frac{Q}{V} = \frac{Q}{-\int_l \mathbf{E} \cdot d\mathbf{l}} \tag{13.110}$$

13.12.2 Calculations of Capacitance

Parallel-plate capacitor   Consider a parallel-plate capacitor shown in Figure 13.17.

1) Let the conductive plates carry charges $+Q$ and $-Q$ or, equivalently, they have surface charge densities as shown in Figure 13.18.

Figure 13.17 Parallel-plate capacitor.

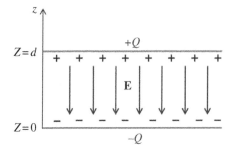

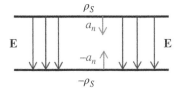

Figure 13.18 Parallel-plate capacitor.

2) Determine **E** from Gauss's law. We previously obtained:

$$\mathbf{E} = \mathbf{E} = \frac{\rho_S}{2\varepsilon}\mathbf{a}_n + \frac{-\rho_S}{2\varepsilon}(-\mathbf{a}_n) = \frac{\rho_S}{\varepsilon}\mathbf{a}_n = -\frac{\rho_S}{\varepsilon}\mathbf{a}_z = -\frac{Q}{\varepsilon A}\mathbf{a}_z \tag{13.111}$$

where A is the area of each plate and ε is the dielectric constant of the medium between the plates.

3) Determine the voltage between the plates.

$$V = -\int_l \mathbf{E}\cdot d\mathbf{l} = -\int_{z=0}^{z=d}\left(-\frac{Q}{\varepsilon A}\mathbf{a}_z\right)\cdot(dz\mathbf{a}_z)$$

$$= \frac{Q}{\varepsilon A}\int_{z=0}^{z=d}dz = \frac{Qd}{\varepsilon A} \tag{13.112}$$

4) Obtain the capacitance from

$$C = \frac{Q}{V} = \frac{Q}{-\int_l \mathbf{E}\cdot d\mathbf{l}} = \frac{Q}{\frac{Qd}{\varepsilon A}} \tag{13.113}$$

or

$$C = \frac{\varepsilon A}{d} \tag{13.114}$$

Two-wire transmission line   A cross-section of a two-wire transmission line is shown in Figure 13.19.

Let's model the transmission line as two infinite parallel conductors of radius a separated by a distance s. Under the assumption of the ratio $s/a > 5$, we may assume that the conductors have a uniform surface charge distribution ρ_S [C/m^2] distributed on the periphery along their length.

One of the conductors carries a positive charge distribution, while the other carries an equal but a negative distribution. The voltage due to a line charge distribution between two points at distances ρ_A and ρ_B from the line was previously calculated as

$$V_{AB} = \frac{\rho_L}{2\pi\varepsilon}\ln\left(\frac{\rho_A}{\rho_B}\right) \tag{13.115}$$

Replacing ρ_A with s and ρ_B with a we get

$$V = \frac{\rho_L}{2\pi\varepsilon}\ln\left(\frac{s}{a}\right) \tag{13.116}$$

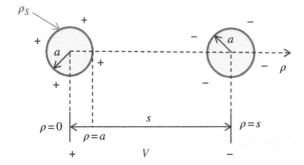

Figure 13.19 Two-wire transmission line.

The total voltage due to both lines of charge, by superposition is twice this result

$$V = \frac{\rho_L}{\pi\varepsilon} \ln\left(\frac{s}{a}\right) \tag{13.117}$$

The per-unit-length capacitance is obtained from

$$c = \frac{\rho_L}{V} = \frac{\rho_L}{\dfrac{\rho_L}{\pi\varepsilon} \ln\left(\dfrac{s}{a}\right)} \tag{13.118}$$

or

$$c = \frac{\pi\varepsilon}{\ln\left(\dfrac{s}{a}\right)} \tag{13.119}$$

Coaxial cable For the coaxial cable shown in Figure 13.20, the voltage between the inner and outer cylinders was determined as

$$V = \frac{\rho_s a}{\varepsilon} \ln\left(\frac{b}{a}\right) \tag{13.120}$$

The surface charge distribution is related to the per-unit-length charge distribution as

$$\rho_l = \rho_s 2\pi a \tag{13.121}$$

thus the voltage becomes

$$V = \frac{\rho_l}{2\pi\varepsilon} \ln\left(\frac{b}{a}\right) \tag{13.122}$$

The per-unit –length capacitance is obtained from

$$c = \frac{\rho_L}{V} = \frac{\rho_L}{\dfrac{\rho_l}{2\pi\varepsilon} \ln\left(\dfrac{b}{a}\right)} \tag{13.123}$$

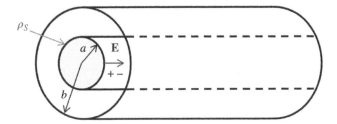

Figure 13.20 Coaxial capacitor.

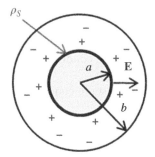

Figure 13.21 Spherical capacitor.

or

$$C = \frac{2\pi\varepsilon}{\ln\left(\dfrac{b}{a}\right)} \tag{13.124}$$

Spherical capacitor For the spherical capacitor shown in Figure 13.21, the voltage between the inner and outer cylinders was determined as

$$V_{AB} = \frac{Q}{4\pi\varepsilon}\left(\frac{1}{a} - \frac{1}{b}\right) \tag{13.125}$$

The capacitance is obtained from

$$C = \frac{Q}{V} = \frac{Q}{\dfrac{Q}{4\pi\varepsilon}\left(\dfrac{1}{a} - \dfrac{1}{b}\right)} \tag{13.126}$$

or

$$C = \frac{4\pi\varepsilon}{\dfrac{1}{a} - \dfrac{1}{b}} \tag{13.127}$$

Capacitance of an isolated sphere Let the outer sphere extend to infinity, i.e. $b \to \infty$. The capacitance of an isolated sphere of radius a then becomes

$$C = 4\pi\varepsilon a \tag{13.128}$$

This capacitance is often referred to as an absolute capacitance. This result is a very useful result in EMC, as we shall see in the application section.

13.13   Electric Boundary Conditions

If the electric field exists in a region consisting of two different media, even though it may be continuous in each medium, it may be discontinuous at the boundary between them, as illustrated in Figure 13.22.

Boundary conditions specify how the tangential and normal components of the field in one medium are related to the components of the field across the boundary in another medium.

We will derive a general set of boundary conditions, applicable at the interface between any two dissimilar media, be they two different dielectrics, or a conductor and a dielectric.

Even though these boundary conditions will be derived for electrostatic conditions, they will be equally valid for time-varying electromagnetic fields.

In each medium we will decompose the electric field intensity $\mathbf{E}$ and electric flux density $\mathbf{D}$ into two orthogonal components:

$$\mathbf{E} = \mathbf{E}_t + \mathbf{E}_n \tag{13.129a}$$

$$\mathbf{D} = \mathbf{D}_t + \mathbf{D}_n \tag{13.129b}$$

This is shown in Figure 13.23.

To determine the boundary conditions, we will use Maxwell's equations for electrostatic fields:

$$\oint_C \mathbf{E} \cdot d\mathbf{l} = 0 \tag{13.130}$$

$$\oint_S \mathbf{D} \cdot d\mathbf{S} = Q_{enc} \tag{13.131}$$

Let's consider the closed path *abcd* shown in Figure 13.24.

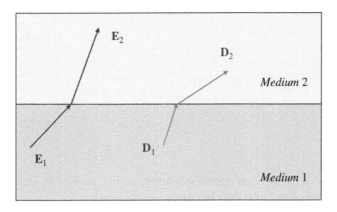

Figure 13.22 Discontinuity at the boundary between two media.

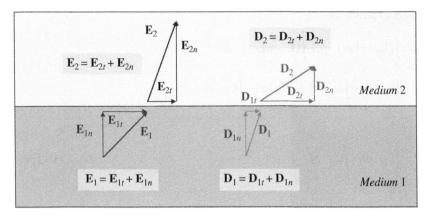

Figure 13.23 Decomposition into the normal and tangential components.

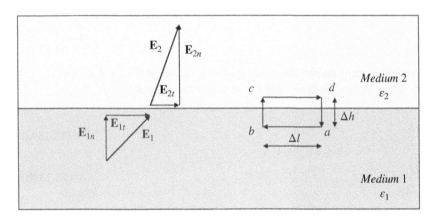

Figure 13.24 Evaluating boundary conditions.

We will apply Eq. (13.133) along this closed path. First, we will break the closed-loop integral in Eq. (13.133) into the integrals along the individual segments:

$$\oint_C \mathbf{E} \cdot d\mathbf{l} = \int_a^b \mathbf{E} \cdot d\mathbf{l} + \int_b^c \mathbf{E} \cdot d\mathbf{l} + \int_c^d \mathbf{E} \cdot d\mathbf{l} + \int_d^a \mathbf{E} \cdot d\mathbf{l} \tag{13.132}$$

Note that the integrals in Eq. (13.132) hold for any length of the integration path. That is, we can let any segment length go to zero, and the right hand-side of Eq. (13.132) will still be true. If we let $\Delta h \to 0$ then the contributions to the line integral by the segments bc and da go to zero and we have

$$\oint_C \mathbf{E} \cdot d\mathbf{l} = \int_a^b \mathbf{E}_1 \cdot d\mathbf{l} + \int_c^d \mathbf{E}_2 \cdot d\mathbf{l} \tag{13.133}$$

Now, since

$$\mathbf{E}_1 = \mathbf{E}_{1t} + \mathbf{E}_{1n} \tag{13.134a}$$

$$\mathbf{E}_2 = \mathbf{E}_{2t} + \mathbf{E}_{2n} \tag{13.134b}$$

We rewrite Eq. (13.133) as

$$\oint_C \mathbf{E} \cdot d\mathbf{l} = \int_a^b (\mathbf{E}_{1t} + \mathbf{E}_{1n}) \cdot d\mathbf{l} + \int_c^d (\mathbf{E}_{2t} + \mathbf{E}_{2n}) \cdot d\mathbf{l}$$

$$= \int_a^b \mathbf{E}_{1t} \cdot d\mathbf{l} + \int_a^b \mathbf{E}_{1n} \cdot d\mathbf{l} + \int_c^d \mathbf{E}_{2t} \cdot d\mathbf{l} + \int_c^d \mathbf{E}_{2n} \cdot d\mathbf{l} \tag{13.135}$$

or

$$\oint_C \mathbf{E} \cdot d\mathbf{l} = \int_a^b \mathbf{E}_{1t} \cdot d\mathbf{l} + \int_c^d \mathbf{E}_{2t} \cdot d\mathbf{l} \tag{13.136}$$

leading to

$$0 = \oint_C \mathbf{E} \cdot d\mathbf{l} = \int_a^b \mathbf{E}_{1t} \cdot d\mathbf{l} + \int_c^d \mathbf{E}_{2t} \cdot d\mathbf{l} = -E_{1t}\Delta l + E_{2t}\Delta l \tag{13.137}$$

or

$$E_{1t} = E_{2t} \tag{13.138}$$

Thus is a very important result: *the tangential component of the electric field is continuous (is the same) across the boundary between any two media.*

Since

$$E_{1t} = \frac{D_{1t}}{\varepsilon_1} \tag{13.139a}$$

$$E_{2t} = \frac{D_{2t}}{\varepsilon_2} \tag{13.139b}$$

We obtain the boundary condition on the electric flux density as

$$\frac{D_{1t}}{\varepsilon_1} = \frac{D_{2t}}{\varepsilon_2} \tag{13.140}$$

To obtain the boundary conditions on the normal components let's consider the closed cylindrical surface shown in Figure 13.25.

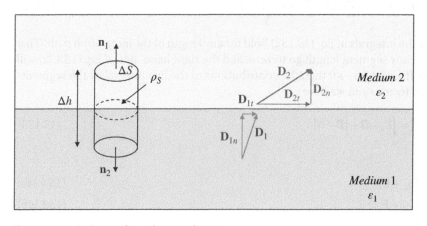

Figure 13.25 Evaluating boundary conditions.

Let's apply the second of the two Maxwell's equations

$$\oint_S \mathbf{D} \cdot d\mathbf{S} = Q_{enc} \tag{13.141}$$

First, we will break the closed-surface integral into three integrals as

$$\oint_S \mathbf{D} \cdot d\mathbf{S} = \int_{side} \mathbf{D} \cdot d\mathbf{S} + \int_{bottom} \mathbf{D}_1 \cdot d\mathbf{S} + \int_{top} \mathbf{D}_2 \cdot d\mathbf{S} \tag{13.142}$$

By letting $\Delta h \to 0$, the contributions to the total flux by the side surface goes to zero.

$$\oint_S \mathbf{D} \cdot d\mathbf{S} = \int_{bottom} \mathbf{D}_1 \cdot d\mathbf{S} + \int_{top} \mathbf{D}_2 \cdot d\mathbf{S} \tag{13.143}$$

Now, since

$$\mathbf{D}_1 = \mathbf{D}_{1t} + \mathbf{D}_{1n} \tag{13.144a}$$

$$\mathbf{D}_2 = \mathbf{D}_{2t} + \mathbf{D}_{2n} \tag{13.144b}$$

We rewrite Eq. (13.143) as

$$\begin{aligned} \oint_S \mathbf{D} \cdot d\mathbf{S} &= \int_{bottom} (\mathbf{D}_{1t} + \mathbf{D}_{1n}) \cdot d\mathbf{S} \mathbf{n}_2 + \int_{top} (\mathbf{D}_{2t} + \mathbf{D}_{2n}) \cdot d\mathbf{S} \mathbf{n}_1 \\ &= \int_{bottom} \mathbf{D}_{1n} \cdot d\mathbf{S} \mathbf{n}_2 + \int_{top} \mathbf{D}_{2n} \cdot d\mathbf{S} \mathbf{n}_1 \end{aligned} \tag{13.145}$$

or

$$\oint_S \mathbf{D} \cdot d\mathbf{S} = \int_{bottom} \mathbf{D}_{1n} \cdot d\mathbf{S} \mathbf{n}_2 + \int_{top} \mathbf{D}_{2n} \cdot d\mathbf{S} \mathbf{n}_1 \tag{13.146}$$

Even if each of the two media happens to have volume charge densities, the only charge remaining in the collapsed cylinder is that distributed on the boundary. Thus,

$$Q = \rho_S \int_{boundary} d\mathbf{S} = \mathbf{D}_1 \cdot \mathbf{n}_2 \int_{bottom} d\mathbf{S} + \mathbf{D}_2 \cdot \mathbf{n}_1 \int_{top} d\mathbf{S} \tag{13.147}$$

or

$$\rho_S = \mathbf{D}_1 \cdot \mathbf{n}_2 + \mathbf{D}_2 \cdot \mathbf{n}_1 \tag{13.148}$$

Since

$$\mathbf{n}_1 = -\mathbf{n}_2 \tag{13.149}$$

we arrive at the boundary condition on the electric flux density as

$$\rho_S = D_{2n} - D_{1n} \tag{13.150}$$

The corresponding condition on the electric field intensity is

$$\rho_S = \varepsilon_2 E_{2n} - \varepsilon_1 E_{1n} \tag{13.151}$$

If no free charge exists on the boundary between the two media, then $\rho_S = 0$, and the boundary conditions become

$$D_{1n} = D_{2n} \tag{13.152}$$

$$\varepsilon_1 E_{1n} = \varepsilon_2 E_{2n} \tag{13.153}$$

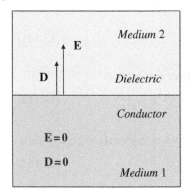

Figure 13.26 Dielectric–conductor boundary.

A very important application of the boundary conditions in EMC is when one medium is a dielectric and the other is a conductor. This is shown in Figure 13.26.

Inside the prefect conductor the fields are zero

$$\mathbf{E} = 0 \tag{13.154a}$$

$$\mathbf{D} = 0 \tag{13.154b}$$

Since

$$\mathbf{E}_1 = 0 \tag{13.155}$$

it follows that

$$\mathbf{E}_{1t} = 0 \tag{13.156a}$$

$$\mathbf{E}_{1n} = 0 \tag{13.156b}$$

Since the tangential component of **E** field must be continuous across the boundary it follows

$$\mathbf{E}_{2t} = 0 \tag{13.157}$$

and since

$$D_{2t} = \varepsilon_2 E_{2t} \tag{13.158}$$

we have

$$\mathbf{D}_{2t} = 0 \tag{13.159}$$

The tangential components of both the **E** vector and **D** vector are zero, but these vectors themselves, in general, are not zero.

*This means that both **E** and **D** vectors are perpendicular to the surface of the prefect conductor.*

Now recall the boundary condition on the normal component of the electric flux density given by Eq. (13.151), repeated here

$$\rho_S = \varepsilon_2 E_{2n} - \varepsilon_1 E_{1n} \tag{13.160}$$

Utilizing Eq. (13.166a) we obtain

$$\rho_S = \varepsilon_2 E_{2n} \tag{13.161}$$

or

$$\rho_S = D_{2n} \tag{13.162}$$

In a vector form we have

$$\mathbf{D}_2 = \varepsilon_2 \mathbf{E}_2 = \mathbf{n}\rho_S \tag{13.163}$$

Electric field points directly away from the conductor surface when ρ_S is positive and directly toward the conductor surface when ρ_S is negative.

13.14 EMC Applications

13.14.1 Electrostatic Discharge (ESD)

When electric charges are separated, an electric field is created. The most important consequence of this in EMC is electrostatic discharge (ESD). ESD can cause component damage, system reset, or signal integrity issues.

The separation of charge may take place when two initially neutral insulating materials, shown in Figure 13.27, come in contact with each other, as shown in Figure 13.28, and subsequently are separated, as shown in Figure 13.29.

When the two materials are in contact, some charges may be transferred between them; upon separation some of these transferred charges may not return to the original material. Effectively, the initially uncharged materials may become charged; one positively, one negatively, as shown in Figure 13.30. Consequently, when the materials are

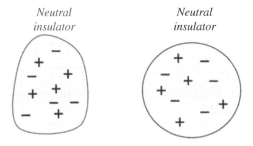

Figure 13.27 Two initially neutral insulating materials separated from each other.

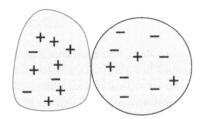

Figure 13.28 Two initially neutral insulating materials in contact with each other.

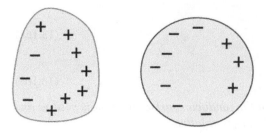

Figure 13.29 Insulators are separated after the contact.

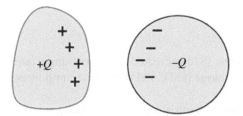

Figure 13.30 Net charge on each insulator after the contact.

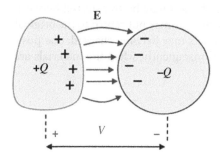

Figure 13.31 Creation of a capacitor.

separated, a capacitor is created with an electric field between the surfaces and a voltage difference between them, as shown in Figure 13.31.

Figure 13.32 shows a *triboelectric list*, i.e. a list of materials that have a greater tendency of giving up electrons (becoming more positive) or attracting electrons (and becoming more negative).

The further apart on the list the materials are, the greater the resulting charge Q and voltage V. The charge, the voltage, and the capacitance are related by

$$C = \frac{Q}{V} \quad \Rightarrow \quad Q = CV \tag{13.164}$$

As the materials are separated, the charge remains constant but the capacitance decreases, causing the voltage between them to increase. When this voltage reaches a high enough level, an electric breakdown may occur in the air separating the materials. This electric breakdown manifests itself as a lightning bolt and intense current; this phenomenon is referred to as *electrostatic discharge* (ESD).

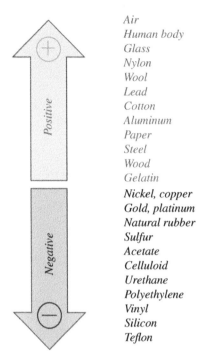

Air
Human body
Glass
Nylon
Wool
Lead
Cotton
Aluminum
Paper
Steel
Wood
Gelatin

Nickel, copper
Gold, platinum
Natural rubber
Sulfur
Acetate
Celluloid
Urethane
Polyethylene
Vinyl
Silicon
Teflon

Figure 13.32 Triboelectric list.

Charged
object

Neutral
conductor

Figure 13.33 Charged object approaches a conductor.

The ESD effect just described involved the charge transfer between two insulating materials. The ESD event can also occur when a charged object (insulator or conductor) approaches a conductor as shown in Figure 13.33.

The initially neutral conductor remains neutral as a whole; the charge, however, is separated. The charge with the opposite polarity to that of a charged object will be exposed on the surface closest to the object, creating the equal but opposite charge on the surface furthest from the object. This is shown in Figure 13.34.

If the conductor with the induced charge is momentarily connected to another conductor (or ground), while still in the vicinity of the charged object, the negative separated charge will be removed from it, as shown in Figure 13.35 (Ott, p. 584.).

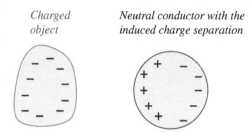

Figure 13.34 Charged object in the vicinity of a neutral conductor.

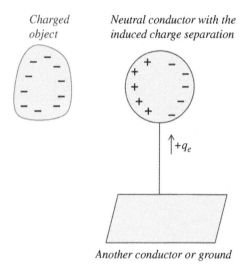

Figure 13.35 Momentary contact with another conductor.

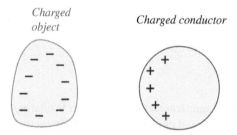

Figure 13.36 Charged conductor.

When the momentary contact with another conductor is removed, while in the vicinity of the charge object, the initially neutral conductor will now be charged, even though it has never touched the charged object. This is shown in Figure 13.36.

When this charged conductor is moved close to another conductor (grounded or not), an electrostatic discharge can occur from one conductor to another. This is illustrated in Figure 13.37.

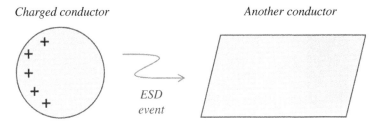

Figure 13.37 ESD event.

Example 13.11 ESD event: walking on a carpet
Walking on a carpet with leather shoe soles can generate voltages as high as 25 kV. Consider the scenario shown in Figure 13.38, where initially uncharged dielectrics (carpet and shoe) come into contact (Paul, 2006, p. 846).

When separated both the carpet and the shoe become charged, as shown in Figure 13.39. Subsequently, the negative shoe charge induces a positive charge on the sole of the foot (conductor), as shown in Figure 13.40.

Positive charge on the sole of the foot causes negative charge to move to the upper parts of the body (finger), as shown in Figure 13.41. As the finger approaches another conducting surface (door knob or electronic component) electrons will be pushed away from the surface closest to the finger, as shown in Figure 13.42. As the finger approaches the charge separation between the finger and the conductor surface creates an intense electrostatic field and voltage. An ESD event takes place: dielectric breakdown of the air occurs, an arc is created, and the discharge current flows through the conductor. This is shown in Figure 13.43.

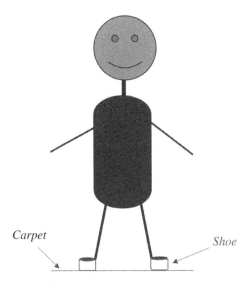

Figure 13.38 Initially uncharged dielectrics come into contact.

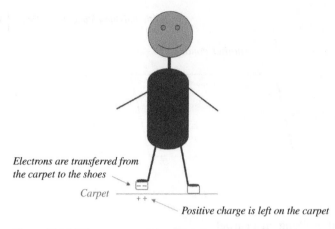

Electrons are transferred from the carpet to the shoes

Carpet

Positive charge is left on the carpet

Figure 13.39 When separated the dielectrics become charged.

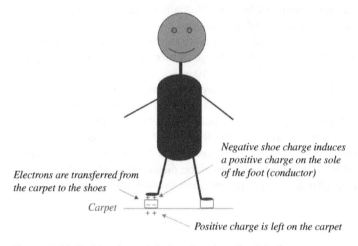

Negative shoe charge induces a positive charge on the sole of the foot (conductor)

Electrons are transferred from the carpet to the shoes

Carpet

Positive charge is left on the carpet

Figure 13.40 Positive charge is induced on the sole of the foot.

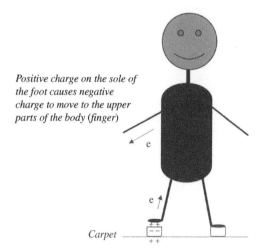

Positive charge on the sole of the foot causes negative charge to move to the upper parts of the body (finger)

Carpet

Figure 13.41 Negative charge moves to the upper parts of the body.

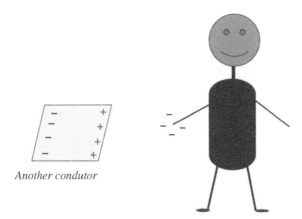

Figure 13.42 Charge separation in an adjacent conductor.

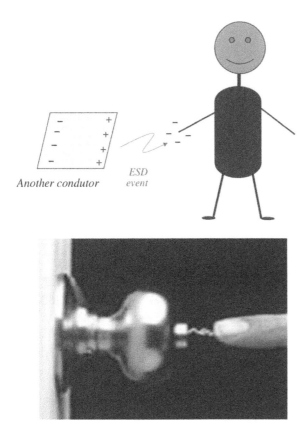

Figure 13.43 ESD event.

During the ESD arc formation the speed of approach is critical. Faster approach results in a physically shorter arc. Thus, for the same voltage difference, a faster approach results in a higher density of voltage per arc length. This results in a larger current and a faster current rise time.

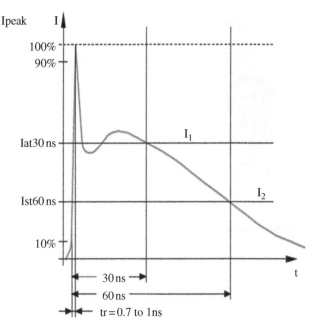

Figure 13.44 Typical shape of the ESD event.

Figure 13.44 shows the typical shape of the ESD event. Typical ESD characteristics are:

- rise time: $20\,\text{ps} < t_r < 70\,\text{ns}$
- spike width (if spike occurs): $0.5\,\text{ns} < t_w < 10\,\text{ns}$
- total duration: $100\,\text{ns} < t_t < 2\,\mu\text{s}$
- peak current: $1\,\text{A} < I_p < 200\,\text{A}$ (or more)

13.14.2 Human-Body Model

Recall: in Section 13.12 we obtained the capacitance of an isolated sphere as

$$C = 4\pi\varepsilon a \tag{13.165}$$

In free space

$$\varepsilon = \varepsilon_0 = \frac{1}{36\pi} \times 10^{-9} \quad \frac{\text{F}}{\text{m}} \tag{13.166}$$

Substituting Eq. (13.166) into Eq. (13.165) we get

$$C = \frac{1}{9} \times 10^{-9} a = 0.111 \times 10^{-9} a = 111 \times 10^{-12} a \tag{13.167}$$

or

$$C = 111a \quad \text{pF} \tag{13.168}$$

where the radius a is in meters. If we model the body as a sphere of radius $1\,\text{m}$, its capacitance would equal to $111\,\text{pF}$.

Equation (13.168) can be used to estimate the absolute capacitance of objects other than a sphere. We first determine the surface area of the object and then calculate the radius of a sphere with the same surface area. Then we use Eq. (13.168) (Ott, 2009, p. 585).

A human body has a surface area approximately equivalent to an area of a 0.5 m radius sphere. Therefore, the absolute capacitance of the human body is

$$C = 111 \times 0.5 = 55.5 \cong 50 \quad \text{pF} \tag{13.169}$$

Using this value, we can now create the *human body model* which serves as the basis for the ESD testing in EMC.

Because of the proximity of other objects to the human body, in addition to the absolute capacitance of the human body, an additional capacitance must be taken into account when determining the total capacitance of a human and the surroundings.

To create the human body model for ESD, we start with the absolute capacitance of 50 pF. In addition to this capacitance we have an additional capacitance between each foot and ground: 50 pF per foot (total 100 pF). Because of the presence of the adjacent objects, an additional capacitance of 50–100 pF may also exist (Ott, 2009, p. 587). This is shown in Figure 13.45.

Thus, the human body capacitance can vary from about 50–250 pF. The equivalent circuit of the human body for ESD is shown in Figure 13.46.

The body capacitance C is first charged up to a voltage V, and then it is discharged through the body resistance R. This body resistance limits the discharge current i. The body resistance can vary from about 500 Ω to 10 kΩ. The body capacitance limits the discharge current rate.

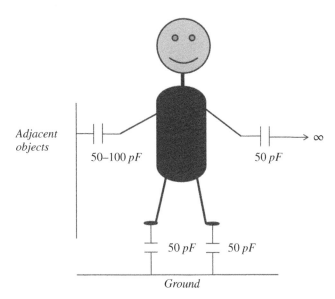

Figure 13.45 Human body capacitance.

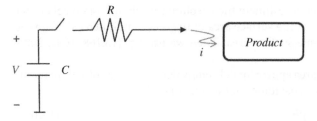

Figure 13.46 Human body circuit model.

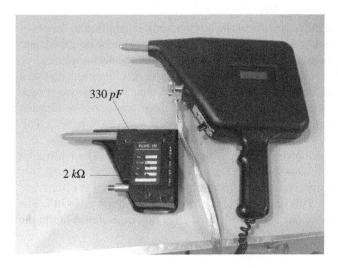

Figure 13.47 An ESD gun and a cartridge.

The most common circuit model of human body consists of 150 pF and 330 Ω (Standard EN 61000-4-2). Typical RC combinations are

$$R = 330\Omega, \quad C = 150\text{pF}$$
$$R = 330\Omega, \quad C = 330\text{pF}$$
$$R = 2000\Omega, \quad C = 150\text{pF} \tag{13.170}$$
$$R = 2000\Omega, \quad C = 330\text{pF}$$

Figure 13.47 shows an ESD gun together with an RC cartridge.

13.14.3 Capacitive Coupling and Shielding

When two conductive bodies are in the vicinity of each other, separated by a dielectric, effectively a capacitive structure is created. We often model this effect as the mutual capacitance.

Note: In the following discussion the conducting structures are electrically short and modeled as lumped parameter circuits.

Consider two circuits: generator circuits (conductor 1) and the receptor circuit (conductor 2) shown in Figure 13.48.

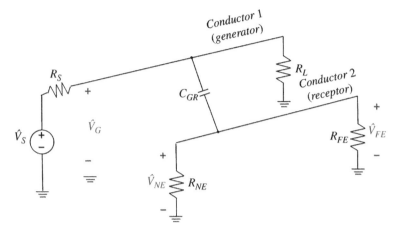

Figure 13.48 Model of a capacitive coupling between the circuits.

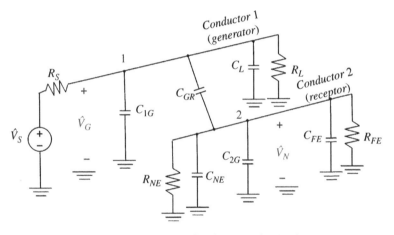

Figure 13.49 Model of a capacitive coupling between the circuits.

C_{GR} represents the mutual capacitance between the two circuits. This mutual capacitance gives rise to a capacitive coupling between these two circuits when the source of the electric field, $\hat{V}_G$ is time-varying. This time-varying source, $\hat{V}_G$, has the potential to induce the near-field noise voltage, $\hat{V}_{NE}$, and the far-field noise voltage, $\hat{V}_{FE}$, in the receptor circuit.

Note: Often in practical circuits, in addition to a capacitive coupling, we have an inductive coupling resulting in the near-end voltage, $\hat{V}_{NE}$, not being equal to the far-end voltage, $\hat{V}_{FE}$. Since in this section we focus on the capacitive coupling only, these two noise voltages will be equal and denoted simply by $\hat{V}_N$.

The more detailed lumped-parameter circuit model of the capacitive coupling is shown in Figure 13.49.

This model is described by:

$\hat{V}_G$ – source of interference (generator circuit)
$\hat{V}_N$ – capacitively induced noise voltage (receptor circuit)

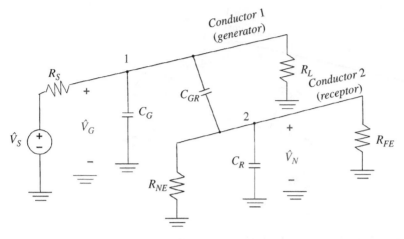

Figure 13.50 Simplified circuit model.

C_{GR} – mutual capacitance between the generator and receptor circuits
C_{1G} – capacitance between the generator circuit and ground
C_L – load capacitance in the generator circuit
C_{2G} – capacitance between the receptor circuit and ground
C_{NE} – near-end load capacitance in the receptor circuit
C_{FE} – far-end load capacitance in the receptor circuit

This circuit model can be simplified to that shown in Figure 13.50.
    In this model:

C_G – total capacitance between the generator circuit and ground
C_R – total capacitance between the receptor circuit and ground

The circuit shown in Figure 13.50 can be represented by that shown in Figure 13.51.
    It is apparent that the capacitance C_G has no effect on the noise voltage, $\hat{V}_N$, and
therefore the circuit in Figure 13.51 can be further simplified to that in Figure 13.52.
    The parallel combination of R and C_R results in an impedance of

$$R \| C_R = \frac{R \dfrac{1}{j\omega C_R}}{R + \dfrac{1}{j\omega C_R}} = \frac{R}{j\omega R C_R + 1} \tag{13.171}$$

The voltage divider produces

$$\hat{V}_N = \frac{\dfrac{R}{j\omega R C_R + 1}}{\dfrac{1}{j\omega C_{GR}} + \dfrac{R}{j\omega R C_R + 1}} \hat{V}_G = \frac{R}{\dfrac{j\omega R C_R + 1}{j\omega C_{GR}} + R} \hat{V}_G$$

$$= \frac{j\omega C_{GR} R}{j\omega R C_R + 1 + j\omega C_{GR} R} \hat{V}_G = \frac{j\omega C_{GR} R}{j\omega R (C_{GR} + C_R) + 1} \hat{V}_G \tag{13.172}$$

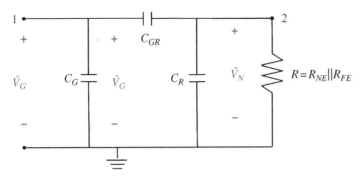

Figure 13.51 Circuit representation.

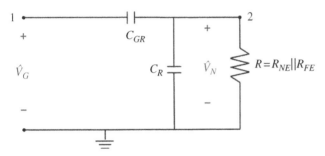

Figure 13.52 Equivalent circuit.

And thus the noise voltage induced in the receptor circuit due to the capacitive coupling is

$$\hat{V}_N = \frac{j\omega \dfrac{C_{GR}}{C_{GR}+C_R}}{j\omega + \dfrac{1}{R(C_{GR}+C_R)}}\hat{V}_G \qquad (13.173)$$

The Bode plot of this voltage is shown in Figure 13.53.
Equation (13.150) can be written as

$$\hat{V}_N = \frac{j\omega \dfrac{C_{GR}}{C_{GR}+C_R}}{j\omega + \dfrac{1}{R(C_{GR}+C_R)}}\hat{V}_G = \frac{j\omega C_{GR}R}{j\omega R(C_{GR}+C_R)+1}\hat{V}_G \qquad (13.174)$$

When

$$j\omega R(C_{GR}+C_R)\ll 1 \quad \Leftrightarrow \quad \omega \ll \frac{1}{R(C_{GR}+C_R)} \qquad (13.175)$$

Equation (13.174) simplifies to

$$\hat{V}_N = j\omega RC_{GR}\hat{V}_G \qquad (13.176)$$

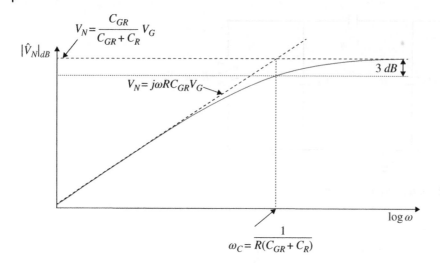

Figure 13.53 Capacitively coupled noise voltage, $\hat{V}_N$.

Inequality (13.175) can be written as

$$R \ll \frac{1}{\omega\left(C_{GR} + C_R\right)} \tag{13.177}$$

In most practical cases this is true (Ott, 2009, p. 46). On the other hand, when

$$j\omega R\left(C_{GR} + C_R\right) \gg 1 \quad \Leftrightarrow \quad \omega \gg \frac{1}{R\left(C_{GR} + C_R\right)} \tag{13.178}$$

Equation (13.174) simplifies to

$$\hat{V}_N = \frac{C_{GR}}{C_{GR} + C_R}\hat{V}_G \tag{13.179}$$

Inequality (13.178) can be also written as

$$R \gg \frac{1}{\omega\left(C_{GR} + C_R\right)} \tag{13.180}$$

Let's return to Eq. (13.176), valid for frequencies much lower than the corner frequency

$$\omega \ll \omega_C = \frac{1}{R\left(C_{GR} + C_R\right)} \tag{13.181}$$

Equation (13.176) can be written as

$$\hat{V}_N = j\omega R C_{GR}\hat{V}_G = R\left(j\omega C_{GR}\hat{V}_G\right) = R\hat{I}_N \tag{13.182}$$

This equation clearly shows why (for electrically small circuits) we model the capacitive coupling as a shunt current source, as shown in Figure 13.54.

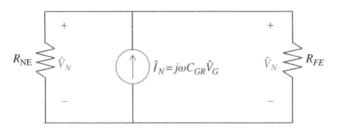

Figure 13.54 Capacitive coupling modeled as a current source.

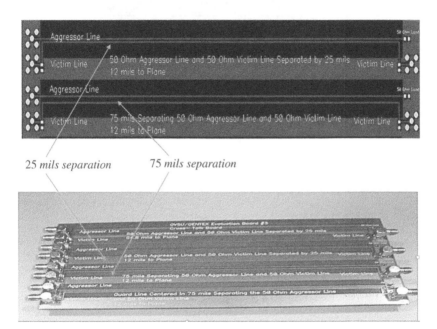

25 *mils separation*   75 *mils separation*

Figure 13.55 Reducing capacitive coupling by moving conductors further apart.

The mutual capacitance, C_{RG}, and the noise voltage, $\hat{V}_N$, can be reduced by moving the conductors further apart, as shown in Figure 13.55.

The mutual capacitance C_{RG} and the noise voltage $\hat{V}_N$ can also be reduced by shielding the receptor circuit. This is shown in Figure 13.56.

Let's investigate the effect of the shield around the receptor circuit on the noise voltage. Figure 13.57 shows the receptor circuit without the shield, while Figure 13.58 shows the receptor circuit with the shield.

Note that the shield is grounded (this makes it effective, as we shall see) and the receptor circuit extends beyond the shield (this corresponds to a practical application).

The model in Figure 13.58 is described by:

C_G – total capacitance between the generator circuit and ground
C_R – total capacitance between the receptor circuit and ground
C_{GR} – mutual capacitance between the generator and receptor circuits

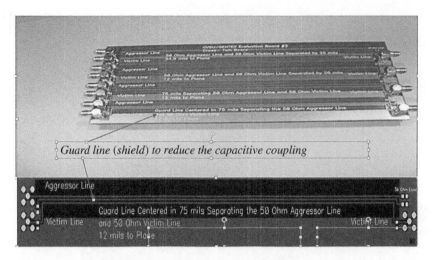

Figure 13.56 Reducing capacitive coupling by shielding the receptor circuit.

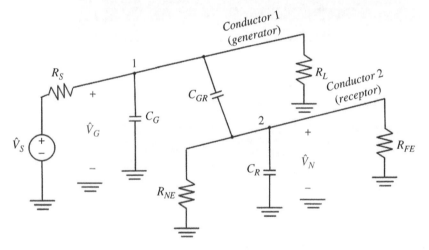

Figure 13.57 Capacitive coupling without a shield around the receptor circuit.

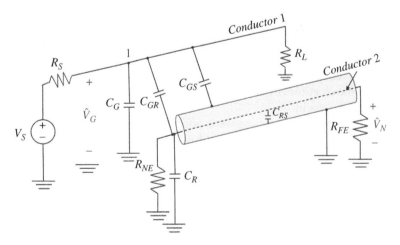

Figure 13.58 Capacitive coupling with a shield around the receptor circuit.

C_{GS} – mutual capacitance between the generator circuit and the shield
C_{RS} – mutual capacitance between the receptor circuit and the shield

The circuit model of the configuration without a shield was presented in Figure 13.52, repeated here as Figure 13.59.

Figure 13.60 shows the circuit model of the configuration shown in Figure 13.58 (with a shield). Note that the capacitance C_{GS} (between the shield and ground) has no effect on the noise voltage, $\hat{V}_N$, (just like the capacitance C_G).

Recall the expressions for the noise voltage without the shield:

$$\hat{V}_N = j\omega RC_{GR}\hat{V}_G, \quad \omega << \frac{1}{R(C_{GR}+C_R)} \tag{13.183a}$$

$$\hat{V}_N = \frac{C_{GR}}{C_{GR}+C_R}\hat{V}_G, \quad \omega >> \frac{1}{R(C_{GR}+C_R)} \tag{13.183b}$$

Comparing the circuits in Figures 13.56 and 13.57 and looking at the Eqs (13.183), we write the expressions for the noise voltage for the case of a shield receptor as

$$\hat{V}_N = j\omega RC_{GR}\hat{V}_G, \quad \omega << \frac{1}{R(C_{GR}+C_R+C_{RS})} \tag{13.184a}$$

$$\hat{V}_N = \frac{C_{GR}}{C_{GR}+C_R+C_{RS}}\hat{V}_G, \quad \omega >> \frac{1}{R(C_{GR}+C_R+C_{RS})} \tag{13.184b}$$

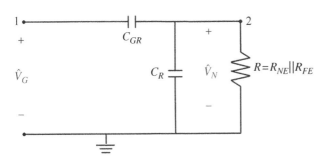

Figure 13.59 Circuit model without a shield.

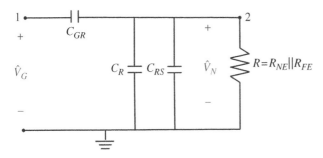

Figure 13.60 Circuit model with a shield.

The expressions for the noise voltage in Eqs (13.183a) and (13.184a) look identical. The difference is that C_{GR} with a shield is much smaller than C_{GR} without a shield.

References

Paul, C.R., *Introduction to Electromagnetic Compatibility*, 2nd ed., John Wiley and Sons, New York, 2006.

Rao, N.N., *Elements of Engineering Electromagnetics*, 6th ed., Pearson Prentice Hall, Upper Saddle River, NJ, 2004.

Sadiku, M.N.O., *Elements of Electromagnetics*, 5th ed., Oxford University Press, New York, 2010.

Ott, H.W., *Electromagnetic Compatibility Engineering*, John Wiley and Sons, Hoboken, NJ, 2009.

14

Static and Quasi-Static Magnetic Fields

In the study of static electric fields, we learned that static distributions of charge led to the definition of the two fundamental vectors:

- electric field intensity **E**
- electric flux density **D**

In the study of static magnetic fields, we will learn that a steady movement of charge (dc current) leads to the definition of the remaining two fundamental vectors:

- magnetic field intensity **H**
- magnetic flux density **B**

Static electric and static magnetic fields, and their corresponding vectors, can be studied independently. In the time-varying case, the fields are no longer independent (hence the name electromagnetic), and all four vectors are involved in the field description.

14.1 Magnetic Flux Density

Recall that the *electric field intensity* **E** at a point in space has been defined in terms of the electric force $\mathbf{F}_e$ acting on a test charge when placed at that point:

$$\mathbf{E} = \frac{\mathbf{F}_e}{q} \tag{14.1}$$

We could refer to Eq. (14.1) as an *explicit* definition of the electric field intensity. Equivalently, Eq. (14.1) could be written as

$$\mathbf{F}_e = q\mathbf{E} \tag{14.2}$$

and we could refer to it as an implicit definition of the electric field intensity.

In a similar manner, we define the *magnetic flux density* **B** at a point in space in terms of the magnetic force $\mathbf{F}_m$ that would be exerted on a charged particle passing with a velocity **u** through that point (Sadiku, 2010, p. 332),

$$\mathbf{F}_m = q\mathbf{u} \times \mathbf{B} \tag{14.3}$$

Foundations of Electromagnetic Compatibility with Practical Applications, First Edition. Bogdan Adamczyk.
© 2017 John Wiley & Sons Ltd. Published 2017 by John Wiley & Sons Ltd.

Equation (14.3) constitutes an implicit definition of the magnetic flux density. The unit of magnetic flux density is Wb/m^2.

14.2   Magnetic Field Intensity

In the previous section we defined the magnetic flux density to denote the presence of a magnetic field in space. We now define the last important EM vector, the *magnetic field intensity* **H** as

$$\mathbf{B} = \mu\mathbf{H} \tag{14.4}$$

where μ. the permeability of the medium. Relationship (14.4) is valid in a linear and isotropic medium.

Equivalently, Eq. (14.4) can be expressed as

$$\mathbf{H} = \frac{\mathbf{B}}{\mu} \quad \left[\frac{A}{m}\right] \tag{14.5}$$

Note that the relationship (14.4) between the **B** and **H** vectors is analogous the one for the **D** and **E** vectors in static electric field:

$$\mathbf{D} = \varepsilon\mathbf{E} \tag{14.6}$$

14.3   Biot–Savart Law

There are two fundamental laws governing magnetostatic fields:

1) Biot–Savart's law
2) Ampere's law

Like Coulomb's law, the Biot–Savart law is the general law of magnetostatics. Just as Gauss's law is a special case of Coulomb's law, Ampere's law is a special case of the Biot–Savart law, and is easily applied in problems involving symmetrical current distributions.

Consider Figure 14.1, where a steady current flows through a thin wire. With this current, we associate a differential current element, $Id\mathbf{l}$. This current element will produce a differential magnetic flux density $d\mathbf{B}$, at an observation point P that is distance R from it.

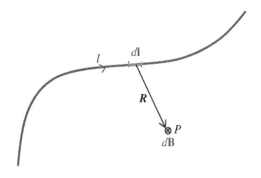

Figure 14.1 Magnetic field due to a current element.

The Biot–Savart law states the differential magnetic field $d\mathbf{B}$ generated by a steady current I flowing through a differential length $d\mathbf{l}$ is given by (Rao, 2004, p. 49)

$$d\mathbf{B} = \frac{\mu_0 I}{4\pi} \frac{d\mathbf{l} \times \mathbf{a}_R}{R^2} \quad \left[\frac{\mathrm{Wb}}{\mathrm{m}^2} = \mathrm{T} \right] \tag{14.7}$$

where R is the distance from the current element and $\mathbf{a}_R$ is the unit vector from the current element to the observation point. The direction of the vector $d\mathbf{B}$ conforms to the right-hand rule (and is into the page in Figure 14.1).

The constant μ_0 is the permeability of free space and is

$$\mu_0 = 4\pi \times 10^{-7} \quad \left[\frac{\mathrm{H}}{\mathrm{m}} \right] \tag{14.8}$$

To determine the total magnetic field $\mathbf{B}$ due to the current-carrying conductor, we need to sum up the contributions due to all the current elements making up the conductor. Hence, the Biot–Savart law becomes

$$\mathbf{B} = \frac{\mu_0}{4\pi} \int_l \frac{I d\mathbf{l} \times \mathbf{a}_R}{R^2} \tag{14.9}$$

14.4 Current Distributions

Electric current can be distributed as a line current, a surface current, or a volume current, as shown in Figure 14.2.

The differential source elements for all three distributions are related by

$$I d\mathbf{l} = K dS = J dv \quad [\mathrm{Am}] \tag{14.10}$$

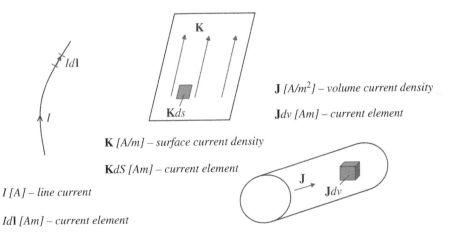

Figure 14.2 Various current distributions.

In terms of the distributed current sources, the Biot–Savart law becomes

$$\mathbf{B} = \frac{\mu_0}{4\pi} \int_S \frac{\mathbf{K} \times \mathbf{a}_R}{R^2} dS \tag{14.11}$$

or

$$\mathbf{B} = \frac{\mu_0}{4\pi} \int_v \frac{\mathbf{J} \times \mathbf{a}_R}{R^2} dv \tag{14.12}$$

Equivalently, the Biot–Savart law can be used to obtain the magnetic field intensity H, for the line, surface, or volume current distributions:

$$\mathbf{H} = \int_l \frac{I d\mathbf{l} \times \mathbf{a}_R}{4\pi R^2} \tag{14.13}$$

$$\mathbf{H} = \int_S \frac{\mathbf{K} \times \mathbf{a}_R}{4\pi R^2} dS \tag{14.14}$$

$$\mathbf{H} = \int_v \frac{\mathbf{J} \times \mathbf{a}_R}{4\pi R^2} dv \tag{14.15}$$

14.5 Ampere's Law

Recall that in electrostatics we could use Coulomb's law to obtain the field due to any charge distribution. The calculations were much easier using Gauss's law when symmetry in the charge distribution was present.

An analogous situation exists in magnetic fields. To obtain the fields due to any current distribution we could use the Biot–Savart's law, but the calculations are much easier using Ampere's law when there is symmetry in the current distribution.

Ampere's law states that the line integral of $\mathbf{H}$ about any closed path is equal the net current enclosed by that path.

$$\oint_l \mathbf{H} \cdot d\mathbf{l} = I_{enc} \tag{14.16}$$

where the enclosed current is often expressed as

$$I_{enc} = \int_S \mathbf{J} \cdot d\mathbf{S} \tag{14.17}$$

Figure 14.3 illustrates the Ampere's law.

Positive current and the direction of the integration path are related by the right-hand rule, as shown in Figure 14.3.

Figure 14.3 Illustration of the Ampere's law.

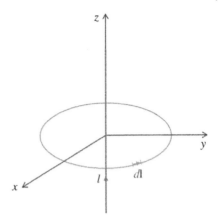

Let's return to Ampere's law in Eq. (14.16). According to the Stokes theorem we have

$$\oint_l \mathbf{H} \cdot d\mathbf{l} = \int_S (\nabla \times \mathbf{H}) \cdot d\mathbf{S} \tag{14.18}$$

Comparing the right-hand sides of Eqs (14.17) and (14.18) we get

$$\int_S \mathbf{J} \cdot d\mathbf{S} = \int_S (\nabla \times \mathbf{H}) \cdot d\mathbf{S} \tag{14.19}$$

leading to

$$\nabla \times \mathbf{H} = \mathbf{J} \tag{14.20}$$

Equations (14.16) and (14.20) constitute another pair of the Maxwell equations.

14.6 Applications of Ampere's Law

In evaluating the line integral in Ampere's law we are free to choose any closed path enclosing the current. When symmetry in the current distribution exists, we choose the path that mirrors the symmetry exhibited by the current distribution.

On such a path, **H** and **B** vectors are either tangential or normal to it, while constant in magnitude. Such a path is called an *Amperian path*. This allows us to write the line integral as

$$\oint_c \mathbf{H} \cdot d\mathbf{l} = \oint_c Hdl = H \oint_c dl \tag{14.21}$$

where the closed integral is simply the length of the contour.

We now apply Ampere's law to determine **H** for some symmetrical current distributions as we did for Gauss's law. We will consider an infinite line current and an infinitely long coaxial line.

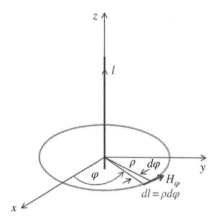

Figure 14.4 Magnetic field due to an infinite line of current.

Example 14.1 Magnetic field of an infinite line of current
Determine the magnetic field due to an infinite line of current, as shown in Figure 14.4.

To determine **H** at an observation point at a distance ρ from the line, we choose an *Amperian path* as shown in Figure 14.4. According to Ampere's law we have

$$I = \oint_c \mathbf{H} \cdot d\mathbf{l} = \oint_c H_\varphi \mathbf{a}_\varphi \cdot \rho d\varphi \mathbf{a}_\varphi = H_\varphi \rho \oint_c d\varphi = 2\pi \rho H_\varphi \tag{14.22}$$

or

$$H\varphi = \frac{I}{2\pi\rho} \tag{14.23}$$

Since

$$\mathbf{H} = H_\phi \mathbf{a}_\phi \tag{14.24}$$

we obtain the magnetic field intensity vector as

$$\mathbf{H} = \frac{I}{2\pi\rho} \mathbf{a}_\varphi \tag{14.25}$$

Example 14.2 Magnetic field around an infinite coaxial line
Determine the magnetic field around an infinite coaxial line shown in Figure 14.5. The inner cylinder carries a current I. The outer cylinder carries the same current oppositely directed.
Ampere's law produces

$$\oint_l \mathbf{H} \cdot d\mathbf{l} = I_{enc} = I + (-I) = 0 \tag{14.26}$$

Figure 14.5 Coaxial line carrying a current *I*.

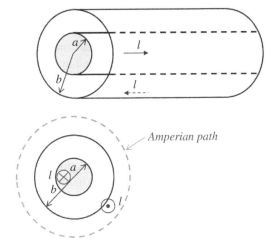

Thus

$$\oint_l \mathbf{H} \cdot d\mathbf{l} = 0 \tag{14.27}$$

for any closed path surrounding the coaxial line. This can only happen when

$$\mathbf{H} = 0, \quad \rho > b \tag{14.28}$$

Thus, *no magnetic field exists outside the (ideal) coaxial cable.*

14.7   Magnetic Flux

The *magnetic flux* through a surface is defined as

$$\Psi = \int_S \mathbf{B} \cdot d\mathbf{S} \quad (Wb) \tag{14.29}$$

The *magnetic flux line*, or *magnetic field line* is the path to which **B** is tangential at every point in a magnetic field.

It is the line along which the needle of a magnetic compass will orient itself if placed in the magnetic field. For example, the magnetic flux lines due to a straight long wire are as shown in Figure 14.6.

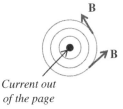

Current out of the page

Figure 14.6 Magnetic field lines.

14.8 Gauss's Law for Magnetic Field

Recall Gauss's law for electric field: the electric flux passing through a closed surface equals the total net charge enclosed by that surface.

$$\oint_S \mathbf{D} \cdot d\mathbf{S} = Q \tag{14.30}$$

The charge Q is the source of the lines of electric flux, and these lines begin and terminate on positive and negative charge, respectively.

There are no known isolated magnetic charges. For that reason, the magnetic flux lines always close form closed paths, and therefore the magnetic flux through a closed surface is zero.

This is Gauss's law for the magnetic field:

$$\oint_S \mathbf{B} \cdot d\mathbf{S} = 0 \tag{14.31}$$

Applying the divergence theorem to Eq. (14.31) we have

$$\oint_S \mathbf{B} \cdot d\mathbf{S} = \int_v \nabla \cdot \mathbf{B} \, dv = 0 \tag{14.32}$$

Equation (14.32) holds for any closed surface, and the volume defined by it. This can only happen when

$$\nabla \cdot \mathbf{B} = 0 \tag{14.33}$$

Equations (14.31) and (14.33) constitute another pair of Maxwell's equations.

14.9 Maxwell's Equations for Static Fields

Let's summarize the four Maxwell equations for static electric and static magnetic fields. Each of the equations can be written in either differential or integral form.

$$\nabla \times \mathbf{E} = 0 \tag{14.34a}$$

$$\oint_l \mathbf{E} \cdot d\mathbf{l} = 0 \tag{14.34b}$$

$$\nabla \times \mathbf{H} = 0 \tag{14.35a}$$

$$\oint_l \mathbf{H} \cdot d\mathbf{l} = I = \int_S \mathbf{J} \cdot d\mathbf{S} \tag{14.35b}$$

$$\nabla \cdot \mathbf{D} = \rho_v \tag{14.36a}$$

$$\oint_S \mathbf{D} \cdot d\mathbf{S} = Q = \int_v \rho_v \, dv \tag{14.36b}$$

$$\nabla \cdot \mathbf{B} = 0 \qquad\qquad (14.37a)$$

$$\oint_S \mathbf{B} \cdot d\mathbf{S} = 0 \qquad\qquad (14.37b)$$

14.10 Vector Magnetic Potential

The concept of a vector magnetic potential is extremely useful in studying radiation from antennas.

To define the vector magnetic potential we use one of the Maxwell's equations

$$\nabla \cdot \mathbf{B} = 0 \qquad\qquad (14.38)$$

and a vector identity

$$\nabla \cdot (\nabla \times \mathbf{A}) = 0 \qquad\qquad (14.39)$$

The vector magnetic potential is defined (implicitly) by

$$\mathbf{B} = \nabla \times \mathbf{A} \qquad\qquad (14.40)$$

Note that this definition is in agreement with (i.e. does not violate) the Maxwell equation (14.38). The following discussion illustrates the usefulness of the concept of vector magnetic potential.

Recall that if the current distribution is known, we can calculate the magnetic field intensity **H** from the Biot–Savart's law as

$$\mathbf{H} = \int_l \frac{I d\mathbf{l} \times \mathbf{a}_R}{4\pi R^2} \qquad\qquad (14.41a)$$

$$\mathbf{H} = \int_S \frac{\mathbf{K} \times \mathbf{a}_R}{4\pi R^2} dS \qquad\qquad (14.41b)$$

$$\mathbf{H} = \int_v \frac{\mathbf{J} \times \mathbf{a}_R}{4\pi R^2} dv \qquad\qquad (14.41c)$$

The integration involved in these calculations is, in general, very difficult because of the cross product and the unit vector calculations.

If the current distribution is known, we can also calculate the vector magnetic potential **A** from

$$\mathbf{A} = \int_l \frac{\mu_0 I d\mathbf{l}}{4\pi R} \qquad\qquad (14.42a)$$

$$\mathbf{A} = \int_S \frac{\mu_0 \mathbf{K} dS}{4\pi R} \qquad\qquad (14.42b)$$

$$A = \int_v \frac{\mu_0 J \, dv}{4 \pi R}$$

(14.42c)

Then we obtain the magnetic flux density from

$$B = \nabla \times A$$

(14.43a)

or the magnetic field intensity from

$$H = \frac{1}{\mu} \nabla \times A$$

(14.43b)

This two-step process of obtaining **B** or **H** is easier than the direct process using Eqs (14.41) because the integrations in Eq. (14.42) are easier to perform than those in Eqs (14.41). The differentiation operation in Eq. (14.43) is well defined and can be easily performed.

14.11 Faraday's Law

Consider an open surface that has a closed loop contour c surrounding it (think of the mouth of a balloon) shown in Figure 14.7. The "balloon" can be inflated or deflated to create different surfaces but the contour c needs to stay unchanged.

This contour can be a conducting wire or an imaginary contour of non-conducting material (free space). Magnetic flux passing through the open surface bounded by this contour gives rise to an electric field.

Faraday's law states that

$$\oint_c E \cdot dl = -\frac{d}{dt} \int_S B \cdot dS$$

(14.44)

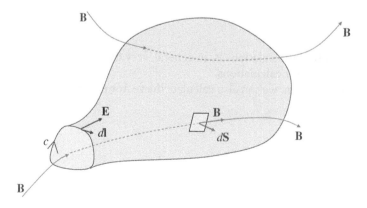

Figure 14.7 Open surface defined by a contour *c*.

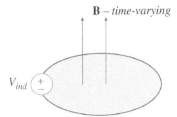

B – *time-varying*

V_{ind}

Figure 14.8 Induced voltage inserted in the loop.

The line integral in Eq. (14.44) is often referred to as an electromotive force

$$V_{emf} = \oint_c \mathbf{E} \cdot d\mathbf{l} \qquad (14.45)$$

The surface integral in Eq. (14.44) is the magnetic flux crossing the contour

$$\Psi = \int_S \mathbf{B} \cdot d\mathbf{S} \qquad (14.46)$$

Using the notation in Eqs (14.45) and (14.46) Faraday's law in Eq. (14.44) can be alternatively expressed as

$$V_{emf} = -\frac{d\Psi}{dt} \qquad (14.47)$$

This form clearly shows that the induced voltage is directly proportional to the rates of change of the magnetic flux. If the loop is electrically small, this induced voltage can be anywhere in the loop as shown in Figure 14.8.

The *magnitude* of this voltage is

$$V_{ind} = \frac{d\Psi}{dt} \qquad (14.48)$$

The *polarity* of this voltage is determined from Lentz's law explained next.

The original magnetic field **B** *gives rise to the induced magnetic field* **B**$_{ind}$*. According to Lentz's law the induced magnetic field* **B**$_{ind}$ *opposes the change in the original magnetic field* **B**.

To facilitate the understating of Lentz's law, let's consider several scenarios shown in Figure 14.9.

As shown in Figure 14.9, the original magnetic field **B** can be either pointing up or down, and can be either increasing or decreasing. Let's investigate each case separately, and apply Lentz's rule to determine the direction of the induced magnetic field.

Case 1 – The original field **B** is pointing up and increasing (Figure 14.10)
The induced field **B**$_{ind}$ opposes this change. Thus, the induced field **B**$_{ind}$ is pointing down.

Case 2 – The original field **B** is pointing up and decreasing (Figure 14.11)
The induced field **B**$_{ind}$ opposes this change. Thus, the induced field **B**$_{ind}$ is pointing up.

Case 3 – The original field **B** is pointing down and increasing (Figure 14.12)

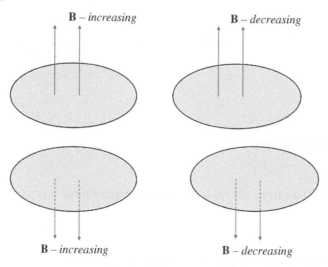

Figure 14.9 Original magnetic field.

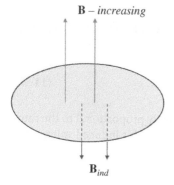

Figure 14.10 **B** field pointing up and increasing.

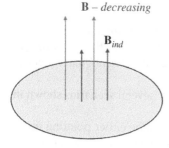

Figure 14.11 **B** field pointing up and decreasing.

The induced field $\mathbf{B}_{ind}$ opposes this change. Thus, the induced field $\mathbf{B}_{ind}$ is pointing up.

Case 4 – The original field **B** is pointing up and decreasing (Figure 14.13)

The induced field $\mathbf{B}_{ind}$ opposes this change. Thus, the induced field $\mathbf{B}_{ind}$ is pointing down.

The knowledge of the direction of the induced field allows us to determine the direction of the induced current (using the right-hand rule). This is shown in Figure 14.14.

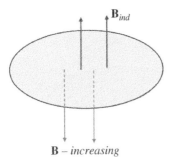

B – *increasing*

Figure 14.12 B field pointing down and increasing.

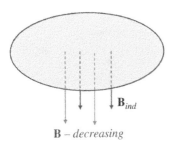

B – *decreasing*

Figure 14.13 B field pointing down and decreasing.

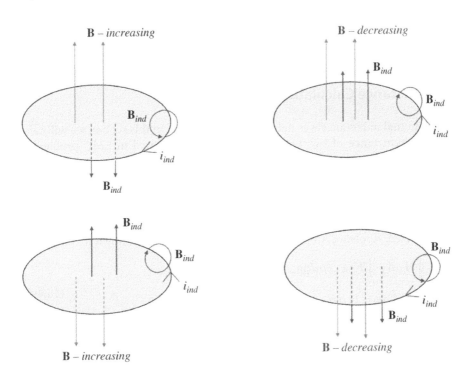

Figure 14.14 Induced current direction.

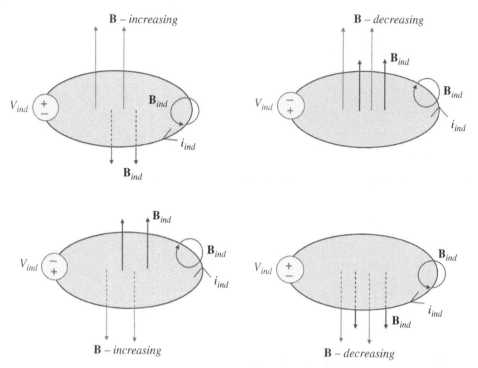

Figure 14.15 Induced voltage polarity.

Since the induced current flows out of the positive terminal of the induced voltage the polarity of the induced voltage is easily determined, as shown in Figure 14.15.

14.12 Inductance Calculations of Structures

The self and mutual inductance was defined in Section 10.1. In this section we will calculate the self inductance of two typical structures encountered in EMC problems: coaxial cable and two parallel wires.

Example 14.3 Inductance of a coaxial cable
Consider the coaxial cable shown in Figure 14.16.

The inner cylinder carries a current I. The outer cylinder carries the same current oppositely directed. Applying Ampere's law for the Amperian path shown we have

$$I_{enc} = \oint_l \mathbf{H} \cdot d\mathbf{l} = \oint_l H_\varphi \mathbf{a}_\varphi \cdot \rho \, d\varphi \mathbf{a}_\varphi$$
$$= H_\varphi \rho \oint_l d\varphi = 2\pi \rho H_\varphi \tag{14.49}$$

and thus

$$H_\varphi = \frac{I}{2\pi\rho}, \quad a \le \rho \le b \tag{14.50}$$

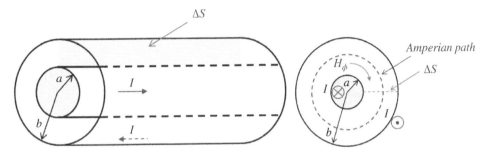

Figure 14.16 Coaxial cable.

The magnetic flux crossing the surface ΔS is

$$
\Psi = \int_S \mathbf{B} \cdot d\mathbf{S} = \int\limits_{\rho=az=0}^{\rho=b}\int\limits_{0}^{l} \frac{\mu I}{2\pi\rho} \mathbf{a}_\varphi \cdot d\rho dz \mathbf{a}_\varphi
$$

$$
= \frac{\mu I l}{2\pi} \int\limits_{\rho=a}^{\rho=b} \frac{1}{\rho} = \frac{\mu I l}{2\pi} \ln\frac{b}{a}
$$

(14.51)

and the inductance of the length l of the cable is

$$
L = \frac{\Psi}{I} = \frac{\mu l}{2\pi} \ln\frac{b}{a} \quad (H)
$$

(14.52)

while the inductance per unit length is

$$
\frac{L}{l} = \frac{\mu}{2\pi} \ln\frac{b}{a} \quad \left(\frac{H}{m}\right)
$$

(14.53)

Example 14.4 Inductance of two parallel wires
Consider the two parallel wires shown in Figure 14.17.
The magnetic field generated by the current in each wire is

$$
H_\varphi = \frac{I}{2\pi\rho}, \quad a \le \rho \le s - a
$$

(14.54)

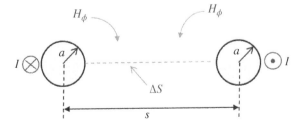

Figure 14.17 Two parallel wires.

The magnetic flux due to both wires is

$$\Psi = 2\int_S \mathbf{B}_\varphi \cdot d\mathbf{S} = 2 \int_{\rho=a}^{\rho=s-a} \int_{z=0}^{l} \frac{\mu I}{2\pi\rho} \mathbf{a}_\varphi \cdot d\rho dz \mathbf{a}_\varphi$$

$$= \frac{\mu Il}{\pi} \int_{\rho=a}^{\rho=s-a} \frac{1}{\rho} = \frac{\mu Il}{\pi} \ln \frac{s-a}{a} \cong \frac{\mu Il}{\pi} \ln \frac{s}{a} \qquad (14.55)$$

The inductance of the section of length l is

$$L = \frac{\Psi}{I} = \frac{\mu l}{\pi} \ln \frac{s}{a}, \quad \frac{s}{a} > 5 \quad [\text{H}] \qquad (14.56)$$

while the inductance per unit length is

$$\frac{L}{l} = \frac{\mu_0}{\pi} \ln \frac{s}{a}, \quad \frac{s}{a} > 5 \quad \left[\frac{\text{H}}{\text{m}}\right] \qquad (14.57)$$

14.13 Magnetic Boundary Conditions

If the magnetic field exists in a region consisting of two different media, even though it may be continuous in each medium, it may be discontinuous at the boundary between them, as illustrated in Figure 14.18.

Boundary conditions specify how the tangential and normal components of the field in one medium are related to the components of the field across the boundary in another medium.

We will derive a general set of boundary conditions, applicable at the interface between any two dissimilar media, be they two different dielectrics, or a conductor and a dielectric.

Even though these boundary conditions will be derived for electrostatic conditions, they will be equally valid for time-varying electromagnetic fields.

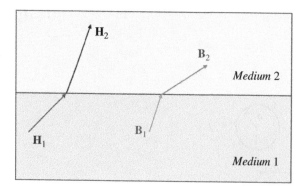

Figure 14.18 Discontinuity at the boundary between two media.

In each medium we will decompose the magnetic field intensity **H** and magnetic flux density **B** into two orthogonal components:

$$\mathbf{H} = \mathbf{H}_t + \mathbf{H}_n \tag{14.58a}$$

$$\mathbf{B} = \mathbf{B}_t + \mathbf{B}_n \tag{14.58b}$$

This is shown in Figure 14.19.

To determine the boundary conditions, we will use Maxwell's equations for magneto-static fields:

$$\oint_L \mathbf{H} \cdot d\mathbf{l} = I \tag{14.59}$$

$$\oint_S \mathbf{B} \cdot d\mathbf{S} = 0 \tag{14.60}$$

Let's consider the closed path *abcd* shown in Figure 14.20. We will apply Eq. (14.59) along this closed path. First, we will break the closed-loop integral in Eq. (14.59) into the integrals along the individual segments:

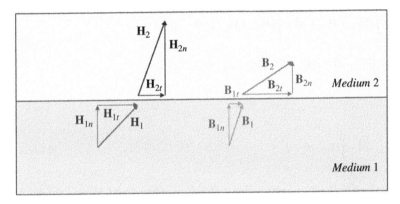

Figure 14.19 Decomposition into the normal and tangential components.

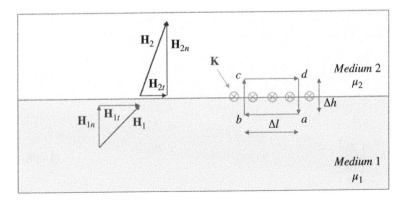

Figure 14.20 Evaluating boundary conditions.

$$I = \oint_C \mathbf{H} \cdot d\mathbf{l} = \int_a^b \mathbf{H} \cdot d\mathbf{l} + \int_b^c \mathbf{H} \cdot d\mathbf{l} + \int_c^d \mathbf{H} \cdot d\mathbf{l} + \int_d^a \mathbf{H} \cdot d\mathbf{l} \qquad (14.61)$$

Note that the integrals in Eq. (14.41) hold for any length of the integration path. That is, we can let any segment length go to zero and the right hand-side of Eq. (14.61) will still be true. If we let $\Delta h \to 0$ then the contributions to the line integral by the segments bc and da go to zero and we have

$$I = \oint_C \mathbf{H} \cdot d\mathbf{l} = \int_a^b \mathbf{H}_1 \cdot d\mathbf{l} + \int_c^d \mathbf{H}_2 \cdot d\mathbf{l} \qquad (14.62)$$

Now, since

$$\mathbf{H}_1 = \mathbf{H}_{1t} + \mathbf{H}_{1n} \qquad (14.63a)$$

$$\mathbf{H}_2 = \mathbf{H}_{2t} + \mathbf{H}_{2n} \qquad (14.63b)$$

we can rewrite Eq. (14.62) as

$$\begin{aligned}
I = \oint_C \mathbf{H} \cdot d\mathbf{l} &= \int_a^b (\mathbf{H}_{1t} + \mathbf{H}_{1n}) \cdot d\mathbf{l} + \int_c^d (\mathbf{H}_{2t} + \mathbf{H}_{2n}) \cdot d\mathbf{l} \\
&= \int_a^b \mathbf{H}_{1t} \cdot d\mathbf{l} + \int_a^b \mathbf{H}_{1n} \cdot d\mathbf{l} + \int_c^d \mathbf{H}_{2t} \cdot d\mathbf{l} + \int_c^d \mathbf{H}_{2n} \cdot d\mathbf{l}
\end{aligned} \qquad (14.64)$$

or

$$\oint_C \mathbf{H} \cdot d\mathbf{l} = \int_a^b \mathbf{H}_{1t} \cdot d\mathbf{l} + \int_c^d \mathbf{H}_{2t} \cdot d\mathbf{l} = I \qquad (14.65)$$

As $\Delta h \to 0$, the surface of the loop approaches a thin line of length Δl. Hence, the total current I flowing through this line is

$$I = K\Delta l \qquad (14.66)$$

Therefore, Eq. (14.65) becomes

$$K\Delta l = \int_a^b \mathbf{H}_{1t} \cdot d\mathbf{l} + \int_c^d \mathbf{H}_{2t} \cdot d\mathbf{l} = -H_{1t}\Delta l + H_{2t}\Delta l \qquad (14.67)$$

or

$$H_{2t} - H_{1t} = K \quad \left[\frac{A}{m}\right] \qquad (14.68)$$

Thus the tangential component of the magnetic field intensity is discontinuous across the boundary between two media.

Since

$$H_{1t} = \frac{B_{1t}}{\mu_1} \tag{14.69a}$$

$$H_{2t} = \frac{B_{2t}}{\mu_2} \tag{14.69b}$$

we obtain the boundary condition on the magnetic flux density as

$$\frac{B_{2t}}{\mu_2} - \frac{B_{1t}}{\mu_1} = K \tag{14.70}$$

When $K = 0$, i.e. the boundary is free of current or the media are not conductors (for K is the free current density), we have

$$\mathbf{H}_{1t} = \mathbf{H}_{2t} \tag{14.71}$$

$$\frac{\mathbf{B}_{1t}}{\mu_1} = \frac{\mathbf{B}_{2t}}{\mu_2} \tag{14.72}$$

This is a very important result that we will utilize when discussing the electromagnetic wave shielding.

To obtain the boundary conditions on the normal components let's consider the closed cylindrical surface shown in Figure 14.21.

Let's apply the second of the two Maxwell's equations

$$\oint_S \mathbf{B} \cdot d\mathbf{S} = 0 \tag{14.73}$$

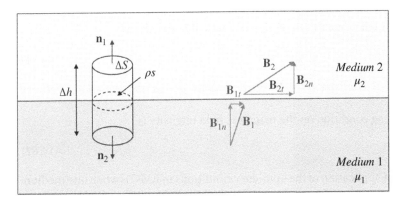

Figure 14.21 Evaluating boundary conditions.

First, we will break the closed surface integral into three integrals as

$$\oint_S \mathbf{B} \cdot d\mathbf{S} = \int_{side} \mathbf{B} \cdot d\mathbf{S} + \int_{bottom} \mathbf{B}_1 \cdot d\mathbf{S} + \int_{top} \mathbf{B}_2 \cdot d\mathbf{S} \tag{14.74}$$

By letting $\Delta h \to 0$, the contributions to the total flux by the side surface goes to zero.

$$\oint_S \mathbf{B} \cdot d\mathbf{S} = \int_{bottom} \mathbf{B}_1 \cdot d\mathbf{S} + \int_{top} \mathbf{B}_2 \cdot d\mathbf{S} \tag{14.75}$$

Now, since

$$\mathbf{B}_1 = \mathbf{B}_{1t} + \mathbf{B}_{1n} \tag{14.76a}$$

$$\mathbf{B}_2 = \mathbf{B}_{2t} + \mathbf{B}_{2n} \tag{14.76b}$$

we rewrite Eq. (14.75) as

$$\oint_S \mathbf{B} \cdot d\mathbf{S} = \int_{bottom} (\mathbf{B}_{1t} + \mathbf{B}_{1n}) \cdot d\mathbf{S}\mathbf{n}_2 + \int_{top} (\mathbf{B}_{2t} + \mathbf{B}_{2n}) \cdot d\mathbf{S}\mathbf{n}_1$$
$$= \int_{bottom} \mathbf{B}_{1n} \cdot d\mathbf{S}\mathbf{n}_2 + \int_{top} \mathbf{B}_{2n} \cdot d\mathbf{S}\mathbf{n}_1 \tag{14.77}$$

or

$$\oint_S \mathbf{B} \cdot d\mathbf{S} = \int_{bottom} \mathbf{B}_{1n} \cdot d\mathbf{S}\mathbf{n}_2 + \int_{top} \mathbf{B}_{2n} \cdot d\mathbf{S}\mathbf{n}_1 \tag{14.78}$$

leading to

$$\mathbf{B}_1 \cdot \mathbf{n}_2 \Delta S + \mathbf{B}_2 \cdot \mathbf{n}_1 \Delta S = 0 \tag{14.79}$$

Since

$$\mathbf{n}_1 = -\mathbf{n}_2 \tag{14.80}$$

we arrive at the boundary condition on the magnetic flux density as

$$\mathbf{B}_{1n} = \mathbf{B}_{2n} \tag{14.81}$$

Thus, *the normal component of* **B** *is continuous across the boundary between two media.*

The corresponding condition on the magnetic field intensity is

$$\mu_1 \mathbf{H}_{1n} = \mu_2 \mathbf{H}_{2n} \tag{14.82}$$

A very important application of the boundary conditions in EMC is when one medium is a dielectric and the other is a conductor. The boundary conditions for this case are summarized in Figure 14.22 (Paul, 2006, p. 905).

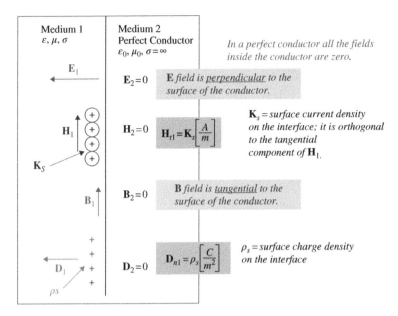

Figure 14.22 Dielectric-conductor boundary.

14.14 EMC Applications

14.14.1 Current Probes

Figure 14.23 shows some typical current probes used in EMC measurements and testing.

Electric current can be measured by connecting a current probe directly to a spectrum analyzer, as shown in Figure 14.24, or by using a preamplifier as shown in Figure 14.25.

A current probe is essentially a transformer, as shown in Figure 14.26. When the probe is clamped around a conductor, the conductor is the primary winding and the probe's windings are the secondary. The current in the conductor produces a magnetic field that is concentrated in, and circulates around, the core of the probe. By Faraday's law, this circulating magnetic field induces V_{ind}, which is measured by a spectrum analyzer.

The probe is calibrated so that the voltage measurement by the probe, V_{ind}, can be translated into the current measurement flowing in the conductor (over the specified frequency range). Typically, the probe's output voltage is specified with the probe terminated in $\hat{Z}_{in} = 50\,\Omega$, as shown in Figure 14.27.

During the calibration process the current of known magnitude and frequency is passed through the probe and the corresponding induced voltage is measured at that frequency.

Then, at each frequency, the ratio of that voltage to current can be calculated:

$$\hat{Z}_T = \frac{\hat{V}}{\hat{I}} \ [\Omega] \tag{14.83}$$

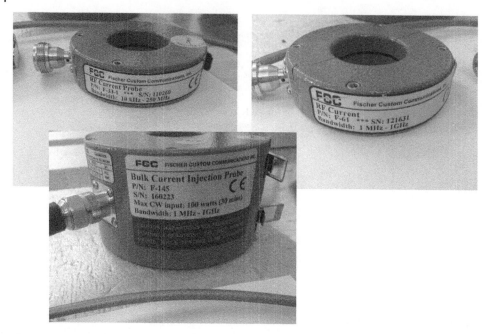

Figure 14.23 Current probes used in EMC.

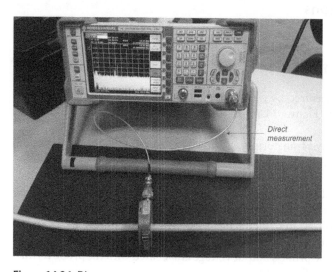

Figure 14.24 Direct current measurement.

This quantity is referred to as the *transfer impedance* of the probe. The unknown current measured by the current probe can then calculated from

$$\hat{I} = \frac{\hat{V}_{probe}}{\hat{Z}_T} \quad [\text{A}]$$

(14.84)

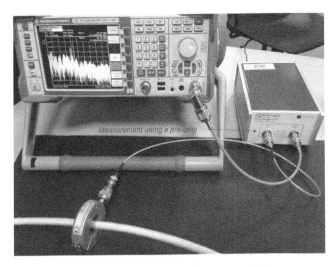

Figure 14.25 Current measurement using a preamplifier.

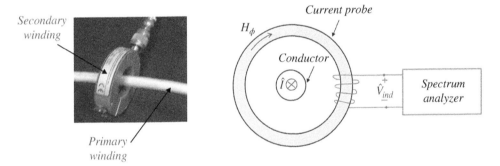

Figure 14.26 Current probe is a transformer.

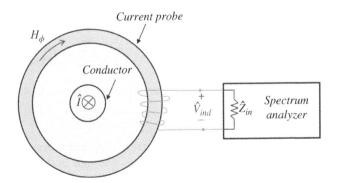

Figure 14.27 Current probe terminated in 50 Ω.

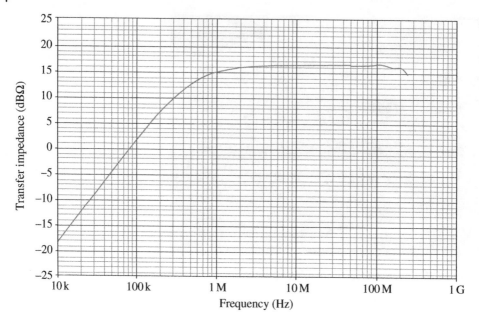

Figure 14.28 Current probe calibration chart.

The transfer impedance is specified in dBΩ instead of the values in Ω.

$$\left|\hat{Z}_T\right| = 20\log\left|\frac{\hat{V}}{\hat{I}}\right| \quad [\text{dB}\Omega] \tag{14.85}$$

Therefore,

$$\left|\hat{Z}_T\right|_{\text{dB}\Omega} = \left|\hat{V}\right|_{\text{dB}\mu\text{V}} - \left|\hat{I}\right|_{\text{dB}\mu\text{A}} \tag{14.86}$$

This allows a direct determination of the unknown current measured by the current probe by a simple subtraction (instead of division):

$$20\log\left|\hat{I}\right| = 20\log\left|\hat{V}_{probe}\right| - 20\log\left|\hat{Z}_T\right| \tag{14.87}$$

and

$$I_{\text{dB}\mu\text{A}} = V_{probe,\text{dB}\mu\text{V}} - Z_{T,\text{dB}\Omega} \tag{14.88}$$

Current probes have an associated calibration chart like the one shown in Figure 14.28.

14.14.2 Magnetic Flux and Decoupling Capacitors

Recall from Section 4.6.3 that when a CMOS gate switches, a current transient is drawn from the power distribution system. This current transient flows through both the power and ground traces. Both of these traces possess (partial) inductance, as shown in Figure 14.29, for a low-to-high transition, and in Figure 14.30 for a high-to-low transition.

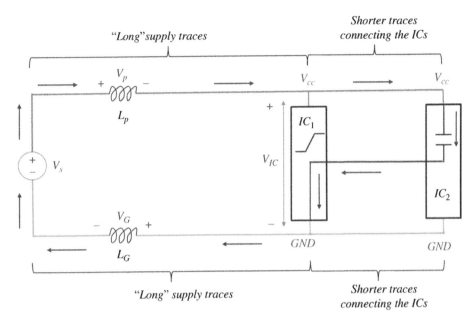

Figure 14.29 CMOS transition from low-to-high.

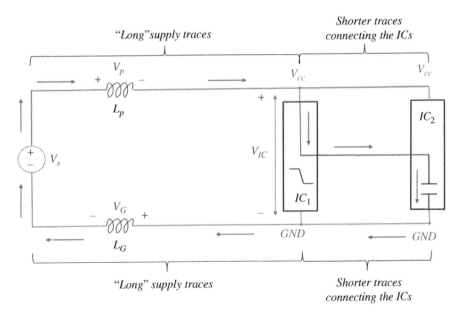

Figure 14.30 CMOS transition from high-to-low.

During such transitions, in addition to causing the signal integrity issues (power rail collapse and ground bounce), the transient current flows in a large loop, resulting in a large magnetic flux crossing that loop (and creating an efficient loop antenna). This is shown in Figure 14.31.

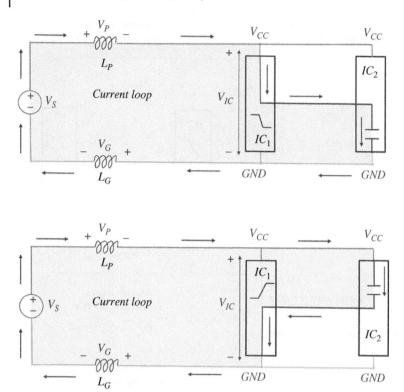

Figure 14.31 Large current loop on CMOS transitions.

One method of reducing this current loop and lowering the flux crossing it is to provide capacitance (bulk decoupling capacitor C_b) between the power and ground conductors near the switching IC. This is shown in Figure 14.32.

The smaller current loop creates a smaller area for the magnetic flux to penetrate, as shown in Figure 14.33.

To further reduce the current loop, we can add a local decoupling capacitor, as shown in Figure 14.34.

This results in the smallest current loop as shown in Figure 14.35.

14.14.3 Magnetic Coupling and Shielding

Consider the two circuits shown in Figure 14.36.

Time-varying current $i_G(t)$ flowing in the generator circuit gives rise to the time-varying magnetic field $\mathbf{H}_G$, which in turn creates the time-varying flux Ψ_{GR} that crosses the adjacent receptor circuit.

According to the Faraday's law this time-varying flux induces a voltage in the receptor circuit. The magnitude of this voltage is

$$v_{ind} = \frac{d\Psi_{GR}}{dt} \tag{14.89}$$

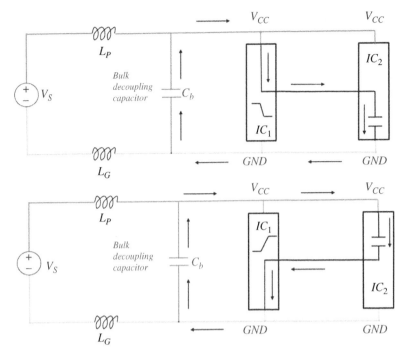

Figure 14.32 Bulk decoupling capacitor effect on the current flow on CMOS transitions.

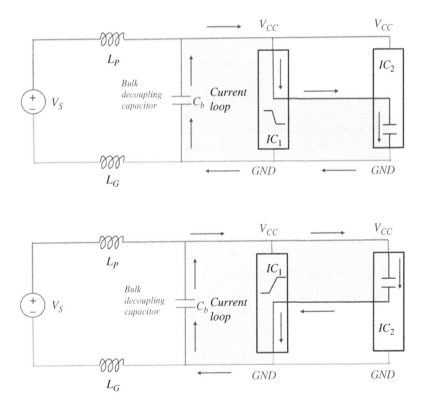

Figure 14.33 Bulk decoupling capacitor effect on the current loop on CMOS transitions.

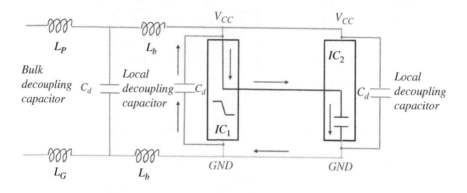

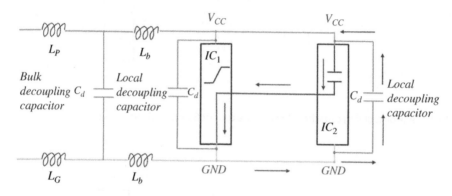

Figure 14.34 Local decoupling capacitor effect on the current flow on CMOS transitions.

The polarity of this voltage is governed by Lentz's law. This induced voltage is shown in Figure 14.37.

To obtain the circuit model of this induced voltage, we introduce the concept of the *mutual inductance* between the generator and the receptor circuit, defined as

$$L_{GR} = \frac{\Psi_{GR}}{i_G} \tag{14.90}$$

From Eq. (14.90) we obtain

$$\Psi_{GR} = L_{GR} i_G \tag{14.91}$$

Substituting Eq. (14.91) into Eq. (14.89) we get

$$v_{ind} = \frac{d\Psi_{GR}}{dt} = \frac{d}{dt}\left(L_{GR} i_G\right) = L_{GR}\frac{di_G}{dt} \tag{14.92}$$

or

$$v_{ind}(t) = L_{GR}\frac{di_G(t)}{dt} \tag{14.93}$$

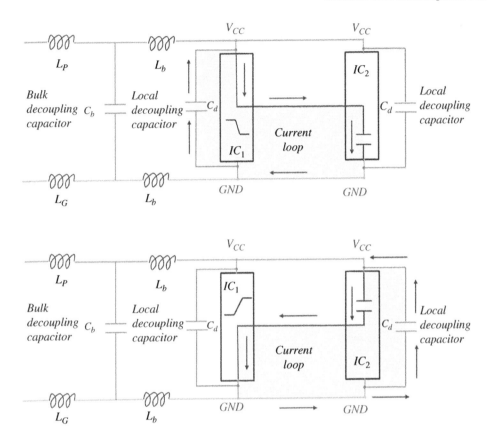

Figure 14.35 Local decoupling capacitor effect on the current loop on CMOS transitions.

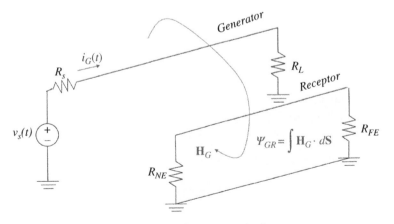

Figure 14.36 Magnetic flux crossing the receptor circuit.

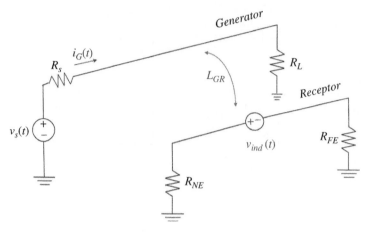

Figure 14.37 Induced voltage in the receptor circuit.

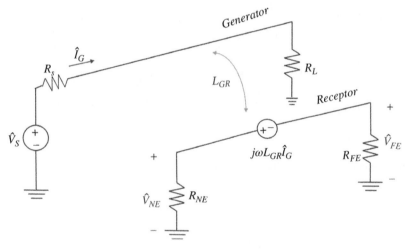

Figure 14.38 Frequency domain circuit model.

When the source voltage driving the generator circuit is sinusoidal, the generator current and the induced voltage in the receptor circuit are also sinusoidal. The circuit analysis is then carried out in the sinusoidal steady-state in the frequency domain, where the time functions are replaced by the corresponding phasors.

$$v_{ind}(t) \leftrightarrow \hat{V}_{ind}$$
$$i_G(t) \leftrightarrow \hat{I}_G \qquad (14.94)$$
$$\frac{di_G(t)}{dt} \leftrightarrow j\omega\hat{I}_G$$

Thus, the frequency (phasor) domain induced voltage in the receptor circuit is

$$\hat{V}_{ind} = j\omega L_{GR}\hat{I}_G \qquad (14.95)$$

This voltage is shown in Figure 14.38.

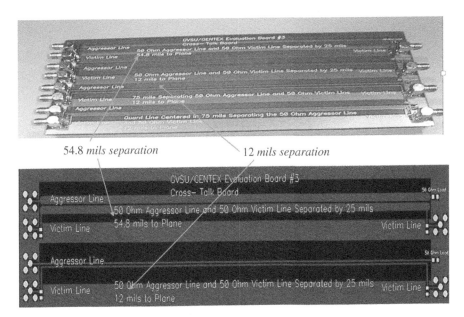

Figure 14.39 Reducing inductive coupling by reducing the area of the receptor circuit.

This induced voltage will have the effect of creating the near-end, $\hat{V}_{NE}$, and far-end, $\hat{V}_{FE}$, noise voltages in the receptor circuit. Applying the voltage divider rule we get

$$\hat{V}_{NE} = \frac{R_{NE}}{R_{NE} + R_{FE}} j\omega L_{GR} \hat{I}_G \qquad (14.96a)$$

$$\hat{V}_{FE} = -\frac{R_{FE}}{R_{NE} + R_{FE}} j\omega L_{GR} \hat{I}_G \qquad (14.96b)$$

The mutual inductance, L_{RG}, and the noise voltage, $\hat{V}_N$, can be lowered by reducing the area of the receptor circuit. In a multilayer PCB, this can be accomplished by moving the ground plane (return path) closer to the signal plane. This is shown in Figure 14.39.

The mutual inductance, L_{RG}, and the noise voltage, $\hat{V}_N$, can also be reduced by shielding the receptor circuit. Let's investigate the effect of the shield around the receptor circuit on the noise voltage. Figure 14.40 shows the receptor circuit with a non-magnetic shield placed around it.

Note that the shield is grounded at both ends (which makes it effective, as we shall see).

Since the shield is non-magnetic, it has no effect on the magnetic properties of the medium between the generator and receptor circuit. Thus, the magnetic flux produced by the current, $\hat{I}_G$, in the generator wire still crosses the receptor-ground circuit and induces a noise voltage in the receptor circuit. This is modeled by the mutual inductance L_{GR}.

Magnetic flux produced by the current, $\hat{I}_G$, in the generator wire also crosses the shield-ground circuit. This flux induces a voltage in the shield circuit that produces a shield current $\hat{I}_S$. This is modeled by the mutual inductance L_{GS}.

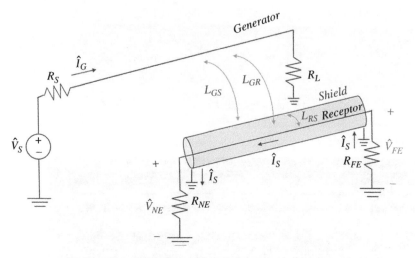

Figure 14.40 Inductive coupling with a shield around the receptor circuit.

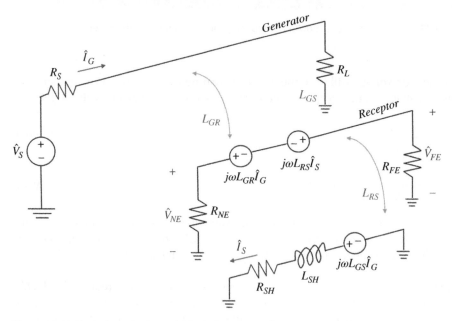

Figure 14.41 Equivalent circuit model for inductive coupling.

The shield current, $\hat{I}_S$, in turn produces a flux that tends to cancel the flux due to the generator wire. This is the essence of the reduction of the magnetic or inductive coupling.

The flux generated by the shield current also crosses the shield-receptor circuit and induces another noise voltage into the receptor circuit. This is modeled by the mutual inductance L_{RS}.

Thus, there are two voltages induced in the receptor circuit, and they are of the opposite polarities, as shown in Figure 14.41.

With the shield present, the near-end, $\hat{V}_{NE}$, and far-end, $\hat{V}_{FE}$, noise voltages in the receptor circuit are

$$\hat{V}_{NE} = \frac{R_{NE}}{R_{NE} + R_{FE}} j\omega \left(L_{GR}\hat{I}_G - L_{RS}\hat{I}_S \right) \tag{14.97a}$$

$$\hat{V}_{FE} = -\frac{R_{FE}}{R_{NE} + R_{FE}} j\omega \left(L_{GR}\hat{I}_G - L_{RS}\hat{I}_S \right) \tag{14.97b}$$

The shield is modeled by the shield resistance, R_{SH}, and shield self inductance, L_{SH}. The shield current equals

$$\hat{I}_S = \frac{j\omega L_{GS}\hat{I}_G}{R_{SH} + j\omega L_{SH}} \tag{14.98}$$

Also, the self inductance of the shield is equal to the mutual inductance between the receptor and the shield (Ott, 2009, p. 59)

$$L_{SH} = L_{RS} \tag{14.99}$$

Also, the mutual inductance between the generator and receptor circuit is equal to the mutual inductance between the generator and the shield (Ott, 2009, p. 61; Paul, 226, p. 655)

$$L_{GR} = L_{GS} \tag{14.100}$$

Substituting Eq. (14.98) into Eqs (14.97) and utilizing Eqs (14.99) and (14.100) we obtain

$$\hat{V}_{NE} = \frac{R_{NE}}{R_{NE} + R_{FE}} j\omega L_{GR}\hat{I}_G \underbrace{\frac{R_{SH}}{R_{SH} + j\omega L_{SH}}}_{\text{effect of shield}} \tag{14.101a}$$

$$\hat{V}_{FE} = -\frac{R_{FE}}{R_{NE} + R_{FE}} j\omega L_{GR}\hat{I}_G \underbrace{\frac{R_{SH}}{R_{SH} + j\omega L_{SH}}}_{\text{effect of shield}} \tag{14.101b}$$

For electrically short lines

$$\hat{I}_G = \frac{\hat{V}_S}{R_S + R_L} \tag{14.102}$$

Utilizing Eq. (14.102) in Eqs (14.101) we get

$$\hat{V}_{NE} = j\omega \left\{ \frac{R_{NE}}{R_{NE} + R_{FE}} \frac{L_{GR}}{R_S + R_L} \underbrace{\frac{R_{SH}}{R_{SH} + j\omega L_{SH}}}_{\text{effect of shield}} \right\} \hat{V}_S \tag{14.103a}$$

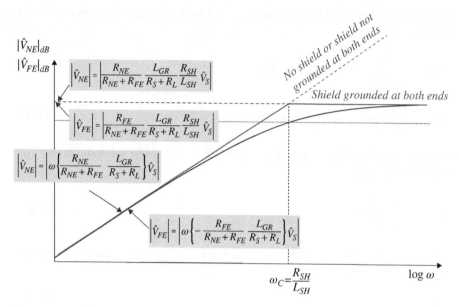

Figure 14.42 Effect of the shield.

$$\hat{V}_{FE} = j\omega \left\{ -\frac{R_{FE}}{R_{NE} + R_{FE}} \frac{L_{GR}}{R_S + R_L} \underbrace{\frac{R_{SH}}{R_{SH} + j\omega L_{SH}}}_{effect\ of\ shield} \right\} \hat{V}_S \tag{14.103b}$$

The effect of the shield is reflected in the shielding factor (SF):

$$SF = \frac{R_{SH}}{R_{SH} + j\omega L_{SH}} \tag{14.104}$$

or

$$SF = \frac{1}{1 + j\dfrac{\omega}{\omega_C}} \tag{14.105}$$

where

$$\omega_C = \frac{R_{SH}}{L_{SH}} \tag{14.106}$$

is the *shield cut-off (break) frequency.*

For frequencies much less than the shield break frequency, the shielding factor can be approximated as

$$SF = \left(\cfrac{1}{1 + j\cfrac{\omega}{\omega_C}} \right)_{\omega \ll \omega_C} \cong 1 \tag{14.107a}$$

while for the frequencies much greater than the shield break frequency it can be approximated as

$$SF = \left(\cfrac{1}{1 + j\cfrac{\omega}{\omega_C}} \right)_{\omega \gg \omega_C} \cong \left(\cfrac{1}{j\cfrac{\omega}{\omega_C}} \right) = \frac{\omega_C}{j\omega} = \frac{R_{SH}}{j\omega L_{SH}} \tag{14.107b}$$

These results are shown in Figure 14.42.

References

Paul, C.R., *Introduction to Electromagnetic Compatibility*, 2nd ed., John Wiley and Sons, New York, 2006.

Rao, N.N., *Elements of Engineering Electromagnetics*, 6th ed., Pearson Prentice Hall, Upper Saddle River, NJ, 2004.

Sadiku, M.N.O., *Elements of Electromagnetics*, 5th ed., Oxford University Press, New York, 2010.

Ott, H.W., *Electromagnetic Compatibility Engineering*, John Wiley and Sons, Hoboken, NJ, 2009.

15

Rapidly Varying Electromagnetic Fields

In the previous chapter we reviewed the static electric fields due to stationary charge distributions, and static magnetic fields due to charges moving at constant speed, i.e. dc currents.

With one exception, the two static fields are independent of each other, allowing us to study them separately. The only time the static fields are linked is in a lossy medium, where the current density J and the electric field intensity E are related by the conductivity of the medium as

$$J = \sigma E \tag{15.1}$$

The E field produces the current density J, which in turn creates the magnetic field.

When the charge distributions and currents vary with time the electric and magnetic fields will also vary with time. When the resulting fields are quasi-static (slowly varying) we can study them separately. When the electric and magnetic fields vary rapidly with time they become coupled – the time-varying electric fields produce the time-varying magnetic fields, and conversely, the time-varying magnetic fields produce the time-varying electric fields.

This field coupling is the key factor in the study of the electromagnetic waves, transmission lines, and antennas, which is the subject of the next three chapters.

15.1 Eddy Currents

In this section we will discuss volume and surface induced electric currents in solid conducting bodies, when exposed to time-varying magnetic field (flux).

These currents (called *eddy currents*) are induced in conductors by a changing magnetic field, due to Faraday's law of induction. Eddy currents flow in closed loops within conductors, in planes perpendicular to the magnetic field that induced them.

The eddy current density, J_{eddy}, is related to the induced electric field intensity E_{ind} by

$$J_{eddy} = \sigma E_{ind} \tag{15.2}$$

where σ is the conductivity of the material.

Figure 15.1 shows the original induced fields and current densities on the surface of a conducting body.

Foundations of Electromagnetic Compatibility with Practical Applications, First Edition. Bogdan Adamczyk.
© 2017 John Wiley & Sons Ltd. Published 2017 by John Wiley & Sons Ltd.

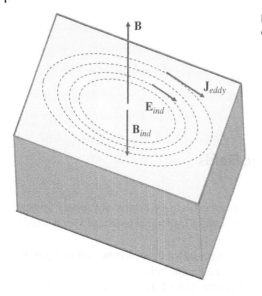

Figure 15.1 Eddy currents on the surface of a conducting body.

As a consequence of eddy currents, electric power is dissipated as heat in a conducting medium due to the resistance of the medium; this is the principle of induction heating.

Another important consequence of eddy currents is the magnetic field that they produce. By Lentz's law, an eddy current creates an induced magnetic field that opposes the change in the magnetic field that created it.

While the induced magnetic field in thin wire circuits is practically always negligible with respect to the original magnetic field, this often is not the case in solid volume conductors.

The induced eddy current density is largest on the surface of the conducting body and decreases exponentially due to the skin effect for harmonically varying magnetic fields.

As we shall see, eddy currents play an important role in magnetic field shielding.

15.2 Charge-Current Continuity Equation

In this section we consider one of the fundamental principles of electromagnetics – the charge-current continuity equation, which is the mathematical expression of the principle of conservation of charge.

For an arbitrary surface S, the *total current* flowing through it can be **defined** as the *flux of the volume current density vector* $\mathbf{J}$ (in units of A/m^2), through the surface S

$$I = \int_S \mathbf{J} \cdot d\mathbf{S} \tag{15.3}$$

For a closed surface S, the *outward flux* of the current density becomes

$$I = \oint_S \mathbf{J} \cdot d\mathbf{S} \tag{15.4}$$

which represents the total outward flow of current. Now, current is also defined as the flow of charge. Thus, *the net current I flowing across S out of v is equal to the negative rate of change of Q*:

$$I = -\frac{dQ}{dt} = -\frac{d}{dt}\int_v \rho_v dv \tag{15.5}$$

where ρ_v is the volume charge density in v. Assuming the volume and the surface to be stationary in time, we obtain

$$\oint_S \mathbf{J} \cdot d\mathbf{S} = -\frac{d}{dt}\int_v \rho_v dv \tag{15.6}$$

By applying the *divergence theorem*, we can convert the surface integral of $\mathbf{J}$ into a volume integral of its divergence, which then gives

$$\oint_S \mathbf{J} \cdot d\mathbf{S} = \int_v \nabla \cdot \mathbf{J} dv = -\frac{d}{dt}\int_v \rho_v dv \tag{15.7}$$

For a stationary volume v, the time derivative operates on ρ_v only. Hence, we can move it inside the integral and express it as a partial derivative of ρ_v:

$$\int_v \nabla \cdot \mathbf{J} dv = -\int_v \frac{\partial \rho_v}{\partial t} dv \tag{15.8}$$

Therefore,

$$\nabla \cdot \mathbf{J} = -\frac{\partial \rho_v}{\partial t} \tag{15.9}$$

or

$$\nabla \cdot \mathbf{J} + \frac{\partial \rho_v}{\partial t} = 0 \tag{15.10}$$

which is known as the *charge-current continuity equation* in differential form. Its integral form can be expressed as (Sadiku, 2010, p. 192)

$$\oint_S \mathbf{J} \cdot d\mathbf{S} + \int_v \frac{\partial \rho_v}{\partial t} dv = 0 \tag{15.11}$$

As we shall see, the continuity equation, together with Maxwell's equations and the constitutive medium relations, provide a complete set of equations needed to describe a general electromagnetic problem.

15.3 Displacement Current

Recall Ampere's law from electrostatics. It states that the line integral of the *static* magnetic field **H** about any closed path must equal the total *conduction current* enclosed by the path. This is the total current bounded by the contour that is *due to free charges*.

$$I_{enc} = \oint_C \mathbf{H} \cdot d\mathbf{l} = \int_S \mathbf{J} \cdot d\mathbf{S} \tag{15.12}$$

Applying the Stokes theorem to Eq. (15.12), we obtain

$$I_{enc} = \oint_c \mathbf{H} \cdot d\mathbf{l} = \int_S (\nabla \times \mathbf{H}) \cdot d\mathbf{S} \tag{15.13}$$

Comparing the right-hand sides of Eqs (15.12) and (15.13) leads to

$$\nabla \times \mathbf{H} = \mathbf{J} \tag{15.14}$$

Equation (15.14) is often referred to as Ampere's law in differential form. It is also one of the Maxwell equations for the static magnetic fields.

Taking the divergence of Eq. (15.14) we obtain

$$\nabla \cdot (\nabla \times \mathbf{H}) = 0 = \nabla \cdot \mathbf{J} \tag{15.15}$$

Since the divergence of the curl of any vector field is identically zero, Eq. (15.15) implies that

$$\nabla \cdot \mathbf{J} = 0 \tag{15.16}$$

However, the charge-current continuity relation requires that

$$\nabla \cdot \mathbf{J} = -\frac{\partial \rho_v}{\partial t} \tag{15.17}$$

which shows that Eq. (15.17) can only be true if

$$\frac{\partial \rho_v}{\partial t} = 0 \tag{15.18}$$

This is an unrealistic limitation. To overcome this difficulty, Maxwell *postulated* the existence of *displacement current density* $\mathbf{J}_d$ which can exist even in a nonconducting and free space medium. After adding the displacement current to the right-hand side of Eq. (15.14), Maxwell obtained

$$\nabla \times \mathbf{H} = \mathbf{J} + \mathbf{J}_d \tag{15.19}$$

where $\mathbf{J}_d$ is yet to be determined. Again taking the divergence of Eq. (15.19), we have

$$\nabla \cdot (\nabla \times \mathbf{H}) = 0 = \nabla \cdot (\mathbf{J} + \mathbf{J}_d) = \nabla \cdot \mathbf{J} + \nabla \cdot \mathbf{J}_d \tag{15.20}$$

or

$$0 = \nabla \cdot \mathbf{J} + \nabla \cdot \mathbf{J}_d \tag{15.21}$$

thus

$$\nabla \cdot \mathbf{J}_d = -\nabla \cdot \mathbf{J} = \frac{\partial \rho_v}{\partial t} \tag{15.22}$$

Now, recall the differential form of the Gauss's law

$$\rho_v = \nabla \cdot \mathbf{D} \tag{15.23}$$

where **D** is the electric flux density. Substituting Eq. (15.23) into Eq. (15.22) we get

$$\nabla \cdot \mathbf{J}_d = \frac{\partial \rho_v}{\partial t} = \frac{\partial}{\partial t}(\nabla \cdot \mathbf{D}) = \nabla \cdot \frac{\partial \mathbf{D}}{\partial t} \tag{15.24}$$

or

$$\mathbf{J}_d = \frac{\partial \mathbf{D}}{\partial t} \tag{15.25}$$

Substituting Eq. (15.25) into Eq. (15.19) results in

$$\nabla \times \mathbf{H} = \mathbf{J} + \frac{\partial \mathbf{D}}{\partial t} \tag{15.26}$$

This is Maxwell's equation for time-varying fields. It shows us that a *time-varying electric field produces a magnetic field*. If we take the surface integral of both sides Eq. (15.26) over an arbitrary open surface S with contour C, we have

$$\int_S (\nabla \times \mathbf{H}) \cdot d\mathbf{S} = \int_S \mathbf{J} \cdot d\mathbf{S} + \int_S \frac{\partial \mathbf{D}}{\partial t} \cdot d\mathbf{S} \tag{15.27}$$

Applying Stokes's theorem to the right-hand side of Eq. (15.27) we obtain

$$\oint_S \mathbf{H} \cdot d\mathbf{l} = \int_S \mathbf{J} \cdot d\mathbf{S} + \int_S \frac{\partial \mathbf{D}}{\partial t} \cdot d\mathbf{S} \tag{15.28}$$

Equation (15.28) can be written in a more general form as

$$\oint_S \mathbf{H} \cdot d\mathbf{l} = \int_S \mathbf{J} \cdot d\mathbf{S} + \frac{d}{dt} \int_S \mathbf{D} \cdot d\mathbf{S} \tag{15.29}$$

Equation (15.29) is known as the *integral form of Maxwell's equation* (Rao, 2004, p. 102). The quantity **H** is the magnetic field intensity vector in A/m. The quantity **J** is the current density vector in A/m^2. The quantity **D** is the electric flux density in C/m^2.

The two terms on the right-hand side of Eq. (15.29) are:

$$I_c = \int_S \mathbf{J} \cdot d\mathbf{S} \quad [A] \tag{15.30}$$

the total *conduction current* that penetrates the surface S bounded by the contour $C.$ – this current is due to free charges – and

$$I_d = \frac{d}{dt} \int_S \mathbf{D} \cdot d\mathbf{S} \quad [A] \tag{15.31}$$

the total *displacement current* that penetrates the surface S bounded by the contour $C.$ This current is due to time-varying electric flux.

As was the case with Faraday's law, any surface shape is suitable so long as contour C bounds it. Only the **J** and **D** that pass through the opening contribute, as shown in Figure 15.2.

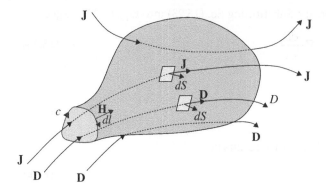

Figure 15.2 Illustration of the Ampere's law – Maxwell's equation.

15.4 EMC Applications

15.4.1 Grounding and Current Return Path

The term *grounding* in our discussion here means that the current return path is through the ground plane (or ground conductor), as illustrated in Figure 15.3.

The ground conductor has a non-zero impedance:

$$Z_G = R_G + j\omega L_G \tag{15.32}$$

At low frequency, the resistance R_G is the dominant factor. At high frequency, the inductance L_G is the dominant factor.

Consider a two-sided PCB with a single trace on top and full copper ground plane on the bottom, shown in Figure 15.4. At points A and B, vias connect the top trace to the ground plane.

The forward current flows on the top trace as shown in Figure 15.5.

How does the return current flow back to the source? The return current has a few options: the direct path from A to B or an alternative path underneath the top trace, or a combination of both, as shown in Figure 15.6.

At low frequencies, the ground current will take the path of least resistance (which corresponds to the path of the lowest impedance). This is shown in Figure 15.7.

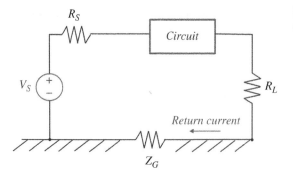

Figure 15.3 "Ground" conductor return current.

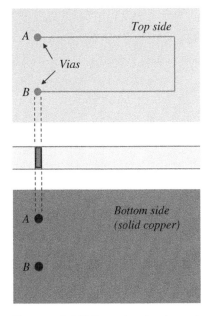

Figure 15.4 PCB illustrating the alternative current return paths.

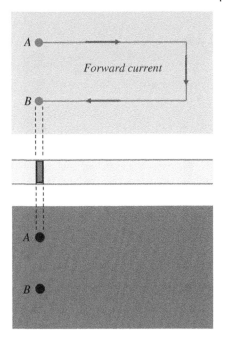

Figure 15.5 Forward current flow.

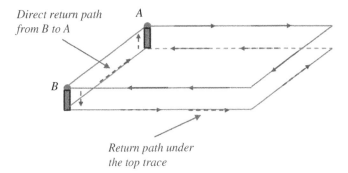

Figure 15.6 Return current alternative paths.

Thus the return current will take the lowest resistance, direct path from A to B, shown in Figure 15.8.

At high frequencies, the return current will take the path of least inductance, which is directly underneath the trace, because this represents the smallest loop area (smallest inductance). This is shown in Figures 15.9 and 15.10.

Let's confirm the above analysis with measurement and simulation results. Figure 15.11 shows the experimental setup used for the return current measurements.

The details of the circuit being investigated are shown in Figure 15.12.

Figure 15.13 presents a circuit diagram of the measurements setup.

The signal from the function generator travels along the center conductor of the coax cable and through the 50 Ω resistor. The return current has two different paths to return

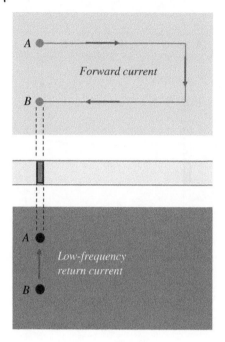

Figure 15.7 Low-frequency return current path.

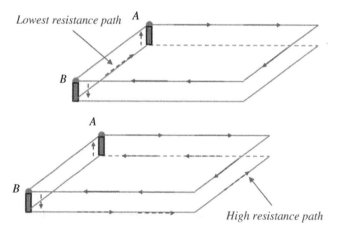

Figure 15.8 Low-frequency current will take the lowest resistance path.

to the source: a direct path over the copper wire or the path through the shield of the coax cable. A current probe is placed over the copper wire, and a sinusoidal signal is generated by the function generator. The frequency of this signal is varied and current through the copper wire is measured.

Figure 15.14 shows the measurement results.

As expected, as the frequency increases more current returns through the shield, as it provides the lower impedance path than the direct copper wire.

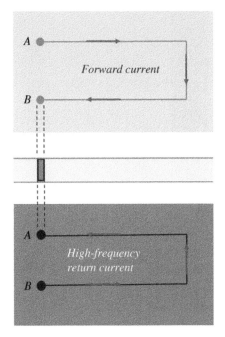

Figure 15.9 High-frequency return current path.

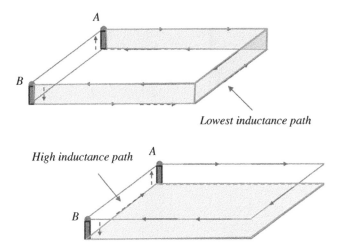

Figure 15.10 High-frequency current will take the lowest inductance path.

Figure 15.15 shows simulation results for the two-sided PCB described and analyzed in this section.

At low frequencies (1–100 kHz) the majority of the return current is through the direct path of least resistance. As the frequency is increased to 500 k–1 MHz, the current splits between the two paths. At high frequencies (10–100 MHz) the majority of the return current flows underneath the top trace through the path of least inductance.

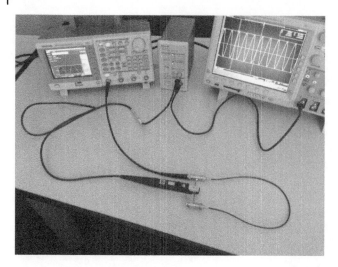

Figure 15.11 Experimental setup for the return current measurements.

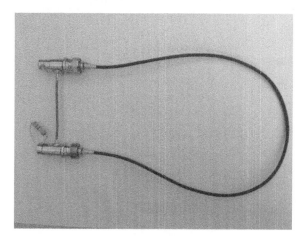

Figure 15.12 Circuit used for the return current measurements.

15.4.2 Common-Impedance Coupling

For common impedance coupling to occur, two circuits must share a current path (with a non-negligible impedance). Before we discuss the common-impedance coupling let's consider a couple of scenarios when the common impedance coupling does not occur.

Consider the circuit shown in Figure 15.16. The current flows from the source to the load, and returns to the source through a zero-impedance ground path.

The voltage at the load (with respect to ground is)

$$\hat{V}_L = R_L \hat{I} \tag{15.33}$$

Now, let's consider the case where the return path has a non-zero ground impedance, as shown in Figure 15.17.

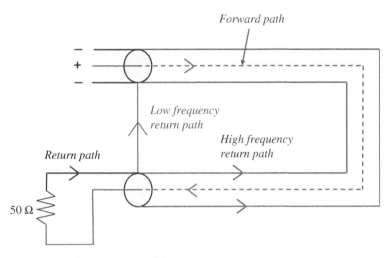

Figure 15.13 Circuit diagram of the measurement setup.

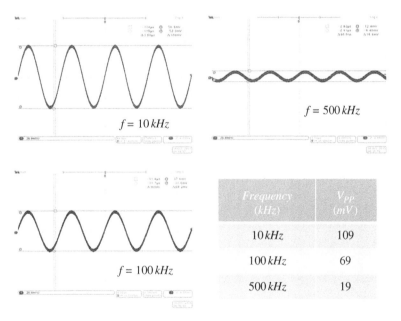

Frequency (kHz)	V_{pp} (mV)
10 kHz	109
100 kHz	69
500 kHz	19

Figure 15.14 Coax cable measurement results.

Now the voltage at the load (with respect to ground is)

$$\hat{V}_L = R_L \hat{I} + \hat{Z}_G \hat{I} \tag{15.34}$$

Obviously the ground impedance, $\hat{Z}_G$, affects the value of the load voltage, but no other circuit influences this value or is impacted by this ground impedance – there is no impedance coupling (since there is no other circuit to be coupled).

Now consider the situation shown in Figure 15.18 where two circuits share the ground return path with zero impedance.

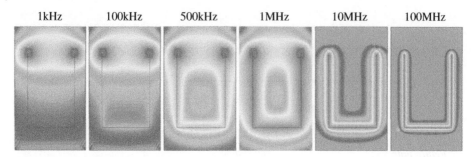

1kHz   100kHz   500kHz   1MHz   10MHz   100MHz

Figure 15.15 Two-sided PCB simulation results.

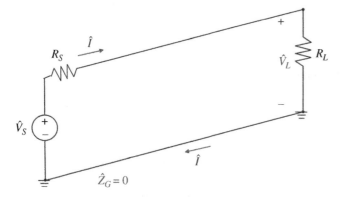

Figure 15.16 Current returns to the source through a zero-impedance ground path.

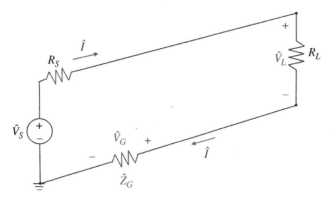

Figure 15.17 Current returns to the source through non-zero impedance ground.

The voltages at the loads are

$$\hat{V}_{L1} = R_{L1}\hat{I}_1 \tag{15.35a}$$

$$\hat{V}_{L2} = R_{L2}\hat{I}_2 \tag{15.35b}$$

Even though the two circuits share the return path, the load voltage of circuit 1, $\hat{V}_{L1}$, is not affected by the return current of circuit 2, $\hat{I}_2$; similarly, the load voltage of circuit 2, $\hat{V}_{L2}$, is not affected by the return current of circuit 1, $\hat{I}_1$.

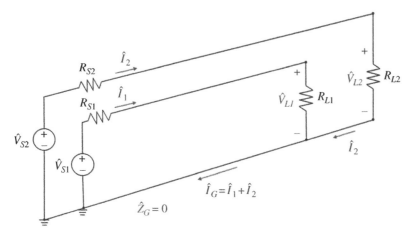

Figure 15.18 Two circuits share a zero-impedance ground path.

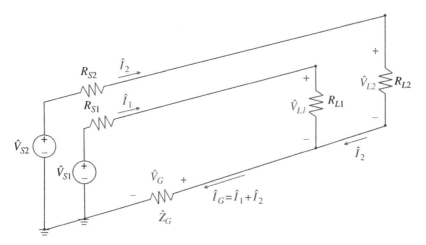

Figure 15.19 Common-impedance coupling circuit

There is no impedance coupling between the circuits (since there is no common impedance shared by both circuits).

Finally, consider the situation shown in Figure 15.19 where two circuits share the ground return path with a non-zero impedance.

The voltages at the loads are

$$\hat{V}_{L1} = R_{L1}\hat{I}_1 + \hat{Z}_G\left(\hat{I}_1 + \hat{I}_2\right)$$ (15.36a)

$$\hat{V}_{L2} = R_{L2}\hat{I}_2 + \hat{Z}_G\left(\hat{I}_1 + \hat{I}_2\right)$$ (15.36b)

Now the load voltage of circuit 1, $\hat{V}_{L1}$, is affected by the return current of circuit 2, $\hat{I}_2$; similarly, the load voltage of circuit 2, $\hat{V}_{L2}$, is affected by the return current of circuit 1, $\hat{I}_1$. This type of coupling is called the *common-impedance coupling*.

References

Rao, N.N., *Elements of Engineering Electromagnetics*, 6th ed., Pearson Prentice Hall, Upper Saddle River, NJ, 2004.

Sadiku, M.N.O., *Elements of Electromagnetics*, 5th ed., Oxford University Press, New York, 2010.

16

Electromagnetic Waves

Recall two of the Maxwell's equations for source-free media:

$$\nabla \times \mathbf{E} = -\mu \frac{\partial \mathbf{H}}{\partial t} \tag{16.1}$$

$$\nabla \times \mathbf{H} = \sigma \mathbf{E} + \varepsilon \frac{\partial \mathbf{E}}{\partial t} \tag{16.2}$$

These equations state that time variations of the magnetic and electric fields give rise to space variations of the electric and magnetic fields, respectively. This interdependence of the space and time variations gives rise to the electromagnetic wave propagation.

In general, electric and magnetic fields have three non-zero components, each of them being a function of all three coordinates and time. That is,

$$\mathbf{E} = \left[E_x(x, y, z, t), E_y(x, y, z, t), E_z(x, y, z, t) \right] \tag{16.3a}$$

$$\mathbf{H} = \left[H_x(x, y, z, t), H_y(x, y, z, t), H_z(x, y, z, t) \right] \tag{16.3b}$$

In the following discussion we will focus on a simple and very useful type of wave: the uniform plane wave. Uniform plane waves not only serve as a building block in the study of electromagnetic waves but also support the study of wave propagation on transmission lines and wave radiation by antennas (Paul, 2006, p. 909).

16.1 Uniform Waves – Time Domain Analysis

To derive the uniform plane wave equations we will use the two Maxwell's equations (16.1) and (16.2). To this end, we first need to make two assumptions: (1) we need to choose the direction of either the electric field intensity vector $\mathbf{E}$ or the magnetic field intensity vector $\mathbf{H}$, and (2) we need to choose the plane in which these two vectors lie.

Let's choose the direction of the electric field intensity vector as

$$\mathbf{E} = \left[E_x(x, y, z, t), 0, 0 \right] \tag{16.4}$$

and let's choose the plane in which both vector lie as the plane parallel to the xy plane (Paul, 2006, p. 445).

Foundations of Electromagnetic Compatibility with Practical Applications, First Edition. Bogdan Adamczyk.
© 2017 John Wiley & Sons Ltd. Published 2017 by John Wiley & Sons Ltd.

Since the wave is uniform in the plane it follows that E_x is not a function of the position in the plane, i.e. it is not a function of x or y,

$$\mathbf{E} = \left[E_x(z,t),0,0 \right] \tag{16.5}$$

and therefore

$$\frac{\partial E_x}{\partial x} = 0 \tag{16.6a}$$

$$\frac{\partial E_x}{\partial y} = 0 \tag{16.6b}$$

In terms of the components, the first Maxwell's equation (16.1) can be written as

$$\nabla \times \mathbf{E} = \left(\frac{\partial E_z}{\partial y} - \frac{\partial E_y}{\partial z} \right) \mathbf{a}_x + \left(\frac{\partial E_x}{\partial z} - \frac{\partial E_z}{\partial x} \right) \mathbf{a}_y + \left(\frac{\partial E_y}{\partial x} - \frac{\partial E_x}{\partial y} \right) \mathbf{a}_z$$
$$= -\mu \frac{\partial H_x}{\partial t} \mathbf{a}_x - \mu \frac{\partial H_y}{\partial t} \mathbf{a}_y - \mu \frac{\partial H_z}{\partial t} \mathbf{a}_z \tag{16.7}$$

Utilizing Eqs (16.5) and (16.6) we get

$$0\mathbf{a}_x + \frac{\partial E_x}{\partial z} \mathbf{a}_y + 0\mathbf{a}_z = -\mu \frac{\partial H_x}{\partial t} \mathbf{a}_x - \mu \frac{\partial H_y}{\partial t} \mathbf{a}_y - \mu \frac{\partial H_z}{\partial t} \mathbf{a}_z \tag{16.8}$$

Thus

$$-\mu \frac{\partial H_x}{\partial t} = 0 \tag{16.9a}$$

$$\frac{\partial E_x}{\partial z} = -\mu \frac{\partial H_y}{\partial t} \tag{16.9b}$$

$$-\mu \frac{\partial H_z}{\partial t} = 0 \tag{16.9c}$$

Since magnetic field intensity is a time-varying quantity, the only way to satisfy equations (16.9a) and (16.9c) is when

$$H_x = 0 \tag{16.10a}$$

$$H_z = 0 \tag{16.10b}$$

Therefore, the magnetic field intensity vector **H** has only a y component

$$\mathbf{H} = \left[0, H_y, 0 \right] \tag{16.11}$$

related to the electric field intensity by Eq. (16.9b). Thus, the **E** and **H** vectors are orthogonal, as shown in Figure 16.1.

Since the wave is uniform in the plane, it follows that H_y is not a function of x or y,

$$\mathbf{H} = \left[0, H_y(z,t),0 \right] \tag{16.12}$$

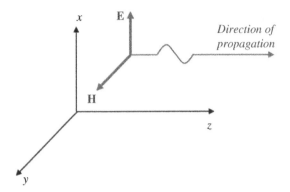

Figure 16.1 Uniform plane wave.

and therefore

$$\frac{\partial H_y}{\partial x} = 0 \tag{16.13a}$$

$$\frac{\partial H_y}{\partial y} = 0 \tag{16.13b}$$

The second Maxwell's equation (16.2) can be written in terms of the components as

$$\nabla \times \mathbf{H} = \left(\frac{\partial H_z}{\partial y} - \frac{\partial H_y}{\partial z} \right) \mathbf{a}_x + \left(\frac{\partial H_x}{\partial z} - \frac{\partial H_z}{\partial x} \right) \mathbf{a}_y + \left(\frac{\partial H_y}{\partial x} - \frac{\partial H_x}{\partial y} \right) \mathbf{a}_z$$

$$= \left(\sigma E_x + \varepsilon \frac{\partial E_x}{\partial t} \right) \mathbf{a}_x + \left(\sigma E_y + \varepsilon \frac{\partial E_y}{\partial t} \right) \mathbf{a}_y + \left(\sigma E_z + \varepsilon \frac{\partial E_z}{\partial t} \right) \mathbf{a}_z \tag{16.14}$$

Utilizing Eqs (16.12) and (16.13) we get

$$-\frac{\partial H_y}{\partial z} \mathbf{a}_x + 0\mathbf{a}_y + 0\mathbf{a}_z = \left(\sigma E_x + \varepsilon \frac{\partial E_x}{\partial t} \right) \mathbf{a}_x + 0\mathbf{a}_y + 0\mathbf{a}_z \tag{16.15}$$

and thus

$$-\frac{\partial H_y}{\partial z} = \sigma E_x + \varepsilon \frac{\partial E_x}{\partial t} \tag{16.16}$$

Therefore, the uniform plane wave is described by a set of coupled partial differential equations:

$$\frac{\partial E_x(z,t)}{\partial z} = -\mu \frac{\partial H_y(z,t)}{\partial t} \tag{16.17a}$$

$$\frac{\partial H_y(z,t)}{\partial z} = -\sigma E_x(z,t) - \varepsilon \frac{\partial E_x(z,t)}{\partial t} \tag{16.17b}$$

This set of equations can be decoupled as follows. Differentiating Eq. (16.17a) with respect to time results in

$$\frac{\partial^2 E_x(z,t)}{\partial z \partial t} = -\mu \frac{\partial^2 H_y(z,t)}{\partial t^2} \tag{16.18a}$$

while differentiating Eq. (16.17b) with respect to z produces

$$\frac{\partial^2 H_y(z,t)}{\partial z^2} = -\sigma \frac{\partial E_x(z,t)}{\partial z} - \varepsilon \frac{\partial^2 E_x(z,t)}{\partial t \partial z} \tag{16.18b}$$

Substituting Eqs (16.17a) and (16.18a) into Eq. (16.18b) gives

$$\frac{\partial^2 H_y(z,t)}{\partial z^2} = -\sigma \left[-\mu \frac{\partial H_y(z,t)}{\partial t} \right] - \varepsilon \left[-\mu \frac{\partial^2 H_y(z,t)}{\partial t^2} \right] \tag{16.19}$$

or

$$\frac{\partial^2 H_y(z,t)}{\partial z^2} = \mu\sigma \frac{\partial H_y(z,t)}{\partial t} + \mu\varepsilon \frac{\partial^2 H_y(z,t)}{\partial t^2} \tag{16.20}$$

This is the first of the decoupled equations. To obtain the second equation, we differentiate Eq. (16.17a) with respect to z and Eq. (16.17b) with respect to time. The result is

$$\frac{\partial^2 E_x(z,t)}{\partial z^2} = -\mu \frac{\partial^2 H_y(z,t)}{\partial t \partial z} \tag{16.21a}$$

$$\frac{\partial^2 H_y(z,t)}{\partial z \partial t} = -\sigma \frac{\partial E_x(z,t)}{\partial t} - \varepsilon \frac{\partial^2 E_x(z,t)}{\partial t^2} \tag{16.21b}$$

Now, substitute Eq. (16.19b) into Eq. (16.19a) to obtain

$$\frac{\partial^2 E_x(z,t)}{\partial z^2} = -\mu \left[-\sigma \frac{\partial E_x(z,t)}{\partial t} - \varepsilon \frac{\partial^2 E_x(z,t)}{\partial t^2} \right] \tag{16.22}$$

or

$$\frac{\partial^2 E_x(z,t)}{\partial z^2} = \mu\sigma \frac{\partial E_x(z,t)}{\partial t} + \mu\varepsilon \frac{\partial^2 E_x(z,t)}{\partial t^2} \tag{16.23}$$

This is the second decoupled wave equation. For source-free and lossless medium ($\sigma = 0$) the wave equations in (16.23) and (16.20) simplify to

$$\frac{\partial^2 E_x(z,t)}{\partial z^2} = \mu\varepsilon \frac{\partial^2 E_x(z,t)}{\partial t^2} \tag{16.24a}$$

$$\frac{\partial^2 H_y(z,t)}{\partial z^2} = \mu\varepsilon \frac{\partial^2 H_y(z,t)}{\partial t^2} \tag{16.24b}$$

Both equations have the same mathematical form, and therefore their solutions will have the same mathematical form. A solution of Eq. (17.24a) is known to be (Rao, 2004, p.174)

$$E_x(z,t) = f\left(t - \frac{z}{v}\right) \tag{16.25}$$

where

$$v = \frac{1}{\sqrt{\mu\varepsilon}} \tag{16.26}$$

and f is an arbitrary twice-differentiable function. Let's verify it, because this verification will reveal a very interesting fact about this solution.

Let

$$\tau = t - \frac{z}{v} \tag{16.27}$$

then

$$E_x(z,t) = f(\tau) = f\left(t - \frac{z}{v}\right) \tag{16.28}$$

Using the chain rule for differentiation, the partial derivatives of E_x with respect to t and z can be expressed as

$$\frac{\partial E_x(z,t)}{\partial t} = \frac{\partial f(\tau)}{\partial \tau}\frac{\partial \tau}{\partial t} = \frac{\partial f(\tau)}{\partial \tau} \tag{16.29a}$$

and

$$\frac{\partial E_x(z,t)}{\partial z} = \frac{\partial f(\tau)}{\partial \tau}\frac{\partial \tau}{\partial z} = -\frac{1}{v}\frac{\partial f(\tau)}{\partial \tau} \tag{16.29b}$$

In a similar manner, we obtain the expressions for the second derivatives

$$\frac{\partial^2 E_x(z,t)}{\partial t^2} = \frac{\partial}{\partial t}\frac{\partial E_x(z,t)}{\partial t} = \frac{\partial}{\partial t}\left[\frac{\partial f(\tau)}{\partial \tau}\right]$$
$$= \frac{\partial^2 f(\tau)}{\partial \tau^2}\frac{\partial \tau}{\partial t} = \frac{\partial^2 f(\tau)}{\partial \tau^2} \tag{16.30a}$$

and

$$\frac{\partial^2 E_x(z,t)}{\partial z^2} = \frac{\partial}{\partial z}\frac{\partial E_x(z,t)}{\partial z} = \frac{\partial}{\partial z}\left[-\frac{1}{v}\frac{\partial f(\tau)}{\partial \tau}\right]$$
$$= -\frac{1}{v}\frac{\partial^2 f(\tau)}{\partial \tau^2}\frac{\partial \tau}{\partial z} = \frac{1}{v^2}\frac{\partial^2 f(\tau)}{\partial \tau^2} \tag{16.30b}$$

Now, substitute Eqs (16.30) into Eq. (16.24a) to obtain

$$\frac{1}{v^2}\frac{\partial^2 f(\tau)}{\partial \tau^2} = \mu\varepsilon\frac{1}{v^2}\frac{\partial^2 f(\tau)}{\partial \tau^2} \tag{16.31}$$

Using Eq. (16.27) in Eq. (16.31) we get

$$\frac{1}{v^2}\frac{\partial^2 f(\tau)}{\partial \tau^2} = \frac{1}{v^2}\frac{\partial^2 f(\tau)}{\partial \tau^2} \tag{16.32}$$

which verifies that Eq. (16.28) is a solution of Eq. (16.24a). In a similar fashion, it can be shown that any twice-differentiable function of the form

$$E_x(z,t) = g\left(t + \frac{z}{v}\right) \tag{16.33}$$

is also a solution of Eq. (16.24a). Thus, the general solution of Eq. (16.24a) is

$$E_x(z,t) = Af\left(t - \frac{z}{v}\right) + Bg\left(t + \frac{z}{v}\right) \tag{16.34}$$

Note that

$$f(z,t) = f\left(t - \frac{z}{v}\right) \tag{16.35a}$$

while

$$f(z + \Delta z, t + \Delta t) = f\left(t + \Delta t - \frac{z + \Delta z}{v}\right) \tag{16.35b}$$

Now, if

$$\Delta z = v\Delta t \tag{16.36}$$

then Eq. (16.35b) becomes

$$f(z + \Delta z, t + \Delta t) = f\left(t + \Delta t - \frac{z + \Delta z}{v}\right)$$
$$= f\left(t + \Delta t - \frac{z + v\Delta t}{v}\right) = f\left(t + \Delta t - \frac{z}{v} - \Delta t\right) = f\left(t - \frac{z}{v}\right) \tag{16.37}$$

Therefore, after a time Δt, the function f retains the same value at a point that is $\Delta z = v\Delta t$ away from the previous position in space (defined by z), as shown in Figure 16.2. This means that an arbitrary function of the form $f(t - z/v)$ represents a *traveling wave* with a velocity

$$v = \frac{\Delta z}{\Delta t} = \frac{1}{\sqrt{\mu\varepsilon}} \tag{16.38}$$

The wave travels in the positive z direction as the time t advances. Similarly, an arbitrary function of the form $g(t + z/v)$ represents a wave with a velocity v in the negative z direction as the time t advances.

The corresponding solution for $H_y(z, t)$ can be obtained as follows. We begin with Eq. (16.9b) repeated here:

$$\frac{\partial E_x}{\partial z} = -\mu\frac{\partial H_y}{\partial t} \tag{16.39}$$

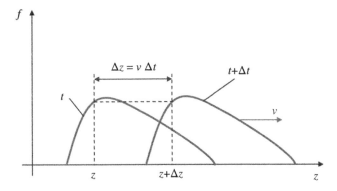

Figure 16.2 A traveling wave.

thus

$$\frac{\partial H_y}{\partial t} = -\frac{1}{\mu}\frac{\partial E_x}{\partial z} \tag{16.40}$$

Using Eq. (16.34) in Eq. (16.40) we get

$$
\begin{aligned}
\frac{\partial H_y}{\partial t} &= -\frac{1}{\mu}\frac{\partial}{\partial z}\left[Af\left(t-\frac{z}{v}\right) + Bg\left(t+\frac{z}{v}\right)\right]\\
&= -\frac{1}{\mu}\left\{ A\left[\frac{\partial}{\partial t}f\left(t-\frac{z}{v}\right)\right]\left[\frac{\partial}{\partial z}\left(t-\frac{z}{v}\right)\right] + B\left[\frac{\partial}{\partial t}g\left(t-\frac{z}{v}\right)\right]\left[\frac{\partial}{\partial z}\left(t+\frac{z}{v}\right)\right]\right\}\\
&\quad -\frac{1}{\mu}\left[A\frac{\partial f\left(t-\frac{z}{v}\right)}{\partial t}\left(-\frac{1}{v}\right) + B\frac{\partial g\left(t-\frac{z}{v}\right)}{\partial t}\left(\frac{1}{v}\right)\right]
\end{aligned}
\tag{16.41}
$$

or

$$
\begin{aligned}
\frac{\partial H_y}{\partial t} &= \frac{1}{\mu v}\left[A\frac{\partial f\left(t-\frac{z}{v}\right)}{\partial t} - B\frac{\partial g\left(t-\frac{z}{v}\right)}{\partial t}\right]\\
&= \frac{\sqrt{\mu\varepsilon}}{\mu}\left[A\frac{\partial f\left(t-\frac{z}{v}\right)}{\partial t} - B\frac{\partial g\left(t-\frac{z}{v}\right)}{\partial t}\right]\\
&= \sqrt{\frac{\varepsilon}{\mu}}\left[A\frac{\partial f\left(t-\frac{z}{v}\right)}{\partial t} - B\frac{\partial g\left(t-\frac{z}{v}\right)}{\partial t}\right]
\end{aligned}
\tag{16.42}
$$

Integrating Eq. (16.42) with respect to time results in

$$H_y(z,t) = \frac{1}{\sqrt{\dfrac{\mu}{\varepsilon}}}\left[Af\left(t-\frac{z}{v}\right) - Bg\left(t+\frac{z}{v}\right)\right] \tag{16.43}$$

or

$$H_y(z,t) = \frac{1}{\eta}\left[Af\left(t - \frac{z}{v}\right) - Bg\left(t + \frac{z}{v}\right) \right] \tag{16.44}$$

where, for a lossless medium, the *intrinsic impedance* of the medium, η, is

$$\eta = \sqrt{\frac{\mu}{\varepsilon}} \quad [\Omega] \tag{16.45}$$

Thus, in summary, the solutions to the wave equations for an arbitrary time-variations of the field are

$$E_x(z,t) = Af\left(t - \frac{z}{v}\right) + Bg\left(t + \frac{z}{v}\right) \tag{16.46a}$$

$$H_y(z,t) = \frac{A}{\eta}f\left(t - \frac{z}{v}\right) - \frac{B}{\eta}g\left(t - \frac{z}{v}\right) \tag{16.46b}$$

16.2 Uniform Waves – Sinusoidal Steady-State Analysis

In the previous section we obtained the solution to the wave equations when the fields were arbitrary functions of time. Of particular interest are the sinusoidal variations of the fields.

Recall the time domain wave equations for the arbitrary variations of the fields:

$$\frac{\partial^2 E_x(z,t)}{\partial z^2} = \mu\sigma \frac{\partial E_x(z,t)}{\partial t} + \mu\varepsilon \frac{\partial^2 E_x(z,t)}{\partial t^2} \tag{16.47a}$$

$$\frac{\partial^2 H_y(z,t)}{\partial z^2} = \mu\sigma \frac{\partial H_y(z,t)}{\partial t} + \mu\varepsilon \frac{\partial^2 H_y(z,t)}{\partial t^2} \tag{16.47b}$$

We wish to obtain the sinusoidal steady-state phasor version of these equations. Recall that differentiation in the time domain corresponds to the multiplication by $j\omega$ in the phasor domain. Thus,

$$\frac{\partial E_x(z,t)}{\partial t} \quad \leftrightarrow \quad j\omega \hat{E}_x(z) \tag{16.48a}$$

$$\frac{\partial H_y(z,t)}{\partial t} \quad \leftrightarrow \quad j\omega \hat{H}_y(z) \tag{16.48b}$$

and

$$\frac{\partial^2 E_x(z,t)}{\partial t^2} = \frac{\partial}{\partial t}\left[\frac{\partial E_x(z,t)}{\partial t} \right] \quad \leftrightarrow \quad (j\omega)(j\omega)\hat{E}_x(z) = -\omega^2 \hat{E}_x(z) \tag{16.49a}$$

$$\frac{\partial^2 H_y(z,t)}{\partial t^2} \quad \leftrightarrow \quad -\omega^2 \hat{H}_y(z) \tag{16.49b}$$

Also

$$\frac{\partial^2 E_x(z,t)}{\partial z^2} \leftrightarrow \frac{d^2 \hat{E}_x}{dz^2} \tag{16.50a}$$

$$\frac{\partial^2 H_y(z,t)}{\partial z^2} \leftrightarrow \frac{d^2 \hat{H}_y}{dz^2} \tag{16.50b}$$

Substituting Eqs (16.48)–(16.50) into Eqs (16.47) we obtain the sinusoidal steady-state wave equations

$$\frac{d^2 \hat{E}_x}{dz^2} = \mu\sigma\left[j\omega\hat{E}_x(z)\right] + \mu\varepsilon\left[-\omega^2 \hat{E}_x(z)\right] \tag{16.51a}$$

$$\frac{d^2 \hat{H}_y}{dz^2} = \mu\sigma\left[j\omega\hat{H}_y(z)\right] + \mu\varepsilon\left[-\omega^2 \hat{H}_y(z)\right] \tag{16.51b}$$

or

$$\frac{d^2 \hat{E}_x}{dz^2} = \left(j\omega\mu\sigma - \omega^2\mu\varepsilon\right)\hat{E}_x(z) \tag{16.52a}$$

$$\frac{d^2 \hat{H}_y}{dz^2} = \left(j\omega\mu\sigma - \omega^2\mu\varepsilon\right)\hat{H}_y(z) \tag{16.52b}$$

Now,

$$j\omega\mu\sigma - \omega^2\mu\varepsilon = j\omega\mu\left(\sigma + j\omega\varepsilon\right) = \hat{\gamma}^2 \tag{16.53}$$

where

$$\hat{\gamma} = \sqrt{j\omega\mu\left(\sigma + j\omega\varepsilon\right)} \tag{16.54}$$

is the *propagation constant*. Using Eq. (16.63) we rewrite the wave equations (16.52) as

$$\frac{d^2 \hat{E}_x}{dz^2} = \hat{\gamma}^2 \hat{E}_x(z) \tag{16.55a}$$

$$\frac{d^2 \hat{H}_y}{dz^2} = \hat{\gamma}^2 \hat{H}_y(z) \tag{16.55b}$$

The solutions of these equations are of the form

$$\hat{E}_x = \hat{A}e^{-\hat{\gamma}z} + \hat{B}e^{\hat{\gamma}z} \tag{16.56a}$$

$$\hat{H}_y = \frac{\hat{A}}{\hat{\eta}}e^{-\hat{\gamma}z} - \frac{\hat{B}}{\hat{\eta}}e^{\hat{\gamma}z} \tag{16.56b}$$

where

$$\hat{\eta} = \sqrt{\frac{j\omega\mu}{\sigma + j\omega\varepsilon}} = \frac{j\omega\mu}{\hat{\gamma}} = \eta\angle\theta_\eta \quad [\Omega] \tag{16.57}$$

is the complex intrinsic impedance of the medium.

Let's verify that the equations (16.56) are indeed the solutions to Eqs (16.55). Differentiating Eqs (16.55) we get

$$\frac{d\hat{E}_x}{dz} = -\hat{\gamma}\hat{A}e^{-\hat{\gamma}z} + \hat{\gamma}\hat{B}e^{\hat{\gamma}z} \tag{16.58a}$$

$$\frac{d\hat{H}_y}{dz} = -\frac{\hat{\gamma}}{\hat{\eta}}\hat{A}e^{-\hat{\gamma}z} - \frac{\hat{\gamma}}{\hat{\eta}}\hat{B}e^{\hat{\gamma}z} \tag{16.58b}$$

Differentiating Eqs (16.58) again results in

$$\frac{d^2\hat{E}_x}{dz^2} = \hat{\gamma}^2\hat{A}e^{-\hat{\gamma}z} + \hat{\gamma}^2\hat{B}e^{\hat{\gamma}z}$$
$$= \hat{\gamma}^2\left(\hat{A}e^{-\hat{\gamma}z} + \hat{B}e^{\hat{\gamma}z}\right) = \hat{\gamma}^2\hat{E}_x \tag{16.59a}$$

$$\frac{d^2\hat{H}_y}{dz^2} = \frac{\hat{\gamma}^2}{\hat{\eta}}\hat{A}e^{-\hat{\gamma}z} - \frac{\hat{\gamma}^2}{\hat{\eta}}\hat{B}e^{\hat{\gamma}z}$$
$$= \hat{\gamma}^2\left(\frac{\hat{A}}{\hat{\eta}}e^{-\hat{\gamma}z} - \frac{\hat{B}}{\hat{\eta}}e^{\hat{\gamma}z}\right) = \hat{\gamma}^2\hat{H}_y \tag{16.59b}$$

which confirms that the Eqs (16.56) is the solution of Eqs (16.55).

The solution in Eqs (16.56) is consistent with that presented by Paul (2006, p. 912). We simply make the notation change from

$$\hat{A} \leftrightarrow \hat{E}_m^+ \tag{16.60a}$$

$$\hat{B} \leftrightarrow \hat{E}_m^- \tag{16.60b}$$

to obtain

$$\hat{E}_x = \hat{E}_m^+ e^{-\hat{\gamma}z} + \hat{E}_m^- e^{\hat{\gamma}z} \tag{16.61a}$$

$$\hat{H}_y = \frac{\hat{E}_m^+}{\hat{\eta}}e^{-\hat{\gamma}z} - \frac{\hat{E}_m^-}{\hat{\eta}}e^{\hat{\gamma}z} \tag{16.61b}$$

Expressing the propagation constant in terms of its real and imaginary parts

$$\hat{\gamma} = \alpha + j\beta \tag{16.62}$$

and the complex intrinsic impedance as

$$\hat{\eta} = \eta\angle\theta_\eta = e^{j\theta_\eta} \tag{16.63}$$

(α is the attenuation constant in Np/m and β is the phase constant in rad/m) we can write the solution in Eqs (16.61) as

$$\hat{E}_x = \hat{E}_m^+ e^{-\alpha z} e^{-j\beta z} + \hat{E}_m^- e^{\alpha z} e^{j\beta z} \tag{16.64a}$$

$$\hat{H}_y = \frac{\hat{E}_m^+}{\eta} e^{-\alpha z} e^{-j\beta z} e^{-j\theta_\eta} - \frac{\hat{E}_m^-}{\eta} e^{\alpha z} e^{j\beta z} e^{-j\theta_\eta} \tag{16.64b}$$

Finally, expressing the complex constant in the exponential form

$$\hat{E}_m^+ = E_m^+ \angle \theta^+ = E_m^+ e^{j\theta^+} \tag{16.65a}$$

$$\hat{E}_m^- = E_m^- \angle \theta^- = E_m^- e^{j\theta^-} \tag{16.65b}$$

we obtain the solution in phasor domain as

$$\hat{E}_x = E_m^+ e^{-\alpha z} e^{-j\beta z} e^{j\theta^+} + E_m^- e^{\alpha z} e^{j\beta z} e^{j\theta^-} \tag{16.66a}$$

$$\hat{H}_y = \frac{E_m^+}{\eta} e^{-\alpha z} e^{-j\beta z} e^{-j\theta_\eta} e^{j\theta^+} - \frac{E_m^-}{\eta} e^{\alpha z} e^{j\beta z} e^{-j\theta_\eta} e^{j\theta^-} \tag{16.66b}$$

which is the form presented by Paul, (2006, Eq. B.65, p. 912).

Examining the equations (16.66), we can immediately write the time domain solution by extracting the magnitudes and phases of the complex expressions and inserting them into the corresponding time domain sinusoids.

$$
\begin{aligned}
E_x = {} & E_m^+ e^{-\alpha z} \cos\left(\omega t - \beta z + \theta^+\right) \\
& + E_m^- e^{\alpha z} \cos\left(\omega t + \beta z + \theta^-\right)
\end{aligned} \tag{16.67a}
$$

$$
\begin{aligned}
H_y = {} & \frac{E_m^+}{\eta} e^{-\alpha z} \cos\left(\omega t - \beta z + \theta^+ - \theta_\eta\right) \\
& - \frac{E_m^-}{\eta} e^{\alpha z} \cos\left(\omega t + \beta z + \theta^- - \theta_\eta\right)
\end{aligned} \tag{16.67b}
$$

In a lossless medium

$$\alpha = 0 \tag{16.68a}$$

$$\hat{\eta} = \eta \angle 0 \tag{16.68b}$$

and with the undetermined constants being real

$$\hat{E}_m^+ = E_m^+ \angle 0 \tag{16.69a}$$

$$\hat{E}_m^- = E_m^- \angle 0 \tag{16.69b}$$

Equations (16.67) become

$$E_x = E_m^+ \cos\left(\omega t - \beta z\right) + E_m^- \cos\left(\omega t + \beta z\right) \tag{16.70a}$$

$$H_y = \frac{E_m^+}{\eta} \cos\left(\omega t - \beta z\right) - \frac{E_m^-}{\eta} \cos\left(\omega t + \beta z\right) \tag{16.70b}$$

Note that we could obtain the solution in Eqs (16.70) by directly employing the time domain solution obtained in the previous section

$$E_x(z,t) = Af\left(t - \frac{z}{v}\right) + Bg\left(t + \frac{z}{v}\right) \tag{16.71a}$$

$$H_y(z,t) = \frac{A}{\eta}f\left(t - \frac{z}{v}\right) - \frac{B}{\eta}g\left(t - \frac{z}{v}\right) \tag{16.71b}$$

Since

$$f\left(t - \frac{z}{v}\right) = \cos\left(\omega t - \frac{z}{v}\right) \tag{16.72a}$$

$$g\left(t - \frac{z}{v}\right) = \cos\left(\omega t - \frac{z}{v}\right) \tag{16.72b}$$

$$f\left(t + \frac{z}{v}\right) = \cos\left(\omega t + \frac{z}{v}\right) \tag{16.72c}$$

$$g\left(t + \frac{z}{v}\right) = \cos\left(\omega t + \frac{z}{v}\right) \tag{16.72d}$$

the solution in Eqs (16.71) becomes

$$E_x(z,t) = A\cos\left(\omega t - \frac{z}{v}\right) + B\cos\left(\omega t + \frac{z}{v}\right) \tag{16.73a}$$

$$H_y(z,t) = \frac{A}{\eta}\cos\left(\omega t - \frac{z}{v}\right) - \frac{B}{\eta}\cos\left(\omega t - \frac{z}{v}\right) \tag{16.73b}$$

Expressing the velocity of propagation as

$$v = \frac{\omega}{\beta} \tag{16.74}$$

and utilizing the substitutions in Eq. (16.60), we can express Eqs (16.73) as

$$E_x = E_m^+ \cos(\omega t - \beta z) + E_m^- \cos(\omega t + \beta z) \tag{16.75a}$$

$$H_y = \frac{E_m^+}{\eta}\cos(\omega t - \beta z) - \frac{E_m^-}{\eta}\cos(\omega t + \beta z) \tag{16.75b}$$

which, of course, agree with Eqs (16.70).

16.3 Reflection and Transmission of Uniform Waves at Boundaries

In the next section we will discuss electromagnetic wave shielding. In order to derive the equations describing this phenomenon we need to understand the reflection and transmission of electromagnetic waves at the boundaries of two media.

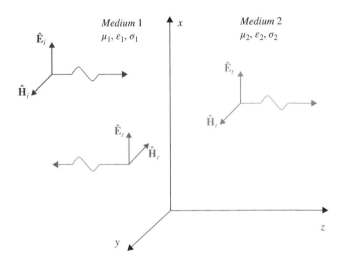

Figure 16.3 Reflection and transmission of a uniform wave at the boundary.

We will consider a normal incidence of a uniform plane wave on the boundary between two media, as shown in Figure 16.3.

When the wave encounters the boundary between two media, reflected and transmitted waves are created (Paul, 2006, p. 472).

The incident wave is described by

$$\hat{\mathbf{E}}_i = \hat{E}_i e^{-\hat{\gamma}_1 z} \mathbf{a}_x = \hat{E}_i e^{-\alpha_1 z} e^{-j\beta_1 z} \mathbf{a}_x \tag{16.76a}$$

$$\hat{\mathbf{H}}_i = \frac{\hat{E}_i}{\hat{\eta}_1} e^{-\hat{\gamma}_1 z} \mathbf{a}_y = \frac{\hat{E}_i}{\eta_1} e^{-\alpha_1 z} e^{-j\beta_1 z} e^{-j\theta_{\eta 1}} \mathbf{a}_y \tag{16.76b}$$

while the reflected wave is expressed as

$$\hat{\mathbf{E}}_r = \hat{E}_r e^{\hat{\gamma}_1 z} \mathbf{a}_x = \hat{E}_r e^{\alpha_1 z} e^{j\beta_1 z} \mathbf{a}_x \tag{16.77a}$$

$$\hat{\mathbf{H}}_r = -\frac{\hat{E}_r}{\hat{\eta}_1} e^{\hat{\gamma}_1 z} \mathbf{a}_y = -\frac{\hat{E}_r}{\eta_1} e^{\alpha_1 z} e^{j\beta_1 z} e^{-j\theta_{\eta 1}} \mathbf{a}_y \tag{16.77b}$$

where the propagation constant and the intrinsic impedance in medium 1 are given by

$$\hat{\gamma}_1 = \sqrt{j\omega\mu_1(\sigma_1 + j\omega\varepsilon_1)} = \alpha_1 + j\beta_1 \tag{16.78a}$$

$$\hat{\eta}_1 = \sqrt{\frac{j\omega\mu_1}{\sigma_1 + j\omega\varepsilon_1}} = \eta_1 \angle \theta_{\eta 1} \tag{16.78b}$$

The transmitted wave is represented as

$$\hat{\mathbf{E}}_t = \hat{E}_t e^{-\hat{\gamma}_2 z} \mathbf{a}_x = \hat{E}_t e^{-\alpha_2 z} e^{-j\beta_2 z} \mathbf{a}_x \tag{16.79a}$$

$$\hat{\mathbf{H}}_t = \frac{\hat{E}_t}{\hat{\eta}_2} e^{-\hat{\gamma}_2 z} \mathbf{a}_y = \frac{\hat{E}_t}{\eta_2} e^{-\alpha_2 z} e^{-j\beta_2 z} e^{-j\theta_{\eta 2}} \mathbf{a}_y \tag{16.79b}$$

where the propagation constant and the intrinsic impedance in medium 2 are given by

$$\hat{\gamma}_2 = \sqrt{j\omega\mu_2\left(\sigma_2 + j\omega\varepsilon_2\right)} = \alpha_2 + j\beta_2 \tag{16.80a}$$

$$\hat{\eta}_2 = \sqrt{\frac{j\omega\mu_2}{\sigma_2 + j\omega\varepsilon_2}} = \eta_2 \angle\theta_{\eta 2} \tag{16.80b}$$

Recall Eq. (13.138), which states that at the boundary of two media, the tangential component of the electric filed intensity is continuous. Thus,

$$\hat{E}_i + \hat{E}_r = \hat{E}_t \quad z=0 \tag{16.81}$$

or

$$\hat{E}_i e^{-\hat{\gamma}_1 z} + \hat{E}_r e^{\hat{\gamma}_1 z} = \hat{E}_t e^{-\hat{\gamma}_2 z} \quad z=0 \tag{16.82}$$

leading to

$$\hat{E}_i + \hat{E}_r = \hat{E}_t \tag{16.83}$$

The boundary condition imposed on the magnetic field, requires that the tangential component of the magnetic field intensity must be continuous. Thus,

$$\hat{H}_i + \hat{H}_r = \hat{H}_t \quad z=0 \tag{16.84}$$

or

$$\frac{\hat{E}_i}{\hat{\eta}_1} e^{-\hat{\gamma}_1 z} - \frac{\hat{E}_r}{\hat{\eta}_1} e^{\hat{\gamma}_1 z} = \frac{\hat{E}_t}{\hat{\eta}_2} e^{-\hat{\gamma}_2 z} \quad z=0 \tag{16.85}$$

leading to

$$\frac{\hat{E}_i}{\hat{\eta}_1} - \frac{\hat{E}_r}{\hat{\eta}_1} = \frac{\hat{E}_t}{\hat{\eta}_2} \tag{16.86}$$

Substituting Eq. (16.83) into Eq. (16.86) results in

$$\frac{\hat{E}_i}{\hat{\eta}_1} - \frac{\hat{E}_r}{\hat{\eta}_1} = \frac{\hat{E}_i + \hat{E}_r}{\hat{\eta}_2} \tag{16.87}$$

or

$$\begin{aligned}
\frac{\hat{E}_i}{\hat{\eta}_1} - \frac{\hat{E}_r}{\hat{\eta}_1} &= \frac{\hat{E}_i}{\hat{\eta}_2} + \frac{\hat{E}_r}{\hat{\eta}_2} \\
\frac{\hat{E}_i}{\hat{\eta}_1} - \frac{\hat{E}_i}{\hat{\eta}_2} &= \frac{\hat{E}_r}{\hat{\eta}_1} + \frac{\hat{E}_r}{\hat{\eta}_2} \\
\hat{E}_i\left(\frac{1}{\hat{\eta}_1} - \frac{1}{\hat{\eta}_2}\right) &= \hat{E}_r\left(\frac{1}{\hat{\eta}_1} + \frac{1}{\hat{\eta}_2}\right) \\
\hat{E}_i\left(\frac{\hat{\eta}_2 - \hat{\eta}_1}{\hat{\eta}_1\hat{\eta}_2}\right) &= \hat{E}_r\left(\frac{\hat{\eta}_2 + \hat{\eta}_1}{\hat{\eta}_1\hat{\eta}_2}\right)
\end{aligned} \tag{16.88}$$

leading to the definition of the *reflection coefficient* at the boundary as

$$\hat{\Gamma} = \Gamma \angle \theta_{\Gamma} = \frac{\hat{E}_r}{\hat{E}_i} = \frac{\hat{\eta}_2 - \hat{\eta}_1}{\hat{\eta}_2 + \hat{\eta}_1} \tag{16.89}$$

Thus the reflected wave is related to the incident wave by

$$\hat{E}_r = \hat{\Gamma}\hat{E}_i \tag{16.90}$$

From Eq. (16.83) we get

$$\hat{E}_r = \hat{E}_t - \hat{E}_i \tag{16.91}$$

Substituting Eq. (16.91) into Eq. (16.87) results in

$$\frac{\hat{E}_i}{\hat{\eta}_1} - \frac{\left(\hat{E}_t - \hat{E}_i\right)}{\hat{\eta}_1} = \frac{\hat{E}_t}{\hat{\eta}_2} \tag{16.92}$$

or

$$\frac{\hat{E}_i}{\hat{\eta}_1} - \frac{\hat{E}_t}{\hat{\eta}_1} + \frac{\hat{E}_i}{\hat{\eta}_1} = \frac{\hat{E}_t}{\hat{\eta}_2}$$

$$\frac{\hat{E}_i}{\hat{\eta}_1} + \frac{\hat{E}_i}{\hat{\eta}_1} = \frac{\hat{E}_t}{\hat{\eta}_2} + \frac{\hat{E}_t}{\hat{\eta}_1}$$

$$\hat{E}_i\left(\frac{1}{\hat{\eta}_1} + \frac{1}{\hat{\eta}_1}\right) = \hat{E}_t\left(\frac{1}{\hat{\eta}_2} + \frac{1}{\hat{\eta}_1}\right) \tag{16.93}$$

$$\hat{E}_i\left(\frac{2}{\hat{\eta}_1}\right) = \hat{E}_t\left(\frac{\hat{\eta}_1 + \hat{\eta}_2}{\hat{\eta}_1\hat{\eta}_2}\right)$$

Leading to the definition of the *transmission coefficient* at the boundary as

$$\hat{T} = T \angle \theta_T = \frac{\hat{E}_t}{\hat{E}_i} = \frac{2\hat{\eta}_2}{\hat{\eta}_2 + \hat{\eta}_1} \tag{16.93}$$

Thus the transmitted wave is related to the incident wave by

$$\hat{E}_r = \hat{T}\hat{E}_i \tag{16.94}$$

16.4   EMC Applications

16.4.1   Electromagnetic Wave Shielding

Metallic shields are often employed in electronic products in order to decrease the radiated emissions or to increase the radiated immunity. This is shown in Figures 16.4 and 16.5, respectively.

The shielding effect can be described using the theory of electromagnetic wave propagation.

Consider a metallic shield of thickness t surrounded on both sides by air (free space), as shown in Figure 16.6.

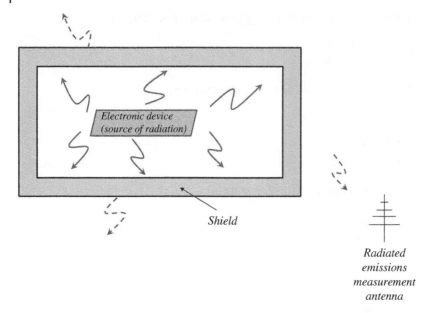

Figure 16.4 Shielding to decrease the radiated emissions.

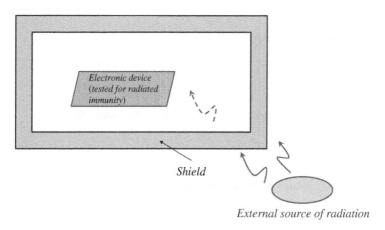

Figure 16.5 Shielding to increase the radiated immunity.

Incident on the left surface of this shield is the uniform plane wave. The incident wave, $(\hat{\mathbf{E}}_i, \hat{\mathbf{H}}_i)$, will be partially reflected, $(\hat{\mathbf{E}}_r, \hat{\mathbf{H}}_r)$, and partially transmitted, $(\hat{\mathbf{E}}_1, \hat{\mathbf{H}}_1)$, through the shield. The transmitted wave, $(\hat{\mathbf{E}}_1, \hat{\mathbf{H}}_1)$, upon arrival at the rightmost boundary will be partially reflected, $(\hat{\mathbf{E}}_2, \hat{\mathbf{H}}_2)$, and partially transmitted, $(\hat{\mathbf{E}}_t, \hat{\mathbf{H}}_t)$ through the shield.

The incident wave is described by

$$\hat{\mathbf{E}}_i = \hat{E}_i e^{-j\beta_0 z} \mathbf{a}_x \tag{16.95a}$$

$$\hat{\mathbf{H}}_i = \frac{\hat{E}_i}{\eta_0} e^{-j\beta_0 z} \mathbf{a}_y \tag{16.95b}$$

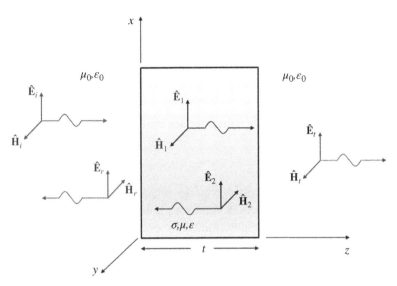

Figure 16.6 Electromagnetic wave shielding.

where

$$\beta_0 = \omega\sqrt{\mu_0\varepsilon_0} \tag{16.96a}$$

$$\eta_0 = \sqrt{\frac{\mu_0}{\varepsilon_0}} \tag{16.96b}$$

The reflected wave is described by

$$\hat{\mathbf{E}}_r = \hat{E}_r e^{j\beta_0 z}\mathbf{a}_x \tag{16.97a}$$

$$\hat{\mathbf{H}}_r = -\frac{\hat{E}_r}{\eta_0}e^{j\beta_0 z}\mathbf{a}_y \tag{16.97b}$$

The wave transmitted through the left interface is described by

$$\hat{\mathbf{E}}_1 = \hat{E}_1 e^{-\hat{\gamma} z}\mathbf{a}_x \tag{16.98a}$$

$$\hat{\mathbf{H}}_1 = \frac{\hat{E}_1}{\hat{\eta}}e^{-\hat{\gamma} z}\mathbf{a}_y \tag{16.98b}$$

where

$$\hat{\gamma} = \sqrt{j\omega\mu(\sigma + j\omega\varepsilon)} = \alpha + j\beta \tag{16.99a}$$

$$\hat{\eta} = \sqrt{\frac{j\omega\mu}{\sigma + j\omega\varepsilon}} \tag{16.99b}$$

The wave reflected at the right interface is described by

$$\hat{\mathbf{E}}_2 = \hat{E}_2 e^{\dot{\gamma}z} \mathbf{a}_x \tag{16.100a}$$

$$\hat{\mathbf{H}}_2 = -\frac{\hat{E}_2}{\hat{\eta}} e^{\dot{\gamma}z} \mathbf{a}_y \tag{16.100b}$$

Finally, the wave transmitted through the right interface is described by

$$\hat{\mathbf{E}}_t = \hat{E}_t e^{-j\beta_0 z} \mathbf{a}_x \tag{16.101a}$$

$$\hat{\mathbf{H}}_t = \frac{\hat{E}_t}{\eta_0} e^{-j\beta_0 z} \mathbf{a}_y \tag{16.101b}$$

The shielding effectiveness, *SE*, can be determined by evaluating the ratio of the transmitted field magnitude to the incident field magnitude (Ott, 2009, p. 244),

$$SE_E = \frac{\left|\hat{E}_t\right|}{\left|\hat{E}_i\right|} \tag{16.102a}$$

$$SE_H = \frac{\left|\hat{H}_t\right|}{\left|\hat{H}_i\right|} \tag{16.102b}$$

Usually, these are expressed in dB, as

$$SE_{E,dB} = 20\log_{10}\frac{\left|\hat{E}_t\right|}{\left|\hat{E}_i\right|} \tag{16.103a}$$

$$SE_{H,dB} = 20\log_{10}\frac{\left|\hat{H}_t\right|}{\left|\hat{H}_i\right|} \tag{16.103b}$$

The *relative* effectiveness of various shield can be determined by the direct field measurements. The experimental setup for **H** field measurements is shown in Figures 16.7 and 16.8.

Figure 16.7 shows the unshielded switched-mode power supply (SMPS) and the H-field probe for the field measurements. Figure 16.8 shows a SMPS with a shield.

The following shields were evaluated:

- phosphorus-bronze 8 mils
- phosphorus-bronze 15 mils
- nickel-silver 8 mils
- cold-rolled-steel 15 mils
- copper tape 3 mils
- cold-rolled-steel w/holes 15 mils

The results are shown in Figures 16.9–16.13.

Figure 16.7 SMPS with no shield.

Figure 16.8 SMPS with a shield.

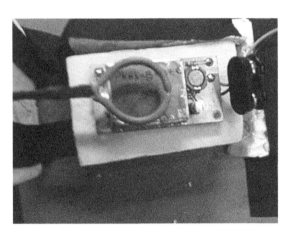

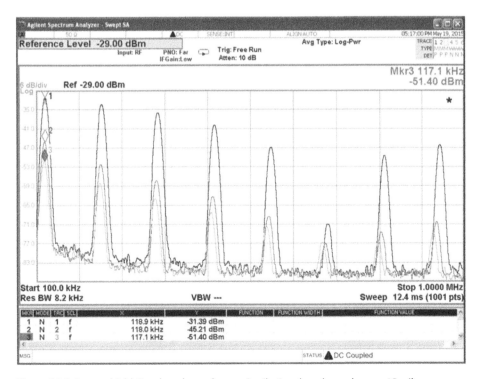

Figure 16.9 1 – no shield; 2 – phosphorus-bronze 8 mils; 3 – phosphorus-bronze 15 mils.

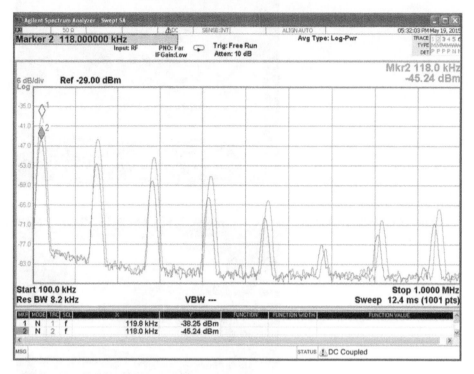

Figure 16.10 1 – nickel-silver 8 mils; 2 – phosphorus-bronze 8 mils.

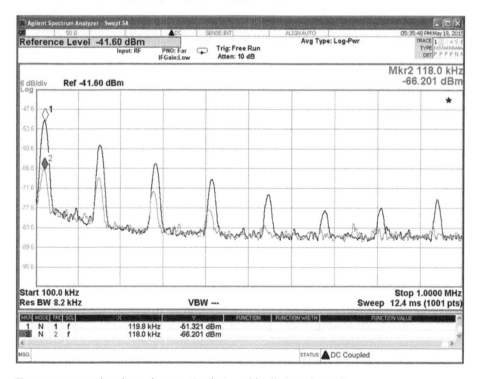

Figure 16.11 1 – phosphorus-bronze 15 mils; 2 – cold-rolled-steel 15 mils.

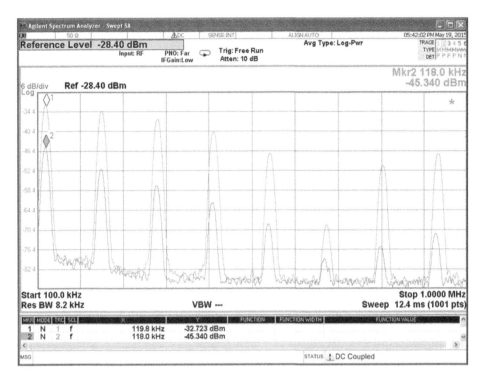

Figure 16.12 1 – no shield; 2 – copper tape 3 mils.

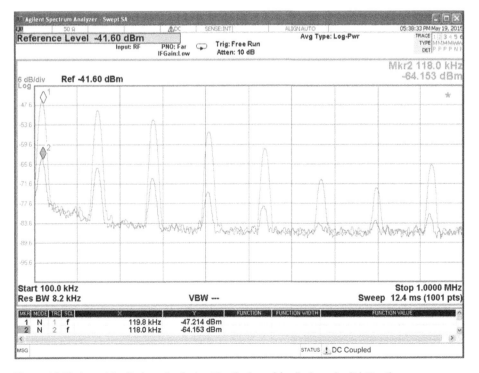

Figure 16.13 1 – cold-rolled-steel w/holes 15 mils; 2 – cold-rolled-steel solid 15 mils.

References

Paul, C.R., *Introduction to Electromagnetic Compatibility*, 2nd ed., John Wiley and Sons, New York, 2006.

Rao, N.N., *Elements of Engineering Electromagnetics*, 6th ed., Pearson Prentice Hall, Upper Saddle River, NJ, 2004.

Ott, H.W., *Electromagnetic Compatibility Engineering*, John Wiley and Sons, Hoboken, NJ, 2009.

17

Transmission Lines

17.1 Transient Analysis

In the previous chapter we reviewed static electric fields due to stationary charge distributions, and static magnetic fields, due to the charges moving at constant speed, i.e. dc currents.

A transmission line can be modeled as a distributed parameter circuit consisting a series of small segments of length Δz, as shown in Figure 17.1

The distributed parameters describing the transmission line are:

r – resistance per-unit-length (Ω/m)
l – inductance per-unit-length (H/m)
g – conductance per-unit-length (S/m)
c – capacitance per-unit-length (F/m)

The transmission line model in Figure 17.1 describes a *lossy* transmission line. To gain an insight into transmission line theory it is very helpful to consider a *lossless* transmission line first. Such a transmission line is shown in Figure 17.2.

To obtain the transmission line equations let's consider a single segment of a lossless transmission line shown in Figure 17.3.

Writing Kirchhoff's voltage law around the outside loop results in

$$-V(z,t)+l\Delta z\frac{\partial I(z,t)}{\partial t}+V(z+\Delta z,t)=0 \tag{17.1}$$

or

$$V(z+\Delta z,t)-V(z,t)=-l\Delta z\frac{\partial I(z,t)}{\partial t} \tag{17.2}$$

Dividing both sides by Δz and taking the limit gives

$$\lim_{\Delta z\to 0}\frac{V(z+\Delta z,t)-V(z,t)}{\Delta z}=-l\frac{\partial I(z,t)}{\partial t} \tag{17.3}$$

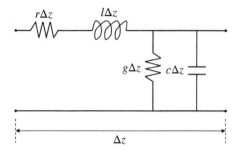

Figure 17.1 Circuit model of a transmission line.

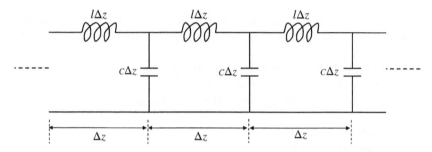

Figure 17.2 Circuit model of a lossless transmission line.

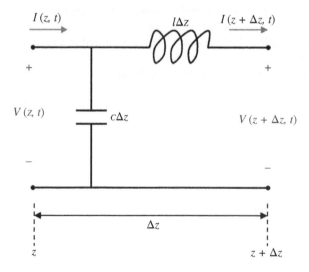

Figure 17.3 Single segment of a lossless transmission line.

or

$$\frac{\partial V(z,t)}{\partial z} = -l\frac{\partial I(z,t)}{\partial t} \qquad (17.4)$$

Writing Kirchhoff's current law at the upper node of the capacitor results in

$$I(z,t) = I(z+\Delta z,t) + c\Delta z\frac{\partial V(z+\Delta z,t)}{\partial t} \qquad (17.5)$$

or

$$I(z+\Delta z,t) - I(z,t) = -c\Delta z\frac{\partial V(z+\Delta z,t)}{\partial t} \qquad (17.6)$$

Dividing both sides by Δz and taking the limit gives

$$\lim_{\Delta z \to 0}\frac{I(z+\Delta z,t) - I(z,t)}{\Delta z} = -\lim_{\Delta z \to 0}c\frac{\partial V(z+\Delta z,t)}{\partial t} \qquad (17.7)$$

or

$$\frac{\partial I(z,t)}{\partial z} = -c\frac{\partial V(z,t)}{\partial t} \qquad (17.8)$$

Equations (17.4) and (17.5) constitute a set of first-order coupled transmission line equations. These equations can be decoupled as flows.

Differentiating Eq. (17.4) with respect to z gives

$$\frac{\partial^2 V(z,t)}{\partial z^2} = -l\frac{\partial^2 I(z,t)}{\partial t\partial z} \qquad (17.9)$$

while differentiating Eq. (17.8) with respect to t gives

$$\frac{\partial^2 I(z,t)}{\partial z\partial t} = -c\frac{\partial^2 V(z,t)}{\partial t^2} \qquad (17.10)$$

Using Eq. (17.10) in eq. (17.9) produces

$$\frac{\partial^2 V(z,t)}{\partial z^2} = lc\frac{\partial^2 V(z,t)}{\partial t^2} \qquad (17.11a)$$

In a similar manner, we can obtain the second transmission line equation as

$$\frac{\partial^2 I(z,t)}{\partial z^2} = lc\frac{\partial^2 I(z,t)}{\partial t^2} \qquad (17.11b)$$

The general solutions to these transmission-line equations are (Rao, 2004. Pg. 372)

$$V(z,t) = V^+\left(t - \frac{z}{v}\right) + V^-\left(t + \frac{z}{v}\right) \qquad (17.12a)$$

$$I(z,t) = I^+\left(t - \frac{z}{v}\right) + I^-\left(t + \frac{z}{v}\right) \qquad (17.12b)$$

where

$$I^+\left(t-\frac{z}{v}\right)=\frac{1}{Z_C}V^+\left(t-\frac{z}{v}\right) \tag{17.13a}$$

$$I^-\left(t+\frac{z}{v}\right)=-\frac{1}{Z_C}V^-\left(t+\frac{z}{v}\right) \tag{17.13b}$$

Z_C is the *characteristic impedance* of the line

$$Z_C=\sqrt{\frac{l}{c}} \tag{17.14}$$

The function $V^+(t-z/v)$ represents a forward-traveling voltage wave traveling in the $+z$ direction, while the function $V^-(t+z/v)$ represents a backward-traveling voltage wave traveling in the $-z$ direction (see Chapter 16 for the detailed explanation).

Similar statements are valid for the current waves. The total solution consists of the sum of forward-traveling and backward-traveling waves.

The velocity of the wave propagation along the line is given by

$$v=\frac{1}{\sqrt{lc}} \tag{17.15}$$

17.1.1 Reflections on Transmission Lines

To simplify the notation in the following discussion, let's rewrite the solution in Eqs (17.12) in a concise form (Rao, 2004, p. 372)

$$V=V^++V^- \tag{17.16a}$$

$$I=I^++I^- \tag{17.16b}$$

From Eqs (17.13) we observe that

$$I^+=\frac{V^+}{Z_C} \tag{17.17a}$$

$$I^-=-\frac{V^-}{Z_C} \tag{17.17b}$$

From Eq. (17.17a) we also note that

$$Z_C=\frac{V^+}{I^+} \tag{17.18}$$

Consider a transmission line of length L driven by a constant voltage source V_S with a source resistance R_S, and terminated by a resistive load R_L, as shown in Figure 17.4.

We assume that no voltage and current exists on the line prior to the switch closing. When the switch closes at $t=0$, forward voltage and current waves originate at $z=0$ and travel toward the load.

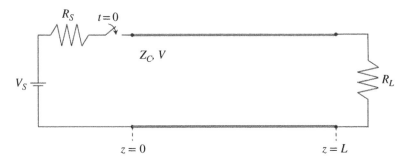

Figure 17.4 Transmission line driven by a constant source and terminated by a resistive load.

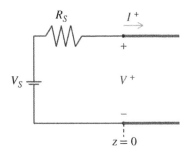

Figure 17.5 Voltage and current forward waves originate at the source.

Let's denote these waves as V^+ and I^+, as shown in Figure 17.5. Writing a KVL for the circuit in Figure 17.5 we get

$$-V_S + R_S I^+ + V^+ = 0 \tag{17.19}$$

Utilizing Eq. (17.18) we obtain

$$-V_S + R_S \frac{V^+}{Z_C} + V^+ = 0 \tag{17.20}$$

or

$$V^+ + \frac{V^+}{Z_C} R_S = V_S$$

$$V^+ \left(1 + \frac{R_S}{Z_C} \right) = V_S \tag{17.21}$$

$$V^+ \left(\frac{R_S + Z_C}{Z_C} \right) = V_S$$

and thus

$$V^+ = \frac{Z_C}{R_S + Z_C} V_S \tag{17.22a}$$

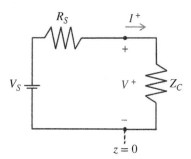

Figure 17.6 Equivalent circuit at $t=0$.

$$I^+ = \frac{V^+}{Z_C} = \frac{V_S}{R_S + Z_C} \qquad (17.22b)$$

Equations (17.22) specify the initial voltage and current values (at location $z=0$) that will propagate towards the load.

Now, consider the circuit shown in Figure 17.6 and look back at Eqs (17.22).

We quickly realize that Eqs (17.22) apply to this circuit. Thus, the circuits shown in Figures 17.6 and 17.5 are equivalent! We may, therefore, say that the source "sees" the transmission line as a resistance equal to the characteristic impedance of the line connected across $z=0$.

Note that the voltage value that propagates towards the load is not equal to the dc voltage of the source, V_S, but is obtained from it using the voltage divider in Eq. (17.22a).

The voltage (and current) wave that originated at the source now travels towards the load, as shown in Figure 17.7.

As this wave travels along the transmission line, the voltage along the line changes from 0 to V^+ and remains at that value (for now).

At the time

$$T = \frac{L}{v} \qquad (17.23)$$

the voltage and current waves reach the load, as shown in Figure 17.8(a).

Applying Ohm's law to the circuit shown in Figure 17.8(a) we get

$$\frac{V^+}{I^+} = R_L \qquad (17.23)$$

We know, however, that the ratio of the forward voltage wave to the forward current wave must be equal to the characteristic impedance of the line

$$\frac{V^+}{I^+} = Z_C \qquad (17.24)$$

This contradiction can only be resolved by postulating the creation of reflected waves at the load, as shown in Figure 17.8(b). We denote these reflected waves as V^- and Γ, respectively.

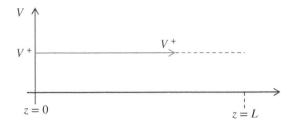

Figure 17.7 Voltage wave travels towards the load.

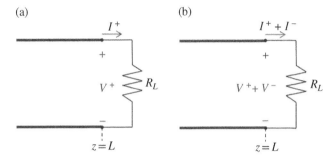

Figure 17.8 Voltage and current waves arrive at the load.

The total voltage across the load and total current through the load are

$$V = V^+ + V^-$$ (17.25a)

$$I = I^+ + I^-$$ (17.25b)

Ohm's law for the circuit in Figure 17.8(b) produces

$$V^+ + V^- = R_L \left(I^+ + I^- \right)$$ (17.26)

We refer to the Eq. (17.26) as the boundary condition at the load. Utilizing Eqs (17.17), repeated here

$$I^+ = \frac{V^+}{Z_C}$$ (17.27a)

$$I^- = -\frac{V^-}{Z_C}$$ (17.27b)

we rewrite Eq. (17.26) as

$$V^+ + V^- = R_L \left(\frac{V^+}{Z_C} - \frac{V^-}{Z_C} \right)$$ (17.28)

or

$$V^+ + V^- = R_L \frac{V^+}{Z_C} - R_L \frac{V^-}{Z_C}$$

$$V^- + R_L \frac{V^-}{Z_C} = R_L \frac{V^+}{Z_C} - V^+$$

$$V^-\left(1 + \frac{R_L}{Z_C}\right) = V^+\left(\frac{R_L}{Z_C} - 1\right) \qquad (17.29)$$

$$V^-\left(\frac{R_L + Z_C}{Z_C}\right) = V^+\left(\frac{R_L - Z_C}{Z_C}\right)$$

and thus the reflected voltage is related to the incident voltage by

$$V^- = V^+ \frac{R_L - Z_C}{R_L + Z_C} \qquad (17.30)$$

The ratio of the reflected (backward-traveling) wave to the incident (forward-traveling) wave is defined as the *voltage reflection coefficient at the load*:

$$\Gamma_L = \frac{V^-}{V^+} = \frac{R_L - Z_C}{R_L + Z_C} \qquad (17.31)$$

Therefore the reflected voltage waveform at the load can be found from the incident wave using the reflection coefficient as

$$V^- = \Gamma_L V^+ \qquad (17.32)$$

The current reflection coefficient at the load is

$$\frac{I^-}{I^+} = \frac{-V^-\big/Z_C}{V^+\big/Z_C} = -\frac{V^-}{V^+} = -\Gamma_L \qquad (17.33)$$

Therefore the reflected current waveform at the load can be found from the incident wave using the reflection coefficient as

$$I^- = -\Gamma_L I^+ \qquad (17.34)$$

The reflected waves now travel back to the source, as shown in Figure 17.9. This wave reaches the source at the time

$$T = \frac{2L}{v} \qquad (17.35)$$

The reflected voltage and current waves reach the load, as shown in Figure 17.10(a). Applying KVL to the circuit shown in Figure 17.10(a) produces

$$-V_S + R_S\left(I^+ + I^-\right) + V^+ + V^- = 0 \qquad (17.36)$$

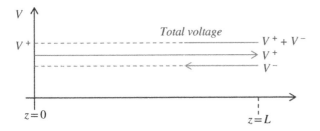

Figure 17.9 Reflected wave travels towards the source.

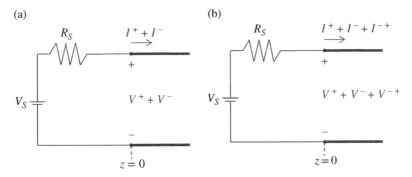

Figure 17.10 Reflected voltage and current waves arrive at the source.

This equation leads to the contradictory conclusions shown next. Utilizing Eqs (17.27), repeated here

$$I^+ = \frac{V^+}{Z_C} \tag{17.37a}$$

$$I^- = -\frac{V^-}{Z_C} \tag{17.37b}$$

we rewrite Eq. (17.36) as

$$-V_S + R_S\left(\frac{V^+}{Z_C} - \frac{V^-}{Z_C}\right) + V^+ + V^- = 0 \tag{17.38}$$

or

$$\frac{R_S}{Z_C}\left(V^+ - V^-\right) + V^+ + V^- = V_S \tag{17.39}$$

Now, using Eq. (17.32), repeated here

$$V^- = \Gamma_L V^+ \tag{17.40}$$

we rewrite Eq. (17.39) as

$$\frac{R_S}{Z_C}\left(V^+ - \Gamma V^+\right) + V^+ + \Gamma V^+ = V_S$$
$$V^+\left(\frac{R_S}{Z_C} - \Gamma\frac{R_S}{Z_C} + 1 + \Gamma\right) = V_S \qquad (17.41)$$

Using Eq. (17.31) repeated here

$$\Gamma_L = \frac{R_L - Z_C}{R_L + Z_C} \qquad (17.42)$$

we get

$$V^+\left(\frac{R_S}{Z_C} - \frac{R_L - Z_C}{R_L + Z_C}\frac{R_S}{Z_C} + 1 + \frac{R_L - Z_C}{R_L + Z_C}\right) = V_S \qquad (17.43)$$

or

$$V^+\left[\frac{R_S\left(R_L + Z_C\right) - \left(R_L - Z_C\right)R_S + Z_C\left(R_L + Z_C\right) + Z_C\left(R_L - Z_C\right)}{Z_C\left(R_L + Z_C\right)}\right] = V_S \qquad (17.44)$$

which reduces to

$$V^+\left[\frac{2\left(R_S + R_L\right)}{\left(R_L + Z_C\right)}\right] = V_S \qquad (17.45)$$

from which

$$V^+ = \frac{R_L + Z_C}{2\left(R_S + R_L\right)}V_S \qquad (17.46)$$

But from Eq. (17.22a), repeated here, we have

$$V^+ = \frac{Z_C}{R_S + Z_C}V_S \qquad (17.47)$$

The only way this inconsistency can be resolved is by postulating the creation of the (re)reflected waves at the source. These reflected waves are denoted as V^{-+} and Γ^{-+}, respectively, and shown in Figure 17.11(b).

The total voltage and current at the source are

$$V = V^+ + V^- + V^{-+} \qquad (17.48a)$$

$$I = I^+ + I^- + I^{-+} \qquad (17.48b)$$

The re-reflected waves now travel back to the load, as shown in Figure 17.11.

KVL applied to the circuit shown in Figure 17.10(b) produces

$$-V_S + R_S\left(I^+ + I^- + I^{-+}\right) + V^+ + V^- + V^{-+} = 0 \qquad (17.49)$$

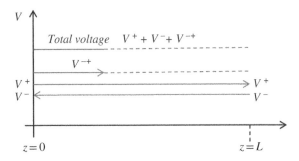

Figure 17.11 Re-reflected wave travels towards the load.

Utilizing

$$I^+ = \frac{V^+}{Z_C} \qquad (17.50\text{a})$$

$$I^- = -\frac{V^-}{Z_C} \qquad (17.50\text{b})$$

$$I^{-+} = \frac{V^{-+}}{Z_C} \qquad (17.50\text{c})$$

in Eq. (17.49) we obtain

$$V^+ + V^- + V^{-+} = V_S - \frac{R_S}{Z_C}\left(V^+ - V^- + V^{-+}\right) \qquad (17.51)$$

Substituting

$$V^+ = V_S \frac{Z_C}{R_S + Z_C} \qquad (17.52)$$

in Eq. (17.51) results in

$$
\begin{aligned}
V_S \frac{Z_C}{R_S + Z_C} + V^- + V^{-+} &= V_S - \frac{R_S}{Z_C}\left(V_S \frac{Z_C}{R_S + Z_C} - V^- + V^{-+}\right) \\
V_S \frac{Z_C}{R_S + Z_C} + V^- + V^{-+} &= V_S - V_S \frac{R_S}{R_S + Z_C} - \frac{R_S}{Z_C}(-V^-) - \frac{R_S}{Z_C}V^{-+} \\
V_S \frac{Z_C}{R_S + Z_C} - V_S + V_S \frac{R_S}{R_S + Z_C} &= -V^- - V^{-+} - \frac{R_S}{Z_C}(-V^-) - \frac{R_S}{Z_C}V^{-+} \\
V_S\left(\frac{Z_C}{R_S + Z_C} + \frac{R_S}{R_S + Z_C} - 1\right) &= -V^- - V^{-+} - \frac{R_S}{Z_C}(-V^-) - \frac{R_S}{Z_C}V^{-+}
\end{aligned}
\qquad (17.53)
$$

Now, the left-hand-side of Eq. (17.53) simplifies to

$$V_S\left(\frac{Z_C}{R_S + Z_C} + \frac{R_S}{R_S + Z_C} - 1\right) = V_S\left(\frac{Z_C + R_S - R_S - Z_C}{R_S + Z_C}\right) = 0 \qquad (17.54)$$

thus

$$0 = -V^- - V^+ - \frac{R_S}{Z_C}\left(-V^-\right) - \frac{R_S}{Z_C}V^+$$

$$V^+ + \frac{R_S}{Z_C}V^+ = +\frac{R_S}{Z_C}V^- - V^-$$

$$V^+\left(1 + \frac{R_S}{Z_C}\right) = \left(\frac{R_S}{Z_C} - 1\right)V^-$$

$$V^+\left(\frac{R_S + Z_C}{Z_C}\right) = \left(\frac{R_S - Z_C}{Z_C}\right)V^- \qquad (17.55)$$

or

$$V^+ = V^- \frac{R_S - Z_C}{R_S + Z_C} \qquad (17.56)$$

The ratio of the re-reflected (forward-traveling) wave to the incident (backward-traveling) wave is defined as the *voltage reflection coefficient at the source*:

$$\Gamma_S = \frac{R_S - Z_C}{R_S + Z_C} \qquad (17.57)$$

Thus,

$$V^+ = \Gamma_S V^- \qquad (17.58)$$

Next, we will discuss some special cases of the reflection coefficient.

Short-circuited line $R_L = 0$ In this case the reflection coefficient is

$$\Gamma = \frac{R_L - Z_C}{R_L + Z_C} = \frac{0 - Z_C}{0 + Z_C} = -1 \qquad (17.59)$$

The reflected voltage is

$$V^- = \Gamma_L V^+ = -V^+ \qquad (17.60)$$

The total voltage at the load is

$$V_{total} = V^+ + V^- = V^+ - V^+ = 0 \qquad (17.61)$$

The reflected voltage is the negative of the incident voltage, and the total voltage across the load is zero. This is consistent with what we would expect across a short circuit.

Open-circuited line $R_L = \infty$ In this case the reflection coefficient is

$$\Gamma = \frac{R_L - Z_C}{R_L + Z_C} = \left.\frac{1 - \dfrac{Z_C}{R_L}}{1 + \dfrac{Z_C}{R_L}}\right|_{R_L \to \infty} = 1 \qquad (17.62)$$

The reflected voltage is

$$V^- = \Gamma_L V^+ = V^+ \qquad (17.63)$$

The total voltage at the load is

$$V_{total} = V^+ + V^- = 2V^+ \qquad (17.64)$$

The reflected voltage wave is equal to the incident wave and the two add up to give the total voltage of double the incident value at the load.

Matched line $R_L = Z_C$ In this case the reflection coefficient is

$$\Gamma = \frac{Z_C - Z_C}{Z_C + Z_C} = 0 \qquad (17.65)$$

The reflected voltage is

$$V^- = \Gamma_L V^+ = 0 \qquad (17.66)$$

The total voltage at the load is

$$V_{total} = V^+ + 0 = V^+ \qquad (17.67)$$

There is no reflection at the load; the total voltage at the load is equal to the incident voltage only.

Example 17.1 Transmission line reflections
Consider the circuit shown in Figure 17.12.
    The experimental setup reflecting this circuit is shown in Figures 17.13. and 17.14.
    A 2 V_{pp} (open-circuit voltage) pulse signal was sent from the function generator along the coaxial cable to the resistive load. The voltages at the source (V_S) and at the load (V_L) were measured using oscilloscope probes. The source was matched to the transmission line and the load resistance was varied as shown in Table 17.1.

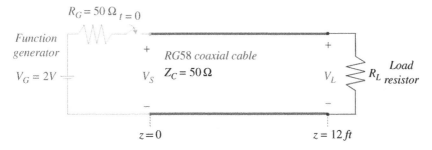

Figure 17.12 Circuit for the load reflection measurements.

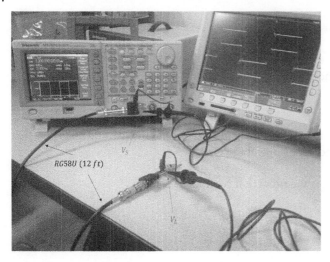

Figure 17.13 Experimental setup for the load reflection measurements.

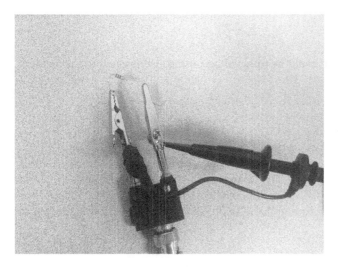

Figure 17.14 Experimental setup – load resistance.

Table 17.1 Resistive load values.

Case 1	$R_L = \infty$ (open circuit)
Case 2	$R_L = 22\ \Omega$
Case 3	$R_L = 47\ \Omega$
Case 4	$R_L = 216\ \Omega$

Case 1 $R_L = \infty$ First, the load was terminated in an open circuit. When the switch closes, the initial voltage wave is created at location $z = 0$. The value of this voltage is

$$V^+ = V_S(z=0) = V_G \frac{Z_C}{R_S + Z_C} = (2)\frac{50}{50+50} = 1\,V$$

and is shown in Figure 17.15. After $t = 18\,ns$ (one-way travel time, T) this waveform arrives at the load. The load reflection coefficient is

$$\Gamma_L = \frac{R_L - Z_C}{R_L + Z_C} = 1$$

The reflected voltage at the load is

$$V^- = \Gamma_L V^+ = (1)(1) = 1\,V$$

The total voltage at the load is

$$V_L = V^+ + V^- = 1+1 = 2\,V$$

as shown in Figure 17.15. The voltage reflected at the load $(V^- = 1\,V)$ travels back to the source and reaches it at $t = 2\,T$. The total voltage at the source at $t = 2\,T$ is

$$V_S = V^+ + V^- = 1+1 = 2\,V$$

Since the source is matched, there is no reflection and the voltage stays at the value of 2 V, as shown in Figure 17.15.

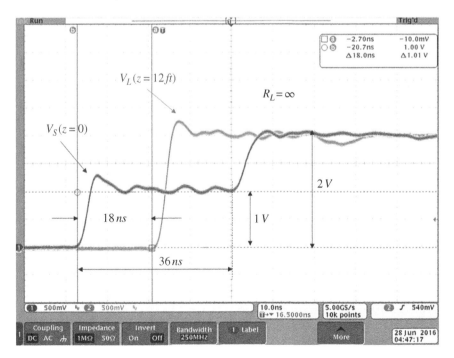

Figure 17.15 Source and load voltages for $R_L = \infty$.

Case 2 $R_L = 22\,\Omega$ The initial voltage at location $z = 0$ is

$$V^+ = V_S(z=0) = 1 \text{ V}$$

After $t = T$ this waveform arrives at the load. The load reflection coefficient is

$$\Gamma_L = \frac{R_L - Z_C}{R_L + Z_C} = \frac{22 - 50}{22 + 50} = -0.39$$

The reflected voltage at the load is

$$V^- = \Gamma_L V^+ = (-0.39)(1) = -0.39 \text{ V}$$

The total voltage at the load is

$$V_L = V^+ + V^- = 1 - 0.39 = 0.61 \text{ V}$$

The voltage reflected at the load ($V^- = -0.39$ V) travels back to the source and reaches it at $t = 2\,T$. The total voltage at the source at $t = 2\,T$ is

$$V_S = V^+ + V^- = 1 - 0.39 = 0.61 \text{ V}$$

Since the source is matched, there is no reflection and the voltage stays at the value of 0.61 V. These results are shown in Figure 17.16.

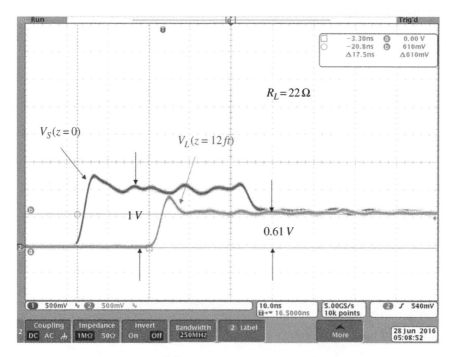

Figure 17.16 Measurement result for $R_L = 22\,\Omega$.

Case 3 $R_L = 47\,\Omega$ The initial voltage at location $z = 0$ is

$$V^+ = V_S(z=0) = 1\ \text{V}$$

After $t = T$ this waveform arrives at the load. The load reflection coefficient is

$$\Gamma_L = \frac{R_L - Z_C}{R_L + Z_C} = \frac{47 - 50}{47 + 50} = -0.03$$

The reflected voltage at the load is

$$V^- = \Gamma_L V^+ = (-0.03)(1) = -0.03\ \text{V}$$

The total voltage at the load is

$$V_L = V^+ + V^- = 1 - 0.03 = 0.97\ \text{V}$$

The voltage reflected at the load ($V^- = -0.03\ \text{V}$) travels back to the source and reaches it at $t = 2\,T$. The total voltage at the source at $t = 2\,T$ is

$$V_S = V^+ + V^- = 1 - 0.03 = 0.97\ \text{V}$$

Since the source is matched, there is no reflection and the voltage stays at the value of 0.97 V. These results are shown in Figure 17.17.

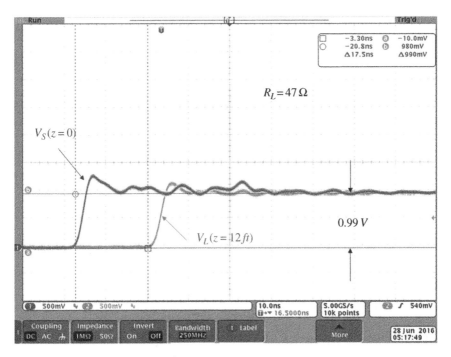

Figure 17.17 Measurement result for $R_L = 47\,\Omega$.

Case 4 $R_L = 216\,\Omega$ The initial voltage at location $z = 0$ is

$$V^+ = V_S(z = 0) = 1\,\text{V}$$

After $t = T$ this waveform arrives at the load. The load reflection coefficient is

$$\Gamma = \frac{R_L - Z_C}{R_L + Z_C} = \frac{216 - 50}{216 + 50} = 0.624$$

The reflected voltage at the load is

$$V^- = \Gamma_L V^+ = (0.624)(1) = 0.624\,\text{V}$$

The total voltage at the load is

$$V_L = V^+ + V^- = 1 + 0.624 = 1.624\,\text{V}$$

The voltage reflected at the load ($V^- = 0.624\,\text{V}$) travels back to the source and reaches it at $t = 2\,T$. The total voltage at the source at $t = 2\,T$ is

$$V_S = V^+ + V^- = 1 + 0.624 = 1.624\,\text{V}$$

Since the source is matched, there is no reflection and the voltage stays at the value of 1.624 V. These results are shown in Figure 17.18.

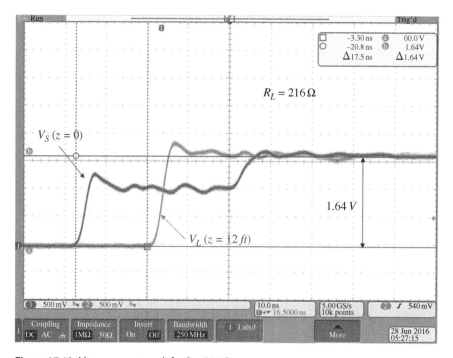

Figure 17.18 Measurement result for $R_L = 216\,\Omega$.

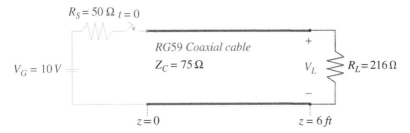

Figure 17.19 Circuit used to create bounce diagram.

17.1.2 Bounce Diagram

Consider the circuit shown in Figure 17.19.

When the switch closes, the forward voltage wave travels towards the load and reaches it at $t = T$ (T = one-way travel time). Since the line and the load are mismatched, a reflection is created and travels back to the source, reaching it at $t = 2T$ (assuming zero-rise time). Since the line and the source are mismatched, another reflection is created, which travels forward to the load, reaching it at $t = 3T$.

This process theoretically continues indefinitely; practically, it continues until the steady-state voltages are reached at the source and at the load. A *bounce diagram* is a plot of the voltage (or current) at the source or the load (or any other location) after each reflection.

Let's create a plot of the voltages at the source and the load for the circuit shown in Figure 17.19.

The initial voltage at the location $z = 0$ is

$$V^+ = V_G \frac{Z_C}{R_S + Z_C} = (10)\frac{75}{50 + 75} = 6 \text{ V}$$

This is shown in Figure 17.20.

The reflection coefficient at the load is

$$\Gamma_L = \frac{R_L - Z_C}{R_L + Z_C} = \frac{216 - 75}{216 + 75} = 0.4845$$

The initial voltage wave of 6 V travels to the load and reaches it at $t = T$ creating a reflection

$$V^- = \Gamma_L V^+ = (0.4845)(6) = 2.907 \text{ V}$$

The total voltage at the load (at $t = T$) is

$$V_L = V^+ + V^- = 6 + 2.907 = 8.907 \text{ V}$$

This is shown in Figure 17.21.

Voltage reflected at the load ($V^- = 2.907$ V) travels back to the source. The reflection coefficient at the source is

$$\Gamma_S = \frac{R_S - Z_C}{R_S + Z_C} = \frac{50 - 75}{50 + 75} = -0.2$$

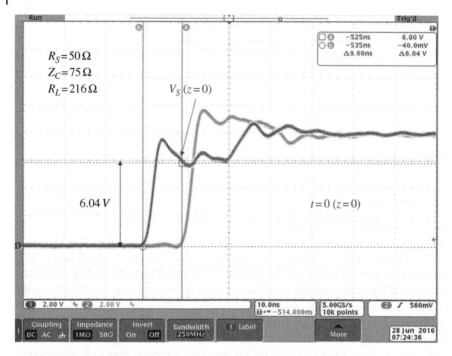

Figure 17.20 Initial voltage wave at $z=0$.

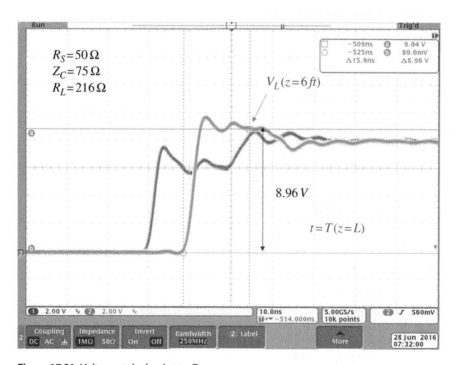

Figure 17.21 Voltage at the load at $t=T$.

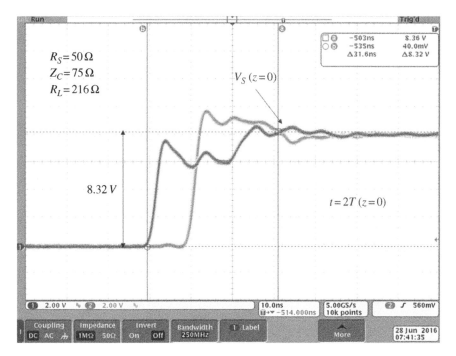

Figure 17.22 Voltage at the source at $t = 2T$.

The re-reflected voltage at the source is

$$V^{-+} = \Gamma_S V^- = (-0.2)(2.907) = -0.5814 \, \text{V}$$

The total voltage at the source at $t = 2T$ is

$$V_S = V^+ + V^- + V^{-+} = 6 + 2.907 - 0.5814 = 8.3256 \, \text{V}$$

This is shown in Figure 17.22.

The voltage reflected at the source ($V^{-+} = -0.5814 \, \text{V}$) travels towards the load where it will create another reflection which will travel towards the source. This process will continue until the steady-state is reached.

The bounce diagram showing the voltages at the source and the load after each reflection is shown in Figure 17.23.

Figure 17.24 shows the voltages at the source ($z = 0$), while Figure 17.25 shows the voltage at the load ($z = L$) during the period $0 \le t < 8T$.

It is apparent that the source and load voltages eventually reach the steady state. Recall that a transmission line can be modeled as a sequence of in-line inductors and shunt capacitors. Under dc conditions (steady-state when driven by a dc source) inductors act as short circuits and capacitors act as open circuits.

Thus in steady state, the circuit in Figure 17.19 is equivalent to the circuit in Figure 17.26, where the transmission line is modeled as an ideal conductor.

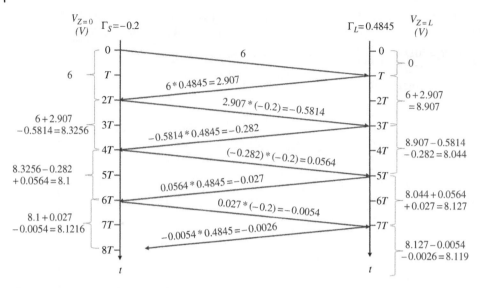

Figure 17.23 Bounce diagram: voltages at the source and the load.

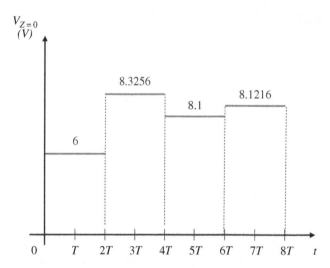

Figure 17.24 Voltage at the source during $0 \le t < 8T$.

The steady state value of the voltage at $z = 0$ is the same as the value at $z = L$ and can be obtained from the voltage divider as

$$V_{SS} = \frac{216}{50+216}(10) = 8.1203 \text{ V}$$

17.1.3 Reflections at an Inductive Load

Consider the circuit shown in Figure 17.27 where the transmission line of length d is terminated by an inductor L.

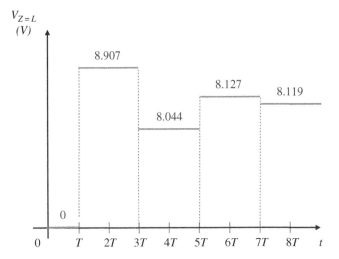

Figure 17.25 Voltage at the load during $0 \le t < 8T$.

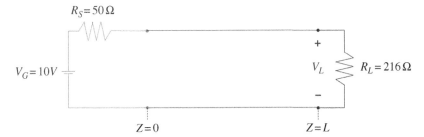

Figure 17.26 Equivalent circuit in steady state.

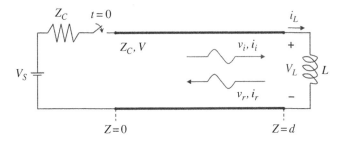

Figure 17.27 Inductive termination of a transmission line.

The source resistance is matched to the characteristic impedance of the line; it is also assumed that the initial current in the inductor is zero

$$i_L\left(0^-\right) = 0 \tag{17.68}$$

When the switch closes at $t = 0$, a wave originates at $z = 0$, with

$$v_i = \frac{V_S}{2} \tag{17.69a}$$

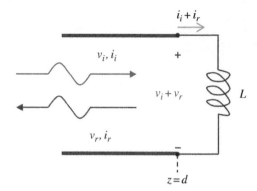

Figure 17.28 Creation of a reflected wave at an inductive load.

$$i_i = \frac{V_S}{2Z_C} \tag{17.69b}$$

and travels towards the load. When this wave arrives at the load (after the time T), the inductor current cannot change instantaneously from zero to the value in (17.69b), KCL is violated at $z = d$, and thus the reflected wave, v_r and i_r is created (Rao, 2004, p. 394) This is shown in Figure 17.28.

The reflected current wave is related to the reflected voltage wave by

$$i_r = -\frac{v_r}{Z_C} \tag{17.70}$$

The voltage–current relationship for an inductor produces

$$v_i + v_r = L\frac{d}{dt}(i_i + i_r) \tag{17.71}$$

or, using Eqs (17.69) and (17.70),

$$\frac{V_S}{2} + v_r = L\frac{d}{dt}\left(\frac{V_S}{2Z_C} - \frac{v_r}{Z_C}\right) \tag{17.72}$$

Since V_S and Z_C are constant, Eq. (17.72) reduces to

$$\frac{V_S}{2} + v_r = -\frac{L}{Z_C}\frac{dv_r}{dt} \tag{17.73}$$

or

$$\frac{L}{Z_C}\frac{dv_r}{dt} + v_r = -\frac{V_S}{2}, \quad v_r(0) = \frac{V_S}{2}, \quad t > T \tag{17.74}$$

This differential equation in v_r was solved in Section 5.4.1. with the result (see Eq. 5.111):

$$v_r(d,t) = -\frac{V_S}{2} + V_S e^{-(Z_C/L)(t-T)}, \quad t > T \tag{17.75}$$

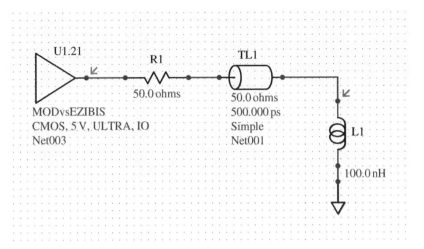

Figure 17.29 HyperLynx circuit model of a transmission line terminated by an inductive load.

The corresponding solution for the reflected current wave is

$$i_r(d, t) = -\frac{v_r(d, t)}{Z_C} = \frac{V_S}{2Z_C} - \frac{V_S}{Z_C} e^{-(Z_C/L)(t-T)}, \qquad t > T \tag{17.76}$$

This reflected voltage wave and reflected current wave travel back to the source; since the source is matched to the line, there is no reflection at the source. The total voltage across the inductor is

$$v(d, t) = v_i(d, t) + v_r(d, t)$$
$$= \begin{cases} 0, & t < T \\ V_S e^{-(Z_C/L)(t-T)}, & t > T \end{cases} \tag{17.77}$$

The total current through the inductor is

$$i(d, t) = i_i(d, t) + i_r(d, t)$$
$$= \begin{cases} 0, & t < T \\ \dfrac{V_S}{Z_C}\left[1 - e^{-(Z_C/L)(t-T)}\right], & t > T \end{cases} \tag{17.78}$$

Figure 17.29 shows the circuit schematic of a transmission line driven by a 5 V CMOS and terminated in an inductive load.

The driver voltage and the voltage across the inductor are displayed in Figure 17.30.

17.1.4 Reflections at a Capacitive Load

Consider the circuit shown in Figure 17.31.

A line of length d is terminated by a capacitor C with zero initial voltage.

$$v_C(0^-) = 0 \tag{17.79}$$

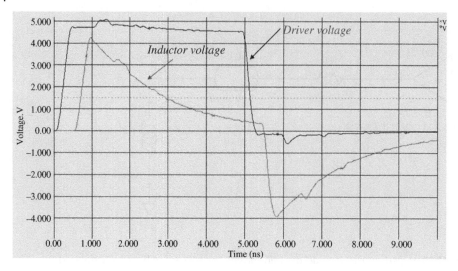

Figure 17.30 Driver voltage and the voltage across the inductor.

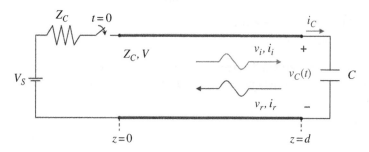

Figure 17.31 Transmission line terminated by a capacitive load.

A constant voltage source with internal resistance equal to the characteristic imped-ance Z_C of the line is connected to the line at $t = 0$. When the switch closes at $t = 0$, a wave originates at $z = 0$, with

$$v_i = \frac{V_S}{2} \tag{17.80a}$$

$$i_i = \frac{V_S}{2Z_C} \tag{17.80b}$$

This wave travels down the line to reach the load end at time T. Upon arriving at the load the reflected voltage and current waves (v_r and i_r) are created. The reflected cur-rent wave is related to the reflected voltage wave by

$$i_r = -\frac{v_r}{Z_C} \tag{17.81}$$

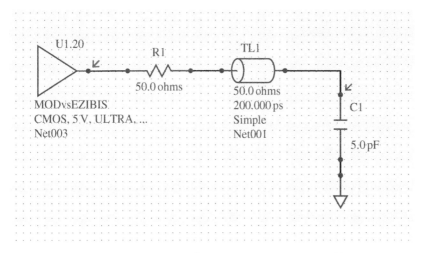

Figure 17.32 HyperLynx circuit model of a transmission line terminated by a capacitive load.

The voltage–current relationship for a capacitor produces

$$i_i + i_r = C\frac{d}{dt}(v_i + v_r) \tag{17.82}$$

or, using Eqs (17.80) and (17.81),

$$\frac{V_S}{2Z_C} - \frac{v_r}{Z_C} = C\frac{d}{dt}\left(\frac{V_S}{2} + v_r\right) \tag{17.83}$$

Since V_S is constant Eq. (17.83) reduces to

$$\frac{V_S}{2} - v_r = CZ_C\frac{dv_r}{dt} \tag{17.84}$$

or

$$CZ_C\frac{dv_r}{dt} + v_r = \frac{V_S}{2}, \quad v_r(0) = \frac{V_S}{2}, \quad t > T \tag{17.85}$$

Equation (17.85) has the same mathematical form as Eq. (17.73). Thus, the solution of Eq. (17.85) will have the same mathematical form as the solution of Eq. (17.73).

Figure 17.32 shows the circuit schematic of a transmission line driven by a 5 V CMOS and terminated in a capacitive load.

The driver voltage and the voltage across the capacitor are displayed in Figure 17.33.

17.1.5 Transmission Line Discontinuity

In this section we will consider the effects of the discontinuity along the transmission line; the discontinuity occurs when the transmission line characteristic impedance changes, as shown in Figure 17.34.

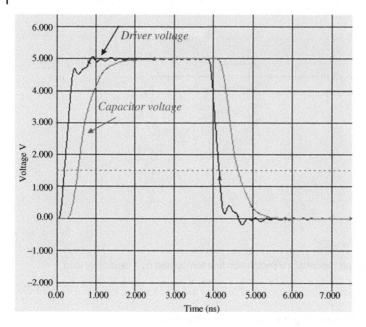

Figure 17.33 Driver voltage and the voltage across the capacitor.

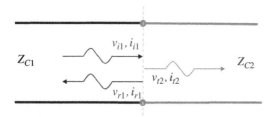

Figure 17.34 Discontinuity along a transmission line.

Let's consider voltage and current waves v_{i1} and i_{i1} traveling on transmission line 1 incident to the junction. Upon their arrival at the junction, the reflected waves v_{r1} and i_{r1}, and the transmitted waves v_{t2} and i_{t2} are created (Paul, 2006, p. 248).

KVL at the junction produces

$$v_{i1} + v_{r1} = v_{t2} \tag{17.86}$$

while the KCL gives

$$i_{i1} + i_{r1} = i_{t2} \tag{17.87}$$

We know that

$$i_{i1} = \frac{v_{i1}}{Z_{C1}} \tag{17.88a}$$

$$i_{r1} = -\frac{v_{r1}}{Z_{C1}} \tag{17.88b}$$

$$i_{t2} = \frac{v_{t2}}{Z_{C2}} \tag{17.88c}$$

Substituting Eq. (17.88) into Eq. (17.87) produces

$$\frac{v_{i1}}{Z_{C1}} - \frac{v_{r1}}{Z_{C1}} = \frac{v_{t2}}{Z_{C2}} \tag{17.89}$$

thus

$$v_{t2} = \frac{Z_{C2}}{Z_{C1}}\left(v_{i1} - v_{r1}\right) \tag{17.90}$$

Using Eq. (17.90) in (17.86) results in

$$v_{i1} + v_{r1} = \frac{Z_{C2}}{Z_{C1}}\left(v_{i1} - v_{r1}\right) \tag{17.91}$$

or

$$v_{i1} + v_{r1} = \frac{Z_{C2}}{Z_{C1}}v_{i1} - \frac{Z_{C2}}{Z_{C1}}v_{r1}$$

$$v_{r1} + \frac{Z_{C2}}{Z_{C1}}v_{r1} = \frac{Z_{C2}}{Z_{C1}}v_{i1} - v_{i1}$$

$$\left(1 + \frac{Z_{C2}}{Z_{C1}}\right)v_{r1} = \left(\frac{Z_{C2}}{Z_{C1}} - 1\right)v_{i1} \tag{17.92}$$

$$\left(\frac{Z_{C1} + Z_{C2}}{Z_{C1}}\right)v_{r1} = \left(\frac{Z_{C2} - Z_{C1}}{Z_{C1}}\right)v_{i1}$$

and thus

$$\frac{v_{r1}}{v_{i1}} = \frac{Z_{C2} - Z_{C1}}{Z_{C2} + Z_{C1}} = \Gamma_{12} \tag{17.93}$$

where Γ_{12} is the voltage reflection coefficient for the wave incident from the left onto the boundary. In terms of the reflection coefficient, the reflected voltage can be expressed as

$$v_{r1} = \Gamma_{12}v_{i1} \tag{17.94}$$

Thus, to the incident wave, the transmission line to the right looks like its characteristic impedance Z_{C2}, as shown in Figure 17.35.

We also define the voltage *transmission coefficient* as the ratio of the transmitted voltage v_{t2} to the incident voltage v_{i1}

$$T_{12} = \frac{v_{t2}}{v_{i1}} \tag{17.95}$$

Since

$$v_{i1} + v_{r1} = v_{t2} \tag{17.96}$$

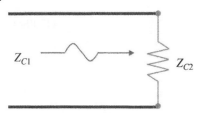

Figure 17.35 Incoming wave sees a termination impedance Z_{C2}.

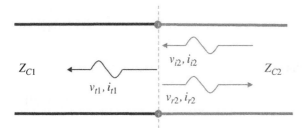

Figure 17.36 Wave incident from the right.

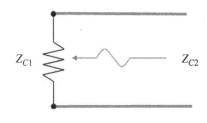

Figure 17.37 Incoming wave sees a termination impedance Z_{C1}.

we have

$$T_{12} = \frac{v_{t2}}{v_{i1}} = \frac{v_{i1} + v_{r1}}{v_{i1}} = 1 + \frac{v_{r1}}{v_{i1}} \tag{17.97}$$

or

$$T_{12} = 1 + \Gamma_{12} \tag{17.98}$$

Utilizing Eq. (17.93) in Eq. (17.98) results in

$$T_{12} = 1 + \frac{Z_{C2} - Z_{C1}}{Z_{C2} + Z_{C1}} = \frac{Z_{C2} + Z_{C1} + Z_{C2} - Z_{C1}}{Z_{C2} + Z_{C1}} \tag{17.99}$$

or

$$T_{12} = \frac{2Z_{C2}}{Z_{C2} + Z_{C1}} \tag{17.100}$$

A similar derivation can be performed for the wave incident on the boundary from the right, as shown in Figure 17.36.

In this case the voltage reflection coefficient is given by

$$\Gamma_{21} = \frac{v_{r1}}{v_{i1}} = \frac{Z_{C2} - Z_{C1}}{Z_{C2} + Z_{C1}} \tag{17.101}$$

Again, to the incident wave, the transmission line to the left looks like its characteristic impedance Z_{C1}, as shown in Figure 17.37.

The voltage transmission coefficient for the wave incident from the right is

$$T_{21} = \frac{2Z_{C1}}{Z_{C1} + Z_{C2}} \qquad (17.102)$$

and the reflection and transmission coefficient for the wave incident from the right are related by

$$T_{21} = 1 + \Gamma_{21} \qquad (17.103)$$

Example 17.2 Transmission line discontinuity
Consider the circuit shown in Figure 17.38.

The experimental setup reflecting this circuit is shown in Figure 17.39.

A 10 V_{pp} (open-circuit voltage) pulse signal was sent from the function generator along the 6 ft long RG58 coaxial cable ($Z_C = 50\,\Omega$) connected to the 6 ft long RG59 coaxial cable ($Z_C = 75\,\Omega$) and terminated with an open circuit. The rise time of the waveform is $t_r = 2.5$ ns.

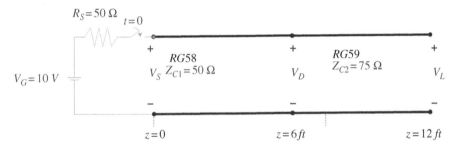

Figure 17.38 Circuit for the reflection measurements.

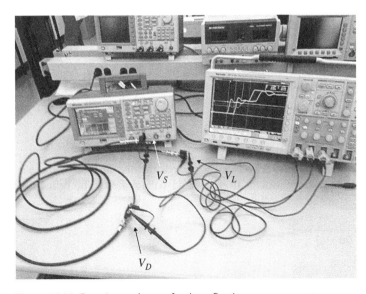

Figure 17.39 Experimental setup for the reflection measurements.

The voltages at the source (V_S), the discontinuity (V_D), and at the load (V_L) were measured using the oscilloscope probes. When the switch closes, the initial voltage wave is created at location $z = 0$. The value of this voltage is

$$v_{i1} = V_S(z=0) = V_G \frac{Z_C}{R_S + Z_C} = (10)\frac{50}{50+50} = 5 \text{ V}$$

After about $t = T = 9$ ns (one-way travel time along RG 58), this waveform arrives at the discontinuity, where it gets reflected and transmitted. The reflection coefficient (from the left) at the discontinuity is

$$\Gamma_{12} = \frac{Z_{C2} - Z_{C1}}{Z_{C2} + Z_{C1}} = \frac{75-50}{75+50} = 0.2$$

The reflected voltage at the discontinuity is

$$v_{r1} = \Gamma_{12}v_{i1} = (0.2)(5) = 1 \text{ V}$$

The total voltage at the discontinuity at $t = T$ is thus

$$V_D = v_{i1} + v_{r1} = 5 + 1 = 6 \text{ V}$$

The reflected voltage wave propagates back towards the source and arrives there about at $t = 2T$ later. The total voltage at the source becomes (after $t = 2T + 2t_r$).

$$V_S = v_{i1} + v_{r1} = 5 + 1 = 6 \text{ V}$$

The incident wave that arrived at the discontinuity at $t = T$ is also transmitted. The voltage transmission coefficient for the wave incident from the left is

$$T_{12} = \frac{2Z_{C2}}{Z_{C2} + Z_{C1}} = \frac{(2)(75)}{75+50} = 1.2$$

The transmitted voltage at $z = 6$ ft is

$$V_D = v_{t2} = T_{12}v_{i1} = (1.2)(5) = 6 \text{ V}$$

which, of course, is the same as the voltage V_D at $t = 0$. These voltages are shown in Figure 17.40.

The transmitted voltage wave travels towards the load where it gets reflected with a load reflection coefficient equal to one (open load). The total voltage at the load is, therefore, at $t = 2T$ is (shown in Figure 17.41):

$$V_L = 6 + 6 = 12 \text{ V}$$

The reflected voltage at the load (6 V) travels back towards the discontinuity, where it gets reflected and transmitted. Let's look at the reflected voltage first. The reflection coefficient (from the right) at the discontinuity is

$$\Gamma_{21} = \frac{Z_{C1} - Z_{C2}}{Z_{C1} + Z_{C2}} = \frac{50-75}{50+75} = -0.2$$

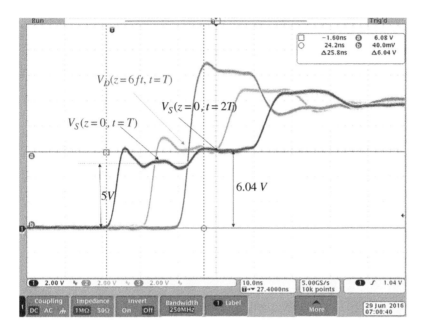

Figure 17.40 Voltages at the source and at the discontinuity at $t = T$ and $t = 2T$.

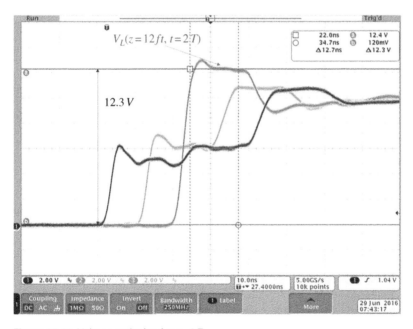

Figure 17.41 Voltage at the load at $t = 2T$.

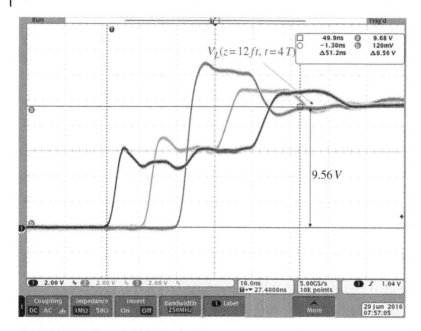

Figure 17.42 Voltage at the load at $t = 4\,T$.

The reflected voltage at the discontinuity is

$$v_{r2} = \Gamma_{21}v_{i2} = (-0.2)(6) = -1.2\ \text{V}$$

This reflected voltage wave propagates towards the load where it gets reflected. The total voltage at the load, at $t = 4\,T$,

$$V_L = 12 - 1.2 - 1.2 = 9.6\ \text{V}$$

This is shown in Figure 17.42.

The incident wave that arrived at the discontinuity (from the right) at $t = 3\,T$ is also transmitted. The voltage transmission coefficient for the wave incident from the right is

$$T_{21} = \frac{2Z_{C1}}{Z_{C2} + Z_{C1}} = \frac{(2)(50)}{75 + 50} = 0.8$$

The transmitted voltage (from right to left) at $z = 6\,\text{ft}$ is

$$v_{t1} = T_{21}v_{i2} = (0.8)(6) = 4.8\ \text{V}$$

Resulting in a total voltage at the discontinuity of

$$V_D = 6 + 4.8 = 10.8\ \text{V}$$

This is shown in Figure 17.43.

Finally, the transmitted voltage of $4.8\,\text{V}$ arrives at the source at $t = 4\,T$, resulting in the source voltage rising to (Figure 17.43).

$$V_S = 6 + 4.8 = 10.8\ \text{V}$$

This process continues until the steady state value (of $10\,\text{V}$ at all locations) is reached.

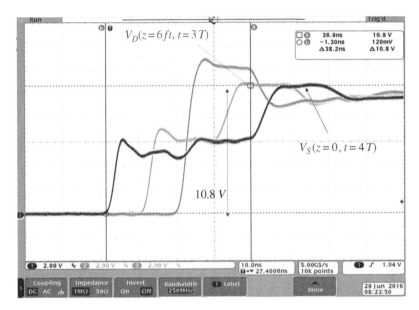

Figure 17.43 Voltage at the discontinuity at $t=3\,T$ and the source at $t=4\,T$.

17.2 Steady-State Analysis

17.2.1 Lossy Transmission Lines

In our discussion so far, we have assumed that the transmission line is lossless. We will now consider the effects of losses. Losses come from two mechanisms: the per-unit-length resistance of the line conductors, r in Ω/m, and the per-unit-length conductance of the surrounding medium, g in S/m.

The per-unit-length equivalent circuit model for a Δz section of the line is shown in Figure 17.44 (Sadiku, 2010, p. 523).

Applying KVL to the outer loop of the circuit in Figure 17.44 results in

$$V(z,t)=r\Delta z I(z,t)+l\Delta z\frac{\partial I(z,t)}{\partial t}+V(z+\Delta z,t) \tag{17.104}$$

or

$$\frac{V(z+\Delta z,t)-V(z,t)}{\Delta z}=-rI(z,t)-l\frac{\partial I(z,t)}{\partial t} \tag{17.105}$$

Taking the limit as $\Delta z \to 0$ we get

$$\frac{\partial V(z,t)}{\partial t}=-rI(z,t)-l\frac{\partial I(z,t)}{\partial t} \tag{17.106}$$

Applying KCL to the upper node in Figure 17.44 gives

$$I(z,t)=I(z+\Delta z,t)+\Delta I \tag{17.107}$$

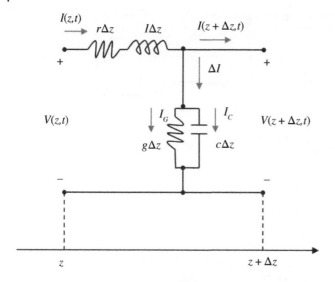

Figure 17.44 The per-unit-length model of a lossy transmission line.

Now, ΔI can be expressed as

$$\Delta I = I_G + I_C = g\Delta z V(z+\Delta z,t) + c\Delta z \frac{\partial V(z+\Delta z,t)}{\partial t} \qquad (17.108)$$

Substituting Eq. (17.108) into Eq. (17.107) results in

$$I(z,t) = I(z+\Delta z,t) + g\Delta z V(z+\Delta z,t) + c\Delta z \frac{\partial V(z+\Delta z,t)}{\partial t} \qquad (17.109)$$

or

$$\frac{I(z+\Delta z,t) - I(z,t)}{\Delta z} = -gV(z,t) - c\frac{\partial V(z+\Delta z,t)}{\partial t} \qquad (17.110)$$

Taking the limit as $\Delta z \rightarrow 0$, we get

$$\frac{\partial I(z,t)}{\partial z} = -gV(z,t) - c\frac{\partial V(z,t)}{\partial t} \qquad (17.111)$$

Equations (17.106) and (17.111) are the coupled transmission line equations. We can decouple them as follows. Differentiating Eq. (17.106) with respect to z gives

$$\frac{\partial^2 V(z,t)}{\partial z^2} = -r\frac{\partial I(z,t)}{\partial z} - l\frac{\partial^2 I(z,t)}{\partial t \partial z} \qquad (17.112)$$

and differentiating Eq. (17.111) with respect to time results in

$$\frac{\partial^2 I(z,t)}{\partial z \partial t} = -g\frac{\partial V(z,t)}{\partial t} - c\frac{\partial^2 V(z,t)}{\partial t^2} \qquad (17.113)$$

Substituting Eq. (17.111) and Eq. (17.113) into Eq. (17.112) gives

$$\frac{\partial^2 V(z, t)}{\partial z^2} = lc\frac{\partial^2 V(z, t)}{\partial t^2} + (lg + rc)\frac{\partial V(z, t)}{\partial t} + rgV(z, t) \tag{17.114}$$

and a similar derivation leads to

$$\frac{\partial^2 I(z, t)}{\partial z^2} = lc\frac{\partial^2 I(z, t)}{\partial t^2} + (lg + rc)\frac{\partial I(z, t)}{\partial t} + rgI(z, t) \tag{17.115}$$

Equations (17.114) and (17.115) are the general transmission line equations. Under sinusoidal excitation, we can rewrite them in a phasor form as

$$\frac{d^2\hat{V}(z)}{dz^2} = -\omega^2 lc\hat{V}(z) + j\omega(lg + rc)\hat{V}(z) + rg\hat{V}(z) \tag{17.116a}$$

$$\frac{d^2\hat{I}(z)}{dz^2} = -\omega^2 lc\hat{I}(z) + j\omega(lg + rc)\hat{I}(z) + rg\hat{I}(z) \tag{17.116b}$$

or

$$\frac{d^2\hat{V}(z)}{dz^2} = (r + j\omega l)(g + j\omega c)\hat{V}(z) \tag{17.117a}$$

$$\frac{d^2\hat{I}(z)}{dz^2} = (r + j\omega l)(g + j\omega c)\hat{I}(z) \tag{17.117b}$$

or

$$\frac{d^2\hat{V}(z)}{dz^2} = \hat{z}\hat{y}\hat{V}(z) \tag{17.118a}$$

$$\frac{d^2\hat{I}(z)}{dz^2} = \hat{z}\hat{y}\hat{I}(z) \tag{17.118b}$$

where

$$\hat{z} = r + j\omega l \tag{17.119a}$$

$$\hat{y} = g + j\omega c \tag{17.119b}$$

are respectively the per-unit-length series impedance and shunt admittance of the transmission line.

Equations (17.118) are often written in the form

$$\frac{d^2\hat{V}(z)}{dz^2} - \hat{\gamma}^2\hat{V}(z) = 0 \tag{17.120a}$$

$$\frac{d^2\hat{I}(z)}{dz^2} - \hat{\gamma}^2\hat{I}(z) = 0 \tag{17.120b}$$

where $\hat{\gamma}$ is the propagation constant of the line, defined by

$$\hat{\gamma} = \sqrt{\hat{z}\hat{y}}\sqrt{\left(r + j\omega l\right)\left(g + j\omega c\right)} = \alpha + j\beta \qquad (17.121)$$

and α is the attenuation constant and β is the phase constant. The general solution of Eqs (17.120) is of the form

$$\hat{V}(z) = \hat{V}^+ e^{-\hat{\gamma}z} + \hat{V}^- e^{\hat{\gamma}z} = \hat{V}^+ e^{-\alpha z} e^{-j\beta z} + \hat{V}^- e^{\alpha z} e^{j\beta z} \qquad (17.122a)$$

$$\hat{I}(z) = \frac{\hat{V}^+}{\hat{Z}_C} e^{-\hat{\gamma}z} - \frac{\hat{V}^-}{\hat{Z}_C} e^{\hat{\gamma}z} = \frac{\hat{V}^+}{\hat{Z}_C} e^{-\alpha z} e^{-j\beta z} - \frac{\hat{V}^-}{\hat{Z}_C} e^{\alpha z} e^{j\beta z} \qquad (17.122b)$$

where the complex *characteristic impedance* is given by

$$\hat{Z}_C = \sqrt{\frac{\hat{z}}{\hat{y}}} = \sqrt{\frac{r + j\omega l}{g + j\omega c}} = Z_C \angle \theta_{Z_C} \qquad (17.123)$$

The solution in Eqs (17.122) consists of the forward and backward traveling waves

$$\hat{V}(z) = \hat{V}_f(z) + \hat{V}_b(z) \qquad (17.124a)$$

$$\hat{I}(z) = \hat{I}_f(z) + \hat{I}_b(z) \qquad (17.124b)$$

where

$$\hat{V}_f(z) = \hat{V}^+ e^{-\alpha z} e^{-j\beta z} \qquad (17.125a)$$

$$\hat{V}_b(z) = \hat{V}^- e^{\alpha z} e^{j\beta z} \qquad (17.125b)$$

$$\hat{I}_f(z) = \frac{\hat{V}^+}{\hat{Z}_C} e^{-\alpha z} e^{-j\beta z} \qquad (17.125c)$$

$$\hat{I}_b(z) = -\frac{\hat{V}^-}{\hat{Z}_C} e^{\alpha z} e^{j\beta z} \qquad (17.125d)$$

When the line is lossless ($\alpha = 0$) the solution in Eq, (17.122) reduces to

$$\hat{V}(z) = \hat{V}^+ e^{-j\beta z} + \hat{V}^- e^{j\beta z} \qquad (17.126a)$$

$$\hat{I}(z) = \frac{\hat{V}^+}{\hat{Z}_C} e^{-j\beta z} - \frac{\hat{V}^-}{\hat{Z}_C} e^{j\beta z} \qquad (17.126b)$$

17.2.2 Standing Waves

Consider the transmission line circuit shown in Figure 17.45.

At any location z, the voltage $\hat{V}(z)$ is the sum of the forward and backward traveling waves $\hat{V}_f$ and $\hat{V}_b$

$$\hat{V}_f = \hat{V}^+ e^{-j\beta z} \qquad (17.127a)$$

$$\hat{V}_b = \hat{V}^- e^{j\beta z} \qquad (17.127b)$$

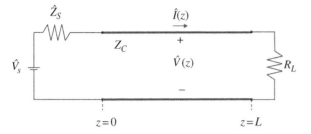

Figure 17.45 Transmission line circuit.

At any location z, we define a complex voltage reflection coefficient as the ratio of the phasor voltages of the backward and forward traveling waves.

$$\hat{\Gamma}(z) = \frac{\hat{V}^- e^{j\beta z}}{\hat{V}^+ e^{-j\beta z}} = \frac{\hat{V}^-}{\hat{V}^+} e^{j2\beta z} \qquad (17.128)$$

From Eq. (17.128) we obtain

$$\hat{V}^- = \hat{\Gamma}(z)\hat{V}^+ e^{-j2\beta z} \qquad (17.129)$$

Substituting Eq. (17.129) into Eqs (17.126) gives

$$\begin{aligned}\hat{V}(z) &= \hat{V}^+ e^{-j\beta z} + \hat{\Gamma}(z)\hat{V}^+ e^{-j2\beta z} e^{j\beta z} \\ &= \hat{V}^+ e^{-j\beta z} + \hat{\Gamma}(z)\hat{V}^+ e^{-j\beta z} \\ &= \hat{V}^+ e^{-j\beta z}\left[1+\hat{\Gamma}(z)\right]\end{aligned} \qquad (17.130)$$

and

$$\begin{aligned}\hat{I}(z) &= \frac{\hat{V}^+}{\hat{Z}_C}e^{-j\beta z} - \frac{\hat{V}^-}{\hat{Z}_C}\hat{\Gamma}(z)\hat{V}^+ e^{-j2\beta z} e^{j\beta z} \\ &= \frac{\hat{V}^+}{\hat{Z}_C}e^{-j\beta z} - \frac{\hat{V}^-}{\hat{Z}_C}\hat{\Gamma}(z)\hat{V}^+ e^{-j\beta z} \\ &= \frac{\hat{V}^+}{\hat{Z}_C}e^{-j\beta z}\left[1-\hat{\Gamma}(z)\right]\end{aligned} \qquad (17.131)$$

Thus, the voltage and current at any location z can be expressed in terms of the voltage reflection coefficient at any location z as

$$\hat{V}(z) = \hat{V}^+ e^{-j\beta z}\left[1+\hat{\Gamma}(z)\right] \qquad (17.132a)$$

$$\hat{I}(z) = \frac{\hat{V}^+}{\hat{Z}_C}e^{-j\beta z}\left[1-\hat{\Gamma}(z)\right] \qquad (17.132b)$$

Evaluating the voltage reflection coefficient in Eq. (17.128) at the load results in

$$\hat{\Gamma}(z=L) = \hat{\Gamma}_L = \frac{\hat{V}^-}{\hat{V}^+}e^{j2\beta L} \qquad (17.133)$$

Returning to Eq. (17.128) we may express the voltage reflection coefficient at any location z as

$$\hat{\Gamma}(z) = \frac{\hat{V}^-}{\hat{V}^+} e^{j2\beta z} = \frac{\hat{V}^-}{\hat{V}^+} e^{j2\beta(z+L-L)}$$

$$= \frac{\hat{V}^-}{\hat{V}^+} e^{j2\beta L} e^{j2\beta(z-L)} = \hat{\Gamma}_L e^{j2\beta(z-L)} \qquad (17.134)$$

Thus, the voltage reflection coefficient at any location z can be expressed in terms of the load reflection coefficient as

$$\hat{\Gamma}(z) = \hat{\Gamma}_L e^{j2\beta(z-L)} \qquad (17.135)$$

Substituting Eq. (17.135) into Eqs (17.132) gives the expressions for the voltage and current at any location z as

$$\hat{V}(z) = \hat{V}^+ e^{-j\beta z} \left[1 + \hat{\Gamma}_L e^{j2\beta(z-L)} \right] \qquad (17.136a)$$

$$\hat{I}(z) = \frac{\hat{V}^+}{Z_C} e^{-j\beta z} \left[1 - \hat{\Gamma}_L e^{j2\beta(z-L)} \right] \qquad (17.136b)$$

The magnitudes of the voltage and current along the line at any distance z *away from the source* are

$$\left| \hat{V}(z) \right| = \left| \hat{V}^+ e^{-j\beta z} \left[1 + \hat{\Gamma}_L e^{j2\beta(z-L)} \right] \right|$$

$$= \left| \hat{V}^+ \right| \left| e^{-j\beta z} \right| \left| \left[1 + \hat{\Gamma}_L e^{j2\beta(z-L)} \right] \right| \qquad (17.137a)$$

$$= \left| \hat{V}^+ \right| \left| \left[1 + \hat{\Gamma}_L e^{j2\beta(z-L)} \right] \right|$$

$$\left| \hat{I}(z) \right| = \left| \frac{\hat{V}^+}{Z_C} e^{-j\beta z} \left[1 - \hat{\Gamma}_L e^{j2\beta(z-L)} \right] \right|$$

$$= \left| \frac{\hat{V}^+}{Z_C} \right| \left| e^{-j\beta z} \right| \left| \left[1 - \hat{\Gamma}_L e^{j2\beta(z-L)} \right] \right| \qquad (17.137b)$$

$$= \left| \frac{\hat{V}^+}{Z_C} \right| \left| \left[1 - \hat{\Gamma}_L e^{j2\beta(z-L)} \right] \right|$$

Thus

$$\left| \hat{V}(z) \right| = \left| \hat{V}^+ \right| \left| \left[1 + \hat{\Gamma}_L e^{j2\beta(z-L)} \right] \right| \qquad (17.138a)$$

$$\left| \hat{I}(z) \right| = \left| \frac{\hat{V}^+}{Z_C} \right| \left| \left[1 - \hat{\Gamma}_L e^{j2\beta(z-L)} \right] \right| \qquad (17.138b)$$

Now, consider the same transmission line but with the distance measured from the load to the source, as shown in Figure 17.46.

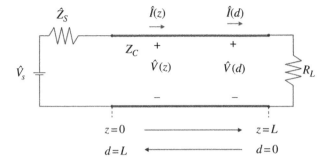

Figure 17.46 Transmission line circuit, distance measured from the load to the source.

The two distance variables are related by

$$d = L - z \tag{17.137a}$$

$$z = L - d \tag{17.137b}$$

In terms of the distance d *away from the load*, the magnitudes of the voltage and current can be expressed as

$$\left|\hat{V}(d)\right| = \left|\hat{V}^+\right|\left[\left|1 + \hat{\Gamma}_L e^{j2\beta(L-d-L)}\right|\right] \tag{17.138a}$$

$$\left|\hat{I}(d)\right| = \left|\frac{\hat{V}^+}{Z_C}\right|\left[\left|1 - \hat{\Gamma}_L e^{j2\beta(L-d-L)}\right|\right] \tag{17.138b}$$

or

$$\left|\hat{V}(d)\right| = \left|\hat{V}^+\right|\left[\left|1 + \hat{\Gamma}_L e^{-j2\beta d}\right|\right] \tag{17.139a}$$

$$\left|\hat{I}(d)\right| = \left|\frac{\hat{V}^+}{Z_C}\right|\left[\left|1 - \hat{\Gamma}_L e^{-j2\beta d}\right|\right] \tag{17.139b}$$

There are four important cases of special interest that we will investigate:

1) The load is a short circuit $\hat{Z}_L = 0$.
2) The load is an open circuit $\hat{Z}_L = \infty$.
3) The load is matched to the transmission line $\hat{Z}_L = Z_C$.
4) Arbitrary resistive load R.

Case 1 – Short-circuited load $\hat{Z}_L = 0$ The load reflection coefficient in the case is

$$\hat{\Gamma}_L = -1 \tag{17.140}$$

Using Eq. (17.140) in Eqs (13.139) gives

$$\left|\hat{V}(d)\right| = \left|\hat{V}^+\right|\left[\left|1 + \hat{\Gamma}_L e^{-j2\beta d}\right|\right]$$

$$= \left|\hat{V}^+\right|\left|e^{-j\beta d}\left(e^{j\beta d} - e^{-j\beta d}\right)\right| = \left|\hat{V}^+\right|\left|e^{-j\beta d} j2\sin(\beta d)\right| \tag{17.141a}$$

$$= \left|\hat{V}^+\right|\left|e^{-j\beta d}\right|\left|2\sin(\beta d)\right| = 2\left|\hat{V}^+\right|\left|\sin(\beta d)\right|$$

and

$$\left|\hat{I}(d)\right| = \frac{\left|\hat{V}^+\right|}{Z_C}\left|1 + e^{-j2\beta d}\right|$$

$$= \frac{\left|\hat{V}^+\right|}{Z_C}\left|e^{-j\beta d}\left(e^{j\beta d} + e^{-j\beta d}\right)\right| = \frac{\left|\hat{V}^+\right|}{Z_C}\left|e^{-j\beta d}2\cos(\beta d)\right| \tag{17.141b}$$

$$= \frac{\left|\hat{V}^+\right|}{Z_C}\left|e^{-j\beta d}\right|\left|2\cos(\beta d)\right| = 2\frac{\left|\hat{V}^+\right|}{Z_C}\left|\cos(\beta d)\right|$$

or

$$\left|\hat{V}(d)\right| = 2\left|\hat{V}^+\right|\left|\sin(\beta d)\right| \tag{17.142a}$$

$$\left|\hat{I}(d)\right| = 2\frac{\left|\hat{V}^+\right|}{Z_C}\left|\cos(\beta d)\right| \tag{17.142b}$$

The phase constant β can be expressed in terms of the wavelength λ as

$$\beta = \frac{\omega}{v} = \frac{2\pi f}{\lambda f} = \frac{2\pi}{\lambda} \tag{17.143}$$

and thus the sine and cosine argument in Eqs (17.142) can be written as

$$\beta d = 2\pi\frac{d}{\lambda} \tag{17.144}$$

Using Eq. (17.144) in Eqs (17.142) produces

$$\left|\hat{V}(d)\right| = 2\left|\hat{V}^+\right|\left|\sin\left(2\pi\frac{d}{\lambda}\right)\right| \tag{17.144a}$$

$$\left|\hat{I}(d)\right| = 2\frac{\left|\hat{V}^+\right|}{Z_C}\left|\cos\left(2\pi\frac{d}{\lambda}\right)\right| \tag{17.144b}$$

The magnitudes of the voltage and current waves for a short-circuited load are shown in Figure 17.47.

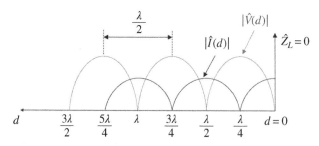

Figure 17.47 Magnitudes of the voltage and current for a short-circuited load.

We observe the following:

1) The voltage is zero at the load and at distances from the load which are multiples of a half wavelength.
2) The current is maximum at the load and is zero at distances from the load that are odd multiples of a quarter wavelength.
3) The corresponding points are separated by one half wavelength.

Case 2 – Open-circuited load $\hat{Z}_L = \infty$ The load reflection coefficient in the case is

$$\hat{\Gamma}_L = 1 \tag{17.145}$$

Using Eq. (17.140) in Eqs (13.139) gives

$$\left|\hat{V}(d)\right| = \left|\hat{V}^+\right|\left|1 + \hat{\Gamma}_L e^{-j2\beta d}\right| = \left|\hat{V}^+\right|\left|1 + e^{-j2\beta d}\right|$$
$$= \left|\hat{V}^+\right|\left|e^{-j\beta d}\left(e^{j\beta d} + e^{-j\beta d}\right)\right| = \left|\hat{V}^+\right|\left|e^{-j\beta d} 2\cos(\beta d)\right| \tag{17.146a}$$
$$= 2\left|\hat{V}^+\right|\left|\cos(\beta d)\right| = 2\left|\hat{V}^+\right|\left|\cos\left(2\pi\frac{d}{\lambda}\right)\right|$$

and

$$\left|\hat{I}(d)\right| = \frac{\left|\hat{V}^+\right|}{Z_C}\left|1 - \hat{\Gamma}_L e^{-j2\beta d}\right| = \frac{\left|\hat{V}^+\right|}{Z_C}\left|1 - e^{-j2\beta d}\right|$$
$$= \frac{\left|\hat{V}^+\right|}{Z_C}\left|e^{-j\beta d}\left(e^{j\beta d} - e^{-j\beta d}\right)\right| \tag{17.146b}$$
$$= \frac{\left|\hat{V}^+\right|}{Z_C}\left|e^{-j\beta d} j2\sin(\beta d)\right| = 2\frac{\left|\hat{V}^+\right|}{Z_C}\left|\sin\left(2\pi\frac{d}{\lambda}\right)\right|$$

or

$$\left|\hat{V}(d)\right| = 2\left|\hat{V}^+\right|\left|\cos\left(2\pi\frac{d}{\lambda}\right)\right| \tag{17.147a}$$

$$\left|\hat{I}(d)\right| = 2\frac{\left|\hat{V}^+\right|}{Z_C}\left|\sin\left(2\pi\frac{d}{\lambda}\right)\right| \tag{17.147b}$$

The magnitudes of the voltage and current waves for an open-circuited load are shown in Figure 17.48.

We observe the following:

1) The current is zero at the load and at distances from the load which are multiples of a half wavelength.
2) The voltage is maximum at the load and is zero at distances from the load that are odd multiples of a quarter wavelength.
3) The corresponding points are separated by one half wavelength.

In both cases, the voltage and current waves do not travel as the time advances, but stay where they are, only oscillating in time between the stationary zeros. In other

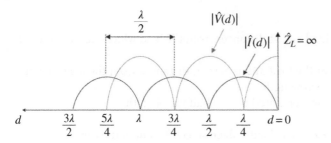

Figure 17.48 Magnitudes of the voltage and current for an open-circuited load.

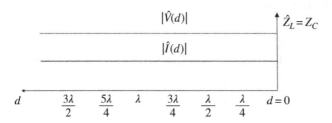

Figure 17.49 Magnitudes of the voltage and current for a matched load.

words, they do not represent a traveling wave in either direction. The resulting wave, which is a superposition of two traveling waves with opposite directions of travel, is thus termed a *standing wave*.

Case 3 – Matched load $\hat{Z}_L = Z_C$ The load reflection coefficient in the case is

$$\hat{\Gamma}_L = 0 \tag{17.148}$$

Using Eq. (17.140) in Eqs (13.139) gives

$$\left|\hat{V}(d)\right| = \left|\hat{V}^+\right|\left|1 + \hat{\Gamma}_L e^{-j2\beta d}\right| = \left|\hat{V}^+\right| \tag{17.149a}$$

$$\left|\hat{I}(d)\right| = \frac{\left|\hat{V}^+\right|}{Z_C}\left|1 - \hat{\Gamma}_L e^{-j2\beta d}\right| = \frac{\left|\hat{V}^+\right|}{Z_C} \tag{17.149b}$$

The magnitudes of the voltage and current waves for matched load are shown in Figure 17.49.

We observe that the voltage and current magnitudes are constant along the line.

Case 4 – Arbitrary resistive load R The magnitudes of the voltage and current waves for matched load are shown in Figure 17.50.

We observe that the locations of the voltage and current maxima and minima are determined by the actual load impedance, but again adjacent corresponding points on each waveform are separated by one half wavelength.

In all cases, except for the matched load, the magnitudes of the voltage and current vary along the line. This variation is quantitatively described by the *voltage standing wave ratio (VSWR)* defined as

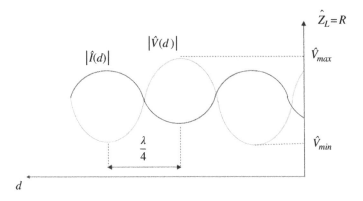

Figure 17.50 Magnitudes of the voltage and current for an arbitrary resistive load.

$$VSWR = \frac{\left|\hat{V}(d)\right|_{max}}{\left|\hat{V}(d)\right|_{min}} \tag{17.150}$$

Let's return to the expressions for the voltage magnitude in Eq. (17.139a), repeated here,

$$\left|\hat{V}(d)\right| = \left|\hat{V}^+\right|\left[\left[1 + \hat{\Gamma}_L e^{-j2\beta d}\right]\right] \tag{17.151}$$

The maximum of this magnitude is

$$\left|\hat{V}(d)\right|_{max} = \left|\hat{V}^+\right|\left[\left[1 + \hat{\Gamma}_L\right]\right] \tag{17.152a}$$

while the minimum of this magnitude is

$$\left|\hat{V}(d)\right|_{min} = \left|\hat{V}^+\right|\left[\left[1 - \hat{\Gamma}_L\right]\right] \tag{17.152b}$$

Substituting Eqs (17.152) into Eq. (17.150) gives an alternative expression for VSWR as

$$VSWR = \frac{1 + \left|\hat{\Gamma}_L\right|}{1 - \left|\hat{\Gamma}_L\right|} \tag{17.153}$$

When the load is short-circuited or open-circuited, we have

$$VSWR = \begin{cases} \infty & \hat{Z}_L = 0 \\ \infty & \hat{Z}_L = \infty \end{cases} \tag{17.154}$$

When the load is matched, we have

$$VSWR = 1 \quad \hat{Z}_L = Z_C \tag{17.155}$$

In general,

$$1 \leq VSWR \leq \infty \tag{17.156}$$

17.3 s Parameters

To characterize high frequency circuits we can use s parameters which relate traveling voltage waves that are incident, reflected, and transmitted when a two-port network in inserted into a transmission line. This is depicted in Figure 17.51.

The incident waves (a_1, a_2) and reflected waves (b_1, b_2) used to define s parameters for a two-port network are shown in Figure 17.52 (Ludwig and Bogdanov, 2009).

The linear equations describing this two-port network in terms of the s parameters are

$$b_1 = s_{11}a_1 + s_{12}a_2 \tag{17.127}$$
$$b_2 = s_{21}a_1 + s_{22}a_2$$

or in a matrix form

$$\begin{bmatrix} b_1 \\ b_2 \end{bmatrix} = \begin{bmatrix} s_{11} & s_{12} \\ s_{21} & s_{22} \end{bmatrix} \begin{bmatrix} a_1 \\ a_2 \end{bmatrix} \tag{17.128}$$

where S is the scattering matrix given by

$$S = \begin{bmatrix} s_{11} & s_{12} \\ s_{21} & s_{22} \end{bmatrix} \tag{17.129}$$

Recall the general solution for the line voltage and current along a transmission line in Eq. (17.122):

$$\hat{V}(z) = \hat{V}^+ e^{-\hat{\gamma}z} + \hat{V}^- e^{\hat{\gamma}z} \tag{17.130a}$$

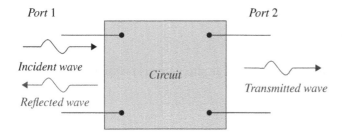

Port 1 *Port 2*

Incident wave

Circuit

Reflected wave

Transmitted wave

Figure 17.51 *s* parameters are related to the traveling waves.

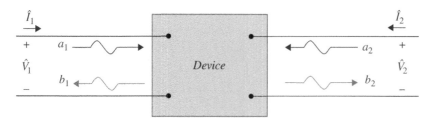

Device

Figure 17.52 Incident and reflected waves.

$$\hat{I}(z)=\frac{\hat{V}^+}{\hat{Z}_C}e^{-\hat{\gamma}z}-\frac{\hat{V}^-}{\hat{Z}_C}e^{\hat{\gamma}z} \qquad (17.130b)$$

at $z=0$ Eqs (17.130) become

$$\hat{V}(0)=\hat{V}^++\hat{V}^- \qquad (17.131a)$$

$$\hat{I}(0)=\frac{\hat{V}^+}{\hat{Z}_C}-\frac{\hat{V}^-}{\hat{Z}_C} \qquad (17.131b)$$

$\hat{V}^+$ and $\hat{V}^-$ in Eq. (17.131) denote the amplitudes of the incident and reflected voltage waves, respectively. From Eq. (17.131b) we get

$$\hat{Z}_C\hat{I}(0)=\hat{V}^+-\hat{V}^- \qquad (17.132)$$

Adding Eqs (17.131a) and (17.132) gives

$$\hat{Z}_C\hat{I}(0)+\hat{V}(0)=2\hat{V}^+ \qquad (17.133)$$

while subtracting Eq. (17.132) from Eq. (17.131a) results in

$$\hat{V}(0)-\hat{Z}_C\hat{I}(0)=2\hat{V}^- \qquad (17.134)$$

Thus, from Eqs (17.133) and (17.134) we obtain

$$\hat{V}^+=\frac{\hat{V}(0)+\hat{Z}_C\hat{I}(0)}{2} \qquad (17.135a)$$

$$\hat{V}^-=\frac{\hat{V}(0)-\hat{Z}_C\hat{I}(0)}{2} \qquad (17.135b)$$

As we shall soon see, it is convenient to use the normalized incident and reflected voltages, instead of the ones in Eq. (17.135):

$$\hat{V}^+=\frac{\hat{V}(0)+\hat{Z}_C\hat{I}(0)}{2\sqrt{Z_C}} \qquad (17.136a)$$

$$\hat{V}^-=\frac{\hat{V}(0)-\hat{Z}_C\hat{I}(0)}{2\sqrt{Z_C}} \qquad (17.136b)$$

We refer to these normalized waves as *power waves*. Using the notation of Figure 17.46, we rewrite Eqs (17.136) at each port as

$$a_1=\frac{\hat{V}_1+\hat{Z}_C\hat{I}_1}{2\sqrt{Z_C}}=\frac{V_{i1}}{\sqrt{Z_C}} \qquad (17.137a)$$

$$b_1=\frac{\hat{V}_1-\hat{Z}_C\hat{I}_1}{2\sqrt{Z_C}}=\frac{V_{r1}}{\sqrt{Z_C}} \qquad (17.137b)$$

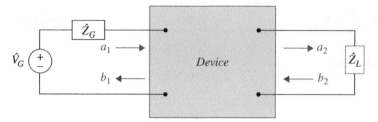

Figure 17.53 Typical two port circuit.

and

$$a_2 = \frac{\hat{V}_2 + \hat{Z}_C \hat{I}_2}{2\sqrt{Z_C}} = \frac{V_{i2}}{\sqrt{Z_C}} \tag{17.138a}$$

$$b_2 = \frac{\hat{V}_2 - \hat{Z}_C \hat{I}_2}{2\sqrt{Z_C}} = \frac{V_{r2}}{\sqrt{Z_C}} \tag{17.138b}$$

Where V_{i1}, V_{i2} are the incident voltage waves, and V_{r1}, V_{r2} are the reflected voltage waves at ports 1 and 2, respectively.

In the typical application of a two-port network, the circuit is driven at port 1 and terminated by a load at port 2, as shown in Figure 17.53.

Returning to Eq. (17.127), repeated here,

$$\begin{aligned} b_1 &= s_{11}a_1 + s_{12}a_2 \\ b_2 &= s_{21}a_1 + s_{22}a_2 \end{aligned} \tag{17.139}$$

We obtain the individual s parameters as

$$s_{11} = \left. \frac{b_1}{a_1} \right|_{a_2=0} \tag{17.140a}$$

Thus, s_{11} is the input port reflection coefficient, when the incident wave at port 2 is zero, which means that port 2 should be terminated in a matched load ($\hat{Z}_L = \hat{Z}_C$) to avoid reflections. This is shown in Figure 17.55.

Also,

$$s_{12} = \left. \frac{b_1}{a_2} \right|_{a_1=0} \tag{17.140b}$$

Thus, s_{12} is the transmission coefficient from port 2 to port 1, with the input port terminated in a matched load, as shown in Figure 17.55.

Also,

$$s_{21} = \left. \frac{b_2}{a_1} \right|_{a_2=0} \tag{17.140c}$$

Thus, s_{21} is the transmission coefficient from port 1 to port 2, with the output port terminated in a matched load, as shown in Figure 17.54.

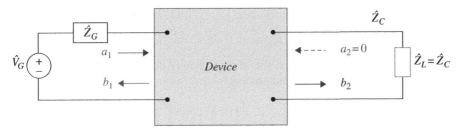

Figure 17.54 Port 2 termination.

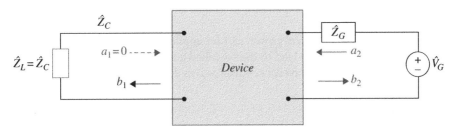

Figure 17.55 Port 1 termination.

Finally,

$$s_{22} = \frac{b_2}{a_2}\bigg|_{a_1=0} \tag{17.140d}$$

Thus, s_{22} is the output port reflection coefficient, when the incident wave on port 1 is zero, as shown in Figure 17.55.

Power waves and generalized scattering parameters Recall: the total voltage and current on a transmission line can be expressed in terms of the incident and reflected voltage wave amplitudes as in (17.131), repeated here:

$$\hat{V}(0) = \hat{V}^+ + \hat{V}^- \tag{17.141a}$$

$$\hat{I}(0) = \frac{\hat{V}^+}{\hat{Z}_C} - \frac{\hat{V}^-}{\hat{Z}_C} \tag{17.141b}$$

The average power delivered to a load can be expressed as

$$P_{AVG} = \frac{1}{2}\text{Re}\{\hat{V}\hat{I}^*\} = \frac{1}{2}\text{Re}\left\{\left[\hat{V}^+ + \hat{V}^-\right]\left[\frac{\hat{V}^+}{\hat{Z}_C} - \frac{\hat{V}^-}{\hat{Z}_C}\right]^*\right\} \tag{17.142}$$

If the line is lossless then its characteristic impedance is real and we have

$$\begin{aligned}
P_{AVG} &= \frac{1}{2}\text{Re}\{\hat{V}\hat{I}^*\} = \frac{1}{2Z_C}\text{Re}\left\{\left[\hat{V}^+ + \hat{V}^-\right]\left[\hat{V}^{+*} - \hat{V}^{-*}\right]\right\} \\
&= \frac{1}{2Z_C}\text{Re}\left\{\left(V^+\right)^2 - \hat{V}^+\hat{V}^{-*} + \hat{V}^{+*}\hat{V}^- + \left(V^-\right)^2\right\}
\end{aligned} \tag{17.143}$$

The middle two terms inside the brackets in Eq. (17.143) are of the form

$$\hat{A} - \hat{A}^* = 2j\text{Im}(A) \tag{17.144}$$

and thus are purely imaginary. Thus, the average power delivered to the load is the difference of the incident and reflected powers.

$$P_{AVG} = \frac{1}{2Z_C}\text{Re}\left\{\left(V^+\right)^2 + \left(V^-\right)^2\right\} = \frac{1}{2Z_C}\left[\left(V^+\right)^2 + \left(V^-\right)^2\right] \tag{17.145}$$

This result is only valid when the line characteristic impedance is real. This result is not valid when the characteristic impedance is complex, as in the case of a lossy transmission line.

The normalized voltage waves or power can be applied to both lossless and lossy lines. These power waves will also lead to the so-called generalized *s* parameters.

Recall the normalized voltage waves, i.e. power waves defined by

$$a_1 = \frac{\hat{V}_1 + \hat{Z}_C\hat{I}_1}{2\sqrt{Z_C}} \tag{17.146a}$$

$$b_1 = \frac{\hat{V}_1 - \hat{Z}_C\hat{I}_1}{2\sqrt{Z_C}} \tag{17.146b}$$

$$a_2 = \frac{\hat{V}_2 + \hat{Z}_C\hat{I}_2}{2\sqrt{Z_C}} \tag{17.146c}$$

$$b_2 = \frac{\hat{V}_2 - \hat{Z}_C\hat{I}_2}{2\sqrt{Z_C}} \tag{17.146d}$$

Let's solve Eqs (17.146a,b) for the voltage and current waves in terms of the power waves amplitudes.

$$a_1 2\sqrt{Z_C} = \hat{V}_1 + \hat{Z}_C\hat{I}_1 \tag{17.147a}$$

$$b_1 2\sqrt{Z_C} = \hat{V}_1 - \hat{Z}_C\hat{I}_1 \tag{17.147b}$$

Adding Eqs (17.147a) and (17.147b) gives

$$a_1 2\sqrt{Z_C} + b_1 2\sqrt{Z_C} = 2\hat{V}_1 \tag{17.148}$$

or

$$\hat{V}_1 = a_1\sqrt{Z_C} + b_1\sqrt{Z_C} = \frac{(a_1 + b_1)Z_C}{\sqrt{Z_C}} \tag{17.149}$$

Subtracting Eq. (17.147b) from Eq. (17.147a) gives

$$a_1 2\sqrt{Z_C} - b_1 2\sqrt{Z_C} = 2\hat{Z}_C\hat{I}_1 \tag{17.150}$$

or

$$a_1 \sqrt{Z_C} - b_1 \sqrt{Z_C} = \hat{Z}_C \hat{I}_1 \tag{17.151}$$

and thus

$$\hat{I}_1 = \frac{a_1 \sqrt{Z_C} - b_1 \sqrt{Z_C}}{\hat{Z}_C} = \frac{a_1 - b_1}{\sqrt{Z_C}} \tag{17.152}$$

Then, using Eqs (17.149) and (17.152) the average power delivered to the load can be expressed as

$$P_{AVG} = \frac{1}{2} \mathrm{Re}\{\hat{V}\hat{I}^*\} = \frac{1}{2} \mathrm{Re}\left\{ \left[\frac{(a_1 + b_1)Z_C}{\sqrt{Z_C}} \right] \left[\frac{a_1 - b_1}{\sqrt{Z_C}} \right]^* \right\} \tag{17.153}$$

or

$$\begin{aligned} P_{AVG} &= \frac{1}{2Z_C} \mathrm{Re}\left\{ \left[(a_1 + b_1)Z_C \right] \left[a_1 - b_1 \right]^* \right\} \\ &= \frac{1}{2} \mathrm{Re}\left\{ |a_1|^2 - a_1 b_1^* + a_1^* b_1 - |b_1|^2 \right\} \end{aligned} \tag{17.154}$$

and thus

$$P_{AVG} = \frac{1}{2}|a_1|^2 - \frac{1}{2}|b_1|^2 \tag{17.155}$$

Similarly, at port 2, we have

$$P_{AVG} = \frac{1}{2}|a_2|^2 - \frac{1}{2}|b_2|^2 \tag{17.156}$$

If the voltages and currents are expressed in terms of the rms values then the average power at port 1 is expressed as

$$P_{AVG} = |a_1|^2 - |b_1|^2 \tag{17.157}$$

while at port 2,

$$P_{AVG} = |a_2|^2 - |b_2|^2 \tag{17.158}$$

where $|a_1|^2$ is the power incident on the input port of the network, $|b_1|^2$ is the power reflected from the input port of the network, $|a_2|^2$ is the power incident on the output port of the network, and $|b_2|^2$ is the power reflected from the output port of the network. Thus, we can relate the s parameters to the powers as follows.

$$\begin{aligned} |s_{11}|^2 &= \frac{|b_1|^2}{|a_1|^2}\bigg|_{a_2=0} \\ &= \frac{\text{Reflected power at port 1}}{\text{Incident power at port 1}} \end{aligned} \tag{17.159a}$$

$$\left|s_{12}\right|^2 = \frac{\left|b_1\right|^2}{\left|a_2\right|^2}\bigg|_{a_1=0}$$

$$= \frac{\text{Transmitted power to port 1}}{\text{Incident power at port 2}} \tag{17.159b}$$

$$\left|s_{21}\right|^2 = \frac{\left|b_2\right|^2}{\left|a_1\right|^2}\bigg|_{a_2=0}$$

$$= \frac{\text{Transmitted power to port 2}}{\text{Incident power at port 1}} \tag{17.159c}$$

$$\left|s_{22}\right|^2 = \frac{\left|b_2\right|^2}{\left|a_1\right|^2}\bigg|_{a_2=0}$$

$$= \frac{\text{Reflected power at port 2}}{\text{Incident power at port 2}} \tag{17.159d}$$

We can express several gains and losses in terms of s parameters, as follows. *Forward power gain* is defined as

$$\text{Forward Power Gain}_{dB} = 10\log\frac{\left|b_2\right|^2}{\left|a_1\right|^2} = 20\log\frac{\left|b_2\right|}{\left|a_1\right|} \tag{17.160a}$$

or

$$\boxed{\text{Forward Power Gain}_{dB} = 10\log\left|s_{21}\right|^2 = 20\log\left|s_{21}\right|} \tag{17.160b}$$

Reverse power gain is defined as

$$\text{Reverse Power Gain}_{dB} = 10\log\frac{\left|b_1\right|^2}{\left|a_2\right|^2} = 20\log\frac{\left|b_1\right|}{\left|a_2\right|} \tag{17.161a}$$

or

$$\boxed{\text{Reverse Power Gain}_{dB} = 10\log\left|s_{12}\right|^2 = 20\log\left|s_{12}\right|} \tag{17.161b}$$

Insertion loss is defined as

$$\text{Insertion Loss}_{dB} = 10\log\frac{\left|a_1\right|^2}{\left|b_2\right|^2} = 20\log\frac{\left|a_1\right|}{\left|b_2\right|} \tag{17.162a}$$

or

$$\boxed{\text{Insertion Loss}_{dB} = 20\log\frac{1}{\left|s_{21}\right|}} \tag{17.162b}$$

Return loss is defined as

$$\text{Return Loss}_{dB} = 10\log\frac{|a_1|^2}{|b_1|^2} = 20\log\frac{|a_1|}{|b_1|} \qquad (17.163a)$$

or

$$\boxed{\text{Return Loss}_{dB} = 20\log\frac{1}{|s_{11}|}} \qquad (17.163b)$$

17.4 EMC Applications

17.4.1 Crosstalk between PCB traces

In this section we will discuss crosstalk between PCB traces and present the simulation and measurement results from a PCB shown in Figure 17.56.

Before presenting the results, let's review the phenomenon of crosstalk.

When two PCB traces (transmission lines) are in the vicinity of one another, a signal propagating along one line can induce a signal on another line, due to capacitive (electric field) and inductive (magnetic field) coupling between the two lines. This phenomenon is called *crosstalk*.

The cross-section of a PCB with the microstrip transmission lines is shown in Figure 17.57.

A PCB of thickness d supports two traces separated by distance s. A ground plane constitutes the reference conductor for the two circuits. This arrangement can be modeled by the circuit shown in Figure 17.58 (Paul, 2006, p. 597).

The generator circuit connects a voltage $V_S(t)$ and its source impedance, R_S, to a load R_L. The adjacent receptor line conductors are terminated by the resistances R_{NE} and R_{FE} at the near end (NE), and the far end (FE), respectively. The signal present on the generator line induces the near-end and far-end coupled crosstalk voltages, V_{NE} and V_{FE}.

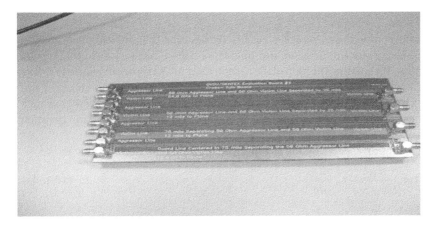

Figure 17.56 PCB used for crosstalk measurements.

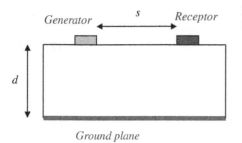

Figure 17.57 Cross-section of a PCB with a microstrip line.

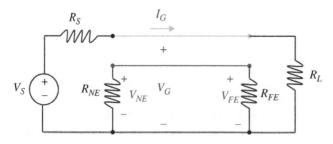

Figure 17.58 Circuit model of a PCB with a microstrip line.

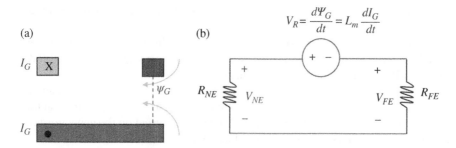

Figure 17.59 Inductive coupling between the circuits (a) field model, (b) circuit model.

In order to determine these voltages, let's briefly describe the physical phenomenon taking place. The current on the generator line, I_G, creates a magnetic field that results in a magnetic flux Ψ_G crossing the loop of the receptor circuit, as shown in Figure 17.59(a).

If this flux is time varying, then according to Faraday's law, it induces a voltage V_R in the receptor circuit. The circuit model of this field phenomenon is represented by a mutual inductance and is shown in Figure 17.59(b). We refer to this interaction between the circuits as the magnetic or inductive coupling.

Using the current divider we obtain the induced near-and far-end voltages as

$$V_{NE}(t) = \frac{R_{NE}}{R_{NE} + R_{FE}} L_m \frac{dI_G}{dt} \tag{17.164a}$$

$$V_{FE}(t) = -\frac{R_{FE}}{R_{NE} + R_{FE}} L_m \frac{dI_G}{dt} \tag{17.164b}$$

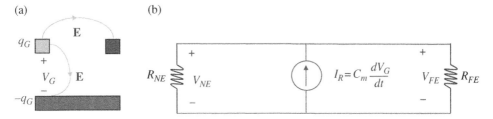

Figure 17.60 Capacitive coupling between the circuits (a) field model, (b) circuit model.

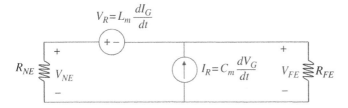

Figure 17.61 Inductive and capacitive coupling circuit model.

Similarly, the voltage between the two conductors of the generator circuit, V_G, has associated with it a charge separation that creates electric field lines, some of which terminate on the conductors of the receptor circuit, as shown in Figure 17.60(a).

If this charge (voltage) varies with time, it induces a current in the receptor circuit. The circuit model of this field phenomenon is represented by a mutual capacitance, and is shown in Figure 17.60(b). We refer to this interaction between the circuits as the electric or capacitive coupling.

Using the voltage divider, we obtain the induced near-and far-end voltages as

$$V_{NE}(t) = \frac{R_{NE}R_{FE}}{R_{NE}+R_{FE}}C_m\frac{dV_G}{dt} \tag{17.165a}$$

$$V_{FE}(t) = \frac{R_{NE}R_{FE}}{R_{NE}+R_{FE}}C_m\frac{dV_G}{dt} \tag{17.165b}$$

Superposition of these two types of coupling results in the circuit model shown in Figure 17.61.

The total induced voltages, by superposition, are given by

$$V_{NE}(t) = \frac{R_{NE}}{R_{NE}+R_{FE}}L_m\frac{dI_G}{dt} + \frac{R_{NE}R_{FE}}{R_{NE}+R_{FE}}C_m\frac{dV_G}{dt} \tag{17.166a}$$

$$V_{FE}(t) = -\frac{R_{FE}}{R_{NE}+R_{FE}}L_m\frac{dI_G}{dt} + \frac{R_{NE}R_{FE}}{R_{NE}+R_{FE}}C_m\frac{dV_G}{dt} \tag{17.166b}$$

If the circuit is electrically small at the highest significant frequency of interest then the generator voltage and current can be obtained from the circuit shown in Figure 17.62.

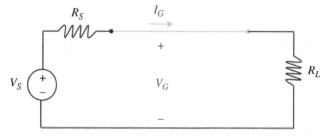

Figure 17.62 Electrically short generator circuit model.

Then,

$$V_G(t) \cong \frac{R_L}{R_S + R_L} V_S(t) \tag{17.167a}$$

$$I_G(t) \cong \frac{1}{R_S + R_L} V_S(t) \tag{17.167b}$$

Substituting Eqs (17.167) into Eqs (17.166) results in

$$V_{NE}(t) = \left[\underbrace{\frac{R_{NE}}{R_{NE} + R_{FE}} L_m \frac{1}{R_S + R_L}}_{\text{Inductive Coupling}} + \underbrace{\frac{R_{NE} R_{FE}}{R_{NE} + R_{FE}} C_m \frac{R_L}{R_S + R_L}}_{\text{Capacitive Coupling}} \right] \frac{dV_S(t)}{dt} \tag{17.168a}$$

$$V_{FE}(t) = \left[-\underbrace{\frac{R_{FE}}{R_{NE} + R_{FE}} L_m \frac{1}{R_S + R_L}}_{\text{Inductive Coupling}} + \underbrace{\frac{R_{NE} R_{FE}}{R_{NE} + R_{FE}} C_m \frac{R_L}{R_S + R_L}}_{\text{Capacitive Coupling}} \right] \frac{dV_S(t)}{dt} \tag{17.168b}$$

Note that the induced crosstalk voltage is proportional to the mutual inductance and capacitance between the two circuits and the derivative of the source voltage.

The crosstalk circuit model in the frequency domain is shown in Figure 17.63

From this equivalent circuit in the frequency domain, or directly from Eqs (17.168) we obtain the near-end and far-end phasor crosstalk voltages as

$$\hat{V}_{NE} = \left[\underbrace{\frac{R_{NE}}{R_{NE} + R_{FE}} L_m \frac{1}{R_S + R_L}}_{\text{Inductive Coupling}} + \underbrace{\frac{R_{NE} R_{FE}}{R_{NE} + R_{FE}} C_m \frac{R_L}{R_S + R_L}}_{\text{Capacitive Coupling}} \right] j\omega \hat{V}_S \tag{17.169a}$$

$$\hat{V}_{FE} = \left[-\underbrace{\frac{R_{FE}}{R_{NE} + R_{FE}} L_m \frac{1}{R_S + R_L}}_{\text{Inductive Coupling}} + \underbrace{\frac{R_{NE} R_{FE}}{R_{NE} + R_{FE}} C_m \frac{R_L}{R_S + R_L}}_{\text{Capacitive Coupling}} \right] j\omega \hat{V}_S \tag{17.169b}$$

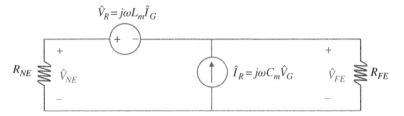

$$\hat{V}_R = j\omega L_m \hat{I}_G$$

Figure 17.63 Inductive and capacitive coupling circuit model in the frequency domain.

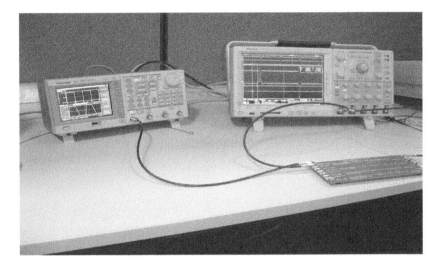

Figure 17.64 Experimental set-up for crosstalk verification.

Table 17.2 Board topologies.

Case	Line separation s [mils]	Distance to Ground Plane d [mils]
1	25	54.8
2	25	12
3	75	12

Observe that the crosstalk induced voltages increase at a rate of 20 dB/decade with frequency.

Figure 17.64 shows the experimental setup to verify the crosstalk derivations (Adamczyk and Teune, 2009). The source is a 1 V_{pp}, 1 MHz trapezoidal pulse with a 50% duty cycle with a 100 ns rise time and 200 ns fall time.

The board with different circuit topologies was investigated and is described in Table 17.2

The load resistors, as well as the near- and far-end resistors were chosen to be 50 Ω. The characteristic impedances of both the generator and receptor circuits were also 50 Ω.

Figures 17.65–17.70 show the generator (aggressor) signal as well as the resulting near- and far-end voltages induced on the receptor (victim) line.

Measured and simulated frequency-domain results are shown in Figure 17.66.

Measured and simulated frequency-domain results are shown in Figure 17.68.

Measured and simulated frequency-domain results are shown in Figure 17.70.

We make the following observation in Case 1, presented in Figure 17.65. Voltage induced on the near end during the rise time is $V_{NE} = 1.54\,\text{mV}$, while the same voltage induced during the fall time is $V_{NE} = 760\,\mu\text{V}$.

Since the value of the rise time is twice that of the fall time, according to Eq. (17.168), the induced voltages should differ in magnitude by a factor of two, which indeed is the case. We also note that the polarities of the two voltages are opposite, which is not unexpected from Eq. (17.168). Similar observations can be made for the voltages induced on the far end.

Furthermore, since the coefficients of coupling for the near-end voltage Eq. (17.168a) are positive, the induced-voltage during the rise is also positive. The far-end voltage is negative during the rise time, indicating that the inductive coupling dominates the capacitive one.

Bringing the ground plane closer to the lines, while keeping the distance between the lines unchanged (Case 2) resulted in the reduction of the induced voltage magnitudes, as shown in Figure 17.66.

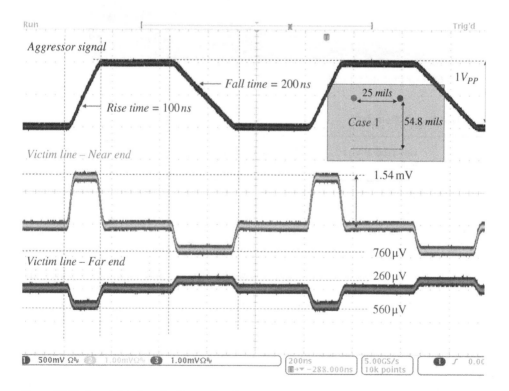

Figure 17.65 Crosstalk induced voltages, Case 1.

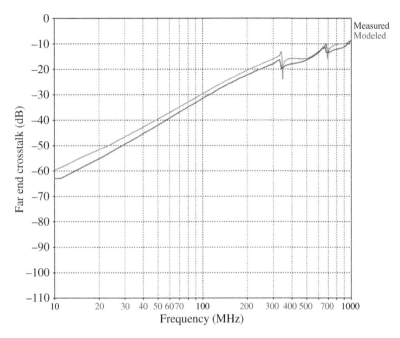

Figure 17.66 FE Crosstalk – Measured and simulated frequency-domain results – Case 1.

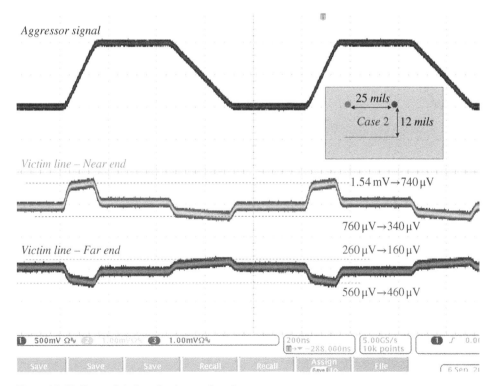

Figure 17.67 Crosstalk induced voltages, Case 2.

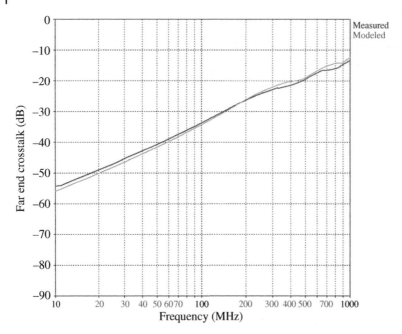

Figure 17.68 FE Crosstalk – Measured and simulated frequency-domain results – Case 2.

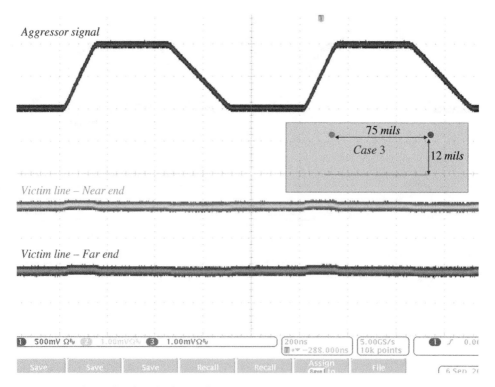

Figure 17.69 Crosstalk induced voltages, Case 3.

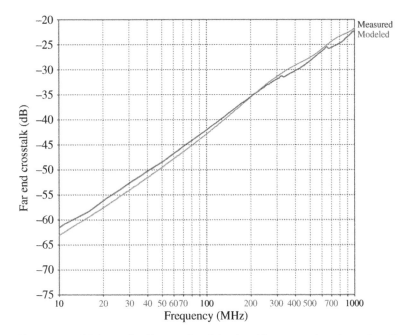

Figure 17.70 FE Crosstalk – Measured and simulated frequency-domain results – Case 3.

Case 3 depicts the scenario where the distance to the ground plane is unchanged from Case 2, but the separation between the lines is increased. This results in negligible induced voltages.

In all cases the induced crosstalk increased at a rate of 20 dB/decade with the frequency.

17.4.2 LISN Impedance Measurement

This section utilizes the s parameter measurements to confirm that the line impedance stabilization network (LISN) meets the CISPR 25 requirements. (Note: See Appendix A for the LISN measurements description.)

CISPR 25 LISN is shown in Figure 17.71.

The CISPR 25 LISN impedance measurement setup is shown in Figure 17.72.

Note that the DC power supply input terminals need to be short circuited. This is accomplished with a shorting bar, as shown in Figure 17.73.

Before the impedance measurement is taken, the calibration process needs to take place to characterize the cable used to connect the LISN to the network analyzer. The calibration involves a calibration kit shown in Figure 17.74, which consists of a short, open, and 50 Ω load.

This calibration kit is used for s_{11} and s_{21} calibration measurements, as shown in Figure 17.75.

Finally, the impedance measurements can be taken as shown in Figure 17.76.

Ideally, the LISN's impedance should be 50 Ω over the entire frequency range of the measurement. According to CISPR 25, the LISN impedance (measured by the network analyzer) in the frequency range of 100 kHz to 100 MHz should be within the specified tolerance band.

Measurement side   *DC input side*

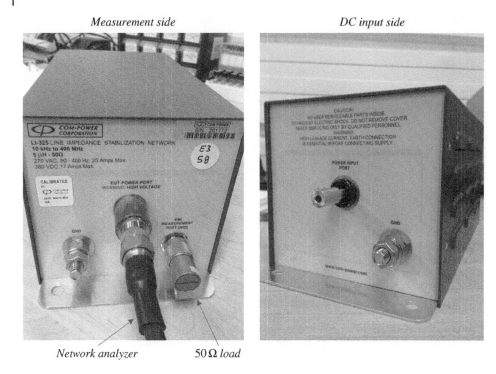

Network analyzer   *50 Ω load*

Figure 17.71 CISPR 25 LISN.

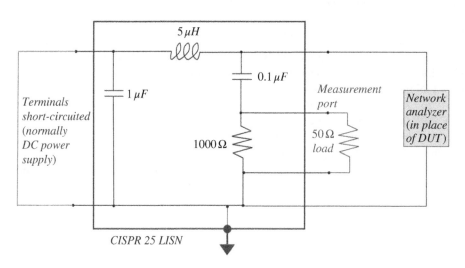

Figure 17.72 CISPR 25 LISN measurement setup.

Figure 17.73 Shorting bar for impedance measurement.

Figure 17.74 Calibration kit.

Let's recall the definition of the s_{11} parameter utilizing Figure 17.77.

$$s_{11} = \frac{b_1}{a_1}\bigg|_{a_2=0} \tag{17.170}$$

Adapting this figure to the LISN impedance measurement setup, we arrive at the Figure 17.78.

The ideal LISN would present 50 Ω impedance to the network analyzer (over the entire frequency range) and thus the reflected voltage wave would equal in magnitude to the incident voltage wave resulting in the ratio.

$$s_{11} = \frac{b_1}{a_1} = 1 = 0_{dB} \tag{17.171}$$

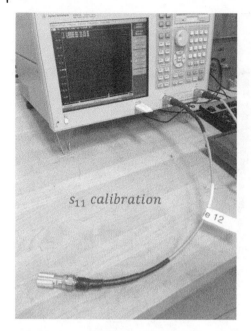
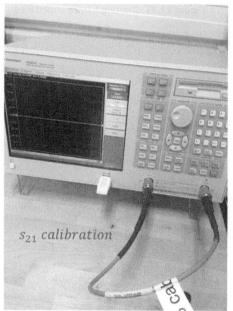

Figure 17.75 Calibration measurements.

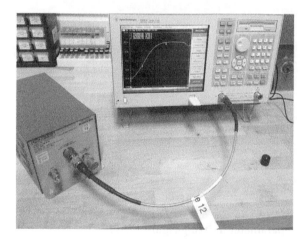

Figure 17.76 Test setup for the impedance measurement.

When both b_1 and a_1 are the voltage magnitudes, we can refer to s_{11} as the (voltage) *reflection coefficient*,

$$s_{11} = \text{Reflection coefficient}\left(b_1 = voltage,\ a_1 = voltage\right) \qquad (17.172)$$

When b_1 is a voltage magnitude and a_1 is the current magnitude (i.e. voltage/resistance ratio calculated internally by the network analyzer when the units of Ω are chosen) then the s_{11} coefficient describes the impedance,

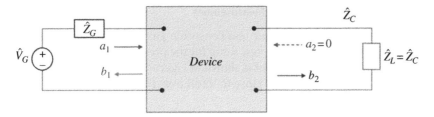

Figure 17.77 Circuit used to define s_{11}.

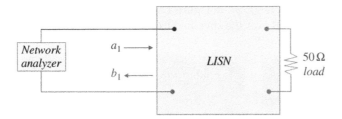

Figure 17.78 Network analyzer measurement of s_{11}.

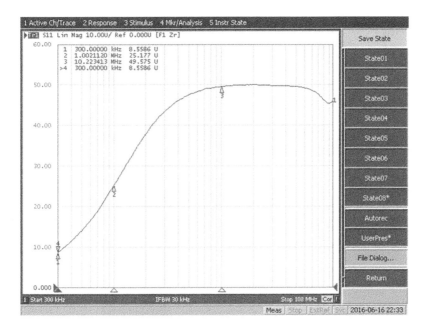

Figure 17.79 s_{11} measurement – LISN impedance.

$$s_{11} = \text{Impedance} \quad \left(b_1 = voltage, a_1 = current \right) \tag{17.173}$$

This is what is displayed when we measure LISN impedance over the frequency (on a linear scale). Figure 17.79 shows the s_{11} measurement of the LISN impedance over the frequency range.

17.4.3 Preamp Gain and Attenuator Loss Measurement

Preamp gain measurement and attenuator loss constitute an s_{21} measurement with a network analyzer. Figure 17.80 shows the test setup for the preamp gain measurement.

The s_{21} gain of the preamp measurement is shown in Figure 17.81.

As expected, the gain of the preamp is close to 32 dB over the range of the specified frequency.

Figure 17.82 shows the test setup for the attenuator loss measurement.

Note: Before this measurement is taken, both cables should be calibrated, as shown in Figure 17.83.

The s_{21} attenuator loss measurement is shown in Figure 17.84.

As shown, the attenuator loss is 10 dB over the specified frequency range.

 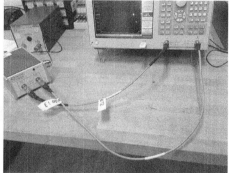

Figure 17.80 Test setup for the preamp gain measurement.

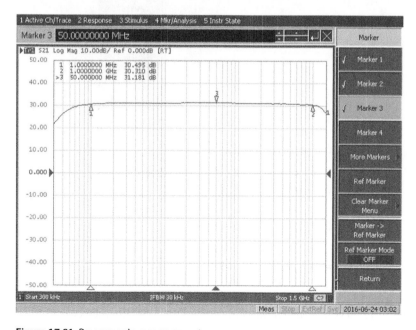

Figure 17.81 Preamp gain measurement.

Figure 17.82 Test setup for the attenuator loss measurement.

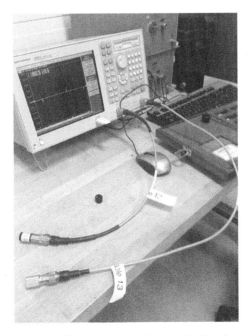
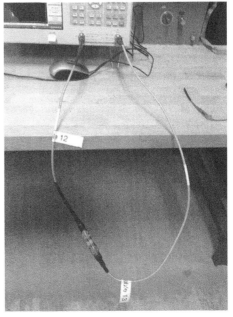

Figure 17.83 s_{11}, s_{22}, and s_{21} (through) calibration.

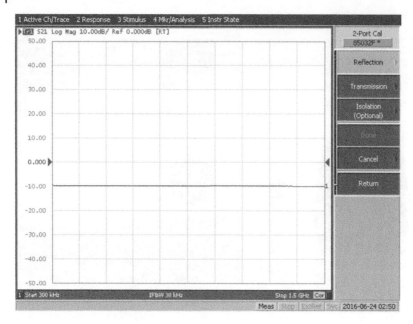

Figure 17.84 Attenuator loss measurement.

References

Adamczyk, B. and Teune, J., "EMC Crosstalk vs. Circuit Topology", ASEE North Central Section Spring Conference, Grand Rapids, MI, 2009.

Ludwig, R. and Bogdanov, G., *RF Circuit Design*, 2nd ed., Pearson, Upper Saddle River, NJ, 2009.

Paul, C.R., *Introduction to Electromagnetic Compatibility*, 2nd ed., John Wiley and Sons, New York, 2006.

Rao, N.N., *Elements of Engineering Electromagnetics*, 6th ed., Pearson Prentice Hall, Upper Saddle River, NJ, 2004.

Sadiku, M.N.O., *Elements of Electromagnetics*, 5th ed., Oxford University Press, New York, 2010.

18

Antennas and Radiation

18.1 Bridge between the Transmission Line and Antenna Theory

In this section we will use the theory of the standing waves on transmission lines discussed in the previous chapter to build a bridge between transmission line theory and the fundamental antenna structure of a dipole antenna.

Consider a standing wave pattern in lossless two-wire transmission line terminated in an open, as shown in Figure 18.1.

When the incident wave arrives at the open-circuited load, it undergoes a complete reflection. The incident and reflected waves combine to create a pure standing wave pattern as shown in Figure 18.1.

The current reflection coefficient at an open-circuited load is −1, and the current in each wire undergoes a 180° phase reversal between adjoining half cycles (this is shown by the reversal of the arrow directions).

The current in a half-cycle of one wire is of the same magnitude but 180° out-of-phase from that in the corresponding half-cycle of the other wire. If the spacing between the two wires is very small ($s \ll \lambda$), the fields radiated by the current of each wire are cancelled by those of the other. Effectively, there is no radiation from this transmission line.

Now, let's flare the terminal section of the transmission line, as shown in Figure 18.2 (Balanis, 2205, p. 18).

It is reasonable to assume that the current distribution is essentially unaltered in form in each of the wires of the transmission line. Since the two wires are no longer parallel and close to each other, the radiated fields do not cancel each other and there is a net radiation from the flared section.

Continuing the flaring process, we arrive at the structure shown in Figure 18.3.

The fields radiated by the two vertical parts of the transmission line will reinforce each other, as long as the total length of the flared section is

$$0 < l < \lambda \tag{18.1}$$

The maximum radiation (in the direction shown in Figure 18.2, i.e. broadside to this antenna) occurs for the total length of both vertical part equal to

$$l = \frac{\lambda}{2} \tag{18.1}$$

Foundations of Electromagnetic Compatibility with Practical Applications, First Edition. Bogdan Adamczyk.

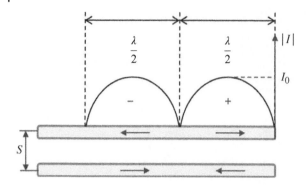

Figure 18.1 Standing wave pattern in a transmission line terminated with an open.

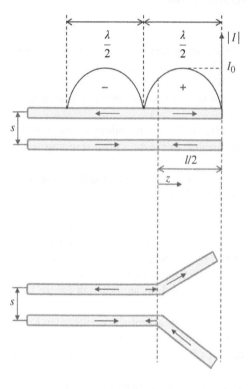

Figure 18.2 Transmission line with terminal section flared.

This is shown in Figure 18.3.

The radiating structure shown in Figure 18.4 is referred to as a half-wave dipole, discussed in the next section. The dipole antenna fields and parameters are derived from the Hertzian (electric) dipole fields, presented next.

18.2 Hertzian Dipole Antenna

In Section 6.7.3 we defined vector magnetic potential as a vector related to the magnetic flux density vector **B** by

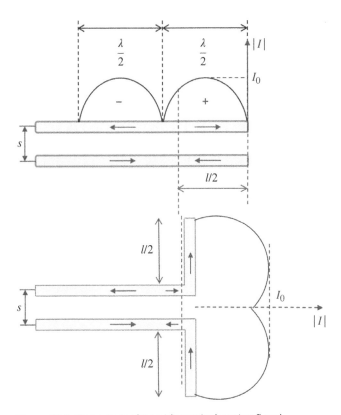

Figure 18.3 Transmission line with terminal section flared.

$$\hat{\mathbf{B}} = \nabla \times \hat{\mathbf{A}} \tag{18.3}$$

Subsequently, we showed that once **A** is known, the electric field intensity **E** can be obtained from

$$\hat{\mathbf{E}} = -j\omega\hat{\mathbf{A}} - j\frac{1}{\omega\mu\varepsilon}\nabla\left(\nabla \cdot \hat{\mathbf{A}}\right) \tag{18.4}$$

and the magnetic field intensity can be obtained from **H** from

$$\hat{\mathbf{H}} = \frac{1}{\mu}\nabla \times \hat{\mathbf{A}} \tag{18.5}$$

Alternatively, once **H** is obtained from Eq. (18.5), **E** can be obtained from **H**:

$$\hat{\mathbf{E}} = \frac{1}{j\omega\varepsilon}\nabla \times \hat{\mathbf{H}} \tag{18.6}$$

In Section 6.7.4 we considered a Hertzian dipole, shown in Figure 18.5. A Hertzian (or electric) dipole consists of a short thin wire of length l, carrying a phasor current $\hat{I}$, positioned symmetrically at the origin of the coordinate system and oriented along the z axis.

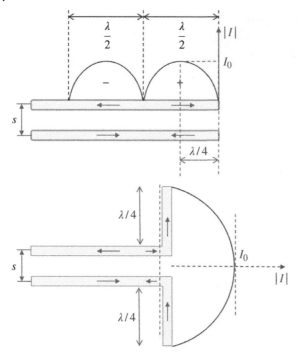

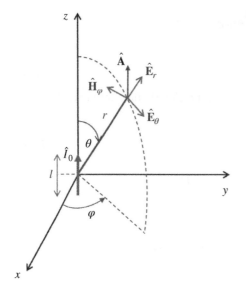

Figure 18.4 Maximum radiation broadside to the antenna.

Figure 18.5 Hertzian dipole.

Utilizing Eqs (18.4)–(18.6) we derived the expressions for the electric and magnetic field intensities at a distance r from a Hertzian dipole as

$$E_r = 2\frac{I_0 l}{4\pi}\eta\beta^2\cos\theta\left[\frac{1}{\beta^2 r^2} - j\frac{1}{\beta^3 r^3}\right]e^{-j\beta r} \tag{18.7a}$$

$$E_\theta = \frac{I_0 l}{4\pi}\eta\beta^2\sin\theta\left(\frac{j}{\beta r} + \frac{1}{\beta^2 r^2} - j\frac{1}{\beta^3 r^3}\right)e^{-j\beta r} \tag{18.7b}$$

$$E_\phi = 0 \tag{18.7c}$$

and

$$H_r = 0 \tag{18.7d}$$

$$H_\theta = 0 \tag{18.7e}$$

$$H_\phi = \frac{I_0 l}{4\pi}\beta^2\sin\theta\left(\frac{j}{\beta r} + \frac{1}{\beta^2 r^2}\right)e^{-j\beta r} \tag{18.7f}$$

The expressions in Eqs (18.7) apply at any distance r from the antenna. They can be simplified at a "large enough" distance for the antenna. To determine what large enough corresponds to, consider a positive number x.

$$\text{if} \quad x < 1 \quad \Rightarrow \quad \frac{1}{x^3} > \frac{1}{x^2} > \frac{1}{x}$$
$$\text{if} \quad x > 1 \quad \Rightarrow \quad \frac{1}{x} > \frac{1}{x^2} > \frac{1}{x^3} \tag{18.8}$$

Thus, for very small x, $(x \ll 1)$, the terms $1/x^2$ and $1/x^2$ will dominate. On the other hand, for large x, $(x \gg 1)$, the terms $1/x^2$ and $1/x^2$ will be negligible compared to the $1/x$ term.

Now let

$$x = \beta r \tag{18.9}$$

Thus, at a small distance from the antenna (referred to as the *near field*)

$$\beta r < 1 \quad \Rightarrow \quad \frac{1}{\beta r} < \frac{1}{\left(\beta r\right)^2} < \frac{1}{\left(\beta r\right)^3} \tag{18.10}$$

At a large distance from the antenna (referred to as the *far field*)

$$\beta r > 1 \quad \Rightarrow \quad \frac{1}{\beta r} > \frac{1}{\left(\beta r\right)^2} > \frac{1}{\left(\beta r\right)^3} \tag{18.11}$$

and the terms $1/(\beta r)^2$ and $1/(\beta r)^3$ will be negligible compared to the $1/\beta r$ term. The *boundary* between the near field and the far field is

$$x = 1 \quad \text{or} \quad \beta r = 1 \quad \Rightarrow \quad r = \frac{1}{\beta} = \frac{1}{2\pi/\lambda} = \frac{\lambda}{2\pi} \cong \frac{1}{6}\lambda \tag{18.12}$$

We should point out that $\lambda/2\pi$ is the boundary between the near and far fields. To be in the far field, we need to be further away from this boundary. How much further? We will address this in the next section.

In the far field, the expressions for the electric and magnetic field intensities at a distance r from a Hertzian dipole are

$$E_r = 0 \tag{18.13a}$$

$$E_\theta = j\frac{I_0 l}{4\pi}\eta\beta\sin\theta\left(\frac{e^{-j\beta r}}{r}\right) \tag{18.13b}$$

$$E_\phi = 0 \tag{18.13c}$$

and

$$H_r = 0 \tag{18.13d}$$

$$H_\theta = 0 \tag{18.13e}$$

$$H_\phi = j\frac{I_0 l}{4\pi}\beta\sin\theta\left(\frac{e^{-j\beta r}}{r}\right) \tag{18.13f}$$

18.3   Far Field Criteria

In this section we will derive the far field criteria for the wire-type and surface-type antennas.

18.3.1   Wire-Type Antennas

Recall that the Hertzian dipole expressions for the θ component of the E field and φ component of the *complete* fields are

$$E_\theta = \frac{I_0 l}{4\pi}\eta\beta^2\sin\theta\left(\frac{j}{\beta r}+\frac{1}{\beta^2 r^2}-j\frac{1}{\beta^3 r^3}\right)e^{-j\beta r} \tag{18.14a}$$

$$H_\phi = \frac{I_0 l}{4\pi}\beta^2\sin\theta\left(\frac{j}{\beta r}+\frac{1}{\beta^2 r^2}\right)e^{-j\beta r} \tag{18.14b}$$

With the radiated wave we associate the wave impedance defined as

$$\hat{Z}_w = \frac{\hat{E}_\theta}{\hat{H}_\phi} \tag{18.15}$$

A far-field criterion for the Hertzian dipole (and other wire-type antennas) is derived from the requirement that the wave impedance in the far field is equal to the intrinsic impedance of free space:

$$\hat{Z}_w = \frac{\hat{E}_\theta}{\hat{H}_\phi} \cong \eta_0 \cong 377\,\Omega \tag{18.16}$$

In free space, $\eta = \eta_0$, $\beta = \beta_0$, and we have

$$\hat{Z}_w = \frac{\hat{E}_\theta}{\hat{H}_\varphi} = \frac{\dfrac{\hat{I}l}{4\pi}\eta_0\beta_0^2\sin\theta\left(j\dfrac{1}{\beta_0 r}+\dfrac{1}{\beta_0^2 r^2}-j\dfrac{1}{\beta_0^3 r^3}\right)e^{-j\beta_0 r}}{\dfrac{\hat{I}l}{4\pi}\beta_0^2\sin\theta\left(j\dfrac{1}{\beta_0 r}+\dfrac{1}{\beta_0^2 r^2}\right)e^{-j\beta_0 r}}$$

$$= \frac{\eta_0\left(j\dfrac{1}{\beta_0 r}+\dfrac{1}{\beta_0^2 r^2}-j\dfrac{1}{\beta_0^3 r^3}\right)}{\left(j\dfrac{1}{\beta_0 r}+\dfrac{1}{\beta_0^2 r^2}\right)} = \eta_0\frac{j\dfrac{1}{(\beta_0 r)}+\dfrac{1}{(\beta_0 r)^2}-j\dfrac{1}{(\beta_0 r)^3}}{j\dfrac{1}{(\beta_0 r)}+\dfrac{1}{(\beta_0 r)^2}}\frac{j(\beta_0 r)^3}{j(\beta_0 r)^3} \quad (18.17)$$

$$= \eta_0\frac{j\dfrac{(\beta_0 r)^3}{(\beta_0 r)}+\dfrac{(\beta_0 r)^3}{(\beta_0 r)^2}-j\dfrac{(\beta_0 r)^3}{(\beta_0 r)^3}}{j\dfrac{(\beta_0 r)^3}{(\beta_0 r)}+\dfrac{(\beta_0 r)^3}{(\beta_0 r)^2}} = \eta_0\frac{-(\beta_0 r)^2+j(\beta_0 r)+1}{-(\beta_0 r)^2+j(\beta_0 r)}$$

Letting,

$$\beta_0 = \frac{2\pi}{\lambda_0} \quad (18.18)$$

we obtain

$$\hat{Z}_w = \frac{\hat{E}_\theta}{\hat{H}_\varphi} = \eta_0\frac{1-\left(\dfrac{2\pi r}{\lambda_0}\right)^2+j\left(\dfrac{2\pi r}{\lambda_0}\right)}{-\left(\dfrac{2\pi r}{\lambda_0}\right)^2+j\left(\dfrac{2\pi r}{\lambda_0}\right)} \quad (18.19)$$

Evaluating this expressions at different distances (in terms of the wavelength) from the antenna we get

$$\hat{Z}_w = 0.707\eta_0\angle-45°, \quad r = \frac{\lambda_0}{2\pi} \quad (18.20a)$$

$$\hat{Z}_w = 375.93\angle-0.01° \cong \eta_0, \quad r = 3\lambda_0 \quad (18.20b)$$

The result in Eq. (81.20b) leads to the *far-field criterion for the wire-type antennas* as

$$d_{far\ field} = 3\lambda_0 \quad (18.21)$$

18.3.2 Surface-Type Antennas

Consider a radiated wave away from a point source, as shown in Figure 18.6.

This wave resembles a spherical wave at distances away from a point source. It is reasonable to assume that in the far field, this spherical wave can be approximated by a uniform plane wave in the vicinity of the receiving antenna.

The approximation criterion can be stated as the allowable difference Δ (expressed in terms of the wavelength) between the ideal plane wave and the actual spherical wave:

$$\Delta = \frac{\lambda}{k} \tag{18.22}$$

Utilizing Figure 18.7, we get

$$d^2 = (d-\Delta)^2 + \left(\frac{D}{2}\right)^2 \tag{18.23}$$

or

$$d^2 = d^2 - 2d\Delta + \Delta^2 + \frac{D^2}{4} \tag{18.24}$$

resulting in

$$-2d\Delta + \Delta^2 + \frac{D^2}{4} = 0 \tag{18.25}$$

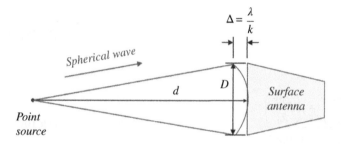

Figure 18.6 Radiating point source.

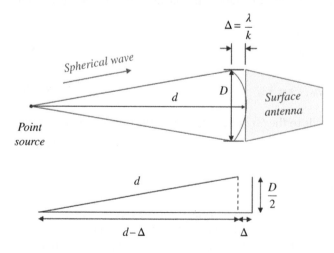

Figure 18.7 Radiating point source geometry.

It is reasonable to make the following assumption:

$$\Delta \ll d \quad \Rightarrow \quad \Delta^2 \ll 2d\Delta \tag{18.26}$$

Then, we may approximate the expression in Eq. (18.25) as

$$2d\Delta \cong \frac{D^2}{4} \tag{18.27}$$

or using Eq. (18.22) as

$$2d\left(\frac{\lambda_0}{k}\right) \cong \frac{D^2}{4} \tag{18.28}$$

resulting in

$$d = \frac{kD^2}{8\lambda_0} \tag{18.29}$$

A reasonable value for k is

$$k = 16 \tag{18.30}$$

Utilizing Eq. (18.30) in Eq. (18.29) results in the *far-field criterion for the surface-type antennas* as

$$d_{far\ field} = \frac{2D^2}{\lambda_0} \tag{18.31}$$

18.4 Half-Wave Dipole Antenna

A half-wave dipole consists of a thin wire fed or excited at the midpoint by a voltage source. The total length of the dipole equals half a wavelength. Each leg of a dipole has a length equal to a quarter of a wavelength, as shown in Figure 18.8.

Often the voltage source is connected to the antenna via transmission line, as shown in Figure 18.9.

The far fields of the half-wave dipole can be obtained by dividing the dipole antenna into infinitesimal dipoles of length dz, as shown in Figure 18.10.

Treating each infinitesimal dipole as a Hertzian dipole, we use the previously derived results

$$dE_\theta = j\frac{\hat{I}(z)dz}{4\pi}\eta\beta_0 \sin\theta'\left(\frac{e^{-j\beta r'}}{r'}\right) \tag{18.32a}$$

$$dH_\varphi = j\frac{\hat{I}(z)dz}{4\pi}\beta_0 \sin\theta''\left(\frac{e^{-j\beta r'}}{r'}\right) \tag{18.32b}$$

where the current distribution is sinusoidal and given by (Paul, 2006, p. 430)

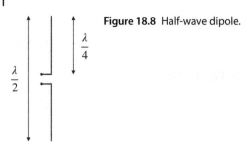

Figure 18.8 Half-wave dipole.

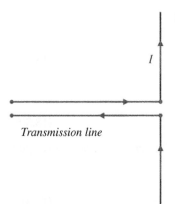

Figure 18.9 Half-wave dipole connected to a transmission line.

Transmission line

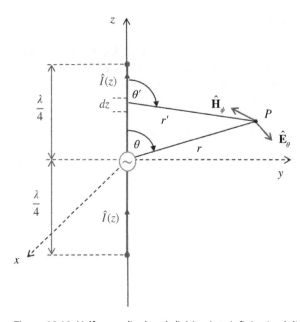

Figure 18.10 Half-wave dipole subdivision into infinitesimal dipoles.

$$\hat{I}(z) = \begin{cases} \hat{I}_0 \sin \beta_0 \left(\dfrac{\lambda}{4} - z \right), & 0 \le z \le \lambda/2 \\[2mm] \hat{I}_0 \sin \beta_0 \left(\dfrac{\lambda}{4} + z \right), & -\lambda/2 \le z \le 0 \end{cases} \tag{18.32a}$$

Adding the contributions from all infinitesimal elements (after some mathematical manipulations) we can obtain the results for the far fields of the half-wave dipole as

$$\hat{E}_\theta = j \frac{\eta_0 \hat{I}_0 e^{-j\beta_0 r}}{2\pi r} F(\theta) \tag{18.33a}$$

$$\hat{H}_\varphi = j \frac{\hat{I}_0 e^{-j\beta_0 r}}{2\pi r} F(\theta) \tag{18.33b}$$

where, the so-called *space factor* is

$$F(\theta) = \frac{\cos \left(\dfrac{\pi}{2} \cos\theta \right)}{\sin\theta} \tag{18.34}$$

The radiation pattern of a half-wave dipole is shown in Figure 18.11.

The electric field is at maximum broadside to the antenna ($\theta = 90°$). In this case, the space factor equals unity.

$$F(\theta) = \frac{\cos \left(\dfrac{1}{2} \pi \cos\theta \right)}{\sin\theta} = \frac{\cos \left(\dfrac{1}{2} \pi \cos 90° \right)}{\sin 90°} = \frac{\cos(0)}{\sin 90°} = \frac{1}{1} = 1 \tag{18.35}$$

Figure 18.11 Radiation pattern of a half-wave dipole.

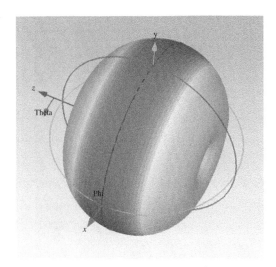

18.5   Quarter-Wave Monopole Antenna

A quarter-wave monopole can be obtained from a half-wave dipole by replacing one of the arms of the dipole by an infinite ground plane, as shown in Figure 18.12.

An infinite ground plane is, of course, not realistic; a practical quarter-wave antenna is shown in Figure 18.13.

The radiation pattern of a quarter-wave monopole above the ground plane is the same as that for the half-wave dipole, as discussed in the next section.

18.6   Image Theory

In image theory, a radiating antenna (actual source) is placed at some distance h from a perfect conducting plane. An image of this antenna (virtual source) is placed below the conducting plane at the same distance h, as shown in Figure 18.14 (Balanis, 2005, p. 185).

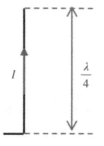

Figure 18.12 Quarter-wave dipole.

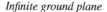

Infinite ground plane

Figure 18.13 A practical monopole antenna.

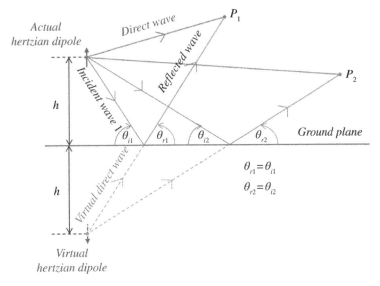

Figure 18.14 Hertzian dipole and its image.

Because of the reflecting ground plane, the total field at an observation point P is the sum of the direct wave and the reflected wave. Obviously, there is no field below the ground plane.

Instead of obtaining the total field by summing the actual direct and the reflected waves, we add the direct waves from the actual source and the direct wave from its image (virtual source) to obtain the same result (above the ground plane). When considering the virtual direct wave, we pretend that the ground plane does not exist and therefore the virtual wave has a direct unobstructed path to the observation point.

Why are we using this approach? Because the calculation of the fields using the actual waves is quite complicated, whereas the calculations using the image theory are quite simple, as we shall see.

Consider the geometry shown in Figure 18.15

The source is an infinitesimal dipole of length l, carrying a constant current I_0. The observation point P is in the far field. Using the previously derived results, the direct component of the E field at the observation point is

$$E_\theta^d = j\eta_0 \frac{\beta_0 I_0 l e^{-j\beta_0 r_1}}{4\pi r_1} \sin\theta_1 \tag{18.36a}$$

The virtual component is

$$E_\theta^v = j\eta_0 \frac{\beta_0 I_0 l e^{-j\beta_0 r_2}}{4\pi r_2} \sin\theta_2 \tag{18.36b}$$

The total field at the observation point is

$$E_\theta = E_\theta^d + E_\theta^v = j\eta_0 \frac{\beta_0 I_0 l e^{-j\beta_0 r_1}}{4\pi r_1} \sin\theta_1 + j\eta_0 \frac{\beta_0 I_0 l e^{-j\beta_0 r_2}}{4\pi r_2} \sin\theta_2 \tag{18.37}$$

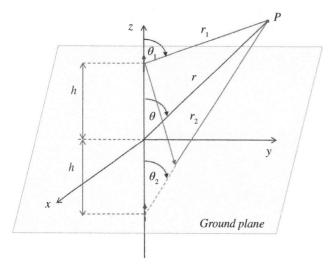

Figure 18.15 Direct waves from the Hertzian dipole and its image.

Using the law of cosines we have

$$r_1 = \sqrt{r^2 + h^2 - 2rh\cos\theta} \tag{18.38a}$$

$$r_2 = \sqrt{r^2 + h^2 - 2rh\cos(\pi - \theta)} \tag{18.38b}$$

In the far field, $r \gg h$ and the Eqs (18.38) can be approximated as

$$r_1 = r - h\cos\theta \tag{18.39a}$$

$$r_2 = r + h\cos\theta \tag{18.39b}$$

Geometrically, Eqs (18.39) represent parallel lines, as shown in Figure 18.16. This is often referred to as the parallel-ray approximation.

Obviously, we have

$$\theta_1 = \theta \tag{18.40a}$$

$$\theta_2 = \theta \tag{18.40b}$$

We will use the approximations in Eqs (18.39) and (18.40) when substituting for r_1 and r_2 in the phase component of the expressions in Eq. (18.37). That is,

$$e^{-j\beta_0 r_1} \cong e^{-j\beta_0(r - h\cos\theta)} \tag{18.41a}$$

$$e^{-j\beta_0 r_2} \cong e^{-j\beta_0(r + h\cos\theta)} \tag{18.41b}$$

When approximating the amplitude components, we may further approximate r_1 and r_2 as

$$r_1 = r \tag{18.42a}$$

$$r_2 = r \tag{18.42b}$$

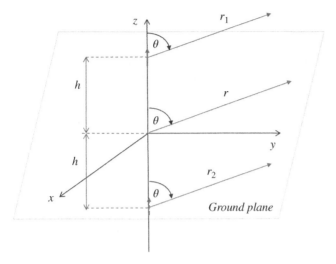

Figure 18.16 Parallel-ray approximation.

Utilizing Eqs (18.41) and (18.42) in Eq. (18.37) we get

$$
\begin{aligned}
E_\theta = E_\theta^d + E_\theta^v &= j\eta_0 \frac{\beta_0 I_0 l e^{-j\beta_0 r_1}}{4\pi r_1} \sin\theta_1 + j\eta_0 \frac{\beta_0 I_0 l e^{-j\beta_0 r_2}}{4\pi r_2} \sin\theta_2 \\
&= j\eta_0 \frac{\beta_0 I_0 l e^{-j\beta_0(r-h\cos\theta)}}{4\pi r} \sin\theta + j\eta_0 \frac{\beta_0 I_0 l e^{-j\beta_0(r+h\cos\theta)}}{4\pi r} \sin\theta \\
&= j\eta_0 \frac{\beta_0 I_0 l}{4\pi r} \sin\theta \left[e^{-j\beta_0(r-h\cos\theta)} + e^{-j\beta_0(r+h\cos\theta)} \right] \\
&= j\eta_0 \frac{\beta_0 I_0 l e^{-j\beta_0 r}}{4\pi r} \sin\theta \left[e^{j\beta_0 h\cos\theta} + e^{-j\beta_0 h\cos\theta} \right] \\
&= j\eta_0 \frac{\beta_0 I_0 l e^{-j\beta_0 r}}{4\pi r} \sin\theta \left[2\cos(\beta_0 h\cos\theta) \right]
\end{aligned}
\tag{18.43}
$$

That is,

$$
E_\theta = \begin{cases} j\eta_0 \dfrac{\beta_0 I_0 l e^{-j\beta_0 r}}{2\pi r} \sin\theta \left[\cos(\beta_0 h\cos\theta) \right], & z \ge 0 \\[2mm] 0, & z < 0 \end{cases}
\tag{18.44}
$$

This result can be extended to the case of a quarter-wave monopole (Paul, Pg. 429).

18.7 Differential- and Common-Mode Currents and Radiation

18.7.1 Differential- and Common-Mode Currents

Consider a typical circuit model shown in Figure 18.17.

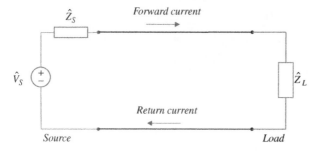

Figure 18.17 A typical circuit model.

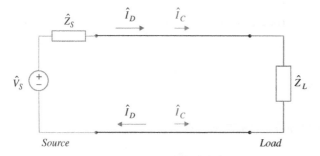

Figure 18.18 A realistic circuit model.

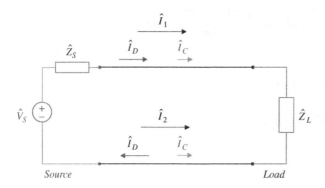

Figure 18.19 Circuit model showing the total currents.

If there were no other circuits or sources or paths of coupling present external to this circuit, the forward current would equal the return current. In virtually any practical circuit a different scenario takes place, as shown in Figure 18.18.

$\hat{I}_D$ is referred to as the *differential-mode (DM) current*, while $\hat{I}_C$ is referred to as the *common-mode (CM) current*. The DM currents are usually the functional currents; they are equal in magnitude and of opposite directions. The CM currents are equal in magnitude and flowing in the same direction. (In the next section, we will show an example of a circuit in which CM currents are created.)

In the analysis of the DM and CM currents we often use the circuit model shown in Figure 18.19, showing, in addition to the DM and CM currents, the *total* currents I_1 and I_2 flowing in the same direction.

These two total currents are related to the DM and CM currents by

$$\hat{I}_1 = \hat{I}_C + \hat{I}_D \qquad (18.45a)$$

$$\hat{I}_2 = \hat{I}_C - \hat{I}_D \qquad (18.45b)$$

Adding and subtracting Eqs (18.45a) and (18.45b) gives

$$\hat{I}_1 + \hat{I}_2 = 2\hat{I}_C \qquad (18.46a)$$

$$\hat{I}_1 - \hat{I}_2 = 2\hat{I}_D \qquad (18.46b)$$

Thus, in terms of the total currents, the DM and CM currents can be expressed as

$$\hat{I}_D = \frac{1}{2}\left(\hat{I}_1 - \hat{I}_2\right) \qquad (18.47a)$$

$$\hat{I}_C = \frac{1}{2}\left(\hat{I}_1 + \hat{I}_2\right) \qquad (18.47b)$$

We are now ready to discuss the radiation from the DM and CM currents.

18.7.2 Radiation from Differential- and Common-Mode Currents

Consider the circuit in Figure 18.20, showing the differential-mode currents and the corresponding radiated **E** fields *in the far field* of these radiating elements.

If we treat the conductors as Hertzian dipoles, or half-wave dipoles, the maximum radiated **E** field is broadside to the antenna ($\theta = 90°$) and in the z direction, as shown.

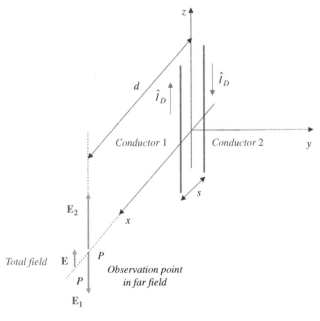

Figure 18.20 Differential-mode radiation.

The radiated fields due to both conductors are of opposite directions, giving a small total radiated field as shown.

The total radiated field at the observation point in the far field can be obtained by treating the two-conductor structure as a small loop antenna (Ott, 2009, p. 465) or by treating each of them as a small dipole antenna and superimposing the fields (Paul, 2006, p. 506).

Now, consider the circuit in Figure 18.21, showing the CM currents and the corresponding radiated **E** fields *in the far field* of these radiating elements.

The radiated fields due to both conductors are of the same direction, thus reinforcing each other to give the total radiated field as shown.

It should be noted that the CM currents could be several orders of magnitude smaller than the DM currents, yet the radiation from them could exceed the regulatory limits.

The total radiated field at the observation point in the far field can be obtained by treating each of the conductors as a small dipole antenna and superimposing the fields (Paul, 2006, p. 515).

We will calculate the total fields using the approach described by Paul (2006, p. 506). In order to calculate the DM and CM radiation, consider the scenario shown in Figure 18.22.

The two linear antennas shown are placed along the x axis, carrying the currents in the z direction. We will determine the radiated field due to both antennas broadside to them (i.e. in the xy plane, or for $\theta = 90°$.

The total radiated electric field at an observation point P in the far field will be the sum of the field of each conductor

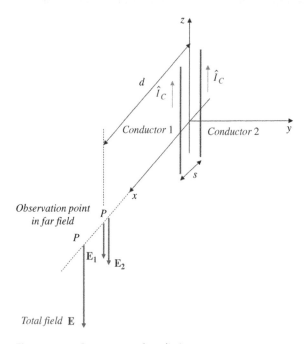

Figure 18.21 Common-mode radiation.

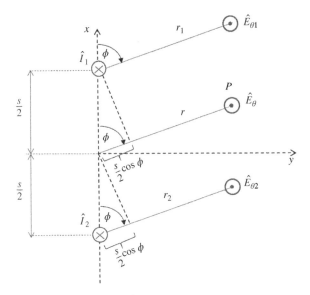

Figure 18.22 Far fields of the two-wire antennas.

$$\hat{E}_\theta = \hat{E}_{\theta,1} + \hat{E}_{\theta,2} \tag{18.48}$$

Treating each conductor as a Hertzian dipole and utilizing Eq. (18.13) we have

$$\hat{E}_{\theta 1} = j\frac{\hat{I}_1 l}{4\pi}\eta_0 \beta_0 \sin\theta \left(\frac{e^{-j\beta_0 r_1}}{r_1}\right) \tag{18.49a}$$

$$\hat{E}_{\theta 2} = j\frac{\hat{I}_2 l}{4\pi}\eta_0 \beta_0 \sin\theta \left(\frac{e^{-j\beta_0 r_2}}{r_2}\right) \tag{18.49b}$$

Thus, the total field in Eq. (18.48) equals

$$\hat{E}_\theta = j\frac{\hat{I}_1 l}{4\pi}\eta_0 \beta_0 \sin\theta \left(\frac{e^{-j\beta_0 r_1}}{r_1}\right) + j\frac{\hat{I}_2 l}{4\pi}\eta_0 \beta_0 \sin\theta \left(\frac{e^{-j\beta_0 r_2}}{r_2}\right)$$

$$= j\frac{l}{4\pi}\eta_0 \beta_0 \sin\theta \left(\frac{\hat{I}_1 e^{-j\beta_0 r_1}}{r_1} + \frac{\hat{I}_2 e^{-j\beta_0 r_2}}{r_2}\right) \tag{18.50}$$

Using the parallel-ray approximation discussed in previous section we have

$$r_1 = r - \frac{s}{2}\cos\phi \tag{18.51a}$$

$$r_2 = r + \frac{s}{2}\cos\phi \tag{18.51b}$$

Substituting Eqs (18.51) into the exponential phase terms in Eq. (18.50) and substituting

$$\hat{r}_1 = r \tag{18.52a}$$

$$\hat{r}_2 = r \tag{18.52b}$$

into the denominators of Eq. (18.50) we obtain

$$
\hat{E}_\theta = j\frac{l}{4\pi}\eta_0\beta_0\sin\theta\left[\frac{\hat{I}_1 e^{-j\beta_0\left(r-\frac{s}{2}\cos\varphi\right)}}{r} + \frac{\hat{I}_2 e^{-j\beta_0\left(r+\frac{s}{2}\cos\varphi\right)}}{r}\right]
$$

$$
= j\frac{l}{4\pi}\eta_0\beta_0\sin\theta\frac{e^{-j\beta_0 r}}{r}\left(\hat{I}_1 e^{+j\beta_0\frac{s}{2}\cos\varphi} + \hat{I}_2 e^{-j\beta_0\frac{s}{2}\cos\varphi}\right) \tag{18.53}
$$

Thus broadside to the antennas ($\sin\theta = 1$) the total radiated field is

$$
\hat{E}_\theta = j\frac{l}{4\pi}\eta_0\beta_0\frac{e^{-j\beta_0 r}}{r}\left(\hat{I}_1 e^{+j\beta_0\frac{s}{2}\cos\varphi} + \hat{I}_2 e^{-j\beta_0\frac{s}{2}\cos\varphi}\right) \tag{18.54}
$$

The maximum radiation will occur in the plane of the wires and on the line perpendicular to the conductors, thus for $\phi = 0°$ or $\phi = 180°$ (Paul, 2006, p. 509). Using $\phi = 180°$ in Eq. (18.54) we obtain

$$
\hat{E}_\theta = j\frac{l}{4\pi}\eta_0\beta_0\frac{e^{-j\beta_0 r}}{r}\left(\hat{I}_1 e^{-j\beta_0\frac{s}{2}} + \hat{I}_2 e^{j\beta_0\frac{s}{2}}\right) \tag{18.55}
$$

Next, we will apply Eq. (18.55) to the DM and CM currents shown in Figures 18.19 and 18.20, respectively.

Differential-mode radiation Letting

$$\hat{I}_1 = \hat{I}_D \tag{18.56a}$$

$$\hat{I}_2 = -\hat{I}_D \tag{18.56b}$$

and replacing the distance r by d (taken from the midpoint between the conductors) in Eq. (18.55) we obtain

$$
\hat{E}_\theta = j\frac{l}{4\pi}\eta_0\beta_0\frac{e^{-j\beta_0 d}}{d}\left(\hat{I}_{D}e^{-j\beta_0\frac{s}{2}} - \hat{I}_{D}e^{j\beta_0\frac{s}{2}}\right)
$$

$$
= j\frac{l}{4\pi}\eta_0\beta_0\frac{\hat{I}_{D}e^{-j\beta_0 d}}{d}\left(e^{-j\beta_0\frac{s}{2}} - e^{j\beta_0\frac{s}{2}}\right)
$$

$$
= j\frac{l}{4\pi}\eta_0\beta_0\frac{\hat{I}_{D}e^{-j\beta_0 d}}{d}j2\sin\left(\beta_0\frac{s}{2}\right) \tag{18.57}
$$

$$
= \frac{l}{2\pi}\eta_0\beta_0\frac{\hat{I}_{D}e^{-j\beta_0 d}}{d}\sin\left(\beta_0\frac{s}{2}\right)
$$

Utilizing

$$\beta_0 = \frac{2\pi}{\lambda_0} \qquad (18.58)$$

in Eq. (18.57) we obtain

$$\hat{E}_\theta = \frac{l}{2\pi} \eta_0 \frac{2\pi}{\lambda_0} \frac{\hat{I}_D e^{-j\beta_0 d}}{d} \sin\left(\frac{2\pi}{\lambda_0} \frac{s}{2}\right)$$
$$= \frac{l\eta_0}{\lambda_0} \frac{\hat{I}_D e^{-j\beta_0 d}}{d} \sin\left(\pi \frac{s}{\lambda_0}\right) \qquad (18.59)$$

For electrically small spacing between the line, i.e.

$$s \ll \lambda_0 \quad \text{or} \quad \frac{s}{\lambda_0} \ll 1 \qquad (18.60)$$

we use the approximation

$$\sin\left(\pi \frac{s}{\lambda_0}\right) \cong \pi \frac{s}{\lambda_0} \qquad (18.61)$$

Using Eq. (18.61) in Eq. (18.59) results in

$$\hat{E}_\theta = \frac{l\eta_0}{\lambda_0} \frac{\hat{I}_D e^{-j\beta_0 d}}{d} \pi \frac{s}{\lambda_0} \qquad (18.62)$$

Now,

$$\lambda_0 = \frac{v_0}{f} = \frac{3 \times 10^8}{f} \qquad (18.63a)$$

$$\eta_0 = 120\pi \qquad (18.63b)$$

Using Eqs (18.63) in Eq. (18.62) produces

$$\hat{E}_\theta = \frac{l\eta_0}{\lambda_0} \frac{\hat{I}_D e^{-j\beta_0 d}}{d} \pi \frac{s}{\lambda_0}$$
$$= \frac{120\pi l \, f}{3 \times 10^8} \frac{\hat{I}_D e^{-j\beta_0 d}}{d} \pi \frac{sf}{3 \times 10^8}$$
$$= \frac{120\pi^2}{9 \times 10^{16}} f^2 ls\hat{I}_D \frac{e^{-j\beta_0 d}}{d} \qquad (18.62)$$
$$= 131.59 \times 10^{-16} f^2 ls\hat{I}_D \frac{e^{-j\beta_0 d}}{d}$$

The magnitude of the total field is

$$E_\theta = 131.59 \times 10^{-16} f^2 \hat{I}_D \frac{ls}{d} \qquad (18.63)$$

This corresponds to the equivalent formulas in Ott (2009, p. 466, Eq. (12–1)) and Paul (2006, p. 510, Eq. (8.12)).

Common-mode radiation Letting

$$\hat{I}_1 = \hat{I}_C \tag{18.64a}$$

$$\hat{I}_2 = \hat{I}_C \tag{18.64b}$$

and replacing the distance r by d (taken from the midpoint between the conductors) in Eq. (18.55) we obtain

$$\hat{E}_\theta = j\frac{l}{4\pi}\eta_0\beta_0\frac{e^{-j\beta_0 d}}{d}\left(\hat{I}_C e^{-j\beta_0\frac{s}{2}} + \hat{I}_C e^{j\beta_0\frac{s}{2}}\right)$$

$$= j\frac{l}{4\pi}\eta_0\beta_0\frac{\hat{I}_C e^{-j\beta_0 d}}{d}\left(e^{-j\beta_0\frac{s}{2}} + e^{j\beta_0\frac{s}{2}}\right) \tag{18.65}$$

$$= j\frac{l}{4\pi}\eta_0\beta_0\frac{\hat{I}_C e^{-j\beta_0 d}}{d}2\cos\left(\beta_0\frac{s}{2}\right)$$

$$= j\frac{l}{2\pi}\eta_0\beta_0\frac{\hat{I}_C e^{-j\beta_0 d}}{d}\cos\left(\beta_0\frac{s}{2}\right)$$

Utilizing

$$\beta_0 = \frac{2\pi}{\lambda_0} \tag{18.66}$$

in Eq. (18.65) we obtain

$$\hat{E}_\theta = j\frac{l}{2\pi}\eta_0\frac{2\pi}{\lambda_0}\frac{\hat{I}_C e^{-j\beta_0 d}}{d}\cos\left(\frac{2\pi}{\lambda_0}\frac{s}{2}\right)$$

$$= j\frac{l\eta_0}{\lambda_0}\frac{\hat{I}_C e^{-j\beta_0 d}}{d}\cos\left(\pi\frac{s}{\lambda_0}\right) \tag{18.67}$$

For electrically small spacing between the line, i.e.

$$s \ll \lambda_0 \quad \text{or} \quad \frac{s}{\lambda_0} \ll 1 \tag{18.68}$$

we use the approximation

$$\cos\left(\pi\frac{s}{\lambda_0}\right) \cong 1 \tag{18.69}$$

Using Eq. (18.69) in Eq. (18.67) results in

$$\hat{E}_\theta = j\frac{l\eta_0}{\lambda_0}\frac{\hat{I}_C e^{-j\beta_0 d}}{d} \tag{18.62}$$

Again,

$$\lambda_0 = \frac{v_0}{f} = \frac{3\times10^8}{f} \tag{18.64a}$$

$$\eta_0 = 120\pi \tag{18.65b}$$

Using Eqs (18.65) in Eq. (18.67) produces

$$
\begin{aligned}
\hat{E}_\theta &= j\frac{l\eta_0}{\lambda_0}\frac{\hat{I}_C e^{-j\beta_0 d}}{d} \\
&= j\frac{120\pi l f}{3\times10^8}\frac{\hat{I}_C e^{-j\beta_0 d}}{d} \\
&= 125.66\times10^{-8} f l\hat{I}_C \frac{e^{-j\beta_0 d}}{d}
\end{aligned}
\tag{18.62}
$$

The magnitude of the total field is

$$E_\theta = 125.66\times10^{-8}\, f\breve{I}_C \frac{l}{d} \tag{18.63}$$

This corresponds to the equivalent formulas in Ott (2009, p. 477, Eq. (12-6)) and Paul (2006, p. 515, Eq. (8.16a)).

18.8 Common Mode Current Creation

In this section we present an example of how the common-mode current is created in a differential signaling circuit. In order to facilitate this discussion, we begin with a simple circuit configuration and augment it to reflect high-speed digital circuits.

18.8.1 Circuits with a Shared Return Path

Consider the circuit shown in Figure 18.23.

The current $I_{forward}$ leaves the source, arrives at the load and the same amount of current, I_{return}, flows back to the source. Note that the load voltage is the same as the source voltage.

The scenario shown in Figure 18.23 is an idealized one. In high-speed digital circuits, we often encounter the arrangement shown in Figure 18.24, where we need to account

Figure 18.23 Load driven by a source.

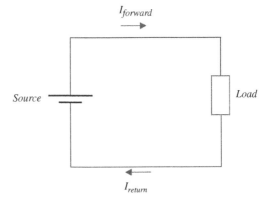

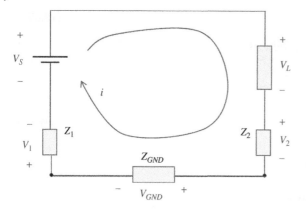

Figure 18.24 Finite impedance in the connections and ground path.

for the finite impedances of the source and load connections to the reference plane, which itself has a non-zero impedance (Johnson and Graham, 2003, p. 365).

Z_1 and Z_2 in Figure 18.24 represent the finite impedances of the IC package pins or balls to make the connection to the ground reference plane, and Z_{GND} represents the finite impedance in the return path.

Writing KVL for the circuit shown in Figure 18.24 we get

$$V_S = V_L + V_2 + V_{GND} + V_1 \tag{18.64}$$

Since

$$V_1 = Z_1 i \tag{18.65a}$$

$$V_1 = Z_1 i \tag{18.65b}$$

Equation (18.64) can be rewritten as

$$V_S = V_L + Z_2 i + V_{GND} + Z_1 i \tag{18.66}$$

or

$$V_L = V_S - Z_2 i - V_{GND} - Z_1 i \tag{18.67}$$

Note that the load voltage is not equal to the source voltage (as is the case in Figure 18.23). The voltage across the load is lower than the source voltage by the various voltage drops along the current loop. Also, if the return path is shared with other circuits, as shown in Figure 18.25, then there is the potential for common impedance coupling between the circuits.

We have considered the finite impedance of the ground connections and Z_{GND} in the ground plane, but a similar scenario occurs in power planes, as shown in Figure 18.26.

Currents flowing in a power plane affect the power voltages in the same way as the currents flowing in the ground plane affect the ground voltages. Whether we are more concerned with the power plane or the ground plane, noise voltage depends on whether the circuit uses the power rails or the ground rail as the internal reference for logic signals (Johnson and Graham, 2003, p. 368).

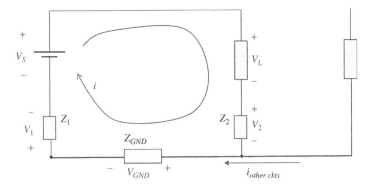

Figure 18.25 Common ground-plane impedance coupling.

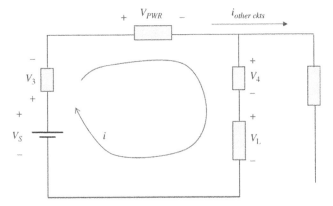

Figure 18.26 Common power-plane impedance coupling.

Another example of common-impedance coupling is shown in Figure 18.27.

The driver for the second circuit, (V_{S2}), shares the same physical IC package as the load for the first circuit. Thus, they share a common ground connection Z_2. As the return current i_2 enters the package through Z_2, it will affect the voltage at the load of the first circuit.

A variation of the circuit shown in Figure 18.27 is shown in Figure 18.28.

The driver for the second circuit, (V_{S2}), shares the same physical IC package as the load for the first circuit but now each circuit has its own ground connection. Obviously, this eliminates the shared common-impedance coupling between the driver and the load in the same package.

Let's assume that no current flows through the ground connections, as shown in Figure 18.29.

Under this assumption, in each circuit, the forward current will be equal in magnitude and opposite in direction to the return current, just like the scenario in the very first, idealized circuit in Figure 18.23.

Also, each of the drivers no longer relies on the reference voltage with respect to the ground plane. The reference voltage, V_S, is simply the voltage difference between the two wires. Finally, each circuit is immune to any type of interference that affects both

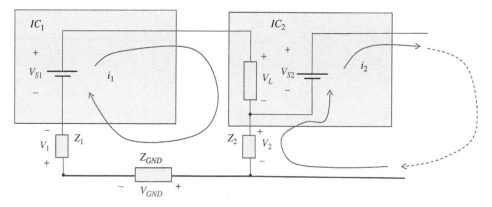

Figure 18.27 Another example of common impedance coupling.

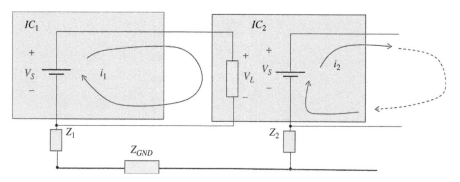

Figure 18.28 A variation of the previous circuit.

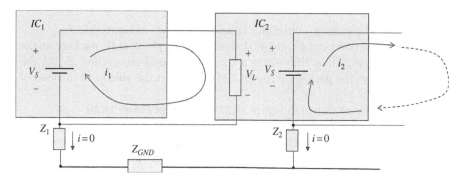

Figure 18.29 No current flows through the ground connections.

wires equally. That is, the potential of each wire is shifted by the same amount, the potential difference between the wires remains the same.

These very desirable characteristics will no longer be true if the currents through the connection impedances Z_1 and Z_2 are non-zero. Thus, in a practical circuit, we would want these currents to be minimized to insignificant levels. Even if this is accomplished, another problem arises in high-speed digital circuits: capacitive coupling to other conducting paths or metallic objects. This is shown in Figure 18.30.

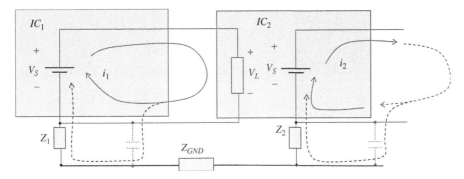

Figure 18.30 Capacitive coupling to other conducting paths or metallic objects.

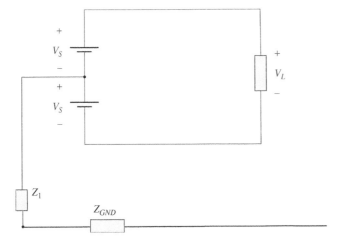

Figure 18.31 Differential signaling circuit.

The return current now has a choice of two paths back to the source: the intended path through the return wire or the parasitic path through the impedance Z_1 or Z_2. This leads us to the topic of differential signaling discussed in the next section.

18.8.2 Differential Signaling

Consider the circuit shown in Figure 18.31.
Writing KVL for the circuit shown gives

$$-V_S - V_S + V_L = 0 \qquad (18.68)$$

or

$$V_L = 2V_S \qquad (18.69)$$

A simple remedy resulting in the load voltage equal the source voltage is shown in Figure 18.32, or in Figure 18.33 with the bottom source polarity reversed.

Thus, the differential signaling results in equal and opposite voltages (with respect to the reference ground) and equal and opposite currents flowing in the forward and return paths. Figure 18.33 shows dc voltages, but of course the same discussion holds for ac voltages.

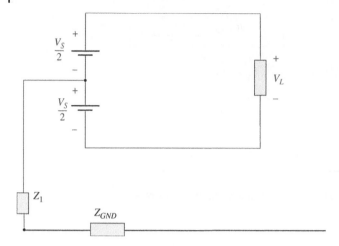

Figure 18.32 A variation of the previous circuit.

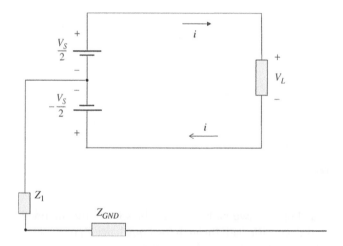

Figure 18.33 Equivalent differential signaling circuit.

18.8.3 Common-Mode Current Creation

Let's add the coupling between the differential pair and the ground plane as in Figure 18.34.

If the layout is symmetrical, then both wires couple equally to the reference ground plane (through Z_F and Z_R), and ac currents induced in the ground plane by one wire will be counteracted by equal and opposite currents induced by the other wire. Thus, there will be no net parasitic current in the reference plane.

If the differential-pair voltages are not precisely complementary, or the stray impedances Z_F and Z_R are not well balanced, then some stray current will flow in the reference plane. This is shown in Figure 18.35.

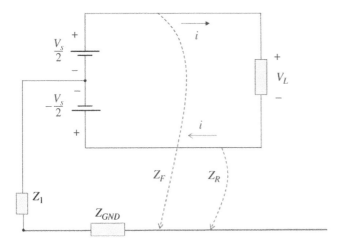

Figure 18.34 Coupling to the ground plane.

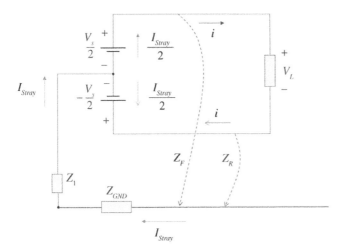

Figure 18.35 Stray current flow.

We refer to this stray current as the common-mode current and designate it in circuits as shown in Figure 18.36.

18.9 Antenna Circuit Model

18.9.1 Transmitting-Mode Model

A physical model of an antenna in a transmitting mode is shown in Figure 18.37.

A corresponding circuit model of an antenna in a transmitting mode is shown in Figure 18.38.

Input impedance of an antenna, $\hat{Z}_{in}$, is the impedance presented by an antenna (to the generator circuit) at its input terminals *A-B*.

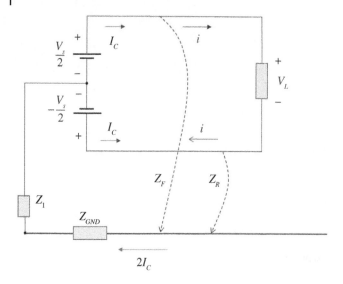

Figure 18.36 Common-mode current flow.

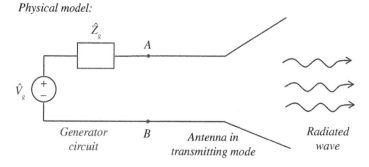

Figure 18.37 Physical model of an antenna in a transmitting mode.

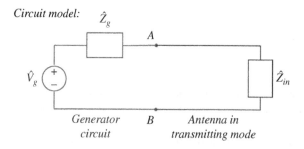

Figure 18.38 Circuit model of an antenna in a transmitting mode.

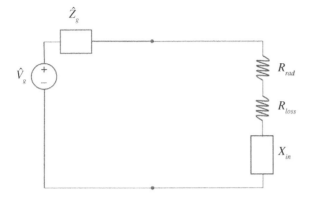

Figure 18.39 Detailed circuit model of an antenna in transmitting mode.

Figure 18.39 shows more details of this circuit model.
The input impedance of an antenna is

$$\hat{Z}_{in} = R_{in} + jX_{in} \quad [\Omega] \tag{18.70}$$

where

$$R_{in} = R_{loss} + R_{rad} \tag{18.71}$$

R_{rad} is the radiation resistance, and X_{in} the radiation reactance of the antenna.
Recall that in Section 6.7.5 (Eq. 6.176), we calculated the radiated power of a Hertzian dipole as

$$P_{rad} = 80\pi^2 \left(\frac{l}{\lambda}\right)^2 \frac{I_0^2}{2} \tag{18.72}$$

The radiation resistance is a fictitious resistance that dissipates the same power as that radiated by the Hertzian dipole when carrying the same current. Thus,

$$P_{rad} = R_{rad} \frac{I_0^2}{2} \tag{18.73}$$

Therefore, the radiation resistance of a Hertzian dipole is

$$R_{rad} = 80\pi^2 \left(\frac{l}{\lambda}\right)^2 \tag{18.74}$$

The radiation resistance of a half-wave dipole is 73 Ω, while the radiation reactance is $j42.5$ Ω (Paul, 2006, p. 435). The values for the quarter-wave monopole are the half of those for the half-wave dipole.
The circuit models (assuming no losses) of a half-wave diploe and a quarter-wave monopole are shown in Figure 18.40, and Figure 18.41, respectively.

18.9.2 Receiving-Mode Model

A physical model of an antenna in a receiving mode is shown in Figure 18.42.

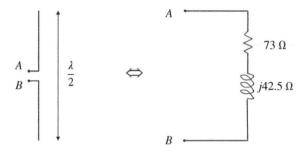

Figure 18.40 Circuit model of a half-wave dipole.

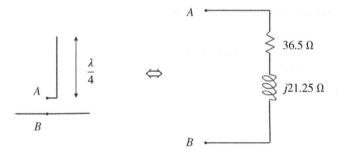

Figure 18.41 Circuit model of a quarter-wave monopole.

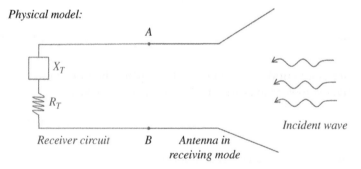

Figure 18.42 Physical model of an antenna in a transmitting mode.

A corresponding circuit model of an antenna in a receiving mode is shown in Figure 18.43.

A visual model of the half-wave dipole receiving antenna operation is shown in Figure 18.44.

An oscillating EM wave (E field shown) arrives at the antenna and is directed along its arms. This E field exerts a force on the electrons in the antenna arms, causing them to move back and forth between the antenna ends, charging them alternatively positive and negative. This oscillating current flows through the receiver resulting in a voltage reading.

Circuit model:

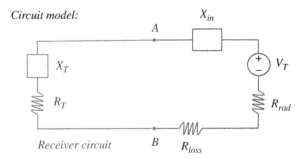

Figure 18.43 Circuit model of an antenna in a receiving mode.

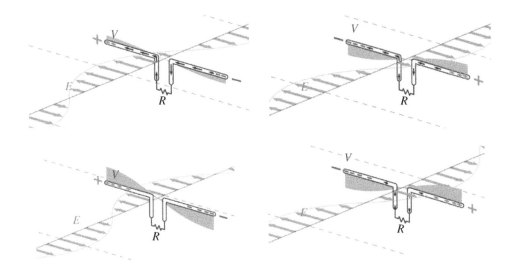

Figure 18.44 Half-wave dipole in a receiving mode.

18.10   EMC Applications

18.10.1   EMC Antenna Measurements

Figure 18.45 shows a log-periodic antenna used for radiated emissions measurements.

The antenna is connected through a high-quality coaxial cable to a receiver as shown in Figure 18.46 (the measurement setup conforms to CISPR 25 requirements).

A simplified test setup is shown in Figure 18.47.

The wave radiating from the equipment under test (EUT) is captured by the measuring antenna, connected through a coax cable to the receiver (spectrum analyzer or EMI receiver).

The voltage measured by this receiver is $\check{V}_{rec}$. In order to relate this voltage reading to the actual electric field measured by the antenna, $\hat{E}_{inc}$, we need the so-called *antenna factor* (supplied by the antenna manufacturer).

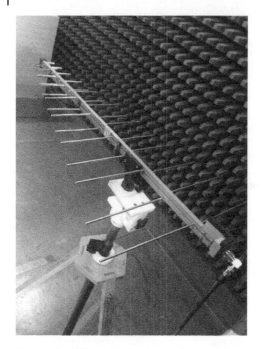

Figure 18.45 Log-periodic antenna for 300–1000 MHz frequency range measurements.

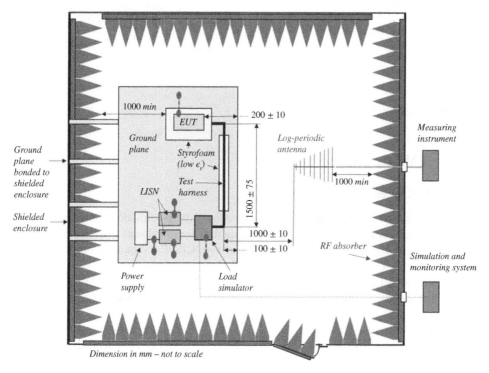

Figure 18.46 Log-periodic antenna connected to a measuring receiver.

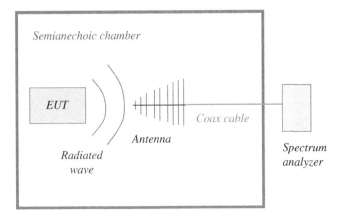

Figure 18.47 Simplified radiated emissions setup.

Antenna factor is defined as

$$AF = \frac{\left|\hat{E}_{inc}\right|}{\left|\hat{V}_{rec}\right|}, \quad \frac{\text{V/m in incident wave}}{\text{V received}} \quad \left(\frac{1}{m}\right) \quad (18.75)$$

That is, the antenna factor is the ratio of the incident electric field at the surface of the measurement antenna to the received voltage at the antenna terminal.

The antenna factor is usually given in dB:

$$AF_{dB} = dB\mu V/m \left(\text{incident wave}\right) - dB\mu V \left(\text{received voltage}\right) \quad (18.76)$$

It is provided by the antenna manufacturer, either as a table or a plot vs frequency. Figure 18.48 shows the antenna factor for a log-periodic antenna.

From Eq. (18.76) we get

$$dB\mu V/m \left(\text{incident wave}\right) = dB\mu V \left(\text{received voltage}\right) + AF_{dB} \quad (18.77)$$

In order to account for the cable loss, we need to modify the above equation to

$$dB\mu V/m \left(\text{incident wave}\right) = dB\mu V \left(\text{received voltage}\right) + AF_{dB} + \text{cable loss}_{dB} \quad (18.78)$$

18.10.2 Antenna VSWR and Impedance Measurements

Consider the model of an antenna system in the receiving mode shown in Figure 18.49.

The spectrum analyzer is matched to the coaxial cable. If the antenna's radiation resistance was 50 Ω over the measurement frequency range then the voltage induced at the base of the antenna would appear at the spectrum analyzer (assuming no cable loss).

If the antenna's resistance differed from 50 Ω then some of the power received by the antenna would be reflected back or reradiated, and the reading at the spectrum analyzer would be lower.

It is therefore very useful to know the impedance of the antenna over its measurement range. One very good indicator of the antenna impedance is obtained by measuring the VSWR of the antenna.

Antenna calibration

Antenna type:		Log periodic
Model:		AL-100
Serial number:		16276
Calibration date:	(mm/dd/yy)	04/11/07
Certificate number:		031121
Frequency	Gain	Factor
MHz	dBi	dB/m
300.0	4.3	15. 4
400.0	5.8	16. 5
500.0	5.8	18. 4
600.0	3.4	22. 4
700.0	5.9	21. 2
800.0	6.0	22. 3
900.0	6.0	23. 3
1000.0	6.2	24. 0
Calibration:		3 meters
Polorization		Horizontal

Figure 18.48 Antenna factor for a log-periodic antenna.

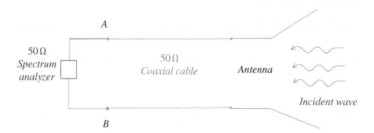

Figure 18.49 Antenna in the receiving mode.

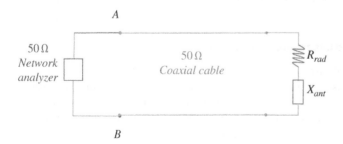

Figure 18.50 Antenna SVWR measurement setup.

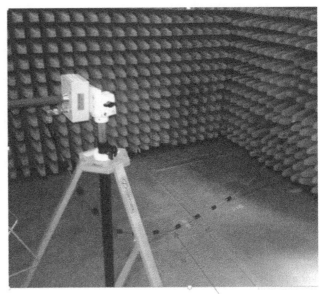

Log-periodic antenna   *Coax cable inside the chamber*

Figure 18.51 Log-periodic antenna measurement setup inside the chamber.

Consider the setup shown in Figure 18.50.

If, in a given frequency range, the antenna's resistance is 50 Ω then the VSWR reading will be 1. The more the impedance of the antenna differs from 50 Ω the higher the VSWR reading.

Figures 18.50 and 18.51 show the actual setup for measuring VSWR of the log-periodic antenna.

The VSWR measurement for this antenna is shown in Figure 18.53, while the impedance measurement is shown in Figure 18.54.

Note that the VSWR is very close to the value of one in the frequency range 300 MHz to 1 GHz, which is the intended frequency range of this antenna.

Note that the impedance measurement is very close to the value of 50 Ω in the intended frequency range of this antenna.

18.10.3 Comb Transmitter Measurements

To evaluate the radiated emissions measurement setup, we often use a comb generator, shown in Figure 18.55.

A comb generator is a transmitting antenna that produces signals of (ideally) the same amplitude, equally spaced in frequency. Figure 18.56 shows the comb generator measurement results in the frequency range 30–300 MHz, which is the intended frequency range for the biconical antenna.

Coax cable outside the chamber   *Preamp*   *EMI receiver*

Figure 18.52 Measurement setup outside the chamber.

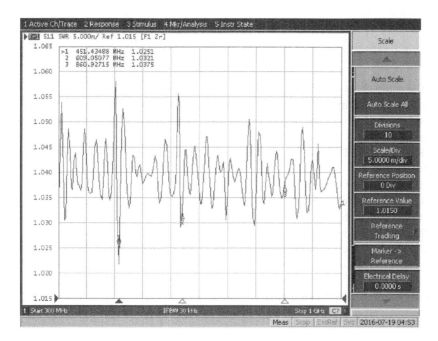

Figure 18.53 Log-periodic antenna VSWR measurement results.

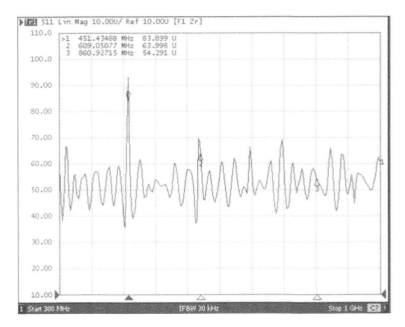

Figure 18.54 Log-periodic antenna impedance measurement results.

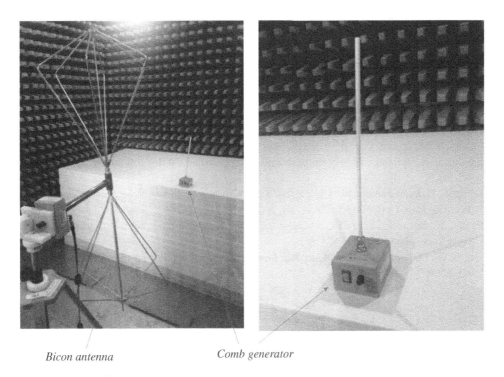

Bicon antenna   *Comb generator*

Figure 18.55 Comb generator measurement setup.

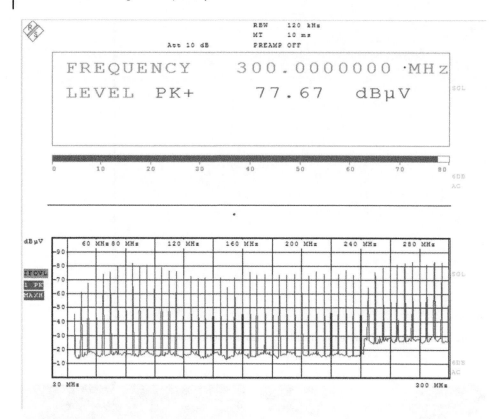

Figure 18.56 Comb generator measurement results.

References

Balanis, C.A., *Antenna Theory Analysis and Design*, 3rd ed., Wiley Interscience, Hoboken. NJ, 2005.

Johnson, H. and Graham, M., *High-Speed Signal Propagation Advanced Black Magic*, Prentice Hall, Upper Saddle River, NJ, 2003.

Paul, C.R., *Introduction to Electromagnetic Compatibility*, 2nd ed., John Wiley and Sons, New York, 2006.

Ott, H.W., *Electromagnetic Compatibility Engineering*, John Wiley and Sons, Hoboken, NJ, 2009.

Appendix A

EMC Tests and Measurements

This appendix presents a description of the basic setups for radiated and conducted emissions, radiated and conducted immunity, and electrostatic discharge (ESD). A representative sample of commercial EMC regulations is used to explain the basics of EMC measurements. Pictures of the typical test setups, equipment, and facilities are presented. Each test is supported by the examples of the real test data, many of them illustrating the "pass" and "fail" results.

This presentation is not intended to review all existing EMC regulations or to discuss the details of each test procedure and the required documentation. The intent is simply to discuss each test and the equipment required to perform it, to the extent needed to gain a basic understanding of and to interpret the test results.

A.1   Introduction – FCC Part 15 and CISPR 22 Standards

EMC standards and regulations have been imposed by various government regulatory bodies and various industries to control allowable emissions from electronic products. In the USA, the Federal Communications Commission (FCC) regulates the use of radio and wire communications. Part 15 of the FCC Rules and Regulations sets forth technical standards and operational requirements for RF devices.

The most widely outside the USA is CISPR 22, which sets limits on the radiated and conducted emissions of information technology equipment, which basically includes all digital devices in the similar meaning as for the FCC (CISPR – Comité International Spécial des Perturbations Radioélectriques – International Special Committee on Radio Interference).

The limits are divided into Class A (commercial devices) and Class B equipment (residential devices), and their meaning is essentially the same as the FCC definitions.

A.1.1   Peak vs Quasi-Peak vs Average Measurements

Most radiated and conducted limits in EMC testing are based on quasi-peak detection mode. Quasi-peak detectors weigh signals according to their repetition rate, which is a way of measuring their "annoyance factor". High amplitude low repetition rate signals could produce the same output as low amplitude high repetition rate signals.

Foundations of Electromagnetic Compatibility with Practical Applications, First Edition. Bogdan Adamczyk.
© 2017 John Wiley & Sons Ltd. Published 2017 by John Wiley & Sons Ltd.

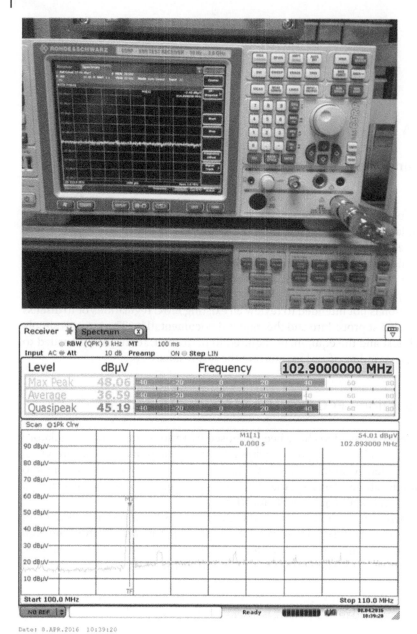

Figure A.1 An EMI receiver and its typical screen output.

As the repetition rate increases, the quasi-peak detector produces a higher voltage output, i.e. a response on spectrum analyzer or EMI receiver. Figure A.1 shows an EMI receiver and its typical screen output.

Quasi-peak detector readings will be less than or equal to the peak detection. An average detector will be less than or equal to the quasi-peak detection. This is shown in Figure A.2.

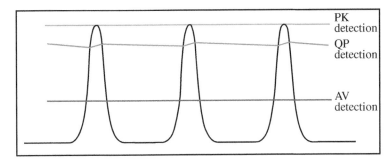

Figure A.2 Relationship between the detectors.

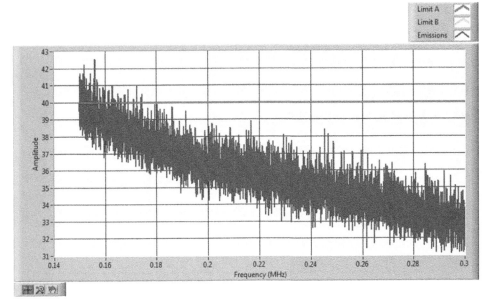

Figure A.3 Peak detector measurement.

Because quasi-peak readings are much slower (by two or three orders of magnitude compared with peak) it is very common to scan initially with the peak detection, and then if this is marginal or fails, switch and run the quasi-peak measurement against the limits.

This approach is illustrated in Figures A.3–A.5 which show the current probe measurements.

Since the peak detector measurement failed, it was followed by the average and the quasi-peak measurements, which passed.

A.1.2   FCC and CISPR 22 Limits

Maybe the easiest way to begin the discussion of EMC regulations and test limits is to start with the FCC and CISPR 22 conducted emission limits, since these two are the same (this is not the case in general).

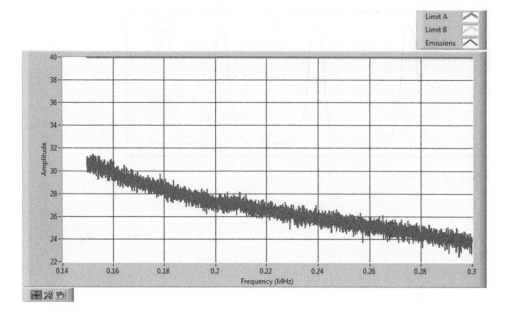

Figure A.4 Average detector measurement.

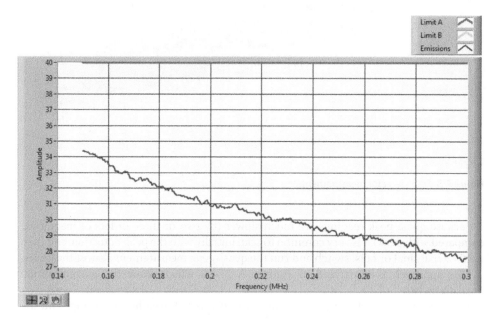

Figure A.5 Quasi-peak detector measurement.

Figure A.6 shows the Class *A* conducted emissions limits, while Figure A.7 shows the limits for Class *B*.

Note that the conducted emission testing is performed in the frequency range of 0.15–30 MHz.

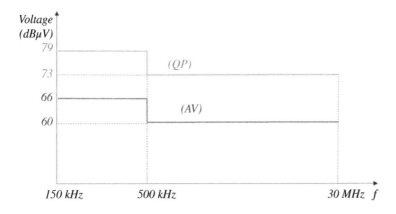

Figure A.6 FCC and CISPR 22 Class *A* conducted emissions limits.

Frequency (MHz)	μV QP (AV)	dBμV QP (AV)
0.15 - 0.5	8912.5 (1995)	79 (66)
0.5 - 30	4467 (1000)	73 (60)

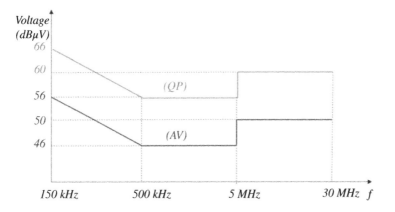

Figure A.7 FCC and CISPR 22 Class *B* conducted emissions limits.

Frequency (MHz)	μV QP (AV)	dBμV QP (AV)
0.15	1995 (631)	66 (56)
0.5	631 (199.5)	56 (46)
0.5 - 5	631 (199.5)	56 (46)
5 - 30	1000 (316)	60 (50)

The FCC and CISPR 22 conducted emissions limits are the same. This is not the case with the radiated emission limits.

CISPR 22 radiated emissions limits are specified at a 10 m distance for both Class *A* and Class *B* devices. FCC radiated emission limits are specified at a 10 m distance for Class *A* devices but at a distance of 3 m for Class *B* devices.

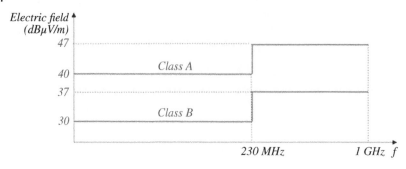

CISPR 22 Class A (measured at 10 m)

Frequency (MHz)	μV/m	dBμV/mQP
30 - 230	100 (QP)	40 (QP)
230 - 1000	224 (QP)	47 (QP)

CISPR 22 Class B (measured at 10 m)

Frequency (MHz)	μV/m	dBμV/mQP
30 - 230	31.6 (QP)	30 (QP)
230 - 1000	70.8 (QP)	37 (QP)

Figure A.8 CISPR 22 Class *B* radiated emissions limits.

Figure A.8 show the CISPR 22 Class *A* and Class *B* radiated emissions limits.

The FCC Class *A* radiated emissions limits are shown in Figure A.9, while the Class *B* radiated emissions limits are shown in Figure A.10.

Finally, Figure A.11 compares the CISPR 22 and FCC radiated emissions limits for Class *A* devices.

A.2 Conducted Emissions

Conducted emissions are the noise currents generated by the EUT (or DUT – device under test) that propagate through the power cord or harness to other components/ systems or power grid.

FCC and CISPR 22 set the limits on the *ac* conducted emissions. CISPR 25 (automotive standard), MIL-STD-461 (military standard) set the limits on the *dc* conducted emissions.

To measure the conducted emissions the artificial network (AN) or the line imped-ance stabilization network (LISN) is used. (LISN looks like a 50 Ω resistor to the EUT and basically acts as an LC low pass filter).

Figure A.12 shows an ac LISN and Figure A.13 shows its schematic.

There are several variations of the dc LISNs. Figures A.14 and A.15 show two of them.

FCC and CISPR 22 require two conducting planes (horizontal and vertical), shown in Figure A.16, and use the *voltage method* to measure the conducted emissions.

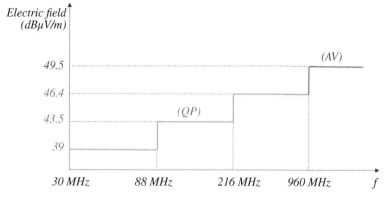

Class A (measured at 10 m)

Frequency (MHz)	μV/m	dBμV/m
30 - 88	90 (QP)	39
88 - 216	150 (QP)	43.5
216 - 960	210 (QP)	46.4
>960	300 (QP)	49.5
>1 GHz	300 (AV)/3000 (PK)	49.5 (AV)/69.5 (PK)

Figure A.9 FCC Class *A* radiated emissions limits.

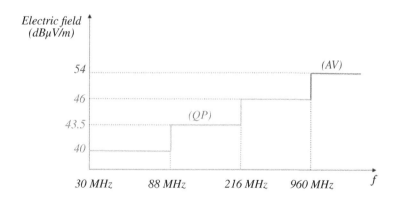

Class B (measured at 3 m)

Frequency (MHz)	μV/m	dBμV/m
30 - 88	100 (QP)	40
88 - 216	150 (QP)	43.5
216 - 960	200 (QP)	46
>960	500 (QP)	54
>1 GHz	500 (AV)/5000 (PK)	54 (AV)/74 (PK)

Figure A.10 FCC Class *B* radiated emissions limits.

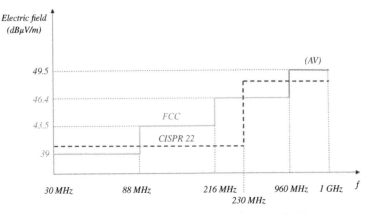

CISPR 22 Class A (measured at 10 m)

Frequency (MHz)	µV/m	dBµV/m QP
30 - 230	100 (QP)	40 (QP)
230 - 1000	224 (QP)	47 (QP)

FCC Class A (measured at 10 m)

Frequency (MHz)	µV/m	dBµV/m
30 - 88	90 (QP)	39
88 - 216	150 (QP)	43.5
216 - 960	210 (QP)	46.4
>960	300 (QP)	49.5
>1 GHz	300 (AV)/3000 (PK)	49.5 (AV)/69.5 (PK)

Figure A.11 Comparison of the CISPR 22 and FCC Class *A* radiated emissions limits.

Figure A.12 FCC/CISPR22 ac LISN.

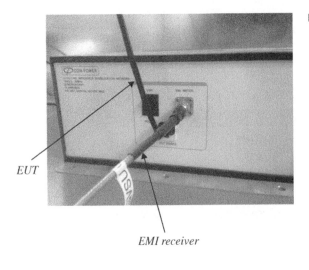

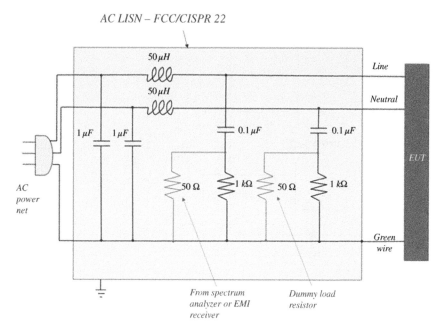

Figure A.13 ac LISN schematic.

Figure A.14 CISPR 25 dc LISN.

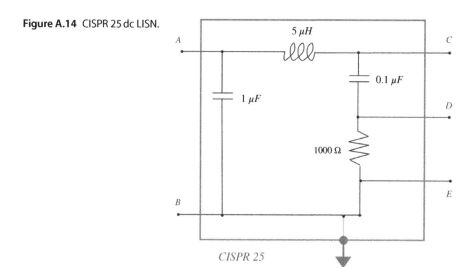

CISPR 25 requires a screen room, shown in Figure A.17, and specifies two methods: *voltage method* and *current probe method*.

A.2.1 FCC and CISPR 22 Voltage Method

The details of the FCC/CISPR 22 conducted emissions voltage method setup are shown in Figure A.18. (for clarity, the vertical ground plane is not shown).

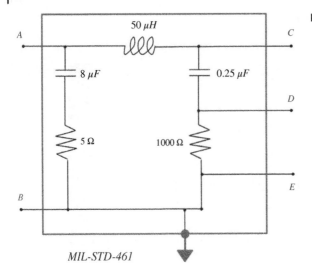

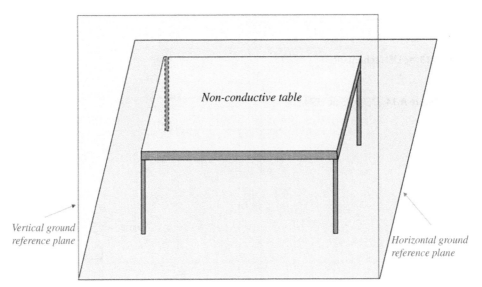

Figure A.16 FCC/CISPR 22 conducted emissions test setup.

Figure A.19 shows a DUT (laptop) positioned on the test table (dimensions not to scale) in a screen room setup for the conducted emissions testing.

The ac conducted emissions are measured on both the line and the neutral lines. An example of the line conducted emissions is shown in Figure A.20, while the emissions on the neutral line are shown in Figure A.21.

A.2.2 CISPR 25 Voltage Method

The details of the CISPR 25 conducted emissions voltage method setup are shown in Figure A.22. and A.23.

Figure A.17 Screen room for CISPR 25 conducted emissions measurements.

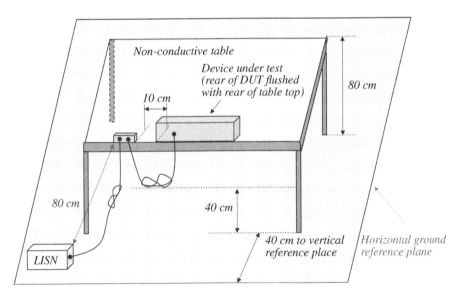

Figure A.18 The details of the FCC/CISPR 22 voltage method setup.

CISPR 25 categorizes the devices into five classes; the classification is based on the physical location of the device in a vehicle and the severity of the exposure to the EM environment. Class 1 limits are the least severe and the Class 5 limits are the most severe, as shown in Tables A.1 and A.2.

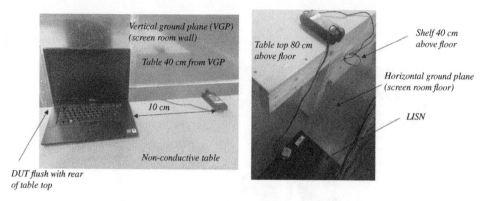

Figure A.19 DUT arrangement in a screen room for conducted emissions testing.

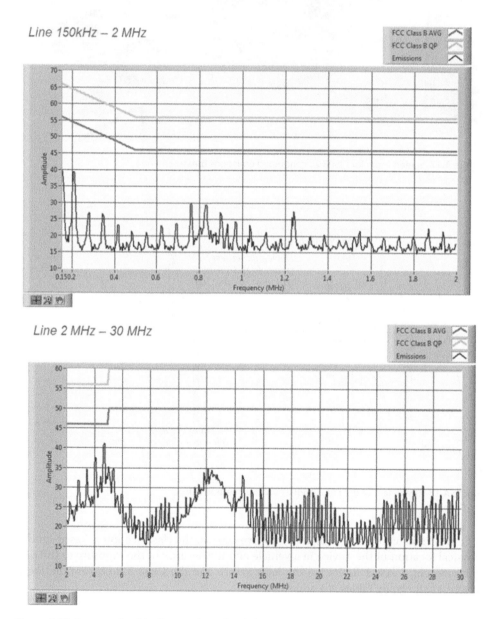

Figure A.20 An example of the line conducted emissions.

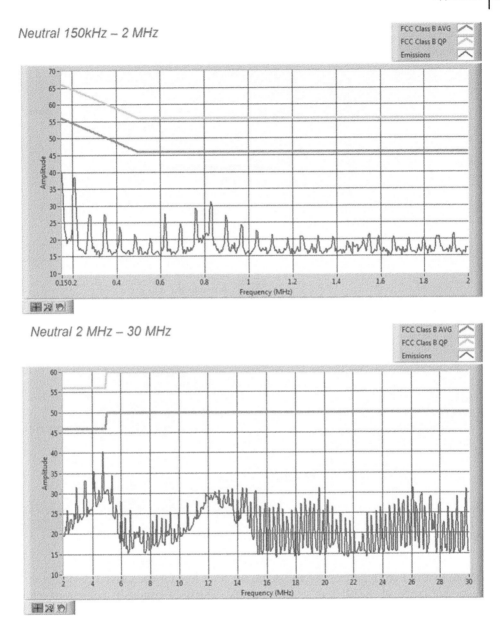

Figure A.21 An example of the neutral conducted emissions.

Note in Table A.1 that the peak limit for Class 3 device in the frequency range of 41–88 MHz is 46 dBµV.

CISPR 25 dc conducted emissions are measured on both the battery and the ground lines in the frequency range 150 kHz to 108 MHz.

Figure A.26 shows an example of the CISPR 25 conducted emissions voltage method (fail) peak detector results on a battery line for a Class 3 device. Figure A.25 shows the (pass) results for the ground line.

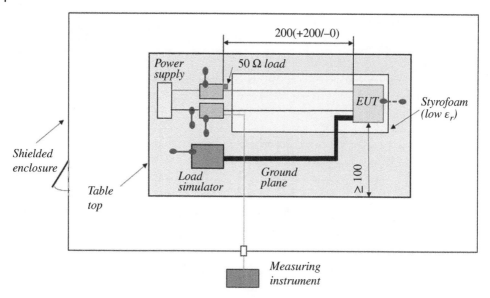

Figure A.22 The details of the CISPR 25 voltage method setup.

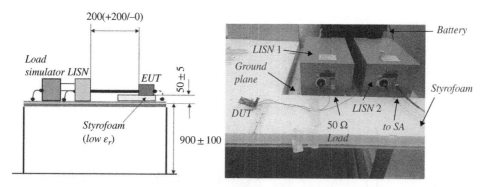

Figure A.23 More details of the CISPR 25 voltage method setup.

Note that in both Figures A.24 and A.25, the limit line for the Class 3 device in the frequency range 41–88 MHz is set at 46 dBμV, which, of course, is consistent with the limits specified in Table A.1.

A.2.3 CISPR 25 Current Probe Method

The details of the CISPR 25 conducted emissions current probe method setup are shown in Figure A.26. and A.27.

Class 1 through Class 5 limits for the current probe method are shown in Tables A.3 and A.4.

Note that in Table A.3 the peak limit for Class 5 device in the frequency range of 41–88 MHz is 0 dBμA. Table A.4 shows that the average limit for Class 5 device in the frequency range of 41–88 MHz is −10 dBμA.

Table A.1 CISPR 25 voltage method – peak and quasi-peak limits.

		Levels (dBμV)									
		Class 1		Class 2		Class 3		Class 4		Class 5	
Service/Band	Frequency MHz	PK	QP	PK	QP	PK	QP	PK	QP	PK	QP
LW	0.15 - 0.3	110	97	100	87	90	77	80	67	70	57
MW	0.53 - 1.8	86	73	78	65	70	57	62	49	54	41
SW	5.9 - 6.2	77	64	71	58	65	52	59	46	53	40
FM	76 - 108	62	49	56	43	50	37	44	31	38	25
TV Band 1	41 - 88	58	–	52	–	46	–	40	–	34	–
CB	26 - 28	68	55	62	49	56	43	50	37	44	31
VHF	30 - 54	68	55	62	49	56	43	50	37	44	31
VHF	68 - 87	62	49	56	43	50	37	44	31	38	25

Table A.2 CISPR 25 voltage method – average detector limits.

		Levels (dBμV)				
		Class 1	Class 2	Class 3	Class 4	Class 5
Service/Band	Frequency MHz	AVG	AVG	AVG	AVG	AVG
LW	0.15 - 0.3	90	80	70	60	50
MW	0.53 - 1.8	66	58	50	42	34
SW	5.9 - 6.2	57	51	45	39	33
FM	76 - 108	42	36	30	24	18
TV Band 1	41 - 88	48	42	36	30	24
CB	26 - 28	48	42	36	30	24
VHF	30 - 54	48	42	36	30	24
VHF	68 - 87	42	36	30	24	18

Figure A.28 shows an example of the CISPR 25 conducted emissions current method peak detector (pass) results for a Class 5 device. Figure A.29 shows the (fail) results for the average detector.

Note that in Figure A.28 the limit line for Class 5 device in the frequency range 41–88 MHz is set at 0 dBμA, while in Figure A.29 the limit line in the same frequency range is at −10 dBμA. This, of course, is consistent with the limits specified in Tables A.3 and A.4.

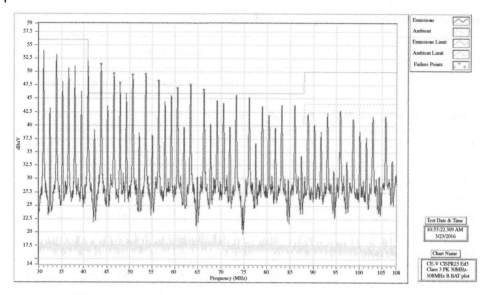

Figure A.24 CISPR 25 conducted emissions voltage method – peak detector results on a battery line for a Class 3 device.

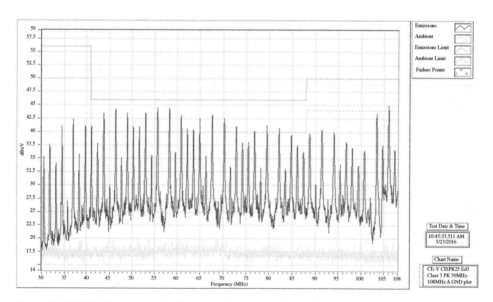

Figure A.25 CISPR 25 conducted emissions voltage method – peak detector results on a ground line for a Class 3 device.

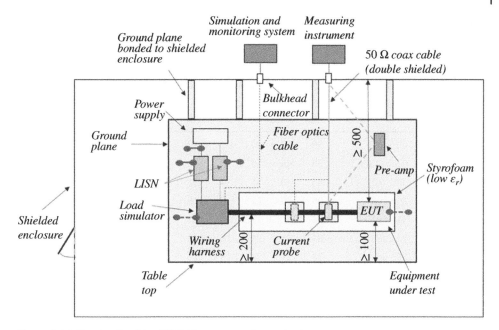

Figure A.26 The details of the CISPR 25 current probe method setup.

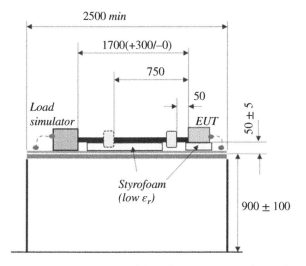

Figure A.27 More details of the CISPR 25 current probe method setup.

Table A.3 CISPR 25 current probe method – peak and quasi-peak limits.

Service/Band	Frequency MHz	Levels (dBμA)									
		Class 1		Class 2		Class 3		Class 4		Class 5	
		PK	QP	PK	QP	PK	QP	PK	QP	PK	QP
LW	0.15 - 0.3	90	77	80	67	70	57	60	47	50	37
MW	0.53 - 1.8	58	45	50	37	42	29	34	21	26	13
SW	5.9 - 6.2	43	30	37	24	31	18	25	12	19	6
FM	76 - 108	28	15	22	9	16	3	10	−3	4	−9
TV Band 1	41 - 88	24	–	18	–	12	–	6	–	0	–
CB	26 - 28	34	21	28	15	22	9	16	3	10	−3
VHF	30 - 54	34	21	28	15	22	9	16	3	10	−3
VHF	68 - 87	28	15	22	9	16	3	10	−3	4	−9

Table A.4 CISPR 25 current probe method – average detector limits.

Service/Band	Frequency MHz	Levels (dBμA)				
		Class 1	Class 2	Class 3	Class 4	Class 5
		AVG	AVG	AVG	AVG	AVG
LW	0.15 - 0.3	70	60	50	40	30
MW	0.53 - 1.8	38	30	22	14	6
SW	5.9 - 6.2	23	17	11	5	−1
FM	76 - 108	8	2	−4	−10	−16
TV Band 1	41 - 88	14	8	2	−4	−10
CB	26 - 28	14	8	2	−4	−10
VHF	30 - 54	14	8	2	−4	−10
VHF	68 - 87	8	2	−4	−10	−16

A.3 Radiated Emissions

CISPR 22 specifies that the measurement of the radiated emissions from products be performed at an OATS (open area test site), shown in Figure A.30, while CISPR 25 requires the measurements to be performed in a semi-anechoic test chamber (also referred to as absorber-lined shielded enclosure – ALSE), shown in Figure A.31.

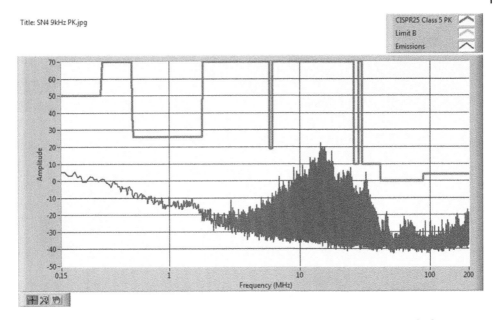

Figure A.28 CISPR 25 conducted emissions current probe method – peak detector results for a Class 5 device.

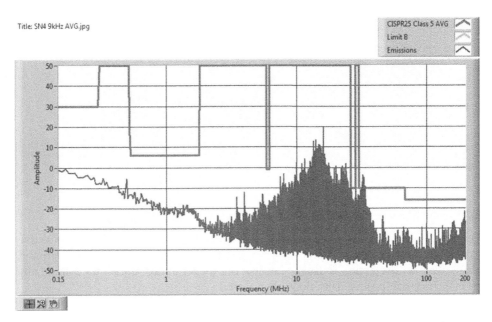

Figure A.29 CISPR 25 conducted emissions current probe method – average detector results for a Class 5 device.

Figure A.30 An open area test site (OATS).

A.3.1 Open-Area Test Site (OATS) Measurements

The ideal OATS is a flat piece of land, free of overhead wires and nearby reflective structures, away from any and all external signals, with a perfectly reflective ground plane. Weather protection is usually needed, but the structure should not contain any metallic material (beams, nails, door hinges, etc.).

Since the OATS should be away from all reflective structures, this requires the control room to be remotely located or located underneath the ground plane. The measurements should be made with a quasi-peak measuring receiver in the frequency range 30 MHz to 1 GHz (peak measurements are permitted).

The test site should be sufficiently large to permit antenna placing at the specified distance. The ground plane should extend at least 1 m beyond the periphery of the EUT

Figure A.31 A semi-anechoic chamber for radiated emissions.

and the largest measuring antenna, and cover the entire area between the EUT and the antenna. The boundary of the area is defined by an ellipse, as shown in Figure A.32.

When performing the radiated emissions measurements at an OATS, it is critical that the ambient measurement is taken first, in order to determine the electromagnetic environment present. Such a measurement is shown in Figure A.33.

After having taken the ambient measurement and identifying any external noise sources present, the DUT emission measurement is taken. Such a measurement is shown in Figure A.34.

A.3.2 Semi-Anechoic Chamber Measurements

CISPR 25 radiated emissions measurements are performed in a semi-anechoic chamber in the frequency range of 150 kHz to 2.5 GHz. In the frequency range of 150 kHz to 30 MHZ a vertical monopole antenna, shown in Figure A.35, is used.

The details of the CISPR 25 radiated emissions setup using a monopole antenna are shown in Figures A.36. and A.37.

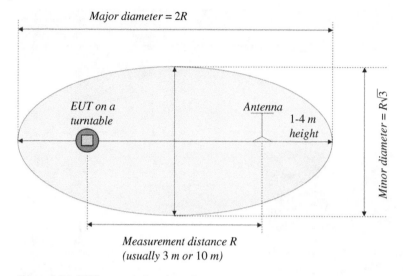

Major diameter = 2R

EUT on a turntable

Antenna 1-4 m height

Minor diameter = R√3

Measurement distance R
(usually 3 m or 10 m)

Figure A.32 OATS ground plane boundary.

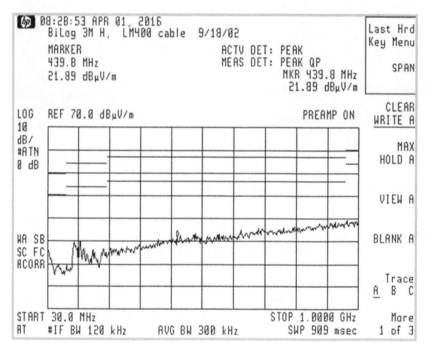

Figure A.33 OATS ambient measurement.

Figure A.38 shows an example of the CISPR 25 radiated emissions, Class 5, peak detector, monopole antenna measurement.

In the frequency range of 30–300 MHZ a biconnical antenna, shown in Figure A.39, is used.

The details of the CISPR 25 radiated emissions setup using a biconnical antenna are shown in Figure A.40. and A.41.

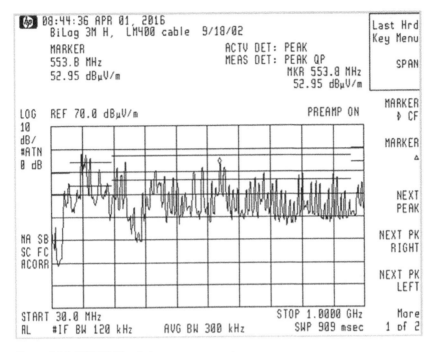

Figure A.34 OATS DUT emissions measurement.

Figure A.35 Monopole antenna for 0.15-30 MHz frequency range measurements.

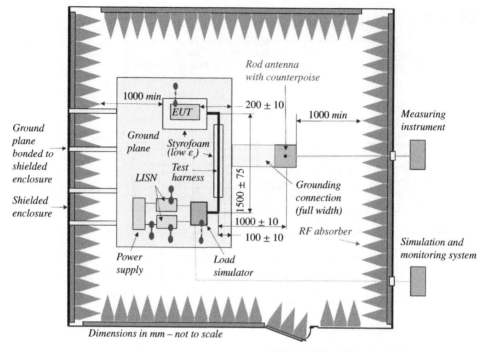

Figure A.36 The details of the CISPR 25 setup with a monopole antenna.

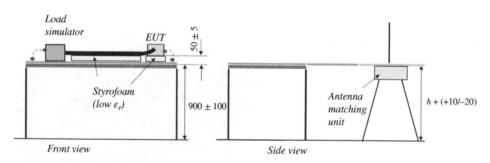

Figure A.37 More details of the CISPR 25 monopole antenna setup.

Figure A.42 shows an example of the CISPR 25 radiated emissions, Class 5, peak detector, biconical antenna measurement.

In the frequency range of 300–1000 MHZ a log-periodic antenna, shown in Figure A.43, is used.

The details of the CISPR 25 radiated emissions setup using a log-periodic antenna are shown in Figure A.44. and A.45.

Figure A.46 shows an example of the CISPR 25 radiated emissions, Class 5, peak detector, log-periodic antenna measurement.

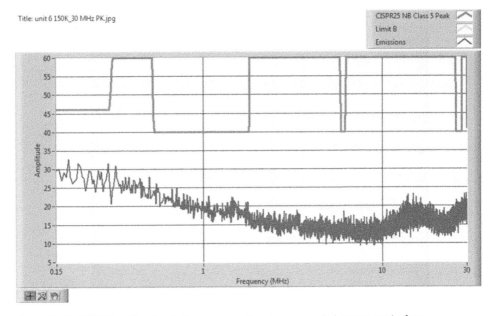

Title: unit 6 150K_30 MHz PK.jpg

Figure A.38 CISPR 25 radiated emissions, monopole antenna – peak detector results for a Class 5 device.

Figure A.39 Biconnical antenna for 30–300 MHz frequency range measurements.

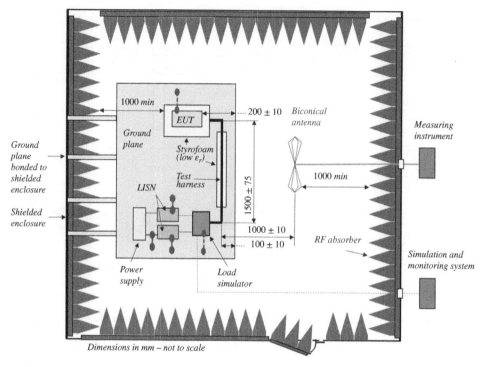

Figure A.40 The details of the CISPR 25 setup with a biconical antenna.

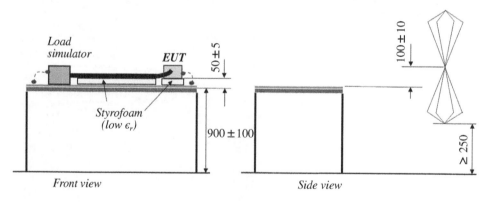

Figure A.41 More details of the CISPR 25 biconical antenna setup.

A.4   Conducted Immunity – ISO 11452-4

ISO 11452-4 specifies the conducted immunity testing using a bulk current injection (BCI) method. BCI is a method of carrying out immunity tests by inducing disturbance signals directly into the wiring harness by means of a current injection probe.

The injection probe is a current transformer through which the wiring harness of the device under test (DUT) is passed, as shown in Figure A.47.

Title: H unit 6 30_300 MHz PK.jpg

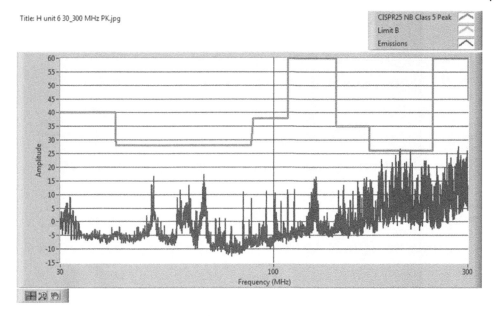

Figure A.42 CISPR 25 radiated emissions, biconical antenna – peak detector results for a Class 5 device.

Figure A.43 Log-periodic antenna for 300–1000 MHz frequency range measurements.

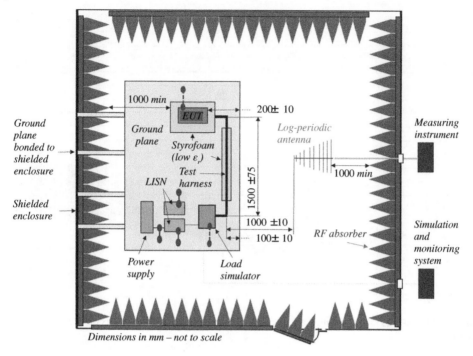

Figure A.44 The details of the CISPR 25 setup with a log-periodic antenna.

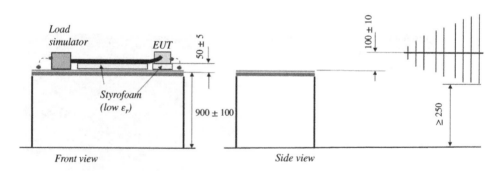

Figure A.45 More details of the CISPR 25 log-periodic antenna setup.

Immunity tests are carried out by varying the test severity level and frequency (1 MHz – 400 MHz) of the induced disturbance. BCI testing is performed in a screen room; the immunity tests require several pieces of additional equipment outside the screen room. These include signal generator, power amplifier, power meter, power sensors, directional coupler, simulation and monitoring system, and computer control. The external equipment pieces and their interconnections are shown in Figure A48.

Figure A.49 shows the location of the equipment external to the screen room used for the immunity testing.

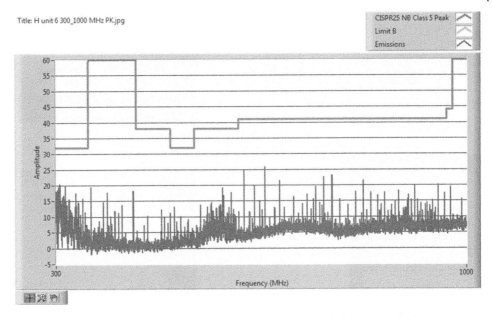

Title: H unit 6 300_1000 MHz PK.jpg

Figure A.46 CISPR 25 radiated emissions, log-periodic antenna – peak detector results for a Class 5 device.

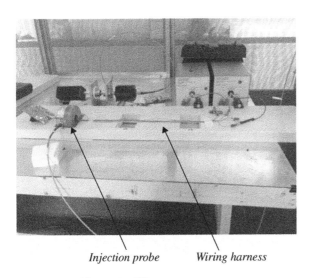

Injection probe　　*Wiring harness*

Figure A.47 ISO 11452-4 BCI test setup.

The measurement setup inside the screen room is shown in Figure A.50.

The internal setup shown in Figure A.50 accommodates two types of BCI test methods specified in ISO 11452-4 the *substitution method* and the *closed-loop method with power limitation*. These will be discussed next.

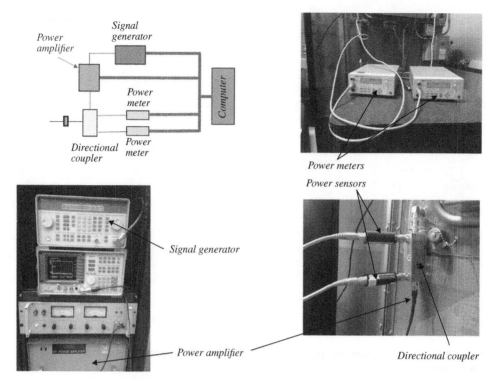

Figure A.48 BCI external equipment.

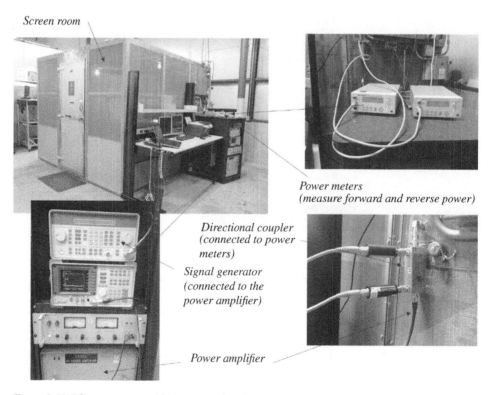

Figure A.49 BCI screen room with the external equipment.

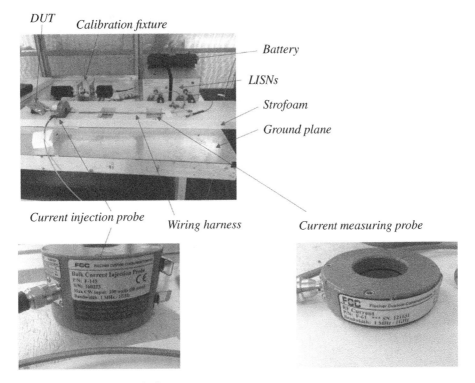

Figure A.50 BCI setup inside the screen room.

A.4.1 Substitution Method

In the substitution method a calibration fixture is used to record the power needed to produce the required current in the 50 Ω load. Then, during the testing, that power is applied over the frequency range.

The details of the ISO 11452-4 conducted immunity testing setup using the substitution method are shown in Figures A.51 and A.52.

ISO 11452-4 specifies five test severity levels as shown in Table A.5.

Note in Table A.5 that in the frequency range of 3–200 MHz the test level I limit is 60 mA while the test level IV limit is 200 mA. This is reflected in the limit lines shown in Figures A.53 and A.54, which show the ISO11542-4 conducted immunity test results using the substitution method.

A.4.2 Closed-Loop Method with Power Limitation

In this method a calibration fixture is used to record the power needed to produce the required current in the 50 Ω environment. Then during the testing, power is applied until the required current is measured or the power limit is reached ($P_{limit} = 4 \times P_{calibration}$).

The details of the ISO 11452-4 conducted immunity testing setup using the closed-loop method with power limitation are shown in Figures A.55 and A.56.

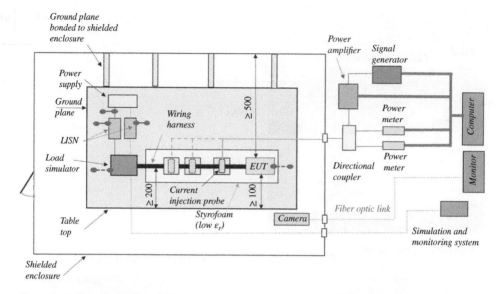

Figure A.51 The details of the ISO 11542–4 setup using the substitution method.

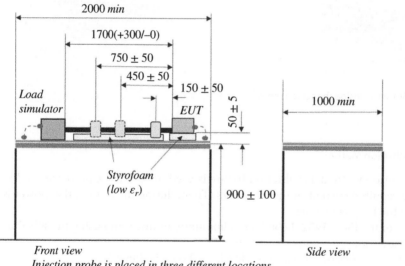

Front view
Injection probe is placed in three different locations.

Side view

Figure A.52 More details of the of the ISO 11542-4 setup using the substitution method.

Table A.5 ISO 11542-4 test severity levels.

Frequency band MHz	Test level I mA	Test level II mA	Test level III mA	Test level IV mA	Test level V mA
1-3	$60 \times f_{(MHz)}/3$	$100 \times f_{(MHz)}/3$	$150 \times f_{(MHz)}/3$	$200 \times f_{(MHz)}/3$	*Specific values agreed between the users of this part of ISO 11452*
3-200	60	100	150	200	
200-400	$60 \times 200/f(MHz)$	$100 \times 200/f(MHz)$	$150 \times 200/f(MHz)$	$200 \times 200/f(MHz)$	

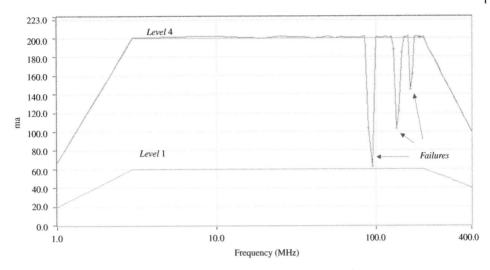

Figure A.53 ISO11452-4 BCI test result (fail) using the substitution method.

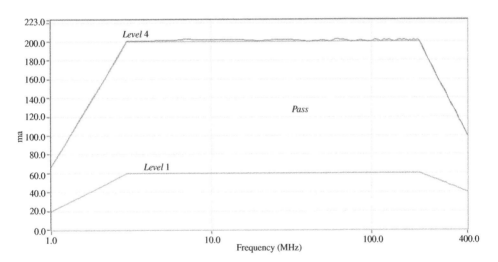

Figure A.54 ISO11452-4 BCI test result (pass) using the substitution method.

A.5 Radiated Immunity

ISO 11452-11 specifies that the measurement of the radiated immunity be performed in a reverberation chamber, while ISO 11452-12 requires the use of a semi-anechoic test chamber (also referred to as an absorber-lined shielded enclosure – ALSE). These chambers are shown in Figure A.57.

A.5.1 Radiated Immunity – ISO 11452-11

ISO 11452-11 requires a reverberation chamber for radiated immunity testing. A reverberation chamber is a shielded highly conductive enclosure. Unlike the semi-anechoic

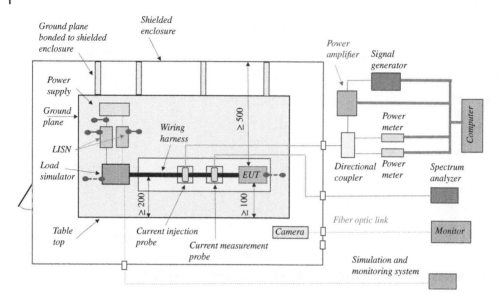

Figure A.55 The details of the ISO 11542-4 setup using the closed-loop method with power limitation.

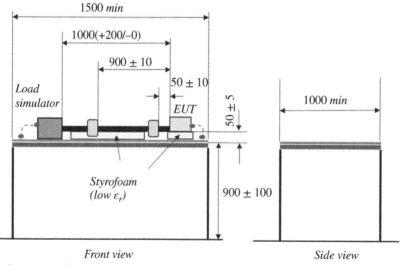

Front view

Side view

Injection probe is placed 900 mm from the DUT.
Mesurement probe is placed 50 mm fro the DUT.

Figure A.56 More details of the of the ISO 11542-4 setup using the closed-loop method with power limitation.

or fully anechoic chamber it is not lined with an absorbing material, as shown in Figure A.58. The metallic walls are highly reflective to the electromagnetic waves.

Just like the conducted immunity tests, this radiated immunity test requires several pieces of additional equipment outside the reverberation chamber. These are shown in Figure A.59.

Reverberation chamber – ISO 11452-11   *Semi-anechoic chamber – ISO* 11452-2

Figure A.57 Reverberation and semi-anechoic chambers for radiated immunity.

Figure A.58 Reverberation chamber is a highly reflective enclosure.

The details of the ISO 11452-11 radiated immunity testing setup are shown in Figure A.60.

Reverberation chamber dimensions should be large compared to the wavelength; the larger the chamber the lower the usable frequency for testing. A mechanical tuner/stirrer, shown in Figure A.61, should have one dimension that is at least one-quarter wavelength at the lowest frequency.

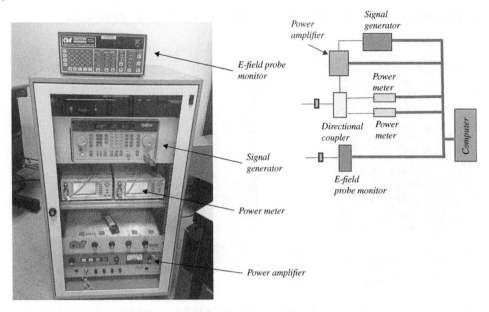

Figure A.59 External equipment required for the reverberation chamber testing.

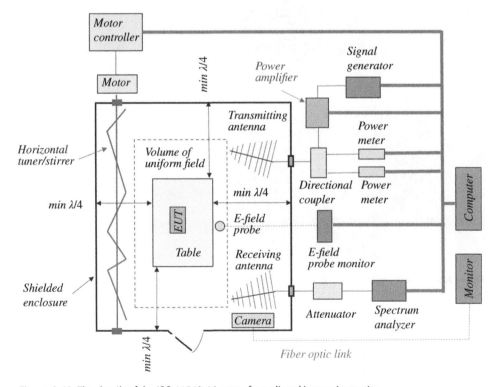

Figure A.60 The details of the ISO 11542-11 setup for radiated immunity testing.

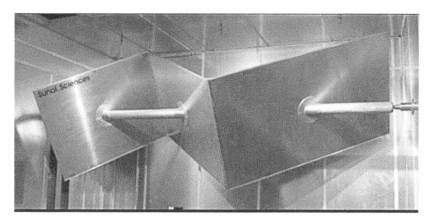

Figure A.61 Mechanical tuner/stirrer.

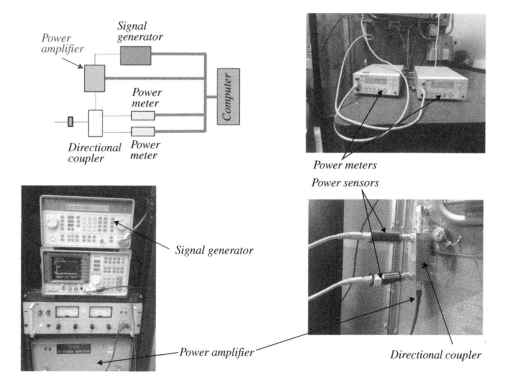

Figure A.62 ISO 11452-2 external equipment.

The mechanical tuner should be shaped asymmetrically to maximize the non-repetitive reflections generated by the transmitting antenna inside the chamber.

A.5.2 Radiated Immunity – ISO 11452-2

ISO 11452-12 requires a semi-anechoic chamber for radiated immunity testing. Just like the radiated immunity test described in the previous section, this test requires a standard immunity equipment external to the chamber, as shown in Figure A.62.

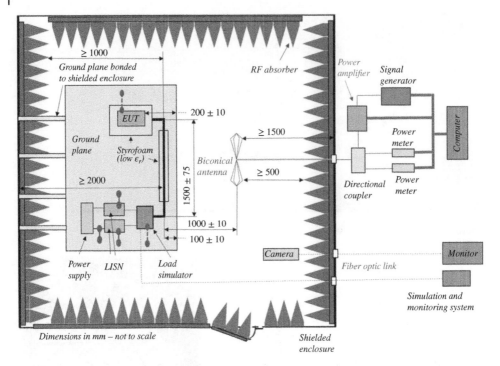

Figure A.63 ISO 11452-2 setup for radiated immunity testing using biconical antenna.

The details of the ISO 11452-2 radiated immunity test setup are shown in Figure A.63 when using a biconical antenna, and in Figure A.64 for log-periodic antenna.

Figures A.65 and A.66 show the ISO11452-4 conducted immunity test results using the substitution method.

A.6   Electrostatic Discharge (ESD)

An ESD test is performed with an ESD gun, as shown in Figure A.67.

Typical RC cartridge combinations are shown in Figure A.68.

The ESD gun reflects the human body circuit model shown in Figure A.69.

The human body model is based on the human body resistance and capacitance (see Chapter 13). The human body model simulates the ESD event when a charged body directly transfers an electrostatic charge to the ESD sensitive device.

ESD specifications define several terms related to the testing methods:

Contact discharge method – a method of testing, in which the electrode of the test generator is held in contact with the EUT, and the discharge actuated by the discharge switch within the generator.

Air discharge method – a method of testing, in which the charged electrode of the test generator is brought close to the EUT, and the discharge actuated by a spark to the EUT.

Direct application – application of the discharge directly to the EUT.

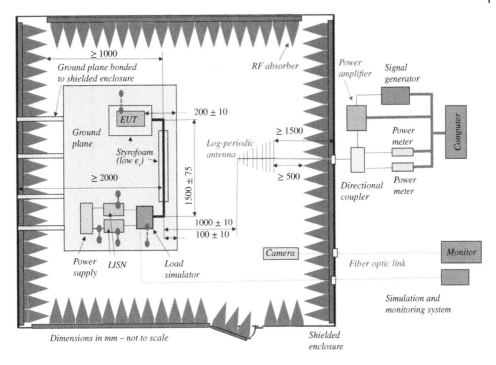

Figure A.64 ISO 11452-2 setup for radiated immunity testing using log-periodic antenna.

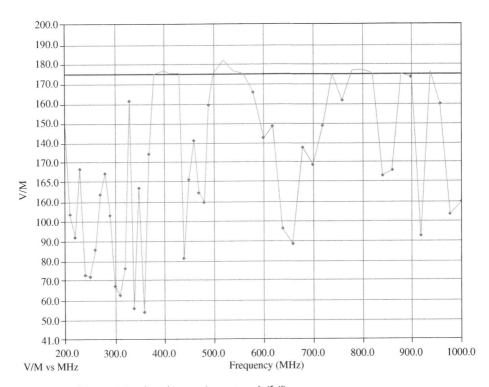

Figure A.65 ISO 11452-2 radiated immunity test result (fail).

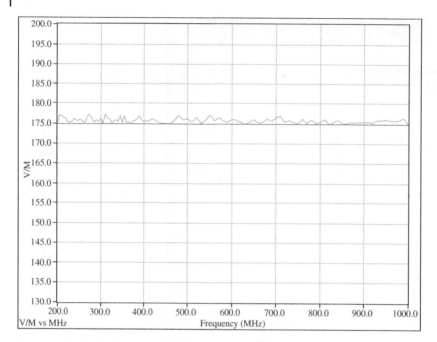

Figure A.66 ISO 11452-2 radiated immunity test result (pass).

ESD gun

Figure A.67 ESD gun and RC cartridge.

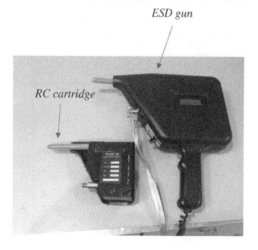

Indirect application – application of the discharge to a coupling plane in the vicinity of the EUT.

These terms are used when specifying the details of the ESD testing, as shown in Tables A.6 and A.7.

An ESD test table-top setup in a screen room is shown in Figure A.70.

The details of the ISO 10605 powered DUT, direct ESD test setup are shown in Figure A.71.

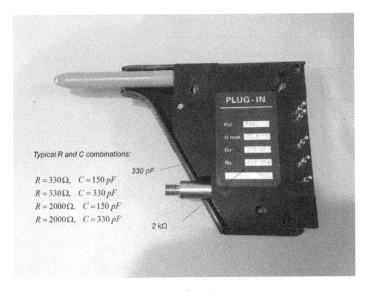

Figure A.68 Typical RC cartridge combinations.

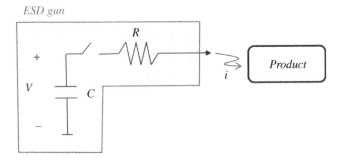

Figure A.69 Human body circuit model.

Table A.6 ISO 10605 – ESD generator parameters.

Parameter	Characteristic
Output voltage range contact discharge mode	*2 kV to 15 kV*
Output voltage range air discharge mode	*2 kV to 25 kV*
Output polarity	*Positive and negative*
Storage capacitances	*150 pF, 330 pF*
Storage resistance	*330 Ω, 2000 Ω*

Table A.7 ISO 61000–4-2 – Test levels and ESD generator parameters.

IEC 61000-4-2 Test Levels and ESD generator parameters

Contact discharge		Air discharge	
Level	Test voltage (kV)	Level	Test voltage (kV)
1	*2*	*1*	*2*
2	*4*	*2*	*4*
3	*6*	*3*	*8*
4	*8*	*4*	*15*
Storage capacitance		**Storage resistance**	
150 pF		*330 Ω*	

Vertical ground plane (IEC 61000-4-2)   *Battery (ISO 10605)*

Figure A.70 ESD table-top test setup.

The details of the ISO 10605 powered DUT, indirect ESD test setup are shown in Figure A.72.

Packaging and handling ISO 10605 test setup details are shown in Figure A.73.

Finally, the details of the ISO 61000-4-2 test setup for table-top equipment are shown in Figure A.74.

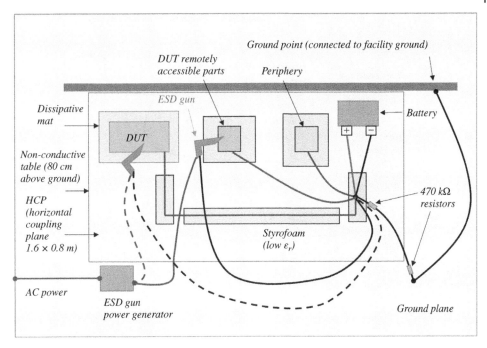

Figure A.71 ISO 10605 powered DUT, direct ESD test setup.

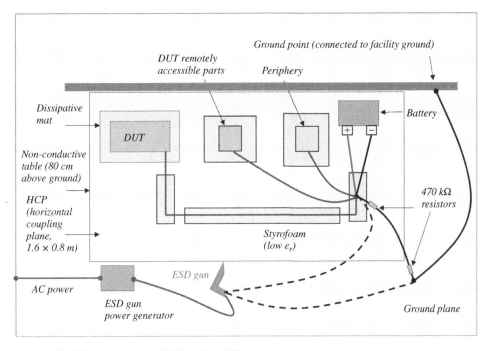

Figure A.72 ISO 10605 powered DUT, indirect ESD test setup.

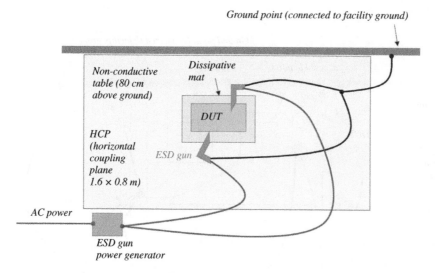

Figure A.73 ISO 10605 packaging and handling, ESD test setup.

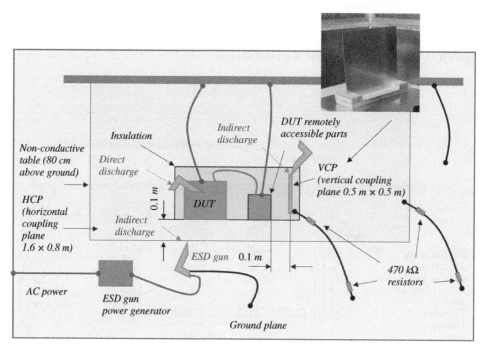

Figure A.74 ISO 61000-4-2 ESD test setup for table-top equipment.

Final remarks about ESD testing
The testing should be performed by direct and indirect application of discharges to the EUT, according to a test plan. This should include:

1) representative operating conditions of the EUT
2) whether the EUT should be tested as table-top or floor-standing
3) the points at which discharges are to be applied

4) at each point, whether contact or air discharges are to be applied
5) the test level to be applied
6) the number of discharges to be applied at each point for compliance testing.

The test results should be classified on the basis of the operating conditions and the functional specifications of the EUT, as in the following, unless different specifications are given by the product committees or product specifications:

1) normal performance within the specification limits
2) temporary degradation or loss of function or performance which is self-recoverable
3) temporary degradation or loss of function or performance which requires operator intervention or system reset
4) degradation or loss of function which is not recoverable due to damage to equipment (components) or software, or loss of data

References

FCC Part 15
CISPR 22
CISPR 25
ISO 11452-2
ISO 11452-4
ISO 11452-11
IEC 61000-4-2
IEC 61000-4-21

Index

Printed and bound by CPI Group (UK) Ltd, Croydon, CR0 4YY

16/04/2025

14658476-0003